Advanced Microwave Circuits and Systems

Advanced Microwave Circuits and Systems

Edited by

Vitaliy Zhurbenko

Advanced Microwave Circuits and Systems,
Edited by Vitaliy Zhurbenko

CBS, Edition 2016

Published by InTech

Janeza Trdine 9, 51000 Rijeka, Croatia

Additional hard copies can be obtained from orders@intechopen.com

Advanced Microwave Circuits and Systems,
Edited by Vitaliy Zhurbenko
p. cm.
ISBN 978-953-307-087-2

Preface

This book is based on recent research work conducted by the authors dealing with the design and development of active and passive microwave components, integrated circuits and systems. It is divided into seven parts. In the first part comprising the first two chapters, alternative concepts and equations for multiport network analysis and characterization are provided. A thru-only de-embedding technique for accurate on-wafer characterization is introduced.

The second part of the book corresponds to the analysis and design of ultra-wideband low-noise amplifiers (LNA). The LNA is the most critical component in a receiving system. Its performance determines the overall system sensitivity because it is the first block to amplify the received signal from the antenna. Hence, for the achievement of high receiver performance, the LNA is required to have a low noise figure with good input matching as well as sufficient gain in a wide frequency range of operation, which is very difficult to achieve. Most circuits demonstrated are not stable across the frequency band, which makes these amplifiers prone to self-oscillations and therefore limit their applicability. The trade-off between noise figure, gain, linearity, bandwidth, and power consumption, which generally accompanies the LNA design process, is discussed in this part.

The requirement from an amplifier design differs for different applications. A power amplifier is a type of amplifier which drives the antenna of a transmitter. Unlike LNA, a power amplifier is usually optimized to have high output power, high efficiency, optimum heat dissipation and high gain. The third part of this book presents power amplifier designs through a series of design examples. Designs undertaken include a switching mode power amplifier, Doherty power amplifier, and flexible power amplifier architectures. In addition, distortion analysis and power combining techniques are considered.

Another key element in most microwave systems is a signal generator. It forms the heart of all kinds of communication and radar systems. The fourth part of this book is dedicated to signal generators such as voltage-controlled oscillators and electron devices for millimeter wave and submillimeter wave applications. This part also covers studies of integrated buffer circuits.

Passive components are indispensable elements of any electronic system. The increasing demands to miniaturization and cost effectiveness push currently available technologies to the limits. Some considerations to meet the growing requirements are provided in the fifth part of this book. The following part deals with circuits based on LTCC and MEMS technologies.

The book concludes with chapters considering application of microwaves in measurement and sensing systems. This includes topics related to six-port reflectometers, remote network analysis, inverse scattering for microwave imaging systems, spectroscopy for medical applications and interaction with transponders in medical sensors.

Editor

Vitaliy Zhurbenko

Contents

1

Mixed-mode S-parameters and Conversion Techniques

Allan Huynh, Magnus Karlsson and Shaofang Gong
Linköping University
Sweden

1. Introduction

Differential signaling in analog circuits is an old technique that has been utilized for more than 50 years. During the last decades, it has also been becoming popular in digital circuit design, when low voltage differential signaling (LVDS) became common in high-speed digital systems. Today LVDS is widely used in advanced electronics such as laptop computers, test and measurement instrument, medical equipment and automotive. The reason is that with increased clock frequencies and short edge rise/fall times, crosstalk and electromagnetic interferences (EMI) appear to be critical problems in high-speed digital systems. Differential signaling is aimed to reduce EMI and noise issues in order to improve the signal quality. However, in traditional microwave theory, electric current and voltage are treated as single-ended and the S-parameters are used to describe single-ended signaling. This makes advanced microwave and RF circuit design and analysis difficult, when differential signaling is utilized in modern communication circuits and systems. This chapter introduces the technique to deal with differential signaling in microwave and millimeter wave circuits.

2. Differential Signal

Differential signaling is a signal transmission method where the transmitting signal is sent in pairs with the same amplitude but with mutual opposite phases. The main advantage with the differential signaling is that any introduced noise equally affects both the differential transmission lines if the two lines are tightly coupled together. Since only the difference between the lines is considered, the introduced common-mode noise can be rejected at the receiver device. However, due to manufacturing imperfections, signal unbalance will occur resulting in that the energy will convert from differential-mode to common-mode and vice versa, which is known as cross-mode conversion. To damp the common-mode currents, a common-mode choke can be used (without any noticeable effect on the differential currents) to prevent radiated emissions from the differential lines. To produce the electrical field strength from microamperes of common-mode current, milliamperes of differential current are needed (Clayton, 2006). Moreover, the generated electric and magnetic fields from a differential line pair are more localized compared to those from single-ended lines. Owing to the ability of noise rejection, the signal swing can be decreased compared to a single-ended design and thereby the power can be saved.

When the signal on one line is independent of the signal on the adjacent line, i.e., an uncoupled differential pair, the structure does not utilize the full potential of a differential design. To fully utilize the differential design, it is beneficial to start by minimizing the spacing between two lines to create the coupling as strong as possible. Thereafter, the conductors width is adjusted to obtain the desired differential impedance. By doing this, the coupling between the differential line pair is maximized to give a better common-mode rejection.

S-parameters are very commonly used when designing and verifying linear RF and microwave designs for impedance matching to optimize gain and minimize noise. Although, traditional S-parameter representation is a very powerful tool in circuit analysis and measurement, it is limited to single-ended RF and microwave designs. In 1995, Bockelman and Einsenstadt introduced the mixed-mode S-parameters to extend the theory to include differential circuits. However, owing to the coupling effects between the coupled differential transmission lines, the odd- and even-mode impedances are not equal to the unique characteristic impedance. This leads to the fact that a modified mixed-mode S-parameters representation is needed. In this chapter, by starting with the familiar concepts of coupling, crosstalk and terminations, mixed-mode S-parameters will be introduced. Furthermore, conversion techniques between different modes of S-parameters will be described.

2.1 Coupling and Crosstalk

Like in single-ended signaling, differential transmission lines need to be correctly terminated, otherwise reflections arise and distortions are introduced into the system. In a system where parallel transmission lines exist, either in differential signaling or in parallel single-ended lines, line-to-line coupling arises and it will cause characteristic impedance variations. The coupling between the parallel single-ended lines is also known as crosstalk and it is related to the mutual inductance (L_m) and capacitance (C_m) existing between the lines. The induced crosstalk or noise can be described with a simple approximation as following

$$V_{noise} = \mathrm{L_m}\frac{\mathrm{dI_{driver}}}{\mathrm{dt}} \tag{1}$$

$$I_{noise} = C_m\frac{dV_{driver}}{dt} \tag{2}$$

where V_{noise} and I_{noise} are the induced voltage and current noises on the adjacent line and V_{driver} and I_{driver} are the driving voltage and current on the active line. Since both the voltage and current noises are induced by the rate of current and voltage changes, extra care is needed for high-speed applications.

The coupling between the parallel lines depends firstly on the spacing between the lines and secondly on the signal pattern sent on the parallel lines. Two signal modes are defined, i.e., odd- and even-modes. The odd-mode is defined such that the driven signals in the two adjacent lines have the same amplitude but a 180 degree of relative phase, which can be related to differential signal. The even-mode is defined such that the driven signals in the two adjacent lines have the same amplitude and phase, which can be related to common-mode noise for a differential pair of signal. Fig. 1 shows the electric and magnetic field lines in the odd- and even-mode transmissions on the two parallel microstrips. Fig. 1a shows that the odd-mode signaling causes coupling due to the electric field between the microstrips, while in the even-mode shown in Fig 1b, there is no direct electric coupling between the lines. Fig. 1c shows that the magnetic field in the odd-mode has no coupling between the two lines while, as shown in Fig. 1d, in the even-mode the magnetic field is coupled between the two lines.

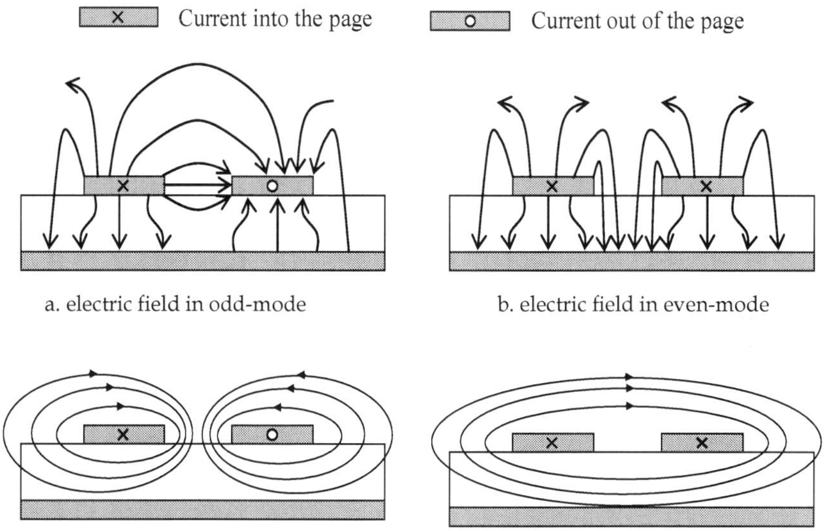

a. electric field in odd-mode b. electric field in even-mode

c. magnetic field in odd-mode d. magnetic field in even-mode

Fig. 1. Odd- and even-mode electric and magnetic fields for two parallel microstrips.

2.2 Odd-mode

The induced crosstalk or voltage noise in a pair of parallel transmission lines can be approximated with Equation 1. For the case of two parallel transmission lines the equation can be rewritten as following

$$V_1 = L_0 \frac{dI_1}{dt} + L_m \frac{dI_2}{dt} \tag{3}$$

$$V_2 = L_0 \frac{dI_2}{dt} + L_m \frac{dI_1}{dt} \tag{4}$$

where L_0 is the equivalent lumped-self-inductance in the transmission line and L_m is the mutual inductance arisen due to the coupling between the lines. Signal propagation in the odd-mode results in $I_1 = -I_2$, since the current is always driven with equal magnitude but in opposite directions. Substituting it into Equations 3 and 4 yeilds

$$V_1 = (L_0 - L_m) \frac{dI_1}{dt} \tag{5}$$

$$V_2 = (L_0 - L_m)\frac{dI_2}{dt} \tag{6}$$

This shows that, due to the crosstalk, the total inductance in the transmission lines reduces with the mutual inductance (L_m).
Similarly, the current noise in the parallel transmission lines can be estimated with Equation 2. For two parallel transmission lines the equation can be rewritten as following

$$I_1 = C_0\frac{dV_1}{dt} + C_m\frac{d(V_1 - V_2)}{dt} \tag{7}$$

$$I_2 = C_0\frac{dV_2}{dt} + C_m\frac{d(V_2 - V_1)}{dt} \tag{8}$$

where C_0 is the equivalent lumped-capacitance between the line and ground, and C_m is the mutual capacitance between the transmission lines arisen due to the coupling between the lines. Signal propagation in odd-mode results in $V_1 = -V_2$. Substituting it into Equations 7 and 8 yields

$$I_1 = (C_0 + 2C_m)\frac{dV_1}{dt} \tag{9}$$

$$I_2 = (C_0 + 2C_m)\frac{dV_2}{dt} \tag{10}$$

Equations 9 and 10 show that, in opposite to the inductance, the total capacitance increases with the mutual capacitance.
The addition of mutual inductance and capacitance shows that the characteristic impedance as well as the phase velocity is directly dependant of the mutual coupling, as shown with the following equations

$$Z_{oo} = \sqrt{\frac{R + j\omega L}{G + j\omega C}} = \sqrt{\frac{R + j\omega(L_0 - L_m)}{G + j\omega(C_0 + 2C_m)}} \tag{11}$$

$$\vartheta_{po} = \frac{1}{\sqrt{LC}} = \frac{1}{\sqrt{(L_0 - L_m)(C_0 + 2C_m)}} \tag{12}$$

where Z_{oo} and v_{po} are the odd-mode impedance and phase velocity, respectively.
Consequently, the total characteristic impedance in the odd-mode reduces due to the coupling or crosstalk between the parallel transmission lines and the phase velocity changes as well.

2.3 Even-mode

In the case of even-mode where the signals are driven with equal magnitude and phase, $V_1 = V_2$ and $I_1 = I_2$, Equations 3, 4, 7 and 8 can be rewritten to the following:

$$V_1 = (L_0 + L_m)\frac{dI_1}{dt} \tag{13}$$

$$V_2 = (L_0 + L_m)\frac{dI_2}{dt} \tag{14}$$

$$I_1 = C_0 \frac{dV_1}{dt} \tag{15}$$

$$I_2 = C_0 \frac{dV_2}{dt} \tag{16}$$

Consequently, in opposite to the odd-mode case, the even-mode wave propagation changes the even-mode impedance (Z_{oe}) and phase velocity (v_{pe}) as shown below:

$$Z_{oe} = \sqrt{\frac{R+j\omega L}{G+j\omega C}} = \sqrt{\frac{R+j\omega(L_0+L_m)}{G+j\omega C_0}} \tag{17}$$

$$\vartheta_{pe} = \frac{1}{\sqrt{LC}} = \frac{1}{\sqrt{(L_0+L_m)(C_0)}} \tag{18}$$

2.4 Terminations

As shown in the previous section, the impedance varies due to the odd- and even-mode transmissions and the coupling between the transmission lines. Fig. 2 shows a graph of the odd- and even-mode impedance change as a function of the spacing between two specific parallel-microstrips. If the loads connected to the parallel lines have a simple termination as commonly used in the single-ended case, reflections will occur due to $Z_{oo} \neq Z_{oe} \neq Z_0$. Fig. 3 shows two termination configurations, i.e., Pi- and T-terminations, which can terminate both the odd- and even-mode signals in coupled parallel transmission lines.

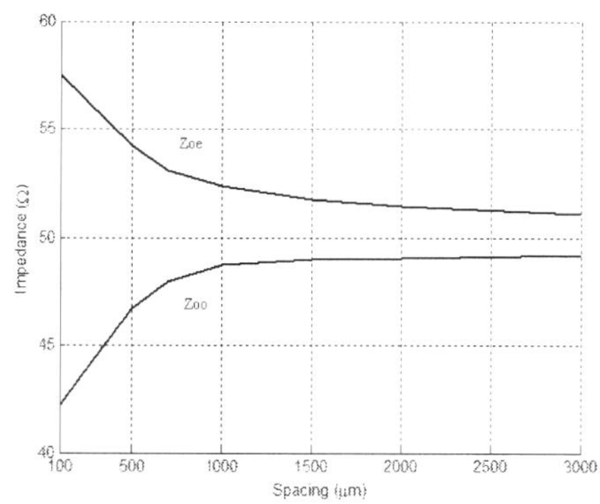

Fig. 2. Variation of the odd- and even-mode impedances as a function of the spacing between two parallel microstrips.

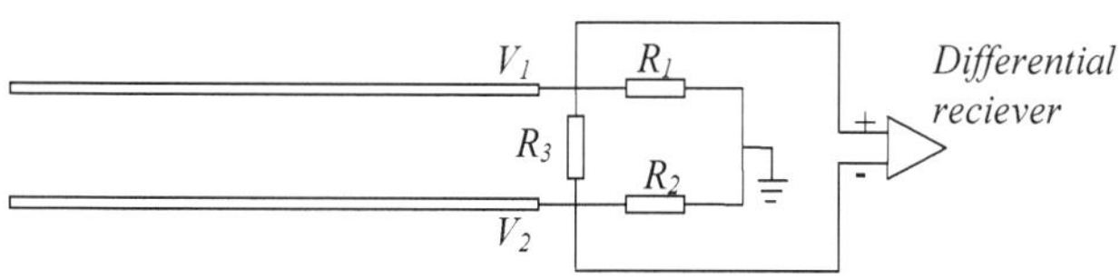

a. Pi-termination

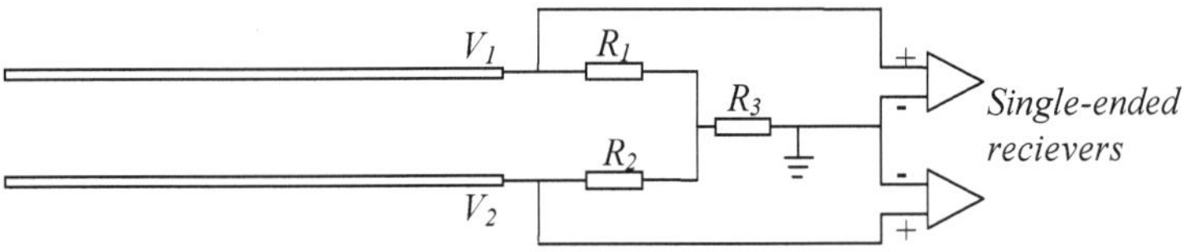

b. T-termination

Fig. 3. Termination configurations for coupled transmission lines.

Fig. 3a shows the Pi-termination configuration. In the odd-mode transmission, i.e., $V_1 = -V_2$ a virtual ground can be imaginarily seen in the middle of R_3 and this forces $R_3/2$ in parallel with R_1 or R_2 equal to Z_{OO}. Since no current flows between the two transmission lines in the even-mode, i.e., $V_1 = V_2$, R_1 and R_2 must thus be equal to Z_{oe}. For a matched differential system with optimized gain and noise, the following expressions need to be fulfilled for a Pi-termination configuration.

$$R_1 = R_2 = Z_{oe} \tag{19}$$

$$R_3 = 2\frac{Z_{oe}Z_{oo}}{Z_{oe}-Z_{oo}} \tag{20}$$

Fig. 3b shows the T-termination configuration. In the odd-mode transmission, i.e., $V_1 = -V_2$, a virtual ground can be seen between R_1 and R_2 and this makes R_1 and R_2 equal to Z_{OO}. In the even-mode transmission, i.e., $V_1 = V_2$ no current flows between the two transmission lines. This makes R_3 to be seen as two $2R_3$ in parallel, as illustrated in Fig. 4. This leads to the conclusion that Z_{oe} must be equal to R_1 or R_2 in serial with $2R_3$. Equations 21 and 22 show the required values of the termination resistors needed for the T-termination configuration in order to get a perfect matched system (Hall et al., 2000).

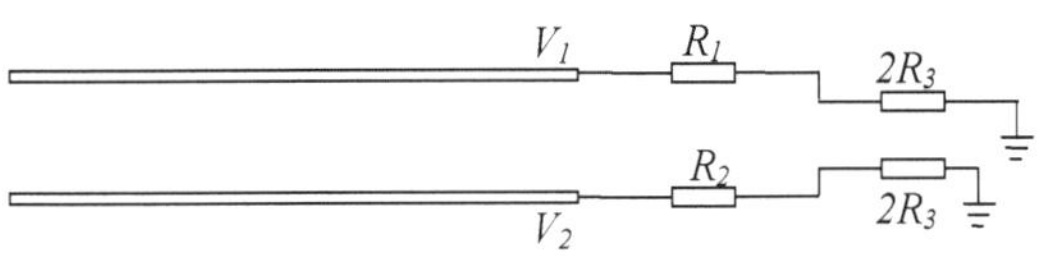

Fig. 4. Equivalent network for T-network termination in even-mode.

$$R_1 = R_2 = Z_{oo} \tag{21}$$

$$R_3 = \frac{1}{2}(Z_{oe} - Z_{oo}) \tag{22}$$

3. S-parameters

Scattering parameters or S-parameters are commonly used to describe an n-port network operating at high frequencies like RF and microwave frequencies. Other well-known parameters often used for describing an n-port network are *Z* (impedance), *Y* (admittance), *h* (hybrid) and *ABCD* parameters. The main difference between the S-parameters and other parameter representations lies in the fact that S-parameters describe the normalized power waves when the input and output ports are properly terminated, while other parameters describe voltage and current with open or short ports. S-parameters can also be used to express other electrical properties like gain, insertion loss, return loss, voltage standing wave ratio, reflection coefficient and amplifier stability.

3.1 Single-ended

The travelling waves used in the transmission line theory are defined with incident normalized power wave a_n and reflected normalized power wave b_n.

$$a_n = \frac{1}{2\sqrt{Z_0}}(V_n + Z_0 I_n) \tag{23}$$

$$b_n = \frac{1}{2\sqrt{Z_0}}(V_n - Z_0 I_n) \tag{24}$$

where index n refers to a port number and Z_0 is the characteristic impedance at that specific port. The normalized power waves are used for the definition of single-ended S-parameters. Fig. 5 shows a sketch of a two-port network with the normalized power wave definitions (Kurokawa, 1965).

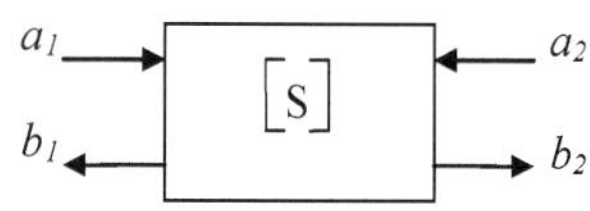

Fig. 5. S-parameters with normalized power wave definition of a two-port network.

The two-port S-parameters are defined as follows.

$$S_{11} = \left.\frac{b_1}{a_1}\right|_{a_2=0} \equiv reflection\ at\ port\ 1 \tag{25}$$

$$S_{12} = \left.\frac{b_1}{a_2}\right|_{a_1=0} \equiv reverse\ voltage\ gain \tag{26}$$

$$S_{21} = \left.\frac{b_2}{a_1}\right|_{a_2=0} \equiv forward\ voltage\ gain \tag{27}$$

$$S_{22} = \left.\frac{b_2}{a_2}\right|_{a_1=0} \equiv reflection\ at\ port\ 2 \tag{28}$$

and in matrix form

$$\begin{Bmatrix} b_1 \\ b_2 \end{Bmatrix} = \begin{bmatrix} S_{11} & S_{12} \\ S_{21} & S_{22} \end{bmatrix} \begin{Bmatrix} a_1 \\ a_2 \end{Bmatrix} \tag{29}$$

When measuring S-parameters, it is important not to have any power wave reflected at port 1 or 2, i.e., $a_1 = 0$ or $a_2 = 0$, as shown in Equations 25-28. Otherwise errors are included in the results. For an n-port network, Equation 29 can be extended to the following expression (Ludwig & Bretchko, 2000):

$$\begin{pmatrix} b_1 \\ b_2 \\ \vdots \\ b_n \end{pmatrix} = \begin{bmatrix} S_{11} & S_{12} & \cdots & S_{1n} \\ S_{21} & S_{22} & \cdots & S_{2n} \\ \vdots & \vdots & \ddots & \vdots \\ S_{n1} & S_{n2} & \cdots & S_{nn} \end{bmatrix} \begin{pmatrix} a_1 \\ a_2 \\ \vdots \\ a_n \end{pmatrix} \tag{30}$$

3.2 Mixed-mode

A two-port single-ended network can be described by a 2x2 S-parameter matrix as shown by Equation 29. However, to describe a two-port differential-network a 4x4 S-parameter matrix is needed, since there exists a signal pair at each differential port. Fig. 6 shows a sketch of the power wave definitions of a two-port differential-network, i.e., a four-port network.

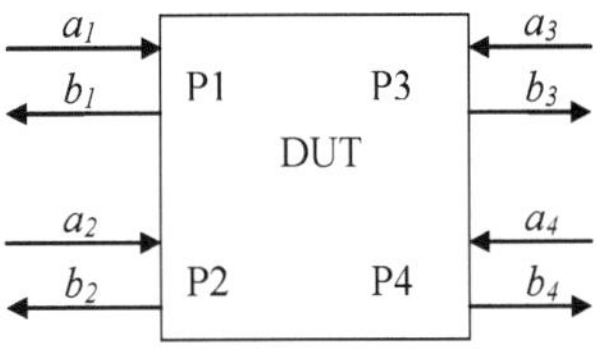

Fig. 6. Power wave definition of a differential two-port network.

Since a real-world differential-signal is composed of both differential- and common-mode signals in general, the single-ended four-port S-parameter matrix does not provide much insight information about the differential- and common-mode matching and transmission. Therefore, the mixed-mode S-parameters must be used. The differential two-port and mixed-mode S-parameters are defined by Equation 31 (Bockelman & Eisenstadt, 1995).

$$\begin{bmatrix} b_{d1} \\ b_{d2} \\ b_{c1} \\ b_{c2} \end{bmatrix} = \begin{bmatrix} \begin{bmatrix} S_{dd11} & S_{dd12} \\ S_{dd21} & S_{dd22} \end{bmatrix} & \begin{bmatrix} S_{dc11} & S_{dc12} \\ S_{dc21} & S_{dc22} \end{bmatrix} \\ \begin{bmatrix} S_{cd11} & S_{cd12} \\ S_{cd21} & S_{cd22} \end{bmatrix} & \begin{bmatrix} S_{cc11} & S_{cc12} \\ S_{cc21} & S_{cc22} \end{bmatrix} \end{bmatrix} \begin{bmatrix} a_{d1} \\ a_{d2} \\ a_{c1} \\ a_{c2} \end{bmatrix} \tag{31}$$

where a_{dn}, a_{cn}, b_{dn} *and* b_{cn} are normalized differential-mode incident-power, common-mode incident-power, differential-mode reflected-power, and common-mode reflected-power at port n. The mixed-mode S matrix is divided into 4 sub-matrixes, where each of the sub-matrixes provides information for different transmission modes.

- S_{dd} sub-matrix: differential-mode S-parameters
- S_{dc} sub-matrix: mode conversion of common- to differential-mode waves
- S_{cd} sub-matrix: mode conversion of differential- to common-mode waves
- S_{cc} sub-matrix: common-mode S-parameters

With mixed-mode S-parameters, characteristics about the differential- and common-mode transmissions and conversions between differential- and common-modes can be found (Bockelman & Eisenstadt, 1995).

3.3 Single-Ended to Mixed-Mode conversion

The best way to measure the mixed-mode S-parameters is to use a four-port mixed-mode vector network analyzer (VNA). In the case where the mixed-mode S-parameters cannot directly be simulated or measured, the single-ended results can first be obtained and then converted into mixed-mode S-parameters by a mathematical conversion. This section shows how it is done for a differential two-port network shown in Fig. 6.

The differential- and common-mode voltages, currents and impedances can be expressed as below, where n is the port number,

$$V_{dn} = V_{2n-1} - V_{2n} \qquad V_{cn} = \frac{V_{2n-1}+V_{2n}}{2} \tag{32}$$

$$I_{dn} = \frac{I_{2n-1}-I_{2n}}{2} \qquad I_{cn} = I_{2n-1} + I_{2n} \tag{33}$$

$$Z_d = \frac{V_d}{I_d} = 2Z_{oo} \qquad Z_c = \frac{V_c}{I_c} = \frac{Z_{oe}}{2} \tag{34}$$

Similar to the single-ended incident- and reflected-powers, the differential- and common-mode incident- and reflected-powers are defined as follows

$$a_{dn} = \frac{1}{2\sqrt{Z_{dn}}}(V_{dn} + Z_{dn}I_{dn}) \tag{35}$$

$$a_{cn} = \frac{1}{2\sqrt{Z_{cn}}}(V_{cn} + Z_{cn}I_{cn}) \tag{36}$$

$$b_{dn} = \frac{1}{2\sqrt{Z_{dn}}}(V_{dn} - Z_{dn}I_{dn}) \tag{37}$$

$$b_{cn} = \frac{1}{2\sqrt{Z_{cn}}}(V_{cn} - Z_{cn}I_{cn}) \tag{38}$$

where a_{dn}, a_{cn}, b_{dn} and b_{cn} are normalized differential-mode incident-power, common-mode incident-power, differential-mode reflected-power, and common-mode reflected-power at port n, respectively. The voltage and current at port n can be expressed by rewriting Equations 35-38 to as follow

$$V_n = \sqrt{Z_0}(a_n + b_n) \tag{39}$$

$$I_n = \frac{1}{\sqrt{Z_0}}(a_n + b_n) \tag{40}$$

Inserting Equations 32-34 and Equations 39-40 into Equations 35-38 and assuming that Z_{oo} = Z_{oe} = Z_0, yields the following results

$$a_{dn} = \frac{a_{2n-1} - a_{2n}}{\sqrt{2}} \qquad a_{cn} = \frac{a_{2n-1} + a_{2n}}{\sqrt{2}} \tag{41}$$

$$b_{dn} = \frac{b_{2n-1} - b_{2n}}{\sqrt{2}} \qquad b_{cn} = \frac{b_{2n-1} + b_{2n}}{\sqrt{2}} \tag{42}$$

As shown by Equations 41 and 42 the differential incident and reflected waves can be described by the single-ended waves. Inserting Equations 41 and 42 into Equation 31, the following expression is obtained.

$$[S^{mm}] = [M][S][M]^{-1} \tag{43}$$

$$where\ [S^{mm}] = \begin{bmatrix} S_{dd11} & S_{dd12} & S_{dc11} & S_{dc12} \\ S_{dd21} & S_{dd22} & S_{dc21} & S_{dc22} \\ S_{cd11} & S_{cd12} & S_{cc11} & S_{cc12} \\ S_{cd21} & S_{cd22} & S_{cc21} & S_{cc22} \end{bmatrix},$$

$$[S] = \begin{bmatrix} S_{11} & S_{12} & S_{13} & S_{14} \\ S_{21} & S_{22} & S_{23} & S_{24} \\ S_{31} & S_{32} & S_{33} & S_{34} \\ S_{41} & S_{42} & S_{43} & S_{44} \end{bmatrix} \ and\ [M] = \frac{1}{\sqrt{2}} \begin{bmatrix} 1 & -1 & 0 & 0 \\ 0 & 0 & 1 & 0 \\ 1 & 1 & 0 & -1 \\ 0 & 0 & 1 & 1 \end{bmatrix}$$

As shown by Equation 43, single-ended S-parameters can be converted into mixed-mode S-parameters with the [M]-matrix (Bockelman & Eisenstadt, 1995).
Note that the conversion method assumes that $Z_{oo} = Z_{oe} = Z_0$. This assumption is only true if the coupling between the differential signals does not exist. Figure 2 clearly shows that if there is a coupling between the transmission lines then $Z_0 \neq Z_{oo} \neq Z_{oe}$. Although this conversion method is widely used today, the weakness of the conversion method has been noticed by people working in the area (Vaz & Caggiano, 2004).
Based on this observation, two new parameters k_{oo} and k_{oe} depending on the coupling between the transmission lines are introduced (Huynh et al., 2007). By this extension, the effect of differential- and common-mode impedances are included in the conversion. Inserting $Z_{oo} = k_{oo}Z_0$ and $Z_{oe} = k_{oe}Z_0$ into Equations 35-38, the following equations can be obtained

$$a_{dn} = \frac{(1+k_{oo})(a_{2n-1} - a_{2n}) + (1-k_{oo})(b_{2n-1} - b_{2n})}{2\sqrt{2k_{oo}}} \tag{44}$$

$$a_{cn} = \frac{(1+k_{oe})(a_{2n-1} + a_{2n}) + (1-k_{oe})(b_{2n-1} + b_{2n})}{2\sqrt{2k_{oe}}} \tag{45}$$

$$b_{dn} = \frac{(1+k_{oo})(b_{2n-1} - b_{2n}) + (1-k_{oo})(a_{2n-1} - a_{2n})}{2\sqrt{2k_{oo}}} \tag{46}$$

$$b_{cn} = \frac{(1+k_{oe})(b_{2n-1} + b_{2n}) + (1-k_{oe})(a_{2n-1} + a_{2n})}{2\sqrt{2k_{oe}}} \tag{47}$$

Inserting Equations 44-47 into Equations 31 results in the following equation

$$[S^{mm}] = ([M_1][S] + [M_2])([M_1] + [M_2][S])^{-1} \tag{48}$$

$$where\ [M_1] = \begin{bmatrix} \frac{1+k_{oo}}{2\sqrt{2k_{oo}}} & -\frac{1+k_{oo}}{2\sqrt{2k_{oo}}} & 0 & 0 \\ 0 & 0 & \frac{1+k_{oo}}{2\sqrt{2k_{oo}}} & -\frac{1+k_{oo}}{2\sqrt{2k_{oo}}} \\ \frac{1+k_{oe}}{2\sqrt{2k_{oe}}} & \frac{1+k_{oe}}{2\sqrt{2k_{oe}}} & 0 & 0 \\ 0 & 0 & \frac{1+k_{oe}}{2\sqrt{2k_{oe}}} & \frac{1+k_{oe}}{2\sqrt{2k_{oe}}} \end{bmatrix}$$

$$and\ [M_2] = \begin{bmatrix} \frac{1-k_{oo}}{2\sqrt{2k_{oo}}} & -\frac{1-k_{oo}}{2\sqrt{2k_{oo}}} & 0 & 0 \\ 0 & 0 & \frac{1-k_{oo}}{2\sqrt{2k_{oo}}} & -\frac{1-k_{oo}}{2\sqrt{2k_{oo}}} \\ \frac{1-k_{oe}}{2\sqrt{2k_{oe}}} & \frac{1-k_{oe}}{2\sqrt{2k_{oe}}} & 0 & 0 \\ 0 & 0 & \frac{1-k_{oe}}{2\sqrt{2k_{oe}}} & \frac{1-k_{oe}}{2\sqrt{2k_{oe}}} \end{bmatrix}$$

As shown by Equation 48, the single-ended S-parameter representation can be converted to mixed-mode S-parameters with the [M_1] and [M_2] matrixes for coupled transmission lines.

4. Conclusions

Owing to the existence of coupling between two parallel transmission lines, the characteristic impedance of the transmission line will change depending on the line spacing and the signal pattern transmitting on the adjacent lines. The defined odd- and even-mode signals can be related to differential- and common-mode signals in the differential transmission technique. Consequently, this leads to the fact that working with highly coupled differential transmission lines one must take the odd- and even-mode impedance variations into account. Otherwise, mismatching will occur and distortions will be introduced into the system.

Furthermore, mixed-mode S-parameters were introduced and how to convert single-ended to mixed-mode S-parameters was also presented. The impedance changes due to the odd- and even-mode signal pattern are included to give a high accuracy when using mixed-mode S-parameters.

What is not included in this chapter is how to find the odd- and even mode impedances in practice. However, one can do it by using a differential time domain reflectormeter (TDR) to find the odd- and even-mode impedances. In fact, companies providing vector network analyzers can provide TDR as an embedded module, so only one vector network analyzer is needed for real measurements.

Moreover, complex odd- and even-mode impedance have not been taken into consideration in this chapter. Further studies can be done for verification of the theory.

5. References

Bockelman D. E. and Eisenstadt W. R., Combined Differential and Common-Mode Scattering Parameters: Theory and Simulation. *IEEE transactions on microwave theory and techniques*, Vol. 43, No. 7, pp.1530-1539, July 1995.

Clayton P.R. *Introduction to Elecctromagnetic Compatibility*, John Wiley & Sons, Inc., ISBN-13:978-0-471-75500-5, Hoboken, New Jersey, United States of America, 2006.

Hall S. H.; Hall G. W. and McCall J.A. *High-Speed digital System Design*, John Wiley & Sons, Inc., ISBN 0-471-36090-2, USA, 2000.

Huynh A.; Håkansson P. and Gong S., "Mixed-mode S-parameter conversion for networks with coupled differential signals" European Microwave Conference, pp. 238-241, July 2007.

Ludwig R. & Bretchko P. *RF Circuit Design Theory and Applications*, Prectice Hall, ISBN 0-13-095323-7, New Jersey, USA, 2000.

Kurokawa K. Power Waves and the Scattering Matrix, *IEEE Transactions of Microwave Theory and Techniques*, Vol., 13, Marsh 1965, pp. 194-202.

Pozar D.M. *Microwave Engineering Third Edition*, John Wiley & Sons Inc., ISBN 0-471-48878-8, USA, 2005.

Vaz K. & Caggiano M. Measurement Technique for the Extraction of Differential S-Parameters from Single-ended S-Parameters. *IEEE 27th International Spring Seminar on Electronics Technology*, Vol. 2, pp. 313-317, May 2004.

2

A thru-only de-embedding method for on-wafer characterization of multiport networks

Shuhei Amakawa, Noboru Ishihara, and Kazuya Masu
Tokyo Institute of Technology
Japan

1. Overview

De-embedding is the process of deducing the characteristics of a device under test (DUT) from measurements made at a distance ((Bauer & Penfield, 1974)), often via additional measurements of one or more dummy devices. This article reviews a simple thru-only de-embedding method suitable for on-wafer characterization of 2-port, 4-port, and $2n$-port networks having a certain symmetry property. While most conventional de-embedding methods require two or more dummy patterns, the thru-only method requires only one THRU pattern.

If the device under measurement is a 2-port and the corresponding THRU pattern has the left/right reflection symmetry, the THRU can be mathematically split into symmetric halves and the scattering matrix for each of them can be determined (Ito & Masu, 2008; Laney, 2003; Nan et al., 2007; Song et al., 2001; Tretiakov et al., 2004a). Once those scattering matrices are available, the effects of pads and leads can be canceled and the characteristics of the device obtained. The method was applied up to 110GHz for characterization of an on-chip transmission line (TL) (Ito & Masu, 2008).

In the case of 4-port devices such as differential transmission lines, 4-port THRU patterns with ground-signal-ground-signal-ground (GSGSG) pads or GSSG pads can often be designed to have the even/odd symmetry in addition to the left/right reflection symmetry. In that case, the scattering matrix for a THRU can be transformed into a block-diagonal form representing two independent 2-ports by an even/odd transformation. Then, the 2-port thru-only de-embedding method can be applied to the resultant two 2-ports. This 4-port thru-only method was applied to de-embedding of a pair of coupled transmission lines up to 50 GHz (Amakawa et al., 2008). The result was found to be approximately consistent with that from the standard open-short method (Koolen et al., 1991), which requires two dummy patterns: OPEN and SHORT.

In the above case (Amakawa et al., 2008), the transformation matrix was known a priori because of the nominal symmetry of the THRU. However, if the 4-port THRU does not have the even/odd symmetry or if the device under measurement is a $2n$-port with $n \geq 3$, the above method cannot be applied. Even if so, the thru-only method can actually be extended to 4-ports without even/odd symmetry or $2n$-ports by using the recently proposed S-parameter-based modal decomposition of multiconductor transmission lines (MTLs) (Amakawa et al., 2009). A $2n$-port THRU can be regarded as nonuniform multiconductor transmission lines,

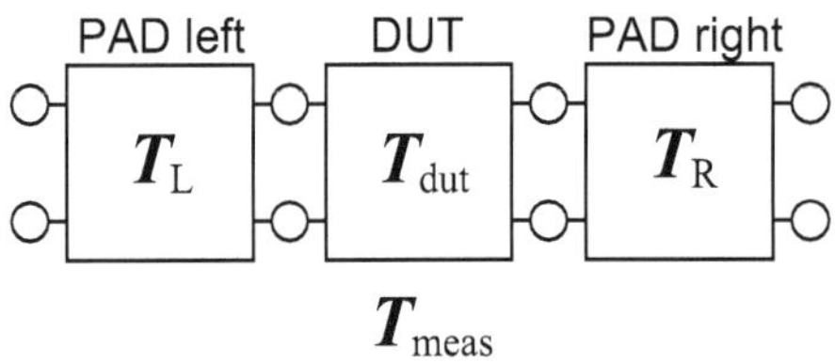

(a) A test pattern with pads and a DUT.

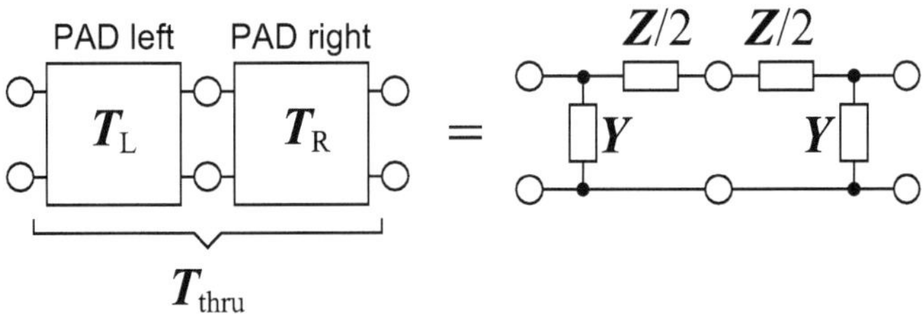

(b) Model of THRU.

Fig. 1. DUT embedded in parasitic networks.

and its scattering matrix can be transformed into a block-diagonal form with 2×2 diagonal blocks, representing n uncoupled 2-ports. The validity of the procedure was confirmed by applying it to de-embedding of four coupled transmission lines, which is an 8-port (Amakawa et al., 2009).

The thru-only de-embedding method could greatly facilitate accurate microwave and millimeter-wave characterization of on-chip multiport networks. It also has the advantage of not requiring a large area of expensive silicon real estate.

2. Introduction

Demand for accurate high-frequency characterization of on-chip devices has been escalating concurrently with the accelerated development of high-speed digital signaling systems and radio-frequency (RF) circuits. Millimeter-wave CMOS circuits have also been becoming a hot research topic.

To characterize on-chip devices and circuits, on-wafer scattering parameter (S-parameter) measurements with a vector network analyzer (VNA) have to be made. A great challenge there is how to deal with parasitics. Since an on-wafer device under test (DUT) is inevitably "embedded" in such intervening structures as probe pads and leads as schematically shown in Fig. 1(a), and they leave definite traces in the S-parameters measured by a VNA, the characteristics of the DUT have to be "de-embedded" (Bauer & Penfield, 1974) in some way from the as-measured data.

While there have been a number of de-embedding methods proposed for 2-port networks, very few have been proposed for 4-port networks in spite of the fact that many important devices, such as differential transmission lines, are represented as 4-ports. In this article, we present a simple 4-port de-embedding method that requires only a THRU pattern (Amakawa

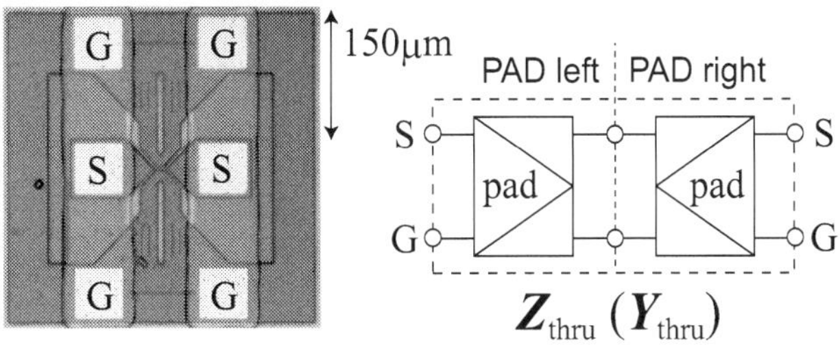

Fig. 2. Micrograph and schematic representation of THRU (Ito & Masu, 2008).

et al., 2008). This method is an extension of a thru-only method for 2-ports. In addition, we also present its extension to $2n$-ports (Amakawa et al., 2009).

The rest of this article starts with a brief description of the thru-only de-embedding method for 2-ports in Section 3. It forms the basis for the multiport method. In Section 4, we explain the mode transformation theory used in the multiport de-embedding method. Section 5 presents an example of performing de-embedding by the thru-only method when the DUT is a 4-port having the even/odd symmetry. Section 6 explains how the mode transformation matrix can be found when the DUT does not have such symmetry or when the DUT is a $2n$-port with $n \geq 3$. Section 7 shows examples of applying the general method. Finally, Section 8 concludes the article.

3. Thru-only de-embedding for 2-ports

Commonly used de-embedding methods usually employ OPEN and SHORT on-chip standards (dummy patterns) (Wartenberg, 2002). De-embedding procedures are becoming increasingly complex and tend to require several dummy patterns (Kolding, 2000b; Vandamme et al., 2001; Wei et al., 2007). The high cost associated with the large area required for dummy patterns is a drawback of advanced de-embedding methods.

Thru-only methods, in contrast, require only one THRU and gaining popularity (Daniel et al., 2004; Goto et al., 2008; Ito & Masu, 2008; Laney, 2003; Nan et al., 2007; Song et al., 2001; Tretiakov et al., 2004a).

In (Ito & Masu, 2008; Laney, 2003; Nan et al., 2007; Song et al., 2001; Tretiakov et al., 2004a), the THRU is modeled by a Π-type equivalent circuit shown in Fig. 1(b). The method of (Goto et al., 2008), on the other hand, was derived from (Mangan et al., 2006), which is related to (Rautio, 1991). It is applicable if the series parasitic impedance Z in Fig. 1(b) is negligible (Goto et al., 2008; Ito & Masu, 2008; Rautio, 1991). In what follows, we will focus on the method of (Ito & Masu, 2008; Laney, 2003; Nan et al., 2007; Song et al., 2001; Tretiakov et al., 2004a).

The THRU pattern used in (Ito & Masu, 2008) is shown in Fig. 2. The 150 μm-pitch ground-signal-ground (GSG) pads are connected with each other via short leads. It turned out that the THRU can be adequately represented by the frequency-independent model shown in Fig. 3. Fig. 4 shows good agreement between the measurement data and the model up to 100 GHz.

The procedure of the thru-only de-embedding method ((Ito & Masu, 2008; Laney, 2003; Nan et al., 2007; Song et al., 2001; Tretiakov et al., 2004a)) is as follows. The 2-port containing the DUT and the THRU a are assumed to be representable by Fig. 1(a) and Fig. 1(b), respectively.

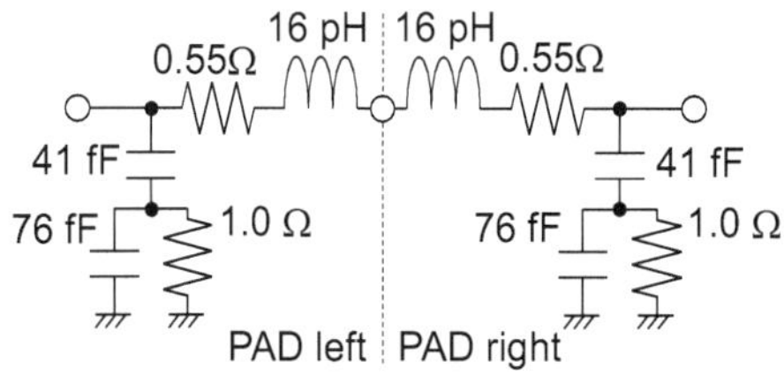

Fig. 3. Lumped-element Π-model of THRU (Ito & Masu, 2008).

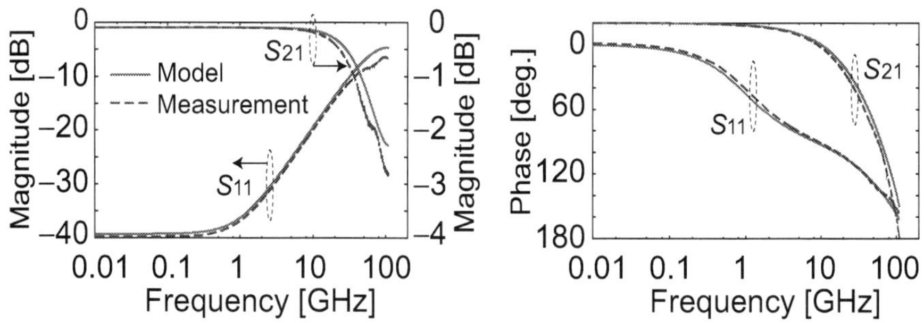

Fig. 4. Measured and modeled (Fig. 3) S-parameters of the THRU pattern. (Ito & Masu, 2008).

In terms of transfer matrices (Mavaddat, 1996), this means that

$$\mathbf{T}_{\mathrm{meas}} = \mathbf{T}_{\mathrm{L}} \mathbf{T}_{\mathrm{dut}} \mathbf{T}_{\mathrm{R}}, \tag{1}$$

$$\mathbf{T}_{\mathrm{thru}} = \mathbf{T}_{\mathrm{L}} \mathbf{T}_{\mathrm{R}}. \tag{2}$$

The S-matrix and T-matrix of a 2-port are related to each other through

$$\mathbf{S} = \begin{bmatrix} S_{11} & S_{12} \\ S_{21} & S_{22} \end{bmatrix} = \frac{1}{T_{11}} \begin{bmatrix} T_{21} & \det \mathbf{T} \\ 1 & -T_{12} \end{bmatrix}, \tag{3}$$

$$\mathbf{T} = \begin{bmatrix} T_{11} & T_{12} \\ T_{21} & T_{22} \end{bmatrix} = \frac{1}{S_{21}} \begin{bmatrix} 1 & -S_{22} \\ S_{11} & -\det \mathbf{S} \end{bmatrix}. \tag{4}$$

Suppose now that the Y-matrix of the THRU is given by

$$\mathbf{Y}_{\mathrm{thru}} = \begin{bmatrix} y_{11} & y_{12} \\ y_{12} & y_{11} \end{bmatrix}. \tag{5}$$

Note that in (5), reciprocity ($y_{21} = y_{12}$) and reflection symmetry ($y_{22} = y_{11}$) are assumed. (5) can be found by converting the measured S-matrix of the THRU into a Y-matrix through (18). If the THRU is split into symmetric halves according to the Π-equivalent in Fig. 1(b),

$$\mathbf{Y}_{\mathrm{L}} = \begin{bmatrix} Y + 2Z^{-1} & -2Z^{-1} \\ -2Z^{-1} & 2Z^{-1} \end{bmatrix} \tag{6}$$

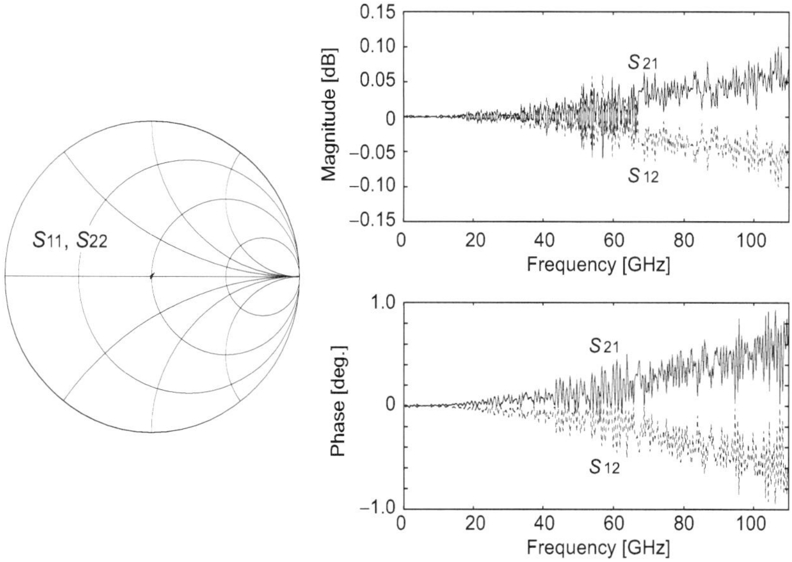

Fig. 5. De-embedded results of the THRU pattern. The thru-only de-embedding method is applied. The maximum magnitude of S_{11} is -33.7 dB (Ito & Masu, 2008).

and

$$\mathbf{Y}_{\mathrm{R}} = \begin{bmatrix} 2Z^{-1} & -2Z^{-1} \\ -2Z^{-1} & Y + 2Z^{-1} \end{bmatrix}, \tag{7}$$

respectively. The parameters in Fig. 1(b) are then given by

$$Y = y_{11} + y_{12}, \tag{8}$$

$$Z = -1/y_{12}. \tag{9}$$

The characteristics of the DUT can be de-embedded as

$$\mathbf{T}_{\mathrm{dut}} = \mathbf{T}_{\mathrm{L}}^{-1}\mathbf{T}_{\mathrm{meas}}\mathbf{T}_{\mathrm{R}}^{-1}. \tag{10}$$

For the procedure to be valid, it is necessary, at least, that the de-embedded THRU that does nothing. That is, S_{11} and S_{22} should be at the center of the Smith chart, and S_{12} and S_{21} are at (1,0). Fig. 5 shows that those do hold approximately. Published papers indicate reasonable success of the thru-only de-embedding method for 2-ports (Ito & Masu, 2008; Laney, 2003; Nan et al., 2007; Song et al., 2001; Tretiakov et al., 2004a).

4. Theory of mode transformation

4.1 General theory

In this section, we explain the theory of S-matrix mode transformation (Amakawa et al., 2008) in preparation for developing thru-only de-embedding for multiports based on the 2-port method explained in the preceding section.

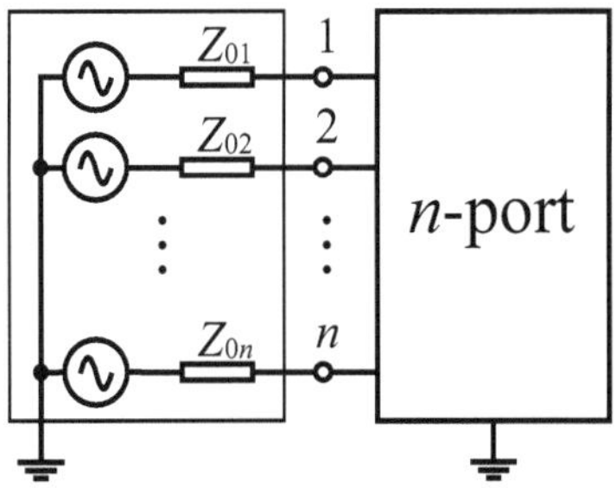

Fig. 6. The right n-port is the network under measurement. The left one is the terminating network, possibly representing a measurement system like a VNA.

A generalized scattering matrix $\mathbf{S}$ of an n-port (Fig. 6) relates the vector, $\mathbf{a}$, of power waves of a given frequency incident upon the n-port to the vector, $\mathbf{b}$, of outgoing power waves (Kurokawa, 1965; Mavaddat, 1996).

$$\mathbf{b} = \mathbf{S}\mathbf{a}, \tag{11}$$

$$\mathbf{a} = \mathbf{R}_0^{-1/2}\mathbf{v}^+ = \frac{1}{2}\mathbf{R}_0^{-1/2}(\mathbf{v} + \mathbf{Z}_0\mathbf{i}), \tag{12}$$

$$\mathbf{b} = \mathbf{R}_0^{-1/2}\mathbf{v}^- = \frac{1}{2}\mathbf{R}_0^{-1/2}(\mathbf{v} - \mathbf{Z}_0^*\mathbf{i}), \tag{13}$$

$$\mathbf{v} = \mathbf{v}^+ + \mathbf{v}^- = \mathbf{Z}\mathbf{i}, \tag{14}$$

$$\mathbf{i} = \mathbf{i}^+ + \mathbf{i}^- = \mathbf{Y}\mathbf{v}. \tag{15}$$

In (12) and (13), * denotes complex conjugate. $\mathbf{Z}_0$ is the reference impedance matrix used to define the generalized S-matrix, and $\mathbf{R}_0 = \Re(\mathbf{Z}_0)$. $\mathbf{Z}_0$ is a diagonal matrix in the conductor domain, in which actual measurements are made with a vector network analyzer (VNA). The kth diagonal element of $\mathbf{Z}_0$ is the reference impedance of the kth port. $\mathbf{Z}_0$ is usually set by a VNA to be a real scalar matrix: $\mathbf{Z}_0 = \mathbf{R}_0 = R_0\mathbf{1}_n$ with $R_0 = 50\,\Omega$. Here $\mathbf{1}_n$ is an $n \times n$ identity matrix. $\mathbf{v}$ and $\mathbf{i}$ in (14) and (15) are, respectively, the port voltage vector and the port current vector in the conductor domain. $\mathbf{Z}$ in (14) is the open-circuit impedance matrix of the n-port under measurement and its inverse is the short-circuit admittance matrix $\mathbf{Y}$ in (15). From (11) and (12), $\mathbf{S}$ and $\mathbf{Z}$ $(= \mathbf{Y}^{-1})$ can be converted to each other by

$$\mathbf{S} = \mathbf{R}_0^{-1/2}\left(\mathbf{Z} - \mathbf{Z}_0^*\right)\left(\mathbf{Z} + \mathbf{Z}_0\right)^{-1}\mathbf{R}_0^{1/2}, \tag{16}$$

$$\mathbf{Z} = \mathbf{R}_0^{1/2}\left(\mathbf{1}_2 - \mathbf{S}\right)^{-1}\left(\mathbf{R}_0^{-1/2}\mathbf{Z}_0^* + \mathbf{S}\mathbf{R}_0^{-1/2}\mathbf{Z}_0\right), \tag{17}$$

$$\mathbf{Y} = \left(\mathbf{R}_0^{-1/2}\mathbf{Z}_0^* + \mathbf{S}\mathbf{R}_0^{-1/2}\mathbf{Z}_0\right)^{-1}\left(\mathbf{1}_2 - \mathbf{S}\right)\mathbf{R}_0^{-1/2}. \tag{18}$$

A network matrix of an n-port (Fig. 6) can be transformed into a different representation by changing the basis sets for voltage and current by the following pair of transformations (Paul, 2008).

$$\mathbf{v} = \mathbf{K}_V\tilde{\mathbf{v}}, \tag{19}$$

$$\mathbf{i} = \mathbf{K}_I\tilde{\mathbf{i}}, \tag{20}$$

subject to

$$\mathbf{K}_\mathrm{V}^\mathrm{T}\mathbf{K}_\mathrm{I}^* = \mathbf{1}_n. \tag{21}$$

Here $^\mathrm{T}$ denotes matrix transposition. $\tilde{\mathbf{v}}$ and $\tilde{\mathbf{i}}$ are, respectively, the port voltage vector and the port current vector in the modal domain. (21) ensures that the power flux remains invariant under the change of bases (Paul, 2008); for example, $\mathbf{a}^\mathrm{T}\mathbf{a}^* = \tilde{\mathbf{a}}^\mathrm{T}\tilde{\mathbf{a}}^*$. (19) and (20) suggest that $\mathbf{K}_\mathrm{V}$ and $\mathbf{K}_\mathrm{I}$ can be expressed in terms of a unitary matrix $\mathbf{K}_\mathrm{U}$ ($\mathbf{K}_\mathrm{U}^\dagger = \mathbf{K}_\mathrm{U}^{-1}$) and a Hermitian matrix $\mathbf{K}_\mathrm{P}$ ($\mathbf{K}_\mathrm{P}^\dagger = \mathbf{K}_\mathrm{P}$) by polar decomposition as

$$\mathbf{K}_\mathrm{V} = \mathbf{K}_\mathrm{U}\mathbf{K}_\mathrm{P}, \quad \mathbf{K}_\mathrm{I} = \mathbf{K}_\mathrm{U}\mathbf{K}_\mathrm{P}^{-1}. \tag{22}$$

Here † denotes conjugate transpose. Since $\mathbf{v}$ and $\mathbf{i}$ are related by the Z-matrix as $\mathbf{v} = \mathbf{Zi}$, impedance matrices undergo the following transformation by the change of bases (19) and (20):

$$\tilde{\mathbf{Z}} = \mathbf{K}_\mathrm{V}^{-1}\mathbf{Z}\mathbf{K}_\mathrm{I} = \mathbf{K}_\mathrm{P}^{-1}\mathbf{K}_\mathrm{U}^\dagger\mathbf{Z}\mathbf{K}_\mathrm{U}\mathbf{K}_\mathrm{P}^{-1}. \tag{23}$$

Likewise,

$$\tilde{\mathbf{Y}} = \mathbf{K}_\mathrm{I}^{-1}\mathbf{Y}\mathbf{K}_\mathrm{V} = \mathbf{K}_\mathrm{P}\mathbf{K}_\mathrm{U}^\dagger\mathbf{Y}\mathbf{K}_\mathrm{U}\mathbf{K}_\mathrm{P}. \tag{24}$$

The reference impedance matrix $\mathbf{Z}_0$ is, in fact, the impedance matrix of the terminating n-port shown in Fig. 6 (with all the signal sources shunted) and is also transformed by (23). If $\mathbf{Z}_0$ is a scalar matrix as is usually the case in the conductor domain,

$$\tilde{\mathbf{Z}}_0 = \mathbf{K}_\mathrm{P}^{-1}\mathbf{Z}_0\mathbf{K}_\mathrm{P}^{-1}. \tag{25}$$

It is only $\mathbf{K}_\mathrm{P}$ (and not $\mathbf{K}_\mathrm{U}$) that affects reference impedances.
If $\mathbf{Z}_0$ is a real scalar matrix, it can be shown that S-matrices undergo the following unitary transformation regardless of the value of $\mathbf{K}_\mathrm{P}$:

$$\tilde{\mathbf{S}} = \mathbf{K}_\mathrm{U}^\dagger\mathbf{S}\mathbf{K}_\mathrm{U}. \tag{26}$$

Mode transformation is particularly useful if the form of $\tilde{\mathbf{S}}$ is diagonal or block-diagonal because it means that the n-port is decoupled into some independent subnetworks. In some cases, the values of $\mathbf{K}_\mathrm{V}$ and $\mathbf{K}_\mathrm{I}$ that give the desired form of $\tilde{\mathbf{S}}$ might be known a priori. Mode transformation is also useful if two or more ports of a network are meant to be excited in a correlated fashion. An example includes differential circuits. In the following, we present a couple of transformations that are used often.

4.2 Even/odd transformation

We define the 2-port even/odd transformation by

$$\mathbf{v} = \begin{bmatrix} V_1 \\ V_2 \end{bmatrix} = \mathbf{K}_{\mathrm{Ve/o}}\mathbf{v}_{\mathrm{e/o}}, \tag{27}$$

$$\mathbf{i} = \begin{bmatrix} I_1 \\ I_2 \end{bmatrix} = \mathbf{K}_{\mathrm{Ie/o}}\mathbf{i}_{\mathrm{e/o}}, \tag{28}$$

$$\mathbf{K}_{\mathrm{Ve/o}} = \mathbf{K}_{\mathrm{Ie/o}} = \frac{1}{\sqrt{2}}\begin{bmatrix} 1 & 0 \\ 0 & -1 \end{bmatrix}, \tag{29}$$

$$\mathbf{v}_{\mathrm{e/o}} = \begin{bmatrix} V_\mathrm{e} \\ V_\mathrm{o} \end{bmatrix} = \frac{1}{\sqrt{2}}\begin{bmatrix} V_1 + V_2 \\ V_1 - V_2 \end{bmatrix}, \tag{30}$$

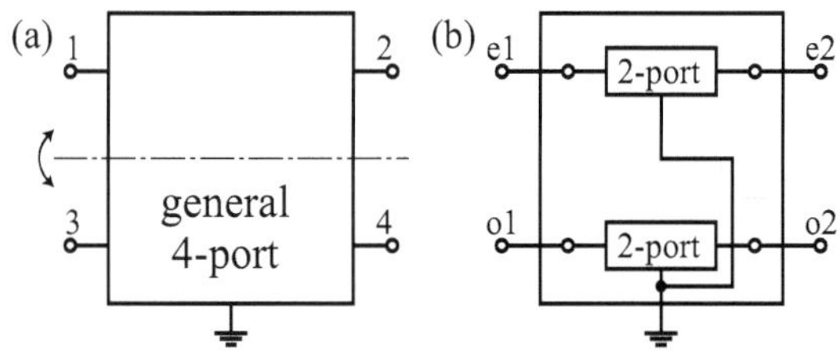

Fig. 7. (a) General 4-port. (b) 4-port consisting of a pair of uncoupled 2-ports.

$$\mathbf{i}_{e/o} = \begin{bmatrix} I_e \\ I_o \end{bmatrix} = \frac{1}{\sqrt{2}} \begin{bmatrix} I_1 + I_2 \\ I_1 - I_2 \end{bmatrix}. \tag{31}$$

Note that $\mathbf{K}_{Ve/o}$ is orthogonal ($\mathbf{K}_{Ve/o}^{T} = \mathbf{K}_{Ve/o}^{-1}$) and, therefore,

$$\mathbf{K}_U = \mathbf{K}_{Ve/o}, \quad \mathbf{K}_P = \mathbf{1}_n. \tag{32}$$

According to (25), the reference impedance matrix in the even/odd domain is given by

$$\mathbf{Z}_{0e/o} = \mathbf{Z}_0. \tag{33}$$

This invariance of the reference impedance matrix is an advantage of the even/odd transformation. This property is consistent with the even mode and the odd mode used in microwave engineering (Pozar, 2005) and transmission line theory (Bakoglu, 1990; Magnusson et al., 2001). From (26),

$$\mathbf{S}_{e/o} = \mathbf{K}_{Ve/o}\mathbf{S}\mathbf{K}_{Ve/o} \tag{34}$$

$$= \frac{1}{2} \begin{bmatrix} S_{11} + S_{21} + S_{12} + S_{22} & S_{11} + S_{21} - S_{12} - S_{22} \\ S_{11} - S_{21} + S_{12} - S_{22} & S_{11} - S_{21} - S_{12} + S_{22} \end{bmatrix}. \tag{35}$$

Extension of the even/odd transformation to 4-ports is straightforward as follows.

$$\mathbf{v} = \begin{bmatrix} V_1 \\ V_2 \\ V_3 \\ V_4 \end{bmatrix} = \mathbf{K}_{Ve/o}\mathbf{v}_{e/o}, \quad \mathbf{i} = \begin{bmatrix} I_1 \\ I_2 \\ I_3 \\ I_4 \end{bmatrix} = \mathbf{K}_{Ie/o}\mathbf{i}_{e/o}, \tag{36}$$

$$\mathbf{K}_{Ve/o} = \mathbf{K}_{Ie/o} = \frac{1}{\sqrt{2}} \begin{bmatrix} 1 & 0 & 1 & 0 \\ 0 & 1 & 0 & 1 \\ 1 & 0 & -1 & 0 \\ 0 & 1 & 0 & -1 \end{bmatrix}, \tag{37}$$

$$\mathbf{v}_{e/o} = \begin{bmatrix} V_{e1} \\ V_{e2} \\ V_{o1} \\ V_{o2} \end{bmatrix} = \frac{1}{\sqrt{2}} \begin{bmatrix} V_1 + V_3 \\ V_2 + V_4 \\ V_1 - V_3 \\ V_2 - V_4 \end{bmatrix}, \quad \mathbf{i}_{e/o} = \begin{bmatrix} I_{e1} \\ I_{e2} \\ I_{o1} \\ I_{o2} \end{bmatrix} = \frac{1}{\sqrt{2}} \begin{bmatrix} I_1 + I_3 \\ I_2 + I_4 \\ I_1 - I_3 \\ I_2 - I_4 \end{bmatrix}. \tag{38}$$

The corresponding port numbering is shown in Fig. 7. Clearly, the exact form of $\mathbf{K}_{Ve/o}$ depends on how the ports are numbered.

Let $\mathbf{S}$ be the conductor-domain 4×4 scattering matrix as measured by a VNA.

$$\mathbf{S} = \begin{bmatrix} \mathbf{S}_{11} & \mathbf{S}_{12} \\ \mathbf{S}_{21} & \mathbf{S}_{22} \end{bmatrix} = \left[\begin{array}{cc|cc} S_{11} & S_{12} & S_{13} & S_{14} \\ S_{21} & S_{22} & S_{23} & S_{24} \\ \hline S_{31} & S_{32} & S_{33} & S_{34} \\ S_{41} & S_{42} & S_{43} & S_{44} \end{array}\right]. \tag{39}$$

From (26) and (32), the S-matrix in the even/odd domain, $\mathbf{S}_{e/o}$, is given by the following orthogonal transformation.

$$\mathbf{S}_{e/o} = \begin{bmatrix} \mathbf{S}_{ee} & \mathbf{S}_{eo} \\ \mathbf{S}_{oe} & \mathbf{S}_{oo} \end{bmatrix} = \mathbf{K}_{Ve/o}\mathbf{S}\mathbf{K}_{Ve/o}, \tag{40}$$

$$\mathbf{S}_{ee} = \begin{bmatrix} S_{e1e1} & S_{e1e2} \\ S_{e2e1} & S_{e2e2} \end{bmatrix} = \frac{1}{2}\left(\mathbf{S}_{11} + \mathbf{S}_{21} + \mathbf{S}_{12} + \mathbf{S}_{22}\right), \tag{41}$$

$$\mathbf{S}_{eo} = \begin{bmatrix} S_{e1o1} & S_{e1o2} \\ S_{e2o1} & S_{e2o2} \end{bmatrix} = \frac{1}{2}\left(\mathbf{S}_{11} + \mathbf{S}_{21} - \mathbf{S}_{12} - \mathbf{S}_{22}\right), \tag{42}$$

$$\mathbf{S}_{oe} = \begin{bmatrix} S_{o1e1} & S_{o1e2} \\ S_{o2e1} & S_{o2e2} \end{bmatrix} = \frac{1}{2}\left(\mathbf{S}_{11} - \mathbf{S}_{21} + \mathbf{S}_{12} - \mathbf{S}_{22}\right), \tag{43}$$

$$\mathbf{S}_{oo} = \begin{bmatrix} S_{o1o1} & S_{o1o2} \\ S_{o2o1} & S_{o2o2} \end{bmatrix} = \frac{1}{2}\left(\mathbf{S}_{11} - \mathbf{S}_{21} - \mathbf{S}_{12} + \mathbf{S}_{22}\right). \tag{44}$$

If the 4-port in question is symmetrical about the horizontal line shown in Fig. 7(a), the off-diagonal submatrices $\mathbf{S}_{eo}$ and $\mathbf{S}_{oe}$ are zero, meaning that the 4-port in the even/odd domain consists of a pair of uncoupled 2-ports as shown in Fig. 7(b). The upper and the lower 2-ports are described by $\mathbf{S}_{ee}$ and $\mathbf{S}_{oo}$, respectively.

4.3 Common/differential transformation

We define the 2-port common/differential transformation by

$$\mathbf{v} = \begin{bmatrix} V_1 \\ V_2 \end{bmatrix} = \mathbf{K}_{Vc/d}\mathbf{v}_{c/d}, \tag{45}$$

$$\mathbf{i} = \begin{bmatrix} I_1 \\ I_2 \end{bmatrix} = \mathbf{K}_{Ic/d}\mathbf{i}_{c/d}, \tag{46}$$

$$\mathbf{K}_{Vc/d} = \begin{bmatrix} 1 & 1/2 \\ 1 & -1/2 \end{bmatrix} = \mathbf{K}_{Ve/o}\begin{bmatrix} \sqrt{2} & 0 \\ 0 & 1/\sqrt{2} \end{bmatrix}, \tag{47}$$

$$\mathbf{K}_{Ic/d} = \left(\mathbf{K}_{Vc/d}^{\dagger}\right)^{-1} = \begin{bmatrix} 1/2 & 1 \\ 1/2 & -1 \end{bmatrix} = \mathbf{K}_{Ie/o}\begin{bmatrix} 1/\sqrt{2} & 0 \\ 0 & \sqrt{2} \end{bmatrix}, \tag{48}$$

$$\mathbf{v}_{c/d} = \begin{bmatrix} V_c \\ V_d \end{bmatrix} = \begin{bmatrix} (V_1 + V_2)/2 \\ V_1 - V_2 \end{bmatrix}, \tag{49}$$

$$\mathbf{i}_{c/d} = \begin{bmatrix} I_c \\ I_d \end{bmatrix} = \begin{bmatrix} I_1 + I_2 \\ (I_1 - I_2)/2 \end{bmatrix}. \tag{50}$$

This definition is consistent with the common mode and the differential mode in analog circuit theory (Gray et al., 2009). The differential mode gives what is interpreted as the signal in

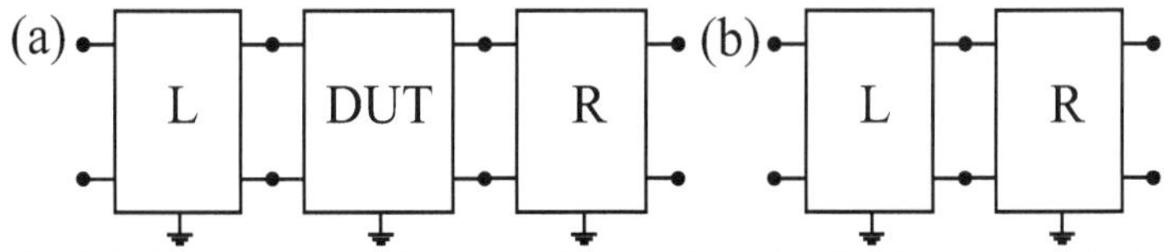

Fig. 8. (a) Model of an as-measured 4-port. The DUT is embedded in between the intervening structures L and R. (b) Model of a THRU dummy pattern.

differential circuits. The common mode describes how the pair would collectively appear when seen from far away.
Since $\mathbf{K}_\mathrm{U} = \mathbf{K}_{\mathrm{Ve/o}}$ and $\mathbf{K}_\mathrm{P} \neq \mathbf{1}_n$ as shown in (47) and (48), according to (25), $\mathbf{Z}_{0\mathrm{c/d}}$ is not equal to $\mathbf{Z}_0$. If, for example,

$$\mathbf{Z}_0 = \begin{bmatrix} 50 & 0 \\ 0 & 50 \end{bmatrix}, \tag{51}$$

then

$$\mathbf{Z}_{0\mathrm{c/d}} = \begin{bmatrix} 25 & 0 \\ 0 & 100 \end{bmatrix}. \tag{52}$$

From (26),

$$\mathbf{S}_{\mathrm{c/d}} = \mathbf{K}_{\mathrm{Ve/o}}\mathbf{S}\mathbf{K}_{\mathrm{Ve/o}} = \mathbf{S}_{\mathrm{e/o}} \tag{53}$$

$$= \frac{1}{2}\begin{bmatrix} S_{11}+S_{21}+S_{12}+S_{22} & S_{11}+S_{21}-S_{12}-S_{22} \\ S_{11}-S_{21}+S_{12}-S_{22} & S_{11}-S_{21}-S_{12}+S_{22} \end{bmatrix}. \tag{54}$$

Extension of the common/differential transformation to 4-ports is also straightforward.

$$\mathbf{v} = \mathbf{K}_{\mathrm{Vc/d}}\mathbf{v}_{\mathrm{c/d}}, \quad \mathbf{i} = \mathbf{K}_{\mathrm{Ic/d}}\mathbf{i}_{\mathrm{c/d}}, \tag{55}$$

$$\mathbf{K}_{\mathrm{Vc/d}} = \begin{bmatrix} 1 & 0 & 1/2 & 0 \\ 0 & 1 & 0 & 1/2 \\ 1 & 0 & -1/2 & 0 \\ 0 & 1 & 0 & -1/2 \end{bmatrix} = \mathbf{K}_{\mathrm{Ve/o}}\begin{bmatrix} \sqrt{2} & 0 & 0 & 0 \\ 0 & \sqrt{2} & 0 & 0 \\ 0 & 0 & 1/\sqrt{2} & 0 \\ 0 & 0 & 0 & 1/\sqrt{2} \end{bmatrix}, \tag{56}$$

$$\mathbf{K}_{\mathrm{Ic/d}} = \begin{bmatrix} 1/2 & 0 & 1 & 0 \\ 0 & 1/2 & 0 & 1 \\ 1/2 & 0 & -1 & 0 \\ 0 & 1/2 & 0 & -1 \end{bmatrix} = \mathbf{K}_{\mathrm{Ve/o}}\begin{bmatrix} 1/\sqrt{2} & 0 & 0 & 0 \\ 0 & 1/\sqrt{2} & 0 & 0 \\ 0 & 0 & \sqrt{2} & 0 \\ 0 & 0 & 0 & \sqrt{2} \end{bmatrix}, \tag{57}$$

$$\mathbf{v}_{\mathrm{c/d}} = \begin{bmatrix} V_{\mathrm{c1}} \\ V_{\mathrm{c2}} \\ V_{\mathrm{d1}} \\ V_{\mathrm{d2}} \end{bmatrix} = \begin{bmatrix} (V_1+V_3)/2 \\ (V_2+V_4)/2 \\ V_1-V_3 \\ V_2-V_4 \end{bmatrix}, \quad \mathbf{i}_{\mathrm{c/d}} = \begin{bmatrix} I_{\mathrm{c1}} \\ I_{\mathrm{c2}} \\ I_{\mathrm{d1}} \\ I_{\mathrm{d2}} \end{bmatrix} = \begin{bmatrix} I_1+I_3 \\ I_2+I_4 \\ (I_1-I_3)/2 \\ (I_2-I_4)/2 \end{bmatrix}. \tag{58}$$

$\mathbf{S}_{\mathrm{c/d}}$ is given by (53) and (40) through (44). This result is consistent with (Bockelman & Eisenstadt, 1995; Yanagawa et al, 1994) except for port ordering.

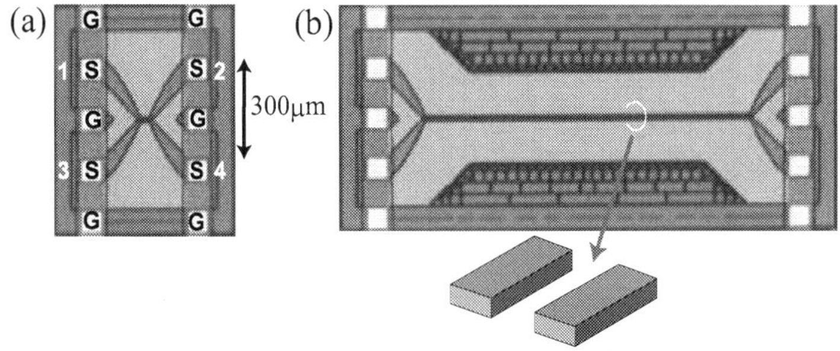

Fig. 9. (a) Micrograph of a THRU with GSGSG pads. (b) A pair of 1 mm-long TLs with the same pads. The line width is 6 μm and the spacing between the lines is 4.6 μm. Nominal differential characteristic impedance is 100 Ω. The technology is a 0.18-μm CMOS process.

5. De-embedding of 4-port with even/odd symmetry

Suppose a 4-port under measurement can be represented as a cascade of three four-ports, as shown in Fig. 8(a). Then, its transfer matrix can be written as $\mathbf{T}_{\text{meas}} = \mathbf{T}_{\text{L}}\mathbf{T}_{\text{dut}}\mathbf{T}_{\text{R}}$. As in the case of a 2-port, if the intervening structures L and R are somehow characterized, the properties of the DUT can be de-embedded by $\mathbf{T}_{\text{dut}} = \mathbf{T}_{\text{L}}^{-1}\mathbf{T}_{\text{meas}}\mathbf{T}_{\text{R}}^{-1}$. The 4-port de-embedding method proposed by (Han et al., 2003) follows this idea. Their method requires that the upper ports and the lower ports of both L and R consist of *uncoupled* two-ports as in Fig. 7(b). This condition may be fulfilled by appropriately configured off-chip systems (Han et al., 2003). However, on-chip THRU patterns (e.g. Fig. 9(a)) can hardly meet this requirement. Note, however, that an on-chip THRU, typically having GSGSG or GSSG probe pads, can often be made symmetrical about the horizontal line shown in Fig. 7(a). In that case, the S-matrix of the THRU can be decomposed into a pair of uncoupled 2-ports (Fig. 7(b)) by the even/odd transformation (40) (Amakawa et al., 2008) or, equivalently, the common/differential transformation (53). Then, each resultant 2-port can be bisected and the matrix representing each half determined as described in Section 3.

The conversion of 4-port $\mathbf{S}$ to/from $\mathbf{T}$ can be done via $\mathbf{S}'$ defined by

$$\mathbf{S}' = \begin{bmatrix} \mathbf{S}'_{11} & \mathbf{S}'_{12} \\ \mathbf{S}'_{21} & \mathbf{S}'_{22} \end{bmatrix} = \left[\begin{array}{cc|cc} S'_{11} & S'_{12} & S'_{13} & S'_{14} \\ S'_{21} & S'_{22} & S'_{23} & S'_{24} \\ \hline S'_{31} & S'_{32} & S'_{33} & S'_{34} \\ S'_{41} & S'_{42} & S'_{43} & S'_{44} \end{array}\right] = \left[\begin{array}{cc|cc} S_{11} & S_{13} & S_{12} & S_{14} \\ S_{31} & S_{33} & S_{32} & S_{34} \\ \hline S_{21} & S_{23} & S_{22} & S_{24} \\ S_{41} & S_{43} & S_{42} & S_{44} \end{array}\right], \tag{59}$$

$$\mathbf{T} = \begin{bmatrix} \mathbf{T}_{11} & \mathbf{T}_{12} \\ \mathbf{T}_{21} & \mathbf{T}_{22} \end{bmatrix} = \left[\begin{array}{cc|cc} T_{11} & T_{12} & T_{13} & T_{14} \\ T_{21} & T_{22} & T_{23} & T_{24} \\ \hline T_{31} & T_{32} & T_{33} & T_{34} \\ T_{41} & T_{42} & T_{43} & T_{44} \end{array}\right] = \begin{bmatrix} \mathbf{S}'^{-1}_{21} & -\mathbf{S}'^{-1}_{21}\mathbf{S}'_{22} \\ \mathbf{S}'_{11}\mathbf{S}'^{-1}_{21} & \mathbf{S}'_{12} - \mathbf{S}'_{11}\mathbf{S}'^{-1}_{21}\mathbf{S}'_{22} \end{bmatrix}, \tag{60}$$

$$\mathbf{S}' = \begin{bmatrix} \mathbf{T}_{21}\mathbf{T}_{11}^{-1} & \mathbf{T}_{22} - \mathbf{T}_{21}\mathbf{T}_{11}^{-1}\mathbf{T}_{12} \\ \mathbf{T}_{11}^{-1} & -\mathbf{T}_{11}^{-1}\mathbf{T}_{12} \end{bmatrix}. \tag{61}$$

We applied the proposed de-embedding method to samples fabricated with a 0.18 μm CMOS process. The frequency ranged from 100 MHz to 50 GHz. The two-step open-short method (Koolen et al., 1991) originally proposed for a two-port was also applied for comparison.

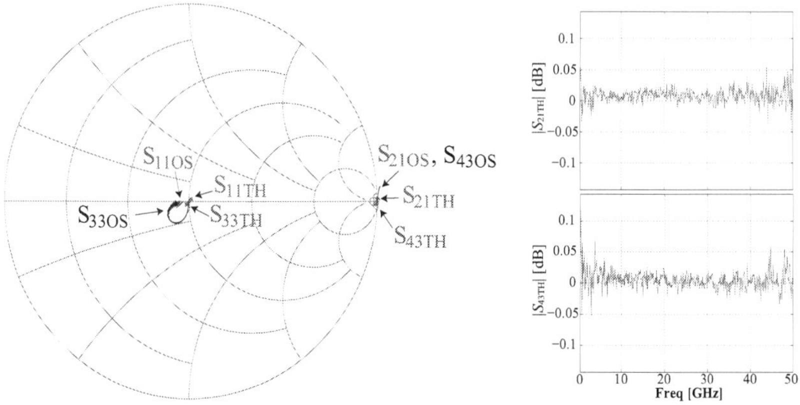

Fig. 10. Characteristics of the THRU (Fig. 9a) after performing the thru-only ($S_{11\mathrm{TH}}$, $S_{33\mathrm{TH}}$, $S_{21\mathrm{TH}}$, $S_{43\mathrm{TH}}$) or open-short ($S_{11\mathrm{OS}}$, $S_{33\mathrm{OS}}$, $S_{21\mathrm{OS}}$, $S_{43\mathrm{OS}}$) de-embedding (Amakawa et al., 2008).

Fig. 10 shows the de-embedded characteristics of a symmetric THRU itself (Fig. 9a). The reflection coefficients obtained by the proposed method ($S_{11\mathrm{TH}}$ and $S_{33\mathrm{TH}}$) stay very close to the center of the Smith chart and the transmission coefficients ($S_{21\mathrm{TH}}$ and $S_{43\mathrm{TH}}$) at its right end as they should. Fig. 11 shows even- and odd-mode transmission coefficients for a pair of 1 mm-long transmission lines shown in Fig. 9(b). A comparatively large difference is seen between the results from the two de-embedding methods for the even mode. One likely cause is the nonideal behavior of the SHORT (Goto et al., 2008; Ito & Masu, 2008). The odd-mode results, on the other hand, agree very well, indicating the immunity of this mode (and the differential mode) to the problem that plague the even mode (and the common mode).

6. Decomposition of a $2n$-port into n 2-ports

The essential used idea in the previous section was to reduce a 4-port problem to two independent 2-port problems by mode transformation. The requirement for it to work was that the 4×4 S matrix of the THRU dummy pattern (a pair of nonuniform TLs) have the even/odd symmetry and left/right symmetry. This development naturally leads to the idea that the same de-embedding method should be applicable to $2n$-ports, where n is a positive integer, provided that the S-matrix of the THRU (n coupled nonuniform TLs) can somehow be block-diagonalized with 2×2 diagonal blocks (Amakawa et al., 2009).

Modal analysis of multiconductor transmission lines (MTLs) have been a subject of intensive study for decades (Faria, 2004; Kogo, 1960; Paul, 2008; Williams et al., 1997). MTL equations are typically written in terms of per-unit-length equivalent-circuit parameters. Experimental characterization of MTLs, therefore, often involves extraction of those parameters from measured S-matrices (Nickel et al., 2001; van der Merwe et al., 1998). We instead directly work with S-matrices. In Section 5, the transformation matrix (37) was known a priori thanks to the even/odd symmetry of the DUT. We now have to find the transformation matrices. As before, we assume throughout that the THRU is reciprocal and hence the associated S-matrix symmetric.

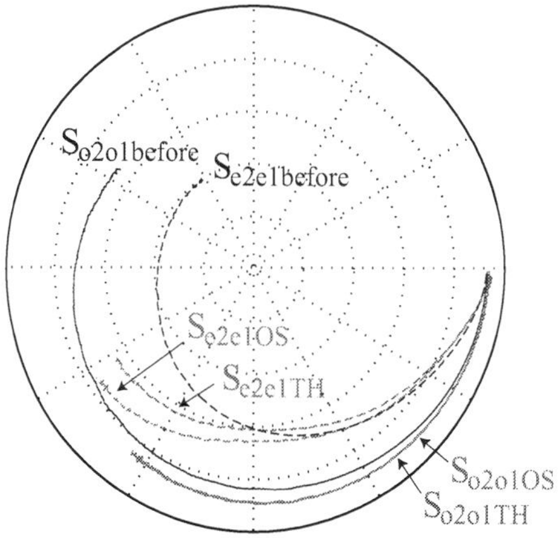

Fig. 11. Even-mode (broken lines) and odd-mode (solid lines) transmission coefficients for a pair of transmission lines (Fig. 9b) before and after de-embedding (thru-only or open-short) (Amakawa et al., 2008).

Our goal is to transform a $2n \times 2n$ scattering matrix $\mathbf{S}$ into the following block-diagonal form:

$$\tilde{\mathbf{S}}' = \begin{bmatrix} \mathbf{S}_{\mathrm{m}1} & & \\ & \ddots & \\ & & \mathbf{S}_{\mathrm{m}n} \end{bmatrix}, \tag{62}$$

where $\mathbf{S}_{\mathrm{m}i}$ are 2×2 submatrices, and the rest of the elements of $\tilde{\mathbf{S}}'$ are all 0. The port numbering for $\tilde{\mathbf{S}}'$ is shown in Fig. 12 with primes. Note that the port numbering convention adopted in this and the next Sections is different from that adopted in earlier Sections. Once the transformation is performed, the DUT can be treated as if they were composed of n uncoupled 2-ports.

This problem is not an ordinary matrix diagonalization problem. The form of (62) results by first transforming $\mathbf{S}$ into $\tilde{\mathbf{S}}$, which has the following form:

$$\tilde{\mathbf{S}} = \left[\begin{array}{c|c} \ddots & \ddots \\ \hline \ddots & \ddots \end{array} \right], \tag{63}$$

and then reordering the rows and columns of $\tilde{\mathbf{S}}$ such that $\mathbf{S}_{\mathrm{m}i}$ in (62) is built from the ith diagonal elements of the four submatrices of $\tilde{\mathbf{S}}$ (Amakawa et al., 2009). The port indices of $\tilde{\mathbf{S}}$ are shown in Fig. 12 without primes. The problem, therefore, is the transformation of $\mathbf{S}$ into $\tilde{\mathbf{S}}$ followed by reordering of rows and columns yielding $\tilde{\mathbf{S}}'$.

In the case of a cascadable $2n$-port, it makes sense to divide the ports into two groups as shown in Fig. 12, and hence the division of $\mathbf{S}$, $\mathbf{a}$, and $\mathbf{b}$ into submatrices/subvectors:

$$\mathbf{b} = \begin{bmatrix} \mathbf{b}_1 \\ \mathbf{b}_2 \end{bmatrix} = \begin{bmatrix} \mathbf{S}_{11} & \mathbf{S}_{12} \\ \mathbf{S}_{21} & \mathbf{S}_{22} \end{bmatrix} \begin{bmatrix} \mathbf{a}_1 \\ \mathbf{a}_2 \end{bmatrix} = \mathbf{S}\mathbf{a}. \tag{64}$$

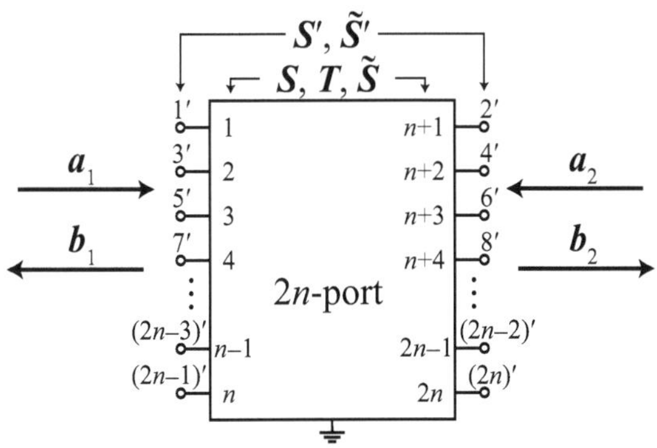

Fig. 12. Port indices for a cascadable $2n$-port. The ports 1 through n of $\mathbf{S}$ constitute one end of the bundle of n lines and the ports $n+1$ through $2n$ the other end.

This was already done in earlier Sections for 4-ports. Since our $2n$-port is reciprocal by assumption, $\mathbf{S}$ is symmetric: $\mathbf{S}^{\mathrm{T}} = \mathbf{S}$. Then, it can be shown that the following change of bases gives the desired transformation.

$$\begin{bmatrix} \mathbf{a}_1 \\ \mathbf{a}_2 \end{bmatrix} = \begin{bmatrix} \mathbf{W}_1 & \\ & (\mathbf{W}_2^{\mathrm{T}})^{-1} \end{bmatrix} \begin{bmatrix} \tilde{\mathbf{a}_1} \\ \tilde{\mathbf{a}_2} \end{bmatrix}, \tag{65}$$

$$\begin{bmatrix} \mathbf{b}_1 \\ \mathbf{b}_2 \end{bmatrix} = \begin{bmatrix} (\mathbf{W}_1^{\mathrm{T}})^{-1} & \\ & \mathbf{W}_2 \end{bmatrix} \begin{bmatrix} \tilde{\mathbf{b}_1} \\ \tilde{\mathbf{b}_2} \end{bmatrix}, \tag{66}$$

where the blanks represent zero submatrices. $\mathbf{W}_1$ and $\mathbf{W}_2$ diagonalize $\mathbf{S}_{21}^{-1}\mathbf{S}_{22}\mathbf{S}_{12}^{-1}\mathbf{S}_{11}$ and $\mathbf{S}_{22}\mathbf{S}_{12}^{-1}\mathbf{S}_{11}\mathbf{S}_{21}^{-1}$, respectively, by similarity transformation:

$$\mathbf{W}_1^{-1}\mathbf{S}_{21}^{-1}\mathbf{S}_{22}\mathbf{S}_{12}^{-1}\mathbf{S}_{11}\mathbf{W}_1 = \mathbf{\Lambda}_1, \tag{67}$$

$$\mathbf{W}_2^{-1}\mathbf{S}_{22}\mathbf{S}_{12}^{-1}\mathbf{S}_{11}\mathbf{S}_{21}^{-1}\mathbf{W}_2 = \mathbf{\Lambda}_2, \tag{68}$$

where $\mathbf{\Lambda}_1$ and $\mathbf{\Lambda}_2$ are diagonal matrices. $\mathbf{W}_1$ and $\mathbf{W}_2$ can be computed by eigenvalue decomposition. The derivation is similar to (Faria, 2004). $\tilde{\mathbf{S}}$ is thus given by

$$\tilde{\mathbf{S}} = \begin{bmatrix} \mathbf{W}_1^{\mathrm{T}}\mathbf{S}_{11}\mathbf{W}_1 & \mathbf{W}_1^{\mathrm{T}}\mathbf{S}_{12}(\mathbf{W}_2^{\mathrm{T}})^{-1} \\ \mathbf{W}_2^{-1}\mathbf{S}_{21}\mathbf{W}_1 & \mathbf{W}_2^{-1}\mathbf{S}_{22}(\mathbf{W}_2^{\mathrm{T}})^{-1} \end{bmatrix}. \tag{69}$$

7. Multiport de-embedding using a THRU

Suppose, as before, that the device under measurement and the THRU can be represented as shown in Fig. 13. Here the DUT is MTLs. In terms of the transfer matrix $\mathbf{T}$ defined by

$$\begin{bmatrix} \mathbf{a}_1 \\ \mathbf{b}_1 \end{bmatrix} = \mathbf{T} \begin{bmatrix} \mathbf{b}_2 \\ \mathbf{a}_2 \end{bmatrix} = \begin{bmatrix} \mathbf{T}_{11} & \mathbf{T}_{12} \\ \mathbf{T}_{21} & \mathbf{T}_{22} \end{bmatrix} \begin{bmatrix} \mathbf{b}_2 \\ \mathbf{a}_2 \end{bmatrix}, \tag{70}$$

$$\mathbf{T} = \begin{bmatrix} \mathbf{T}_{11} & \mathbf{T}_{12} \\ \mathbf{T}_{21} & \mathbf{T}_{22} \end{bmatrix} = \begin{bmatrix} \mathbf{S}_{21}^{-1} & -\mathbf{S}_{21}^{-1}\mathbf{S}_{22} \\ \mathbf{S}_{11}\mathbf{S}_{21}^{-1} & \mathbf{S}_{12} - \mathbf{S}_{11}\mathbf{S}_{21}^{-1}\mathbf{S}_{22} \end{bmatrix}, \tag{71}$$

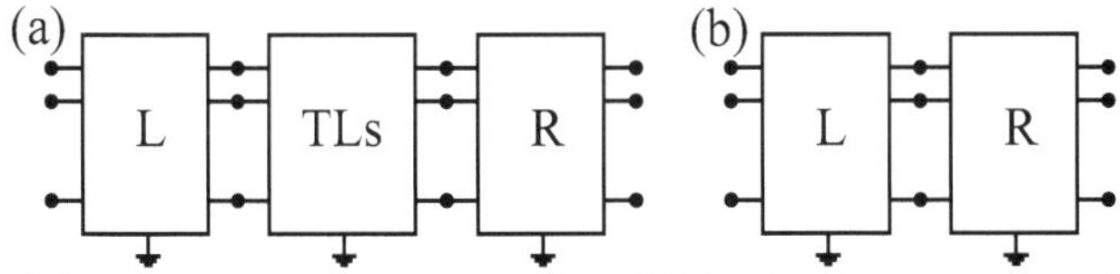

Fig. 13. (a) Model of n coupled TLs measured by a VNA. The TLs sit between the intervening structures L and R. (b) Model of a THRU.

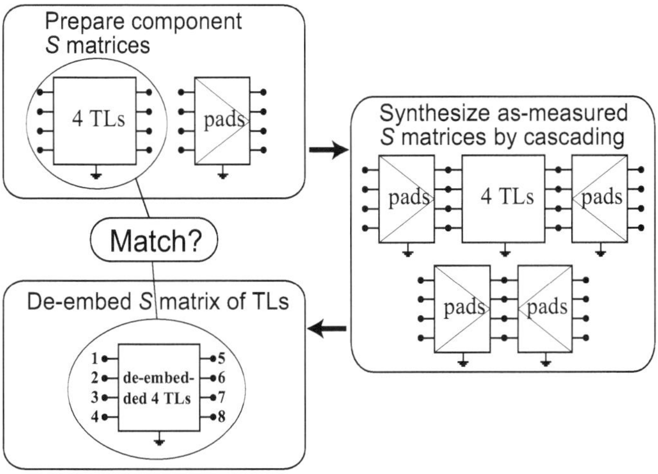

Fig. 14. Flow of validating the de-embedding method.

$$\mathbf{S} = \begin{bmatrix} \mathbf{S}_{11} & \mathbf{S}_{12} \\ \mathbf{S}_{21} & \mathbf{S}_{22} \end{bmatrix} = \begin{bmatrix} \mathbf{T}_{21}\mathbf{T}_{11}^{-1} & \mathbf{T}_{22} - \mathbf{T}_{21}\mathbf{T}_{11}^{-1}\mathbf{T}_{12} \\ \mathbf{T}_{11}^{-1} & -\mathbf{T}_{11}^{-1}\mathbf{T}_{12} \end{bmatrix}. \tag{72}$$

The as-measured T-matrix for Fig. 13(a) is $\mathbf{T}_{\text{meas}} = \mathbf{T}_{\text{L}}\mathbf{T}_{\text{TL}}\mathbf{T}_{\text{R}}$.

In order to de-embed $\mathbf{T}_{\text{TL}}$ from $\mathbf{T}_{\text{meas}}$, the THRU (Fig. 13(b)) is measured, and the result ($\mathbf{T}_{\text{thru}} = \mathbf{T}_{\text{L}}\mathbf{T}_{\text{R}}$) is transformed into the block-diagonal form $\tilde{\mathbf{S}}'_{\text{thru}}$. Since each of the resultant 2×2 diagonal blocks of $\tilde{\mathbf{S}}'_{\text{thru}}$ is symmetric by assumption, the method in Section 3 can be applied to determine $\mathbf{T}_{\text{L}}$ and $\mathbf{T}_{\text{R}}$. Then, the characteristics of the TLs are obtained by $\mathbf{T}_{\text{TL}} = \mathbf{T}_{\text{L}}^{-1}\mathbf{T}_{\text{meas}}\mathbf{T}_{\text{R}}^{-1}$.

Shown in Fig. 14 is the procedure that we followed to validate the thru-only de-embedding method for $2n$-ports (Amakawa et al., 2009). S-parameter files of 1 mm-long 4 coupled TLs and pads were generated by using Agilent Technologies ADS. A cross section of the TLs is shown in Fig. 15. The schematic diagram representing the pads placed at each end of the bundle of TLs is shown in Fig. 16. Figs. 17 and 18 show the characteristics of the "as-measured" TLs and the THRU, respectively. The characteristics of the bare (un-embedded) TLs and the de-embedded results are both shown on the same Smith chart in Fig. 19, but they are indistinguishable, thereby demonstrating the validity of the de-embedding procedure.

We also applied the same de-embedding method to the TLs shown in Fig. 9, analyzed earlier by the even/odd transformation in Section 5 (Amakawa et al., 2008). The numerical values

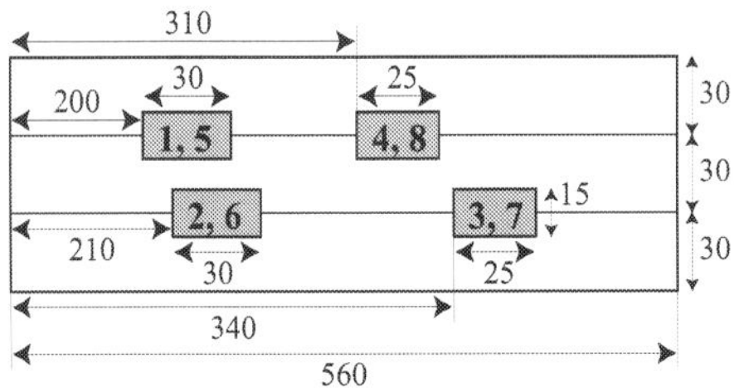

Fig. 15. Schematic cross section of the 1 mm-long 4 coupled TLs (not to scale), labeled with port numbers. Dimensions are in μm. Relative dielectric permittivity is 4. Metal conductivity is $5.9 \times 10^7\ (\Omega \cdot \mathrm{m})^{-1}$. $\tan\delta = 0.04$ (Amakawa et al., 2009).

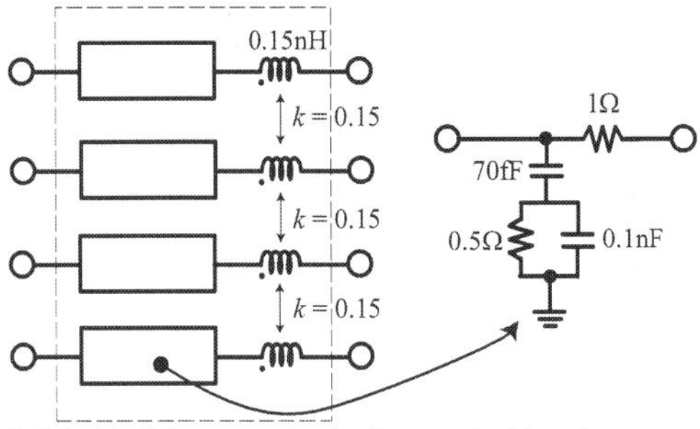

Fig. 16. Model of the left half of the THRU including pads (Amakawa et al., 2009).

of even/odd transformed and fully block-diagonalized S-matrix of the THRU (Fig. 9(a)) at 10 GHz are, respectively,

$$\mathbf{S}'_{\mathrm{e/o}} = \begin{bmatrix} 0.050-0.064\mathrm{j} & 0.828-0.369\mathrm{j} & 0.001-0.000\mathrm{j} & 0.001-0.000\mathrm{j} \\ 0.828-0.369\mathrm{j} & 0.051-0.064\mathrm{j} & -0.000+0.002\mathrm{j} & 0.001+0.000\mathrm{j} \\ 0.001-0.000\mathrm{j} & -0.000+0.002\mathrm{j} & -0.030-0.124\mathrm{j} & 0.904-0.322\mathrm{j} \\ 0.001-0.000\mathrm{j} & 0.001+0.000\mathrm{j} & 0.904-0.322\mathrm{j} & -0.030-0.123\mathrm{j} \end{bmatrix},$$

$$\tilde{\mathbf{S}}' = \begin{bmatrix} 0.051-0.064\mathrm{j} & 0.828-0.369\mathrm{j} & 0.000+0.000\mathrm{j} & 0.000+0.000\mathrm{j} \\ 0.828-0.369\mathrm{j} & 0.050-0.064\mathrm{j} & 0.000+0.000\mathrm{j} & 0.000+0.000\mathrm{j} \\ 0.000+0.000\mathrm{j} & 0.000+0.000\mathrm{j} & -0.030-0.124\mathrm{j} & 0.903-0.324\mathrm{j} \\ 0.000+0.000\mathrm{j} & 0.000+0.000\mathrm{j} & 0.903-0.324\mathrm{j} & -0.032-0.123\mathrm{j} \end{bmatrix}.$$

The reference impedance matrices are $\mathbf{Z}'_{0\mathrm{e/o}} = \tilde{\mathbf{Z}}'_0 = 50 \cdot \mathbf{1}_4$. The upper diagonal block in $\mathbf{S}'_{\mathrm{e/o}}$ is the even-mode S matrix and the lower diagonal block is the odd-mode S matrix. The residual nonzero off-diagonal blocks in $\mathbf{S}'_{\mathrm{e/o}}$, representing the crosstalk between the even and odd modes, were ignored in Section 5 (Amakawa et al., 2008). The transformation (69) can better block-diagonalize the S-matrix.

8. Conclusions

We have reviewed the simple thru-only de-embedding method for characterizing multiport networks at GHz frequencies. It is based on decomposition of a $2n$-port into n uncoupled

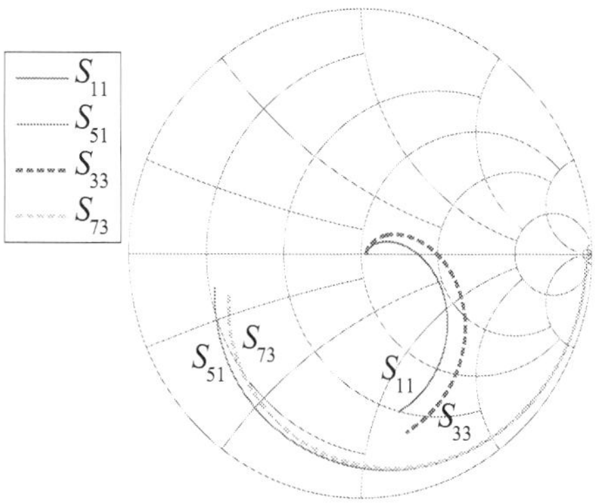

Fig. 17. Characteristics of the 4 coupled TLs from 100 MHz to 40 GHz before de-embedding (Amakawa et al., 2009).

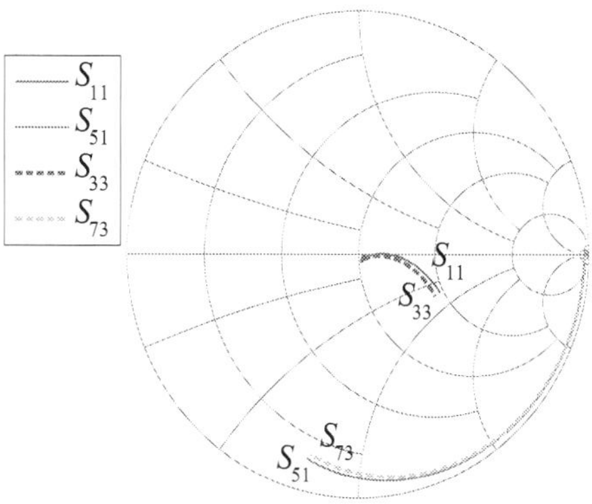

Fig. 18. Characteristics of the THRU from 100 MHz to 40 GHz (Amakawa et al., 2009).

2-ports. After the decomposition, the 2-port thru-only de-embedding method is applied. If the DUT is a 4-port and the THRU pattern has the even/odd symmetry, the transformation matrix is simple and known a priori (Amakawa et al., 2008). If not, the S-parameter-based decomposition proposed in (Amakawa et al., 2009) can be used.

While the experimental results reported so far are encouraging, the validity and applicability of the de-embedding method should be assessed carefully. It is extremely important to ascertain the validity of the 2-port de-embedding method because the validity of the multiport method depends entirely on it. One Fig. 1(b) In particular, hardly any justification has been given for the validity of the Π-equivalent-based bisecting of the THRU (Ito & Masu, 2008; Laney, 2003; Nan et al., 2007; Song et al., 2001; Tretiakov et al., 2004a). There are other possible ways of bisecting the THRU. T-equivalent-based bisection is one example (Kobrinsky et al., 2005). Once the foundations of the 2-port method are more firmly established, the multiport method can be used with greater confidence.

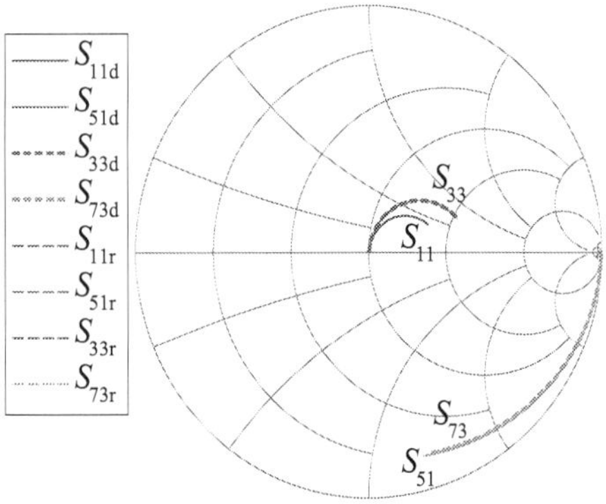

Fig. 19. Reference characteristics of the 4 coupled TLs (with a subscript r) and the de-embedded results. (with a subscript d). Actually, those two are indistinguishable on the Smith chart (Amakawa et al., 2009).

9. Acknowledgments

The authors thank H. Ito, T. Sato, T. Sekiguchi, and K. Yamanaga for useful discussions. This work was partially supported by KAKENHI, MIC.SCOPE, STARC, Special Coordination Funds for Promoting Science and Technology, and VDEC in collaboration with Agilent Technologies Japan, Ltd., Cadence Design Systems, Inc., and Mentor Graphics, Inc.

10. References

Amakawa, S., Ito, H., Ishihara, N., and Masu, K. (2008). A simple de-embedding method for characterization of on chip four-port networks, *Advanced Metallization Conference*, pp. 105–106; *Proceedings of Advanced Metallization Conference 2008 (AMC 2008)*, pp. 99–103, 2009, Materials Research Society.

Amakawa, S., Yamanaga, K., Ito, H., Sato, T., Ishihara, N., and Masu, K. (2009). *S*-parameter-based modal decomposition of multiconductor transmission lines and its application to de-embedding, *International Conference on Microelectronic Test Structures*, pp. 177–180.

Bakoglu, H. B., *Circuits, Interconnections, and Packaging for VLSI*, Addison Wesley, 1990.

Bauer, R. F. and Penfield, Jr, P. (1974). De-embedding and unterminating, *IEEE Transactions on Microwave Theory and Techniques*, vol. 22, no. 3, pp. 282–288.

Bockelman, D. E. and Eisenstadt, W. R. (1995). Combined differential and common-mode scattering parameters: theory and simulation, *IEEE Transactions on Microwave Theory and Techniques*, vol. 43, no. 7, pp. 1530–1539.

Daniel, E. S., Harff, N. E., Sokolov, V., Schreiber, S. M., and Gilbert, B. K. (2004). Network analyzer measurement de-embedding utilizing a distributed transmission matrix bisection of a single THRU structure, *63rd ARFTG Conference*, pp. 61–68.

Faria, J. A. B. (2004). A new generalized modal analysis theory for nonuniform multiconductor transmission lines, *IEEE Transactions on Power Systems*, vol. 19, no. 2, pp. 926–933.

Goto, Y., Natsukari, Y. , and Fujishima, M. (2008). New on-chip de-embedding for accurate evaluation of symmetric devices, *Japanese Journal of Applied Physics*, vol. 47, no. 4, pp. 2812–2816.

P. R. Gray, P. J. Hurst, S. H. Lewis, and R. G. Meyer, *Analysis and Design of Analog Integrated Circuits*, 5th edition, Wiley, 2009.

Han, D.-H., Ruttan, T. G., and Polka, L. A. (2003), Differential de-embedding methodology for on-board CPU socket measurements, *61st ARFTG Conference*, pp.37–43.

Issakov, V., Wojnowski, M., Thiede, A., and Maurer, L. (2009). Extension of thru de-embedding technique for asymmetrical and differential devices, *IET Circuits, Devices & Systems*, vol. 3, no. 2, pp. 91–98.

Ito, H. and Masu, K. (2008). A simple through-only de-embedding method for on-wafer S-parameter measurements up to 110 GHz, *IEEE MTT-S International Microwave Symposium*, pp. 383–386.

Kobrinsky, M. J., Chakravarty, S., Jiao, D., Harmes, M. C., List, S., and Mazumder, M. (2005). Experimental validation of crosstalk simulations for on-chip interconnects using S-parameters, *IEEE Transactions on Advanced Packaging*, vol. 28, no. 1, pp. 57–62.

Kogo, H. (1960). A study of multielement transmission lines, *IRE Transactions on Microwave Theory and Techniques*, vol. 8, no. 2, pp. 136–142.

Kolding, T. E. (1999). On-wafer calibration techniques for giga-Hertz CMOS measurements, *International Conference on Microelectronic Test Structures*, pp.105–110.

Kolding, T. E. (2000a). Impact of test-fixture forward coupling on on-wafer silicon device measurements, *IEEE Microwave Guided Wave Letters*, vol. 10, no. 2, pp. 73–74.

Kolding, T. E. (2000b). A four-step method for de-embedding gigahertz on-wafer CMOS measurements, *IEEE Transactions on Electron Devices*, vol. 47, no. 4, pp. 734–740.

Koolen, M. C. A. M., Geelen, J. A. M., and Versleijen, M. P. J. G. (1991). An improved de-embedding technique for on-wafer high-frequency characterization, *Bipolar Circuits and Technology Meeting*, pp.188–191.

Kurokawa, K. (1965). Power waves and the scattering matrix, *IEEE Transactions on Microwave Theory and Techniques*, vol. 13, no. 2, pp. 194–202.

Laney, D. C. (2003). *Modulation, Coding and RF Components for Ultra-Wideband Impulse Radio*, PhD thesis, University of California, San Diego, San Diego, California.

Magnusson, P. C., Alexander, G. C., Tripathi, V. K., and Weisshaar, A., *Transmission Lines and Wave Propagation*, 4th edition, CRC Press, 2001.

Mangan, A. M., Voinigescu, S. P., Yang, M.-T., and Tazlauanu, M. (2006). De-embedding transmission line measurements for accurate modeling of IC designs, *IEEE Transactions on Electron Devices*, vol. 53, no. 2, pp. 235–241.

Mavaddat, R. (1996). *Network Scattering Parameters*, World Scientific.

Nan, L., Mouthaan, K., Xiong, Y.-Z., Shi, J., Rustagi, S. C., and Ooi, B.-L. (2007) Experimental characterization of the effect of metal dummy fills on spiral inductors, *Radio Frequency Integrated Circuits Symposium*, pp. 307–310.

Nickel, J. G., Trainor, D., and Schutt-Ainé, J. E. (2001). Frequency-domain-coupled microstrip-line normal-mode parameter extraction from *S*-parameters, *IEEE Transactions on Electromagnetic Compatibility*, vol. 43, no. 4, pp. 495–503.

Paul, C. R. (2008). *Analysis of Multiconductor Transmission Lines*, 2nd edition, Wiley-Interscience.

Pozar, D. M., *Microwave Engineering*, 3rd edition, Wiley, 2005.

Rautio, J. C. (1991). A new definition of characteristic impedance, *IEEE MTT-S International Microwave Symposium*, pp. 761–764.

Song, J., Ling, F., Flynn, G., Blood, W., and Demircan, E. (2001). A de-embedding technique for interconnects, *IEEE Topical Meeting on Electrical Performance and Electronic Packaging*, pp. 129–132.

Tretiakov, Y., Vaed, K., Ahlgren, D., Rascoe, J., Venkatadri, S., and Woods, W. (2004a). On-wafer de-embedding techniques for SiGe/BiCMOS/RFCMOS transmission line interconnect characterization, *International Interconnect Technology Conference*, pp. 166–168.

van der Merwe, J., Reader, H. C., and Cloete, J. H. (1998). *S*-parameter measurements yielding the characteristic matrices of multiconductor transmission lines, *IEEE Transactions on Electromagnetic Compatibility*, vol. 40, no. 3, pp. 249–256.

Vandamme, E. P., Schreurs, D. M. M.-P., and van Dinther, C. (2001). Improved three-step de-embedding method to accurately account for the influence of pad parasitics in silicon on-wafer RF test-structures, *IEEE Transactions on Electron Devices*, vol. 48, no. 4, pp. 737–742.

Wartenberg, S. A. (2002). *RF Measurements of Die and Packages*, Artech House.

Wei, X., Niu, G., Sweeney, S. L., Liang, Q., Wang, X., and Taylor, S. S. (2007). A general 4-port solution for 110 GHz on-wafer transistor measurements with or without impedance standard substrate (ISS) calibration, *IEEE Transactions on Electron Devices*, vol. 54, no. 10, pp. 2706–2714.

Williams, D. F., Hayden, L. A., and Marks, R. B. (1997). A complete multimode equivalent-circuit theory for electrical design, *Journal of Research of the National Institute of Standards and Technology*, vol. 102, no. 4, pp. 405–423.

Yanagawa, K., Yamanaka, K., Furukawa, T., and Ishihara, A. (1994). A measurement of balanced transmission line using *S*-parameters, *IEEE Instrumentation and Measurement Technology Conference*, pp.866–869.

3

Current reuse topology in UWB CMOS LNA

TARIS Thierry
University of Bordeaux
France

1. Introduction

In February 2002, the Federal Communications Commission (FCC) gave the permission for the marketing and operation of a new class of products incorporating Ultra Wide Band (UWB) technology. The early applications of UWB technology were primarily radar related, driven by the promise of fine-range resolution that comes with large bandwidth. But the recent 3.1-10.6GHz allocation extends the UWB use to a larger application area for which the specific frequency ranges are reported in the table 1.

Class/Applications	Frequency band for operation at part 1 limit
Communications and measurement systems	3.1 to 10.6GHz (different out-of-band emission limits for indoor and outdoor devices)
Imaging: ground penetrating radar, wall, medical imaging	< 960MHz or 3.1 to 10.6GHz
Imaging: through wall	< 960MHz or 1.99 to 10.6GHz
Imaging: surveillance	1.99 to 10.6GHz

Table I. FCC allocations for each UWB category [1]

In this work we will focus on communication applications for which maximum emissions in the prescribed bands are at an effective isotropic radiated power (EIRP) of −41.3 dBm/MHz or a maximum peak power level of 0dBm/50MHz, and the −10 dB level of the emissions must fall within the prescribed band. Hence UWB signal transmissions must respect the spectrum mask presented in the figure 1 for the 7.5GHz here considered bandwidth.
Unlike narrow band standards located within the 0.9 to 6 GHz range, UWB technology can meet the growing demand for high data rate communications in short range distance with relatively low power consumption. Several Gigabits per second (Gbps) are expected within a 5 meter range [2][3]. However targeting mass market applications its deployment success is first driven by a low cost implementation. To meet this requirement CMOS technology is a promising candidate. From technical point of view the digital part obviously benefits from Moore's law, but scaling of the CMOS devices with increasing f_T and f_{max} also facilitates the processing of large bandwidth analog signals with low power.

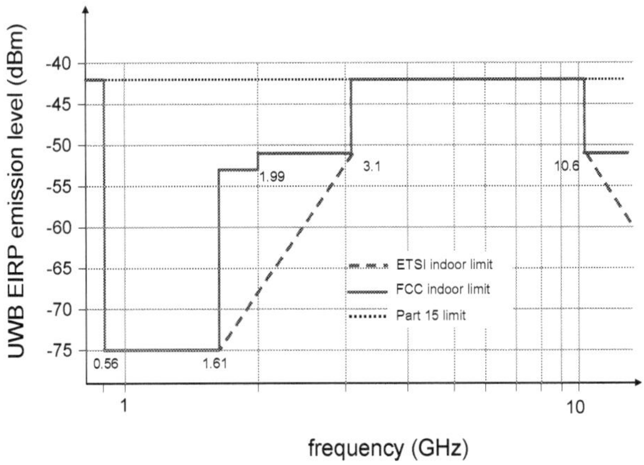

Fig. 1. UWB spectral mask for indoor communication systems [1]

Furthermore full CMOS wireless transceivers are often limited by output power capability of active device. The UWB spectral mask presented in the figure 1 relaxes such constrain enabling the use of CMOS PA. Hence the single chip solution becomes feasible with UWB technology.

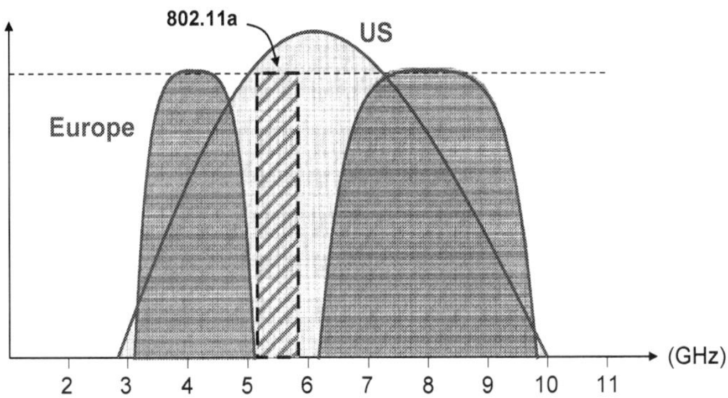

Fig. 2. UWB spectrum redrawing for US (blue) and Europe (red)

UWB allocation shares frequency spectrum with other wireless applications. As well the redrawing of the 3.1-10.6GHz band depends on the local regulation. For example, the entire 7.5GHz bandwidth, red line in the fig. 2, is used in US. In Europe, it is split in two bands, blue lines in the fig. 2: the lower from 3.1 to 5.1 GHz and the upper band from 6.4 to 10.6 GHz because of the 802.11a application which is located within the 5.15 to 5.725 GHz range. Currently, there are two major IEEE UWB radio schemes, i.e., multi-band (MB) OFDM [4]and direct sequence (DS), both dividing the 7.5GHz UWB bandwidth into multiple sub-bands, and use carrier in radio transmission. An alternative UWB radio solution, fig. 3(b),[5], is a fully pulse-based, non-carrier, single-band (7.5GHz), most-digital, all-CMOS UWB transceiver that can take the full advantages of the original impulse UWB radio technology to achieve the required multi-Gbps throughput.

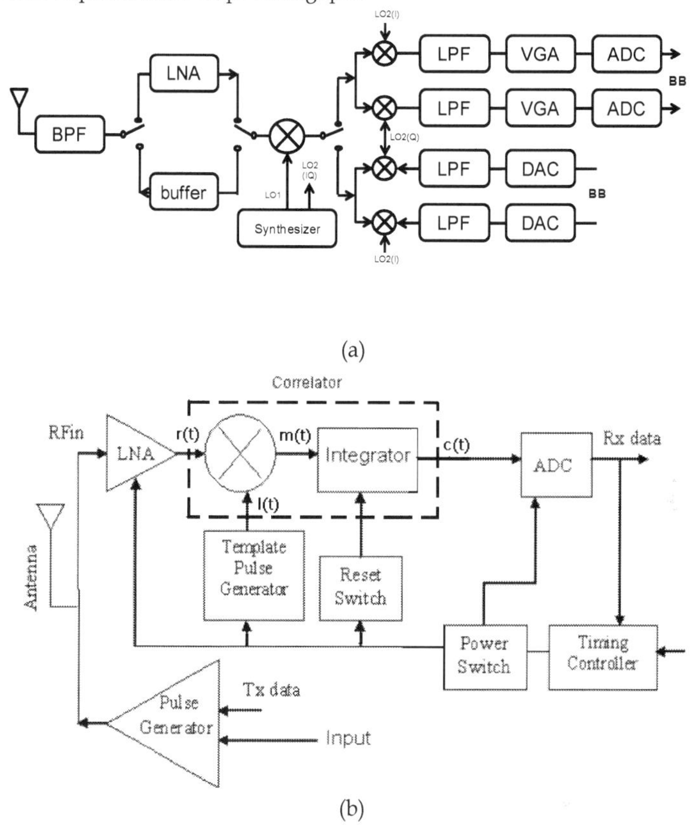

Fig. 3. MB-OFDM receiver [4] *(a)* pulse-based UWB transceiver [5] *(b)*

Whatever the radio scheme that is selected, both kinds of receiver architectures require a Low Noise Amplifier (LNA) after the antenna. This critical building block must exhibit a low return loss, a low noise figure, and a high gain across the entire frequency band of interest. These characteristics are mainly supported by the input matching skills of the circuit. As well the fig. 4 proposes some simplified schematics for the most common UWB LNA topologies. The two first, LC filter cascode [6], fig. 3(a), and transformer feedback [7][8], fig. 4(b), involve the transistor gate to source capacitance (C_{gs}) in the synthesis of a pass-band filter. Both architectures, sensitive to technology modeling, remain the best suited to achieve broadband operation, in term of gain, input matching, and low noise figure under low power consumption. Considering the implementation, the large number of inductor required to set up these techniques induces a huge, and so expensive, silicon area. The two last configurations resistive feedback (FB) [9][10] and common gate (CG) [11], fig. 4(c) and (d) respectively, offer a very compact silicon alternative. However resistive FB amplifier is known for large current consumption and stringent gain-bandwidth trade-off inducing limited optimization. Basic CG topology can perform wideband response at the expense of increased noise figure (NF). The input impedance is so dominated by the trans-conductance once the input gate to source capacitor is cancelled out within the considered bandwidth [12].

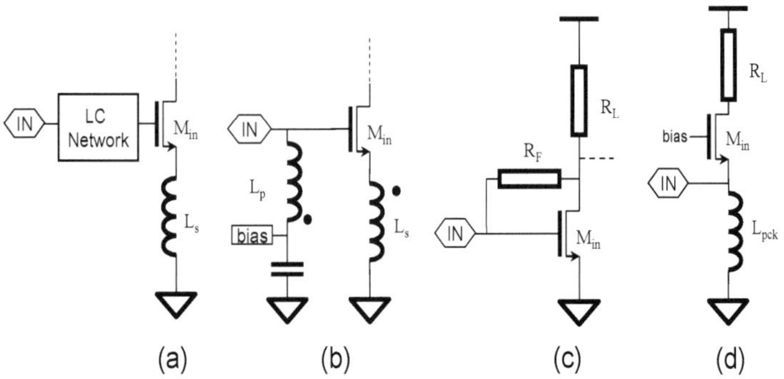

Fig. 4. Four basic topologies of UWB LNA input matching

For a couple of decade CMOS technologies scaling has induced a drastic reduction of the supply voltage reported in the fig. 5. From analog point of view, a meaningful figure of scaling impact is the threshold to supply margin (VDD-VthN). Indeed, it determines the number of devices that can be stacked between the two supply rails thus steering the design methodology and constrains.

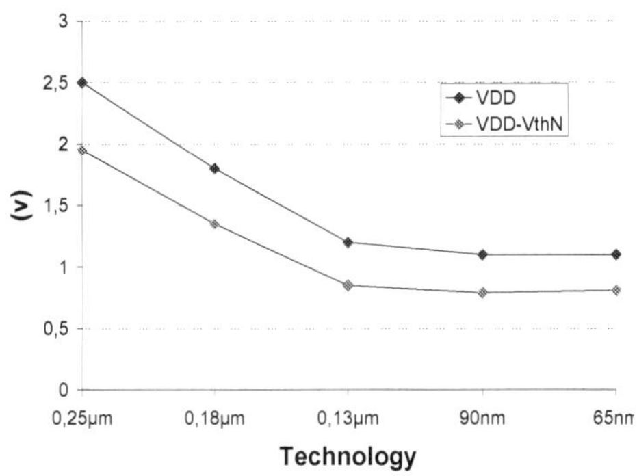

TECHNOLOGY	SUPPLY (VDD)	V_{THN}	VDD-V_{THN}
0.25µm	2.5V	550mV	1.95V
0.18µm	1.8V	450mV	1.35V
0.13µm	1.2V	350mV	0.85V
90nm	1.1V	310mV	0.79V
65nm	1.1V	290mV	0.81V

Fig. 5. CMOS scaling impact on VDD and V_{th}

In RF designs it defines the maximum allowed headroom voltage which is directly connected to the linearity, the gain, and the signal to noise ratio (SNR) of the building blocks. As a matter of consequence UWB LNA performances greatly suffer from voltage scaling in deep sub-micron CMOS technologies. Considering such context this work proposes a new circuit technique making the aforementioned UWB topologies low voltage compatible.

2. Resistive feedback input matching

2.1 Resistive feedback theory

The single stage wide band amplifier, proposed in fig. 4(c), can be studied according to the simplified small signal model depicted in fig. 6. Where g_{mMin}, C_{gsMin} and C_{gdMin} are the transconductance, gate to source and drain to source capacitors of M_{in} respectively. R_L and R_F are load and feedback resistances. We assume that the circuit is connected to a source generator whose the R_s impedance is typically 50 Ω.

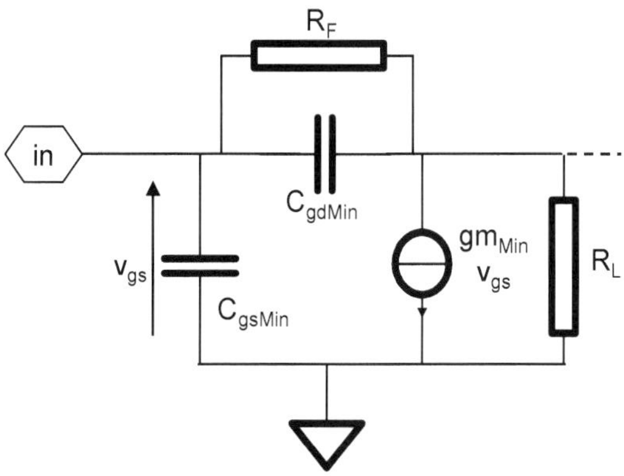

Fig. 6. Simplified model of a resistive feedback input stage

In first approximation, we can derive the analytic expressions of the voltage gain (A_v), noise figure (NF) and input impedance (Z_{in}) as:

$$|A_v| \approx g_{m_{M_{in}}}.(R_L // R_F) \tag{1}$$

$$NF \approx 1 + \frac{2}{3}\frac{1}{g_{m_{M_{in}}} R_S}.(\frac{1}{R_S} + \frac{R_S}{R_F{}^2}) + (\frac{f}{f_{T_{Min}}})^2.\frac{2}{3} g_{m_{M_{in}}} R_S + \frac{R_S}{R_F} \tag{2}$$

$$Z_{in} \approx \frac{R_F + R_L}{1 + |A_v|} \tag{3}$$

In a -3dB bandwidth (BW_{-3dB}) defined as it follows:

$$BW_{-3dB} \approx \frac{1+|A_v|}{R_F C_{gsM_{in}}} \tag{4}$$

With f_{TMin} the cutoff frequency of M_{in}.
From (1) it comes a high voltage gain requires a large g_m and, or, a large load resistor R_L. But the square law of current drain in CMOS transistors implies an expensive DC current to significantly increase the transconductance thus limiting the gain improvement through g_m in wireless applications. Concerning R_L it is limited by biasing conditions since it must keep M_{in} in saturation region. Hence moderate gains are performed with this kind of topology. In expression (2) the larger the R_F is the lower the NF is. But R_F is limited in (3) for a fixed input impedance, typically 50Ω, and contributes to the bandwidth tuning in (4), a roughly 10GHz in UWB applications. So to estimate the capability of such amplifier configuration to address UWB requirements, the expressions (1) to (4) have been drawn, fig. 7, based on 0.13μm CMOS skills.

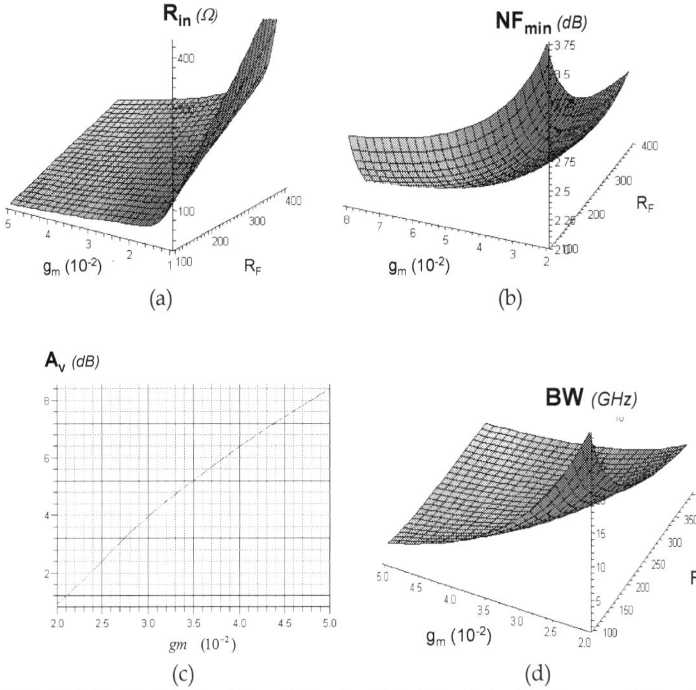

Fig. 7. Re(Zin) (a) NFmin (b) Av (c) and Bandwidth (d) of single stage resistive feedback amplifier with typical 0.13μm CMOS technology skills. VDD=1.6V

Considering a 50Ω input impedance in fig. 7(a), the transconductance, g_m, and the feedback resistance, R_F, must be tuned to 30 mS and 130 Ω respectively. Under such conditions a roughly 3 dB NF_{min}, fig. 7(b), and 4 dB voltage gain, fig. 7(c), are performed. According to

Friis formula [10], this available gain cannot ensures a low NF_{min} to the front end. To overcome this drawback we can either implement a multi-stage LNA either design a mixer performing both a low noise figure and a high gain. The first solution is power hungry and so unexpected for wearable applications. The second puts on the mixer some hard design constrains. Hence these preliminary investigations emphasize that a conventional resistive feedback topology is unsuited for UWB LNA design in a 0.13µm CMOS implementation.

2.2 Current reuse improvement

It has been demonstrated in the previous part that the lack of gain is the bottleneck of resistive feedback amplifier optimization. The current reuse topology relying on active load configuration can complete this missing. Indeed both improving the voltage gain and lowering the design trade-off, this technique makes resistive feedback topology more attractive for ultra wide band purpose.

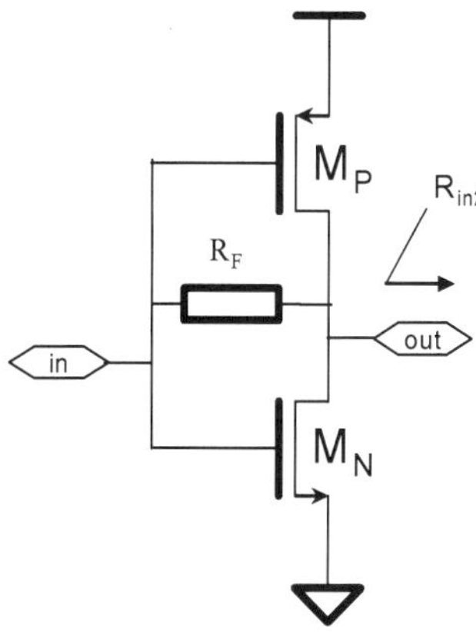

Fig. 8. resistive feedback current reuse configuration [12]

Considering the topology depicted in fig. 8, the overall trans-conductance is now ($g_{mN}+g_{mP}$) that is twice larger the basic resistive feedback one for the same biasing current. Loading the NMOS transistor with the PMOS allows the circuit to operate under lower supply voltage than resistive load configuration. As well the expressions *(1)*, *(2)*, *(3)* and *(4)* remain the same according to the equivalences reported in the Table II:

RESISTIVE FEEDBACK	REUSE FEEDBACK
C_{gsMin}	$C_{gsN}+C_{gsP}$
C_{gdMin}	$C_{gdN}+C_{gdP}$
R_L	$R_F//r_{dsN}//r_{dsP}//R_{in2}$
g_{mMin}	$g_{mN}+g_{mP}$

Table II. resistive to reuse feedback equivalence table

With R_{in2} the input impedance of the following stage. In reuse approach, R_F remains large to lower NF in *(2)* where as g_m boosting achieves a high A_v for a reasonable current consumption in *(1)*. Hence the growth of A_v mitigates the increase of parasitic capacitor value to keep a wide -3dB bandwidth in *(4)*.

Measurement results

The schematic of the LNA is depicted in fig. 9. A 0.7nH input inductor is added to compensate for the parasitic capacitors of M_N and M_P at high frequencies. A resistive feedback buffer with peaking inductor load is coupled with the reuse UWB stage to drive the 50Ω output load.

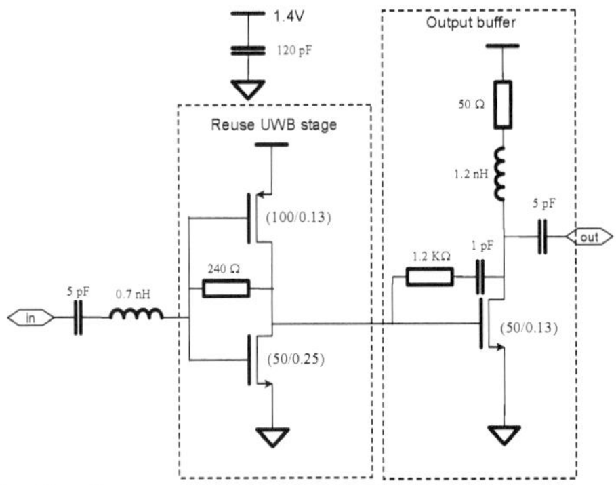

Fig. 9. UWB resistive feedback current reuse LNA [11]

This circuit is implemented in a 6 metal levels 130nm CMOS technology from STM. The measurement results have been performed with a GSG probe bench combined with a HP 8510B network analyzer and a HP 8970 NF-meter.

The reuse input stage, in charge of NF and input matching, depicts, in fig. 10, its ability to achieve a wide band input matching. Indeed S_{11}, in good agreement with simulations, is kept lower than -12dB from 1 to 14.8 GHz.

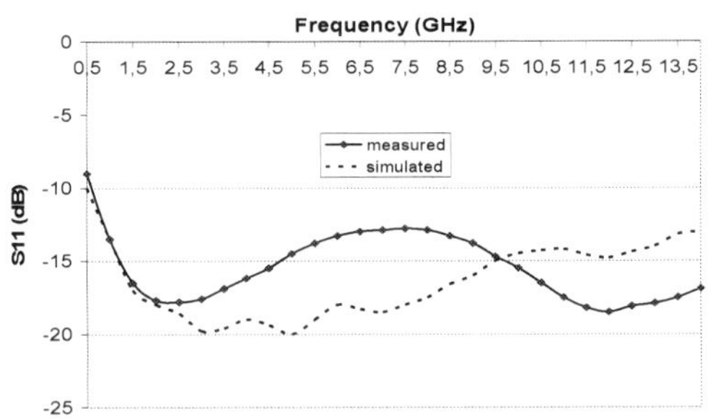

Fig. 10. Measured x Simulated S_{11} parameter

As far as noise is concerned, both NMOS and PMOS of the current reuse stage are sized to provide the lowest NF at 3.6GHz -i.e. a good tradeoff regarding to the 3 to 11 GHz frequency band of interest-. However resistive amplifier is not well suited to synthesize the imaginary part required by the noise matching of MOS transistor, especially over a wide bandwidth. For this reason, such kind of topology remains noisier than its LC ladder counterpart [9]. According to the expression *(2)*, the high f_T of the 130nm MOS technology combined with a large R_F feedback resistor keeps the NF lower than 9dB up to 10 GHz thus addressing UWB requirement. It reaches a 4.45dB minimum at 3.74GHz, as showed in the fig. 11.

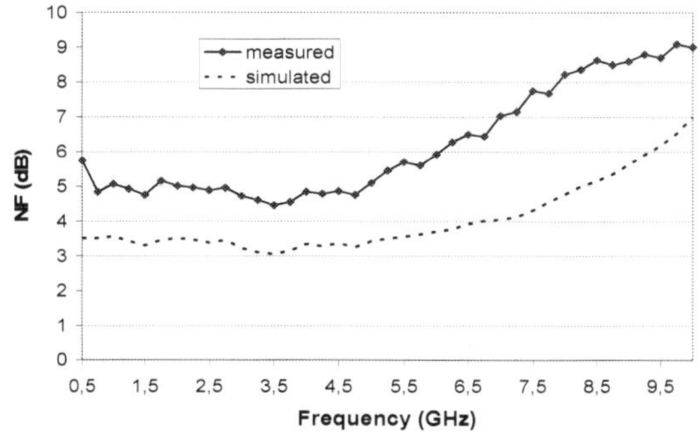

Fig. 11. Measured x Simulated Noise Figure

In fig. 12, an 11.5dB quasi flat band gain is provided from 2 to 9 GHz. This gain is entirely performed by the current reuse stage which consumes a 17mW under 1.4 supply voltage. The buffer does not provide additional gain it both compensates for the roll-off induced by C_{gsMin} in the previous stage, with the 1.2 nH peaking inductor, and delivers the signal to the 50Ω load.

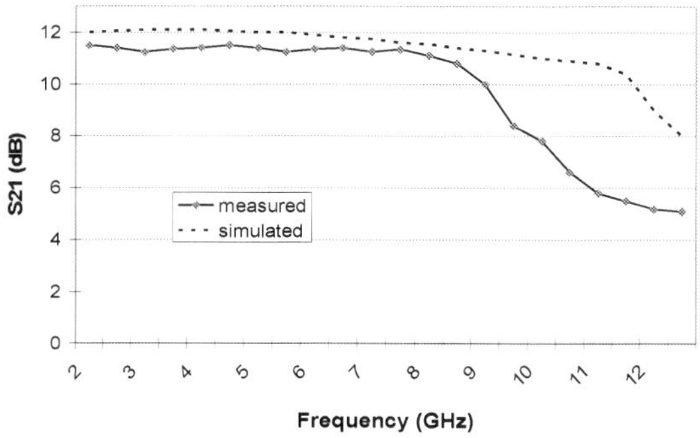

Fig. 12. Measured x Simulated S_{21} parameters

Fig. 13, depicting S_{21} and NF_{min} variations versus supply voltage emphasizes the ability of the current reuse topology to operate under low supply voltage. The sweep range is lower limited to 0.9V since the input matching is not completed below. A roughly 6dB gain 5.5 dB NF_{min} are achieved under 1V. Whereas the gain is voltage dependent, the minimum noise figure is almost constant, close to 4.5dB, until 1.1V supply. For lower supply voltages the gain drop-off mainly contributes to the noise figure growth. According to expression (4) it has been also observed that the bandwidth slightly shrinks with the voltage decreasing, it covers a 2 to 8GHz range under 1V.

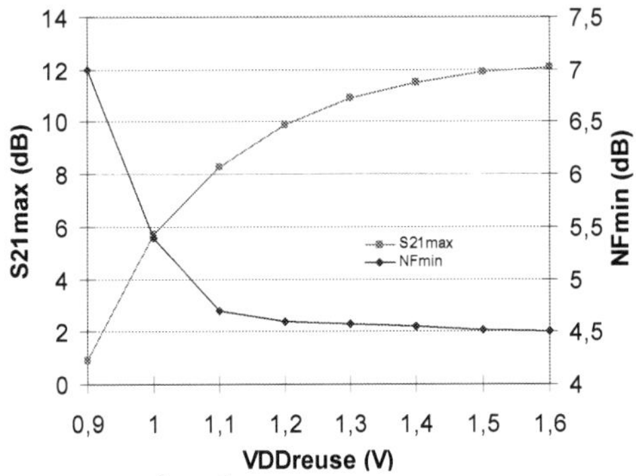

Fig. 13. S_{21} and NF_{min} versus supply voltage

The -1dB input compression point is reported in fig. 14. The supply reduction first limits the allowed voltage swing at the output of the current reuse stage thus degrading the ICP1. For sub-1V operation the serious lack of gain permits larger power processing.

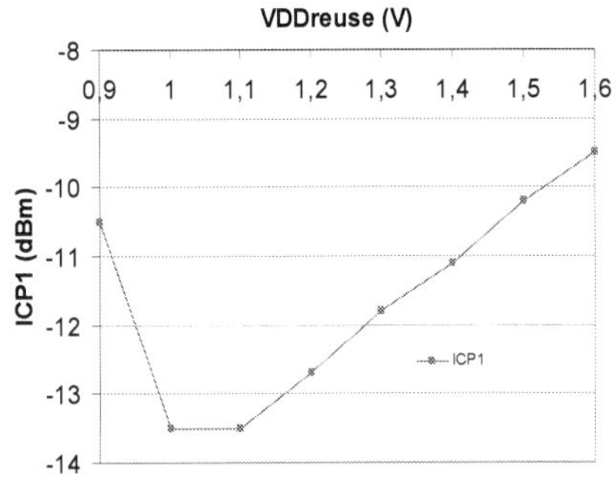

Fig. 14. ICP1 versus supply voltage

3. LC ladder input matching

Another useful technique to perform wide band input matching in UWB LNA is the filter synthesis. This solution is interesting since it achieves a pass band filtering instead of a wide band response like resistive amplifiers do. Out-band blockers are so attenuated thus relaxing the linearity and power consumption constrains of the LNA. The synthesis is based on LC elements, the larger the number of cascaded cells is the wider and better the filtering is. But inductors are silicon expensive, so designers have to deal with input matching level and bulk management. Fourth and sixth order topologies are good trade-off in UWB applications.

3.1 LC ladder theory

The LC network presented in fig. 15(a) is a 6th order pass-band filter whose the impedance response, Zin, is depicted in fig. 15(b). Hence Zin is purely resistive within a frequency range defined as it follows:

$$f_L \sim \frac{1}{2\pi RC_2} \sim \frac{R}{2\pi L_1} \tag{5}$$

$$f_U \sim \frac{1}{2\pi RC_1} \tag{6}$$

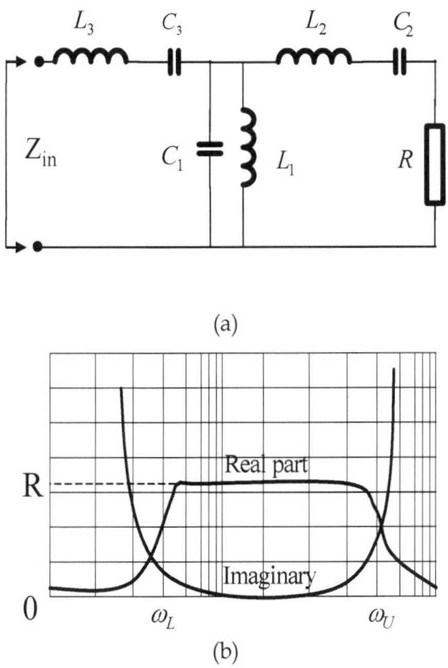

Fig. 15. 6th order pass-band filter form [8] (a) Z_{in} response (b)

This input matching configuration is often implemented in UWB LNA in accordance with the schematic reported in fig. 16. The circuit proposed in fig. 16 has been implemented in a 0.25µm SiGe BiCMOS technology [10].

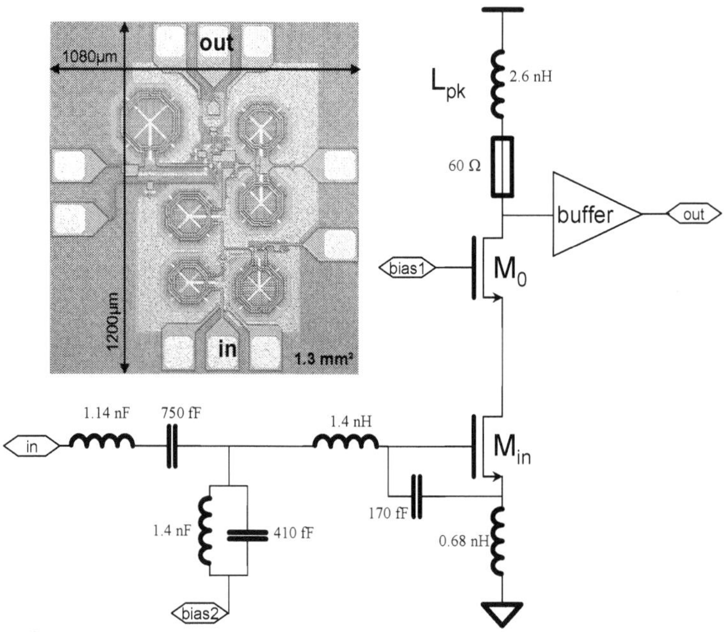

Fig. 16. UWB LC ladder cascode LNA [13]

The roll off induced over the gate to source capacitor of M_{in} is compensated by L_{pk} peaking load inductor to perform a quasi flat band voltage gain, G_v, expressed as:

$$G_v = \frac{g_{mM_{in}}[r_{pk} + jL_{pk}\omega]}{2R_S.jC_{GS\,tot}\omega} \tag{7}$$

Close to the geometric average frequency $(f_L.f_U)^{0.5}$, the minimum noise figure, based on narrow band derivation described in [10], is written as :

$$NF_{\min} = 1 + \frac{2}{\sqrt{5}}\frac{\omega}{g_{\min}}C_{GStot}\sqrt{\gamma.\delta.(1-|c^2|)} \tag{8}$$

With C_{GStot} the combination between M_{in} gate to source and 170 fF external capacitors, g_{mMin} the transconductance of M_{in}, $\gamma = 3/2$, c is the correlation term equals to j0.395, and δ is the coefficient of gate noise equals to 4/3.

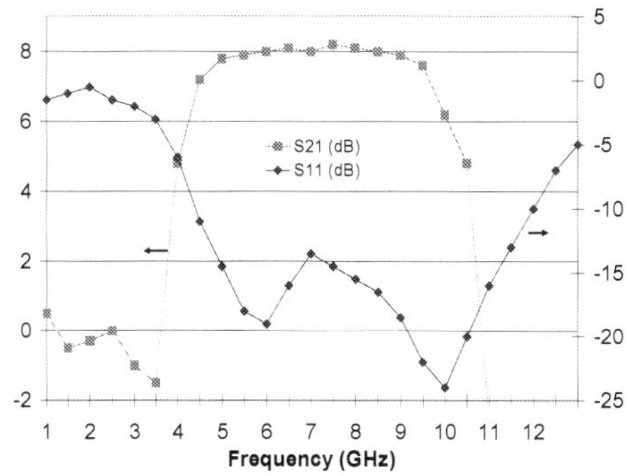

(a)

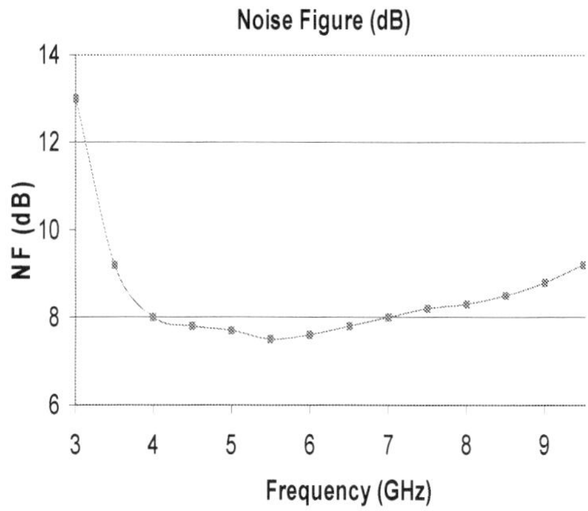

(b)

Fig. 17. S_{21} and S_{11} versus frequency *(a)* NF versus frequency *(b)*

Operating under 2.5V, it achieves a 8,5 dB maximum gain, fig. 17(a), with a 8,5 mW power consumption. S_{11} remains lower than -10dB from 4.2 to 12 GHz thus depicting a very wide band input matching. The NF, fig. 17(b), smoothly increases from 7.5 dB@5.6GHz to 9.5 dB@9.8GHz. The measurement results of the circuit are summarized in Table III.

S_{11} (DB) [BW IN GHZ]	<-10 [4,2-12]
S_{21} (dB) [BW-3dB in GHz]	8.5 [4.4-10.6]
NF (dB)	7.5/9.5
Pdiss core (mW)	8.5
Area (mm²)	1

Table III. 0.25μm UWB LC ladder cascode LNA performances

The 4th order LC ladder LNA topology presented in the fig. 18 has been implemented and simulated in a 0.13μm CMOS technology. Two circuits have been designed according to the table IV, they perform a more than 10dB gain and a good input matching over the targeted bandwidth as showed in fig. 19. Supplied under 1.6V they roughly consume a 5mA.

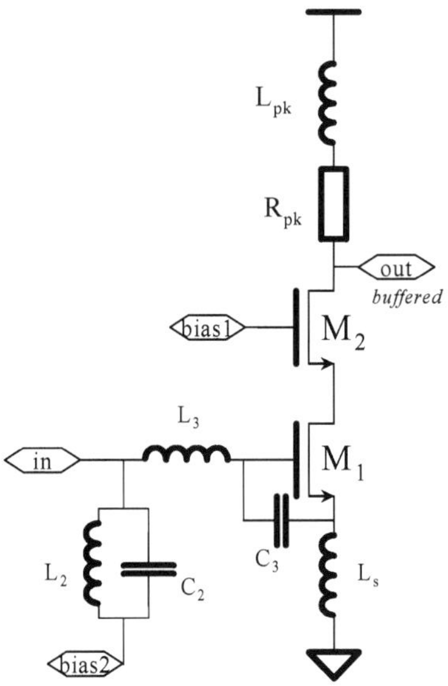

TABLE IV. DEVICE SIZES FOR 3-5GHZ AND 6-10GHZ

	3-5 GHZ	6-10 GHZ
L_2/C_2	*1.9nH/400fF*	*0.6nH/70fF*
L_3/C_3	*2.2nH/520fF*	*1.4nH/130fF*
L_s	*1.1nH*	*1.1nH*
$(W/L)_{M1}$	*40/0.13 μm*	*40/0.13 μm*
$(W/L)_{M2}$	*60/0.13 μm*	*50/0.13 μm*
R_{pk}	*50Ω*	*35Ω*
L_{pk}	*3nH*	*2nH*

Fig. 18. 4th order LC ladder LNA dedicated to 3-5GHz or 6-10GHz bands

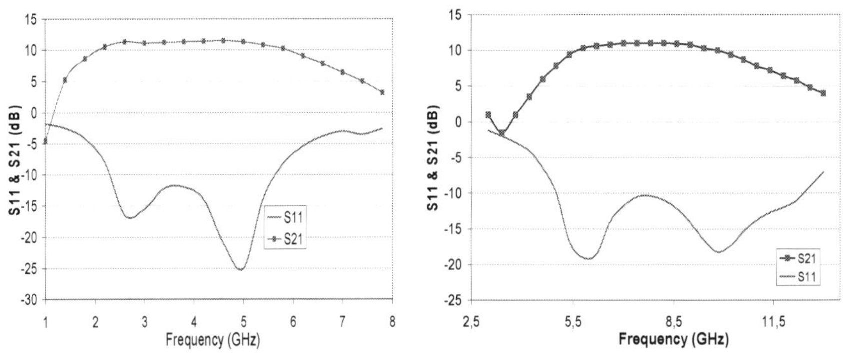

(a) (b)

Fig. 19. Simulated S_{11} and S_{21} of a 4th order LC ladder cascode LNA dedicated to 3-5GHz *(a)* and 6-10GHz *(b)*

A wide band operation requires from active devices a large current consumption to ensure a sufficient high cutoff frequency. This means that the gate overdrive voltage of M_1 and M_2, fig. 18, must be selected within a 0.3 to 0.5V range, according to [17], thus limiting the minimum supply voltage of cascode topology compared to a narrow band design issue. The gain to VDD dependency is drawn in fig. 20 for the 6 to 10GHz circuit. A less than 10dB gain is achieved with the technology nominal voltage, 1.2V, and then it severally decreases for sub 1V operations.

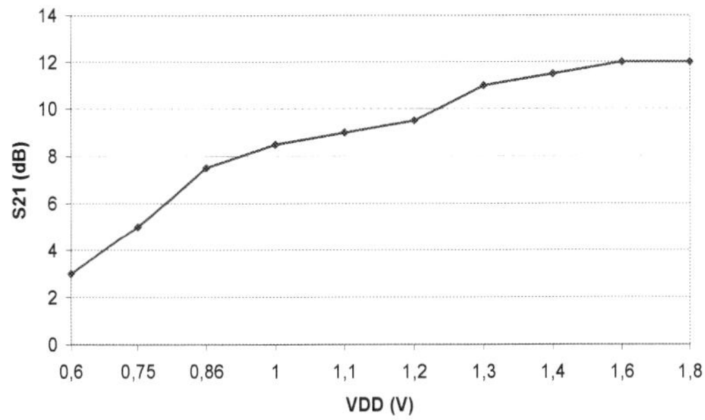

Fig. 20. 4th order LC ladder LNA gain versus supply voltage

So LC ladder network synthesis is a very efficient technique to perform wide band input matching in LNA designs. But the cascode topology would suffer from technology scaling in the near future because of supply voltage reduction. To find an alternative we investigate the combination of LC ladder with current reuse configuration.

3.2 Current reuse improvement

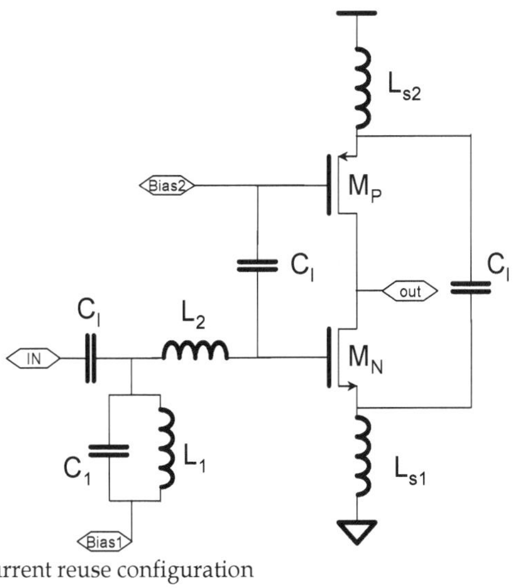

Fig. 21. LC ladder current reuse configuration

The theoretical topology of a LC ladder current reuse LNA is presented in fig. 21. It is here combined with a 4th order LC ladder filter and four inductors are needed for the synthesis. C_l shunting the two sources of the transistors in AC mode, Ls1 and Ls2 are in parallel and achieve the real part of the input impedance. The configuration of active load improves the voltage gain compared to a cascode topology.
The minimum supply voltage, keeping both the PMOS and NMOS transistors in saturation region, is derived in expression (9). It is drawn in the Fig. 22 based on the features of a 0.13µm CMOS technology. W_p is adjusted to 30 µm which implies a C_{gstot} equals to 300fF intended for 6 to 10 GHz range. It points out the ability of current reuse topology to sustain low voltage operation.

$$VDD_{\min} = \sqrt{\frac{2L_P I_D}{\mu_p C_{ox} W_P}} + |V_{TP}| \tag{9}$$

With L_p and W_p the PMOS transistor length and width respectively. μ_P is the hole mobility.

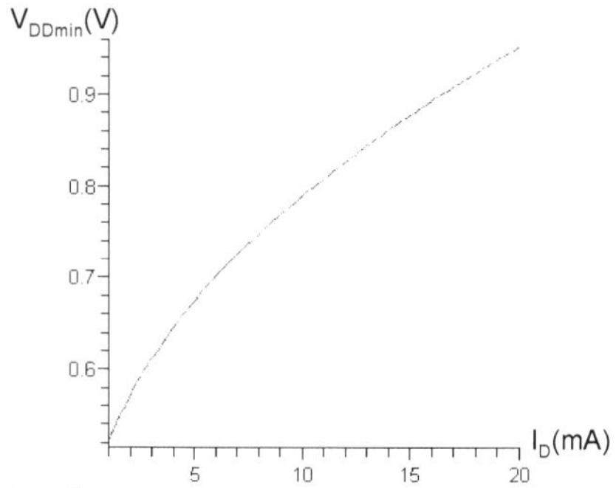

Fig. 22. Minimum supply voltage versus current consumption in a LC ladder current reuse topology. W_P=30µm, L_P=0.13µm, VT_{PMOS}=400mV

The expressions *(10)* and *(11)* derive the noise figure and the voltage gain of the circuit presented in Fig. 21. Driven with the same current both current reuse and cascode topologies perform similar cut-off frequency. Hence comparable NF_{min} and Gv are achieved. Though current reuse topology can support sub 1V operation, low noise and high gain performances require a higher supply voltage since it directly sets the current of the stage, as depicted in fig. 22.

$$F_{\min} = 1 + \frac{2}{\sqrt{5}} \frac{\omega}{g_{mtot}} C_{GStot} \sqrt{\gamma.\delta.(1 - |c^2|)} \tag{10}$$

$$G_v = \frac{g_{m\,tot}.(r_{dsn} // r_{dsp})}{2R_S.j\omega C_{GS\,tot}} \tag{11}$$

With $g_{mtot}=g_{mP}+g_{mN}$ and $C_{gstot}=C_{gsN}+C_{gsP}$

3.3 Measurement results

The simplified schematic of the circuit depicted in fig. 23 was implemented in a 6 metal levels 130nm CMOS technology from STM. This LNA is designed to cover the 6-10GHz upper band of European UWB standard. A DC Feedback (DCFB) loop regulates the output DC voltage of the first stage. The nominal supply is set to 1.6V to ensure both high gain and low noise figure. The buffer is loaded with peaking inductors which compensate for the roll-off occurring in the first stage.

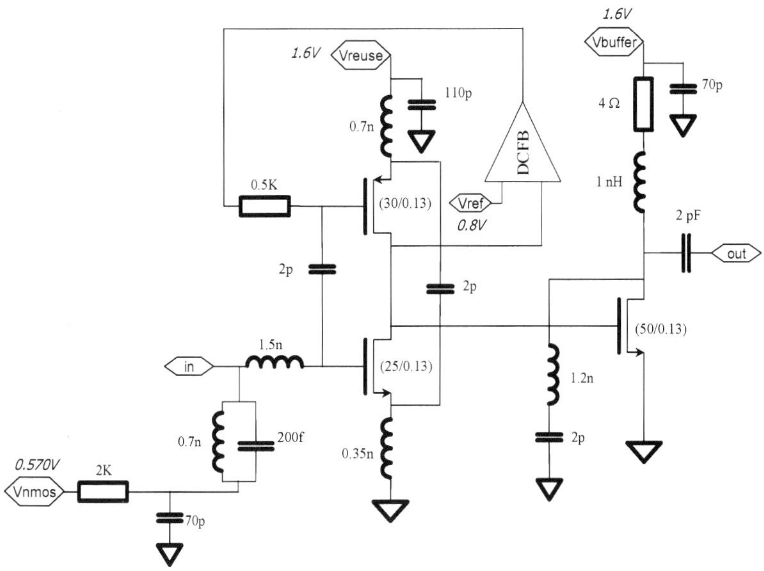

Fig. 23. UWB LC ladder current reuse LNA

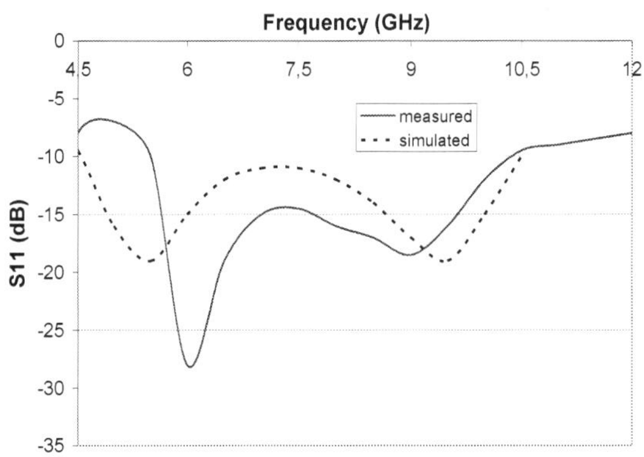

Fig. 24. Measured x Simulated S_{11} parameter

The measurement results have been performed with a GSG probe bench combined with a HP 8510B network analyzer and a HP 8970 NF-meter. The input matching, based on a 4th order pass-band filter, exhibits two minimum at 6 and 10 GHz as depicted in fig. 24. The S_{11} parameter is kept lower than -10 dB from 5.4 to 10.2 GHz. The discrepancy between simulation and measurement forms emphasizes the sensitivity of LC ladder implementation to parasitic elements.
Intended for 6-10GHz, the circuit, operating under 1.6V, achieves a more than 10dB quasi flat band gain from 5.6 to 8.8 GHz with a maximum 12.2dB at 6.5GHz, Fig. 25. A -3dB attenuation is here provided by the buffer. So the current reuse stage performs a roughly 15dB maximum gain for a 3mA current consumption. After retro-simulations it has been underlined extra capacitors, at the output node, strengthen the voltage roll-off of the first stage. As a matter of consequence it is not fully compensated by the peaking load above 9 GHz.

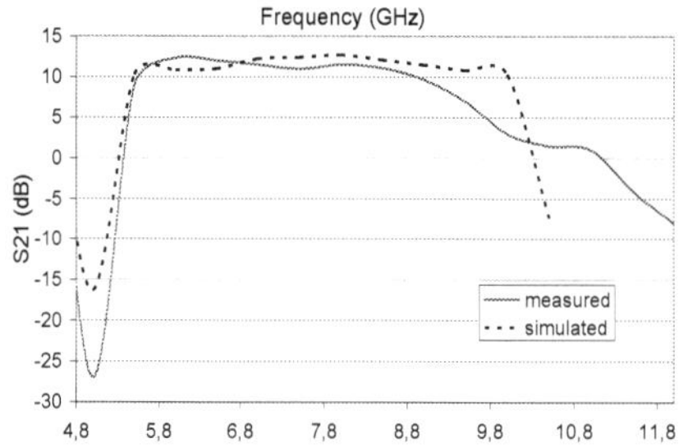

Fig. 25. Measured x Simulated S_{21} parameter

From noise point, M_N and M_P transistors are sized to provide the lowest NF at 7GHz that is a good tradeoff regarding the 6 to 10 GHz frequency band of interest. Likewise input matching, the form of the NF measurements is shifted to low frequency in fig. 26. The LC ladder technique remains lower noise than its resistive feedback counterpart. Indeed the NF is kept constant at 4.5dB from 4.5 GHz to 7.5 GHz, then it smoothly increases until 7dB at 10 GHz.

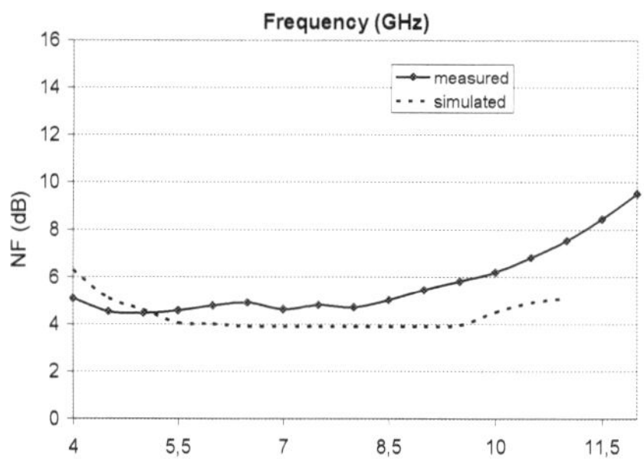

Fig. 26. Measured x Simulated NF

The voltage gain and minimum noise figure versus V_{dd} are drawn in fig. 27. A less than 3dB attenuation is observed where as the supply is reduced from 1.6 to 1V. This behavior is driven by the DC feedback loop which tracks and compensates for the supply variations. The NF_{min} is kept lower than 4.6dB up to 1.1V operation. For sub-1V operation it smoothly increases with respect to the S_{21} drop off. The input matching as well as the circuit bandwidth are sustained until 1V.

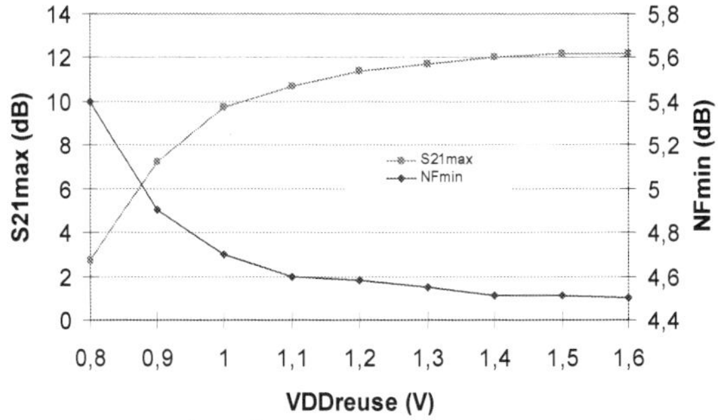

Fig. 27. S_{21} and NF_{min} versus supply voltage

The -1dB input compression point versus supply variation is reported in fig. 28. It first decreases with the voltage reduction. The LNA still performs a more than 10dB gain within a 1 to 1.6V range, fig. 27, so the output voltage range preventing from MOS linear region operation shrinks with the supply lessening thus degrading the ICP1. For sub-1V operation the lack of gain permits large signal processing improving the ICP1.

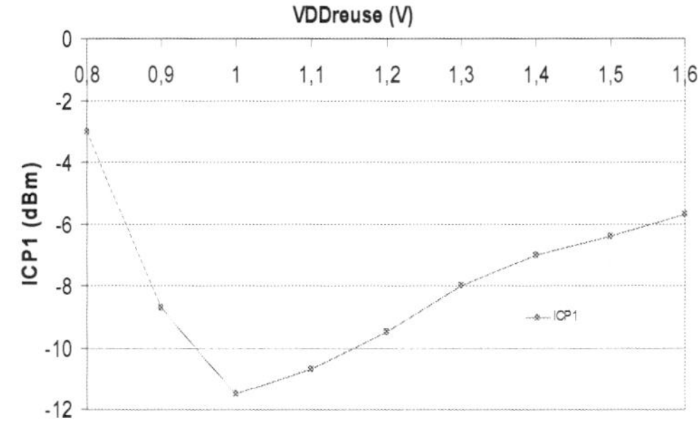

Fig. 24. ICP1 versus supply voltage

4. Design trend discussion

Though UWB systems are dedicated to mass market, the silicon area is a significant matter of concern since it directly impacts the chip cost. The micrographs of the two current reuse LNA reported in this work are depicted in fig. 29. The resistive FB topology, 0.6mm^2, fig. 29(a), allows silicon saving compared to its LC ladder counterpart, 1mm^2, fig. 29(b). However these considerations also need to be discussed at system level. The resistive feedback technique exhibiting a wide band response requires a pre-filtering stage which achieves strong out band rejection.

DESIGN	GAIN [DB] (BW IN GHZ)	NF [DB] (BW IN GHZ)	S11 [DB] (BW IN GHZ)	S22 [DB (BW IN GHZ)	INPUT TYPE	CHIP AREA (ACTIVE)	P_{DC} (WITHOUT BUFFER)
0.13µm CMOS [TW1]*	12.3±2dB (5.6-8.8)	4.4 to 7 (4-10)	<-10dB (5.4-10)	<-10dB (5.2 - 12)	LC ladder	1.2mm²	5mW@1.6V
0.13µm CMOS [TW2]*	11.5±1dB (2-9)	4.45 to 9 (3-10)	<-10dB (1-14.6)	<-10dB (1 - 14.2)	Resistive feedback+L	0.6mm²	17mW@1.4V
0.18µm CMOS [6]	9±1.5dB (2.3-9.3)	4 to 9.2 (3 -10)	<-9.9 (2.6 - 11.7)	<-13 (3 - 10)	LC ladder	1.1mm²	9mW@1.8V
0.25µm BiCMOS [13]	8.5±3dB (4.4 - 10.6)	7.5 to 9.5 (3 - 11)	<-10 (4.2 - 12)	<-9dB (2.8 - 11)	LC ladder	1.3mm²	8.5mW@2.5V
0.18µm CMOS [15]	13.5±3dB (2 - 9)	2.3 to 7.4 (2 - 9)	<-8.8dB (2 - 9)	<-15dB (2 - 10)	Resistive feedback+L	0.9mm²	25mW@1.8V
0.18µm CMOS [16]	9.8±3dB (2 - 4.6)	2.3 to 6 (3 - 5.4)	<-10dB (3 - 6)	<-10dB (2 - 6)	Resistive feedback+LC	0.9mm²	12mW@1.8V

**TW means This Work*

Table V. Measured performance summary and comparison

To do so BWA or SAW filters are mandated thus increasing the device price. LC ladder topologies performing pass-band response help in out band rejection thus lowering the rejection constrains on the pre-filtering stage. The silicon integration of such a filter can be expected lessening the device cost. Hence both approaches must be considered at transistor and system levels with respect to the specifications.

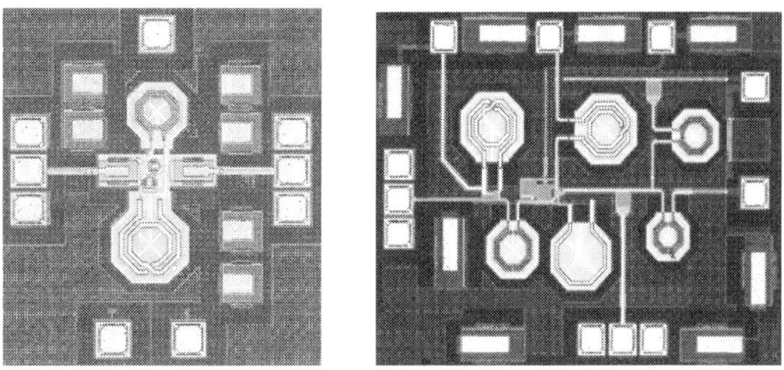

740µm*810µm (a) 1000µm*1200µm (b)

Fig. 29. Chip micrograph of resistive FB current reuse LNA (a) and LC ladder current reuse LNA (b)

The table V summarizes the performances of the circuits and prior published works. It is worth noting UWB devices intended for wireless applications must be low power compatible. The LC ladder references [TW1], [6] and [13] perform comparable performances with resistive architectures [TW2], [15] and [16] for a roughly half power consumption making this technique well suited for RF systems. However it has been demonstrated in section II that the performances of resistive feedback circuits, especially the bandwidth and the power consumption, are technology dependent. We can estimate the implementation of resistive LNA in newer technologies, 65nm or 40nm nodes as instance, would compete with LC ladder approaches.

Furthermore the table V clearly demonstrates that current reuse approach is a promising configuration for ultra wide band LNA. Indeed comparing the circuits, here presented, [TW1] and [TW2] with proposed references [6], [13] and [15] it allows both a lower supply voltage operation and a power saving. Since current reuse configuration is not a wideband dedicated technique, the improvements, here reported, can even be extended to the general purpose of CMOS LNA design under low power and low voltage constrains.

5. Perspectives

About Software Radio

The concept of Software Radio (SR) was first introduced by the US Army. The main goal was to secure communications on hostile battlefield. The project "Speakeasy" was the first stone brought to researches and opened all a new field in the telecommunication industry, military and civil ones. Mitola exposed a defined concept of Software Radio in 1999. It is described as a fully reconfigurable wireless device that adapts its communication radio in response to network and user demands.

The telecommunication industry has seen the opportunity to work on a very promising principle. It is summed up by replacing several receiving chain in mobile phones, each one addressing a standard, through a one-chip-solution. This unique chip would enable the reception of pre-defined and unknown telecommunication standards by self reconfiguration. In a few words the SR device, presented in Fig. 30, is expected to be the ultimate radio system. To compete with advanced multi-standard solutions under development SR solution must cope with stringent constrains among them are:

- a reduced cost of design
- a reduced cost of power consumption
- a reduced cost of human investment

The first assessment implies a CMOS implementation. The second one issues from the huge power consumption of the solutions based on multi RF front end. The last point assumes the focus of the whole engineer staff on a single project would lead to a cost effective development.

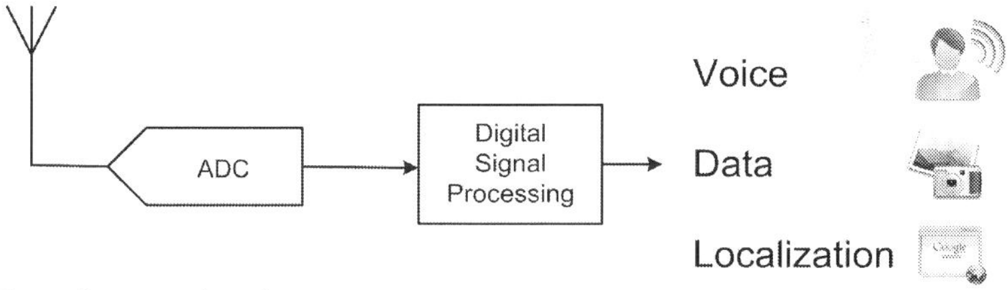

Fig. 30. The software radio solution

Nowadays, some technological bottlenecks make difficult the development of a full SR receiver intended for mobile terminals. Indeed, in Fig. 30, the input signals coming from the antenna have to be converted into digital at least at twice the RF frequencies, a roughly 10GHz, with a high resolution, 16 bits. Such kind of Analog to Digital Convertor (ADC) does not exist and are not expected before 15 years on silicon. So today SR is rather derived into Software Defined Radio (SDR). These systems allow reconfigurations among predefined wireless standards [18][19][20][21]. The solution proposed in the Fig. 31 is based on a conventional heterodyne receiver with reconfigurable building blocks. But the adjusting of some functions like filters and antenna remains intricate and limits the covering of the receiver to a few of applications. The successful of SR implementation needs a new deal in radio schemes.

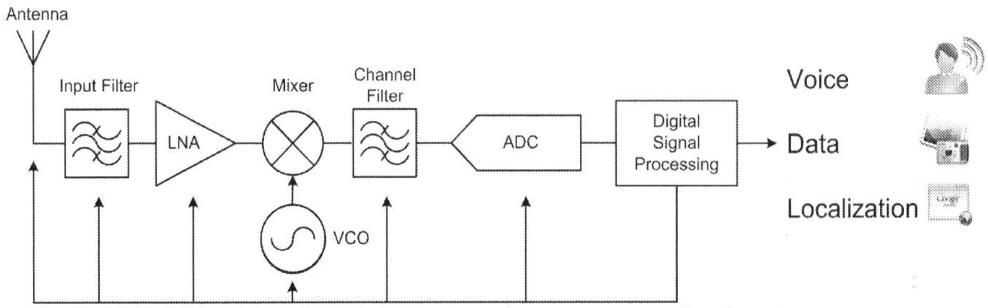

Fig. 31. A software defined radio solution based on conventional radio scheme

Other works on disruptive solution with an intensive research in analog Front End are proposed. Some explore discrete time analog signal processing in order to translate into analog domain the processing work usually done into digital [22]. It enables better trade-off between reconfigurability and power consumption. Among the most promising solution under study is the Sampled Analog Signal Processor (SASP) [23], fig. 32. The SASP uses the principle of an analog discrete Fourier transforms to translate into frequency domain the RF signal [sasp7]. The calculation is based on voltage samples.

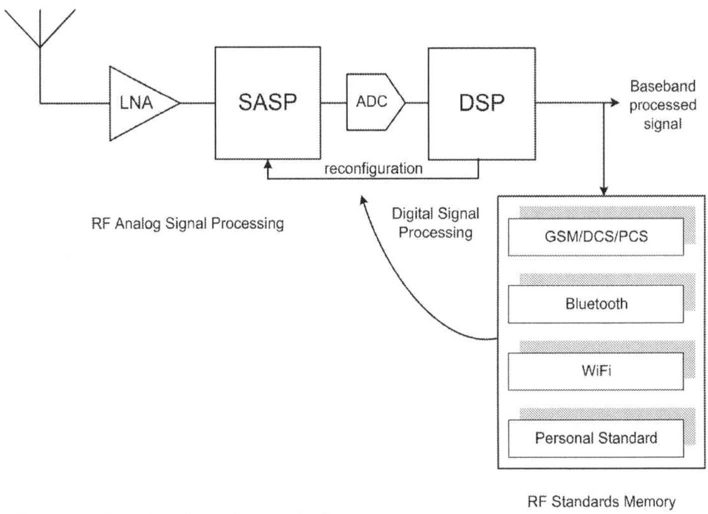

Fig. 32. A software defined radio solution [23]

Once the spectrum containing the information in a 0 to 5GHz range is processed, it is a matter to select the only one (or several) signal envelops to be demodulated. It is composed by few voltage samples among thousands. These samples are converted into digital for a digital signal processing. This system is able to cover any RF standards according to a low power consumption given by analog signal processing and flexibility offered by a DSP. The DSP manages the use of stored standards and reconfigures as needed the parameters of the SASP. The standard memory can be updated and accepts news standards through old standards, Fig. 32.

This brief on next SDR and SR solutions emphasizes the coming revolution in radio scheme of future wireless handsets. Hence the conventional solutions lying on (super) heterodyne receivers will blow up leaving the corner to an embedded analog/digital processing of the RF signal. These novel frontends, like SASP in Fig. 32, will still need two traditional building blocks: an antenna and a LNA.

What would LNA for SR be?

The answer does not depend on the choice between SR or SDR solution. In both cases the amplifier would cover a wide bandwidth. It must features good performances, NF, gain and linearity, making voltage samples suited for analog and digital processing.

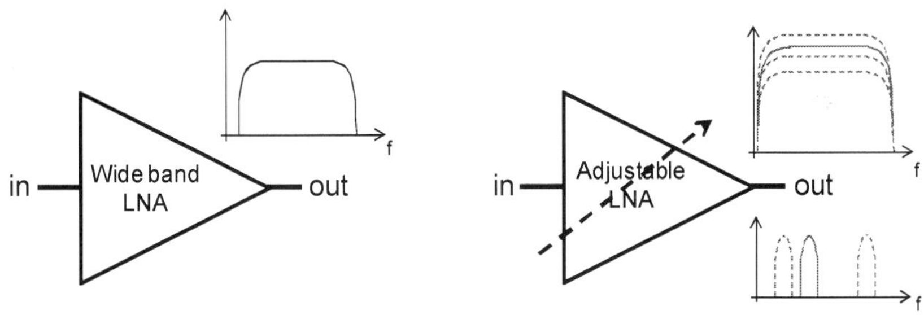

(a) (b)

Fig. 33. Behavioral LNA solution for SR receivers: wide band *(a)* adjustable *(b)*

A first solution, Fig. 33(a), would rely on a wide band LNA achieving the best performances, required by the most stringent standard, over its whole bandwidth. But it is not an optimized approach, in particularly from power saving, since the LNA often provides useless skills regarding the application needs. A smart management is to adjust the LNA performances according to the standard requirements. As reported in expressions (1), (2), (7), (8), (10) and (11), both the NF and gain depend on the current consumption through the transconductance of the input transistors. These features can be so tuned by varying the biasing of the LNA. Considering the linearity, the wide band operation puts on the receiver some stringent constrains from intermodulation rejections. The IIP3 and IIP2 can be improved by a narrow band response of the circuit. So we can draw the behavior of a SR LNA as it follows, Fig. 33(b):

- a wide band operation for signal tracking
- a tunable topology to adjust the performances with the targeted applications
- a narrow band response to fill high IIP_i requirements

As mentioned in [24], the performances of a SASP unit, and later SR devices, are technology dependent. The narrower the transistor is the higher the operating frequency and the resolution is. So each new CMOS generation will directly impact on SR capabilities. This means that the scaling constrains on LNA design, above mentioned, will be further stringent. Based on the investigations reported in this work, the circuit depicted in Fig. 34 proposes a SR LNA topology suited for low voltage supply.

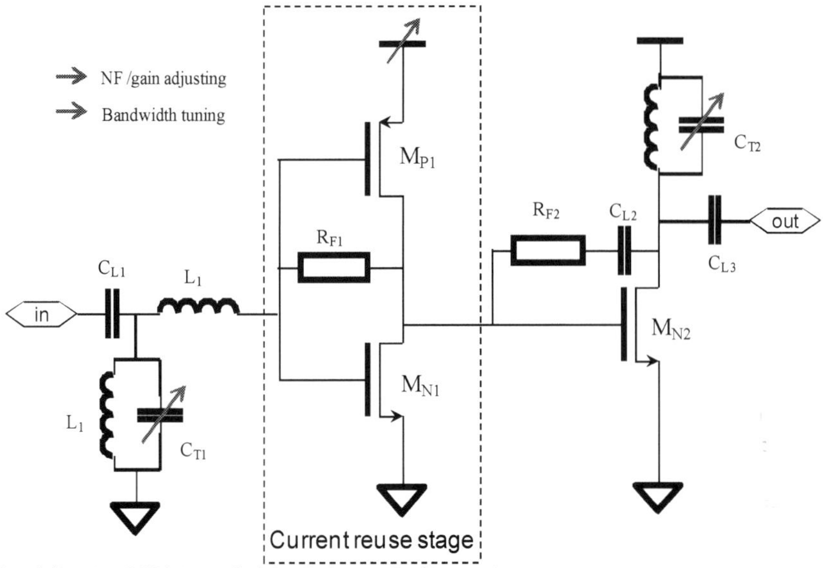

Fig. 34. Adjusting LNA intended for Software Radio receiver

The core of the LNA is a current reuse configuration to be low voltage compatible. The resistive feedback approach ensures a very wide band response for signal tracking mode. The supply control, blue arrow, allows NF and Gv variations according to Fig. 13. The variable capacitors, red arrows, synthesizing filters adjust the bandwidth and operating frequency in narrow band mode.

If full SR systems make useless some building blocks like mixer and VCO, they open a large field of investigations for LNA. The complexity of such next radio solution would focus the interest of front-end designers in the years to come. A first question that they would answer is: what is the best suited LNA behavior for a SR receiver? The circuits will cope with some new concepts of frequency agility, from low to high frequency in narrow and/or wide band operations. Both a system management and a reconfigurable architecture are expected to fill this point. The requirement for adjusting NF and gain characteristics will also challenge the implementation of LNA. Novel circuit techniques are mandated in such case.

6. References

[1]R. A. Scholtz, D.M. Pozar, W. Namgoong (2005) "Ultra-Wideband Radio", journal on applied signal processing (EURASIP), pp.252-272, March2005

[2]G. R. Aiello and G. D. Rogerson, (2003). "Ultra-wideband wireless systems," IEEE Microwave Mag., vol. 4, pp. 36-47, June2003.

[3]S. Roy, J. R. Foerster, V. S. Somayazulu, and D. G. Leeper, (2004). "Ultrawideband radio design: the promise of high-speed, short-range wireless connectivity," Proc. Of the IEEE, vol. 92, pp.295-311, Feb. 2004.

[4]Hui Zheng, Shuzuo Lou, Dongtian Lu, Cheng Shen, Tatfu Chan, and Howard C. Luong (2007). "A 3.1-8.0 GHz MB-OFDM UWB Transceiver in 0.18µm CMOS", IEEE Custom Intergrated Circuits Conference, (CICC 2007), pp.651-653, Sept. 2007, San Francisco, USA

[5]H. Xie et al. (2007). "A broadband CMOS multiplier-based correlator for IR-UWB transceiver SOC", IEEE Radio Frequency Integrated Circuit, RFIC2007, pp. 493-496, Hawai, USA

[6]A. Bevilacqua, A. M Niknejad, (2004). "Ultra-Wideband CMOS LNA for 3.1 to 10.6 GHz Wireless Receivers", IEEE Integrated Solid State Circuit Conference, ISSCC 2004, pp. 382-383, San Francisco, USA

[7]Michael T. Reiha, John R. Long and John J. Pekarik, (2006). "A 1.2 V Reactive-Feedback 3.1-10.6 GHz Ultrawideband Low-Noise Amplifier in 0.13 µm CMOS", IEEE Radio Frequency Integrated Circuit, RFIC 2006, pp.384-385, San Franscisco, USA.

[1]Carlin H.J., Civalleri P.P. (1998). "Wideband Circuit Design" CRC Press LLC, 1998, chap. 6, pp289-295

[8]Chang-Tsung Fu and Chien-Nan Kuo, (2006). "3~11-GHz CMOS UWB LNA Using Dual Feedback for Broadband Matching", IEEE Radio Frequency Integrated Circuit, RFIC 2006, pp.67-70, San Franscisco, USA

[9]D. K. Shaeffer and T. H. Lee, (1997). "A 1.5V, 1.5GHz CMOS Low Noise Amplifier", IEEE Journal of Solid State Circuits, vol 32, pp. 745-759, May 1997.

[10]D. Im, S. Song, H. Kim, K. Lee, (2007). "A wideband CMOS variable-gain Low Noise Amplifier for multistandard terrestrial and cable TV tuner", IEEE Radio Frequency Integrated Circuit, RFIC2007, pp. 621-624, Hawai, USA

[11]B. Razavi, T. Aytur, C. Lam, Fei-Ran Yang, Kuang-Yu Li, Ran-Hong Yan, Han-Chang Kang, Cheng-Chung Hsu, Chao-Cheng Lee, (2005). "A UWB CMOS transceiver", IEEE Journal of Solid-States Circuits, Vol. 40, issue 12, pp2555-2562, Dec 2005

[12]T. Taris, O. Elgharniti, JB. Begueret, E. Kerhervé, (2006). "UWB LNAs using LC ladder and transformers for input matching networks", IEEE International Conference on Electronics, Circuits and Systems, ICECS2006, pp. 323-326, Nice, France.

[13]T. Taris, JB. Begueret, Y. Deval, (2007). "A low voltage current reuse LNA in a 130nm CMOS technology for UWB applications", Microwave Integrated Circuit Conference, EuMIC 2007, pp.307-310, Munich, Germany

[14]C.W. Kim, M.S. Jung and S.G. Lee, (2005). "Ultra-Wideband CMOS low noise amplifier", electronics letters, vol. 41, n°7, pp.384-385

[15]C.-W. Kim, M.-S. Kang, P. Tuan Anh, H.-T. Kim, and S.-G. Lee An (2005). "Ultra-Wideband CMOS Low Noise Amplifier for 3–5-GHz UWB System", IEEE Journal of Solid-States Circuits, Vol. 40, N° 2, pp544-547, March 2005

[16]Willy M. C. Sansen, (2006). "Analog Design Essentials", Springer, 2006, chap. 1

[17] P. K. Prakasam et al., (2008). "Emerging Technologies in Software Defined Receivers," in Proc. IEEE Radio and Wireless Symposium (RWS),pp.719-722, Orlando, US

[18] V. J. Arkesteijn et al., (2002) "An Analogue Front-End Architecture for Software Defined Radio," in Proc. 13th proRISC workshop, November 2002, pp. 165-168.

[19]K. Muhammad et al., (2004) "A Discrete-Time Bluetooth Receiver in a 0.13_m Digital CMOS Process," in Proc. IEEE ISSCC, San Francisco, USA, 2004.

[20]A. A. Abidi, (2006) "Evolution of a software-defined radio receivers rf frontend, in Proc. IEEE Radio Frequency Integrated Circuits Symposium (RFIC), San Francisco, USA, 2006, pp. 027-030.

[21]A. K. Mal et al., (2004) "Sampled Analog Architecture for DCT and DST," in Proc. Int. Symp. Circuits Systems, Vancouver, Canada, May 2004, pp.825-828.

[22]F. Rivet, Y. Deval, D. Dallet, JB. Begueret, P. Cathelin and D. Belot, (2008) "A Disruptive Receiver Architecture Dedicated To Software Defined Radio, in IEEE TCAS-II, vol. 55, n4, pp. 344-348, Apr. 2008.

[23]F. Rivet, Y. Deval, D. Dallet, JB. Begueret and D. Belot, (2008) "65nm CMOS Circuit Design of a Sampled Analog Signal Processor dedicated to RF Applications," in Proc. IEEE NEWCAS'08, Montreal, Quebec, June 2008.

4

Multi-Block CMOS LNA Design for UWBWLAN Transform-Domain Receiver Loss of Orthogonality

Mohamed Zebdi, Daniel Massicotte
Université du Québec à Trois-Rivières
(Québec) CANADA

Christian Jesus B. Fayomi
Université du Québec à Montréal
(Québec) CANADA

1. Introduction

Transform-domain ultra-wideband (UWB) receiver (Hoyos & Sadler, 2006), is a new UWB receiver implementation method, considering a novel approach that utilizes, analog basis expansion of the input signal, followed by parallel sampling of the basis coefficients (Hoyos & Sadler, 2004), to face significant implementation challenges, including achieving sufficient front-end dynamic range, to support desired receiver processing gain, and rejection of large narrowband interferers (NBI), and overcoming channel-induced distortion.

This method enables parallel digital signal processing, and leads to considerable complexity reduction, while still achieve a performance very close to Nyquist rate digital receivers, even when operating at a Sub-Nyquist sampling rate, without significant BER penalty, if the truncation error is negligible compared with the additive noise, or if the incurred SNR degradation can be mitigated by the channel decoder (Hoyos & Sadler, 2006). Since the receiver has N parallel paths, the sampling rate for each path will be N times lower than if, a Nyquist rate time-domain ADC were used to sample the input signal.

If the front-end frequency selectivity of a conventional receiver does not provide the required attenuation of the adjacent frequencies, the remaining unknown and undesired adjacent channel interference, will fold into the signal band. Thus, by using the orthogonality principle in the frequency domain, the transform-domain receiver can select signals with great accuracy, even if strong interferers are nearby.

Unfortunately, there are practical limitations to this solution due, to the limited number of frequency samples that can be taken, because of the complexity in the parallel bank of mixers and integrators needed. This limitation requires us to perform frequency sampling over short time windows, which in turn produces bandwidth expansion due, to convolution in the frequency domain. This frequency expansion implies overlapping (aliasing) in the frequency domain, i.e., loss of orthogonality.

The loss of orthogonality causes the adjacent, as well as in-band, interferers to overlap with the signal of interest producing destructive aliasing. A structure of multipath transform-domain receiver was proposed in (Prakasam et al., 2008). However, this chapter aims at proposing a selective, time-domain UWB, Wireless Local Area Network (WLAN) front-end, employing a direct-sequence architecture (Razavi, 1997), and introducing a multi-block designed Low-Noise Amplifier (LNA), in order to minimise the loss of orthogonality effect in the transform-domain receiver, due to short windowing in parallel sampling of the basis coefficients, during the analog basis expansion of the input signal, while ensuring a better trade-off between, selectivity, linearity, noise figure, and power consumption.

The first topic of this chapter describes the transform-domain receiver architecture, where the analog basis expansion principal of the input signal is briefly introduced. Then, the next topic presents the circuit techniques, including the circuit design and simulations results of the front-end receiver, while the last topic draws the conclusions.

2. Transform-Domain Receiver

The transform-domain receiver architecture (Fig. 1), shows the transmitted bit a_i estimation process, from the set of the coefficient $\Phi_n(t)$,provided by the quantized basis coefficients . An estimate of a , namely $\hat{a}$ is obtained via the linear receiver matrix H as:

$$\hat{a} = H \bullet r \tag{1}$$

In the case of an transform-domain receiver, the linear formulation in (1) includes the Inter-Symbol Interference (ISI) between the received pulses of $\tilde{r}(t)$, so solving the linear filtering problem presented in (1) will perform both signal detection and equalization (Hoyos & Sadler, 2006). Thus, the transform-domain receiver parallelization can be exploited to add robustness to adjacent channel interference, witch would be greatly helpful for the WLAN receivers, occupying the 5-6GHZ frequency band (group #2), in the case of the UWB system standard (Federal Communications Commission, 2002), where the neighboring nodes may be very closes, contrasting with the expensive signal detection and equalization with conventional architectures (Blazquez et al., 2005), (Chen & Chiueh, 2006).

The fundamental difference between the transform-domain receiver front-end and the conventional single ADC architectures is the way sampling of the received signal is performed. Folding of the signal spectrum introduced by the time-domain sampling produces the classical frequency aliasing effect. The adjacent channel interference will fold into the signal band, and may seriously distort the discrete-time representation of the signal, if the front-end frequency selectivity of a conventional receiver does not provide the required attenuation.

Unfortunately, even when we use the orthogonality principal in the frequency domain, witch can select signals with great accuracy, even when strong interferers are nearby, the number of the frequency samples that can be taken is limited, because of the complexity in the parallel bank of mixers and integrators needed (Fig. 1). Thus, loss of orthogonality can be caused by the frequency expansion induced by the convolution in frequency domain, when performing a frequency-domain sampling over short time windows.

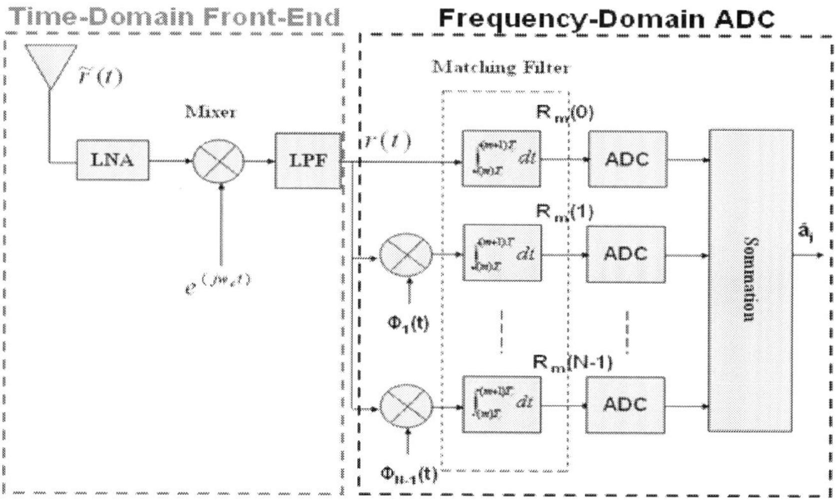

Fig. 1. Mixed-Signal Transform-Domain Receiver Architecture.

Attenuation of the adjacent channel at the front-end of the receiver is critical to avoid catastrophic aliasing after frequency-domain sampling. The aim of this work is to provide a great selective time-domain front-end reception part, for the transform-domain UWB WLAN receiver, in order to minimize the overlapping (aliasing) in the frequency domain caused by the short time windowing limitation.

3. Design Techniques

3.1 Dynamic Feedback LNA

Several proposals were introduced lately for the design of low-power, UWB CMOS LNA (Park, et al., 2005), (Yu et al., 2006), (Shameli & Heydari, 2006), (Yo & Yoo, 2007). Due particularly to its low input impedance, the common-gate input LNA exhibits an excellent linearity, selectivity, and impedance matching over the common-base LNAs (Cusmai & Brandolini, 2006). This makes it the best candidate to integrate into transform-domain receivers, wherein the selectivity compared to the adjacent UWB groups, would be of a major importance.

Unfortunately, the limited small signal gain and relatively high noise figure (NF) values, constitute serious handicaps for their use. Figure 2 depicts the structure of the proposed, multi-block designed LNA, introducing the dynamic feedback architecture. For linearity purpose, the voltage-voltage feedback is the better way to take advantage of the common-gate LNA.

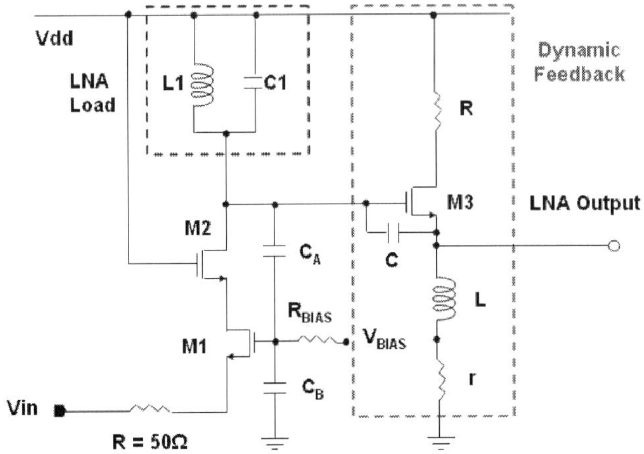

Fig. 2. Dynamic Feedback LNA.

One interesting property of the circuit is its input impedance depends on the loop transfer function. The idea is then synthesizing a single resonance load network, able to simultaneously shape the frequency transfer function, and the input impedance. By inspection of the circuit, and since the marginal wideband gain contribution of the source follower, the input impedance $Z_{in}(j\omega)$ and the frequency transfer function $T(j\omega)$ are given by :

$$Z_{in}(j\omega) = 1/g_m + \alpha . Z_{Load}(j\omega) \tag{2}$$

$$T(j\omega) = \frac{Z_{Load}(j\omega)}{Z_{in}(j\omega)} \tag{3}$$

Where $g_m = g_{m1} // g_{m2}$, is the equivalent transconductance gain introduced by the transistor M_1 and M_2, α the feedback factor, and Z_{Load} the load impedance. From (2), the input impedance has the same frequency dependence as the load impedance. α is chosen so that $(1/g_m) + \alpha . R_p = R_s$, where R_p is the load resistance at the resonance. Thus, the LNA is easily matched to the source resistance R_s at frequency band of interest.

Furthermore, if the load impedance intentionally presents also a series resonance for filtering purposes, the amplifier input impedance is unmatched at resonance frequency. The resulting power reflection is beneficial, contributing to reduce the signal processed at the interferer frequency. The amplifier load impedance Z_{Load} can be designed to filter out 3-5GHz and 6-10GHz interferers, while amplifying 5-6-GHz WLAN UWB signals and the input impedance is matched to the source in UWB WLAN band only.

On the other hand, the dynamic feedback LNA achieves a significant conversion gain improvement, mainly due initially, to the composed open loop dynamic feedback LNA structure, as depicted in the figure 3, where the C_2 represent the equivalent capacitor between the input and the output of the circuit, and secondly to the inductive load of the feedback circuit, which as will be detailed later on, increases the gain by reducing the LNA to a simple second order circuit with zero at the origin, for certain inductance L values. The small signal gain is then, maximized within the frequency band between the two poles.

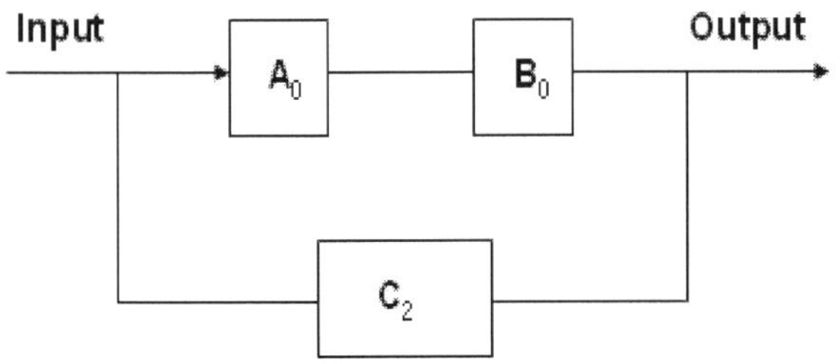

Fig. 3. The Dynamic Feedback LNA Block Diagram.

In the case of the UWB WLAN communication systems, we can maximise the small signal gain by replacing them within the 5-6GHz frequency range. The M_1 transistor is biased with 3.8mA, and the load circuit (L_1C_1) initially centers the small signal gain at 5GHz, the UWB WLAN lower limit. C_A symbolise the M_1-M_3 gate-to-gate capacitance (~200fF), where C_B capacitor still in the order of 700fF. The two poles introduced by the dynamic feedback are mainly dependent on the high frequency capacitors of the transistor M_3 (Razavi, 2001). Equations (4), (5) and (6) show the source follower high frequency transfer function (Razavi, 2006).

$$\frac{V_{out}}{Vin} = \frac{1 + \frac{C_{GS3}S}{g_{m3}}}{(aS^2 + bS + 1)} \tag{4}$$

where:

$$\begin{cases} a = \frac{R_{s3}}{g_{m3}}(C_{GD3}.C_{GS3} + C_{GD3}.C_{SB3} + C_{GS3}.C_{SB3}) \\ b = R_{s3}C_{GD3} + \frac{C_{GD3} + C_{SB3}}{g_{m3}} \end{cases} \tag{5}$$

These equations show that, by introducing two additional poles, a UWB LNA with dynamic feedback is likely to suffer from a chronic instability over all UWB frequency band. In our case, the small signal model of the proposed dynamic feedback circuit is shown in figure 4. Thus, the feedback circuit transfer function can be described as:

$$\frac{V_{out}}{V_{in}}(s) = \frac{r_{03}.Z.(C_2 s + g_{m3})}{[a+b]} \tag{6}$$

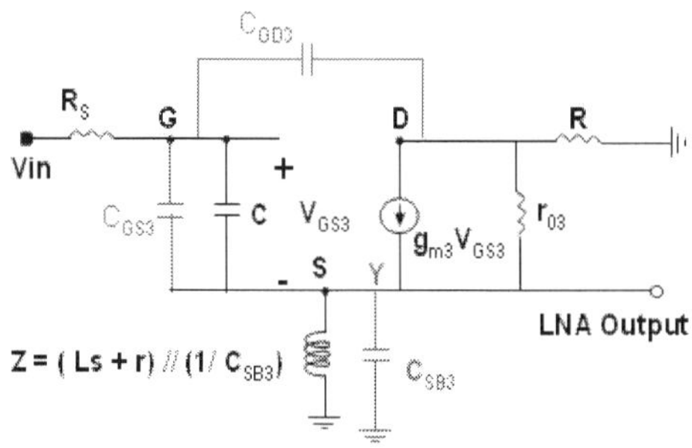

Fig. 4. The Dynamic Feedback Small Signal Model.

$$\text{With}: \begin{cases} a = \{R_s(C_2 + C_{GD3})s + 1\}.(Z + R + r_{03}) \\ b = r_{03}.Z.(C_2 s + g_{m3}).(Z + RR_s C_{GD3} s) \end{cases} \quad (7)$$

$$\text{And}: \begin{cases} C_2 = C + C_{GS3} \\ Z = (Ls + r)//(\dfrac{1}{C_{SB3}}) \end{cases} \quad (8)$$

Accordingly, both the impedance Z, and the coupling capacitor C are introduced for the LNA gain conversion optimization. On the other part, the dynamic feedback based on a source follower circuit, with an inductive output, allows an inductive behaviour of the feedback circuit input impedance (Fig. 5) witch helps improving the small signal gain too, with more idealized voltage-voltage amplifier circuit (Razavi, 2001). Consequently, the dynamic feedback circuit can also make it possible, to minimize the NF, induced by the purely capacitive feedback circuit proposed by (Cusmai & Brandolini, 2006).

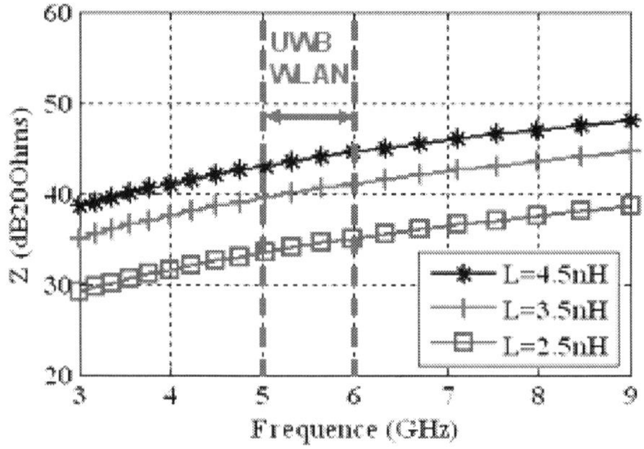

Fig. 5. The Dynamic Feedback Input Impedance.

It's important to note here the marginal dynamic feedback open loop contribution, for the UWB WLAN LNA, in terms of small signal gain (Fig. 6), and NF (Fig. 7), especially at the frequency band of interest. Thus, the small signal gain improvement can be achieved without power noise amplification, witch help improving the NF as needed. The frequency response simulation results suggest that, for a specific inductive load range values L, the closed loop feedback circuit contribution is effectively reduced to a simple zero at the origin (Fig. 8), witch ensures a perfect stability for the LNA circuit, over the entire UWB frequency band. However, for other inductance L range values, the UWB WLAN LNA could become deeply instable, with four poles occupying a larger frequency range (ex. L=1.5nH). The conversion gain can thus be maximized, by introducing an optimum inductance L, and capacitor C values (L=4.5nH, C=0.4pF), until reaching 27dB at 5.65 GHz (Fig. 9), (Fig. 10).

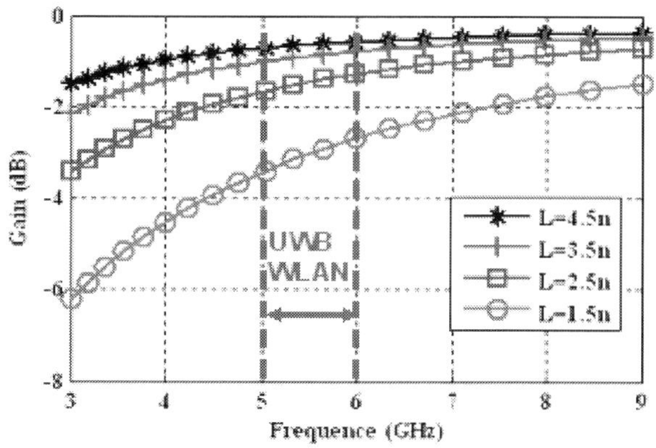

Fig. 6. The Dynamic Feedback Open Loop Gain Contribution.

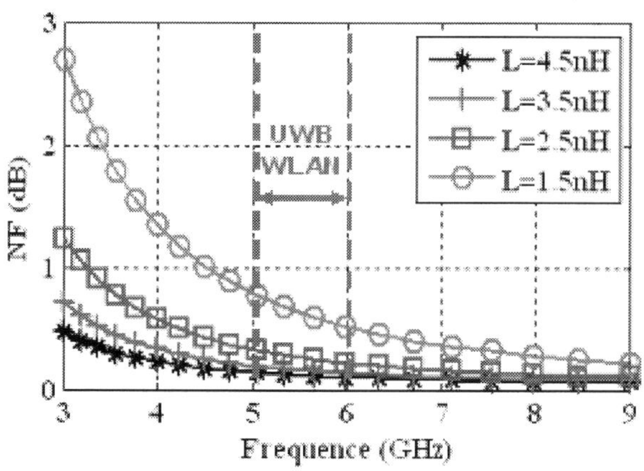

Fig. 7. The Dynamic Feedback Open Loop Noise Figure Contribution.

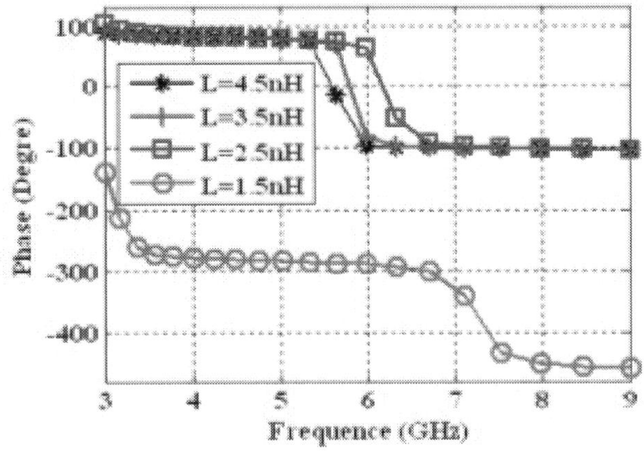

Fig. 8. Dynamic Feedback LNA Phase Simulations.

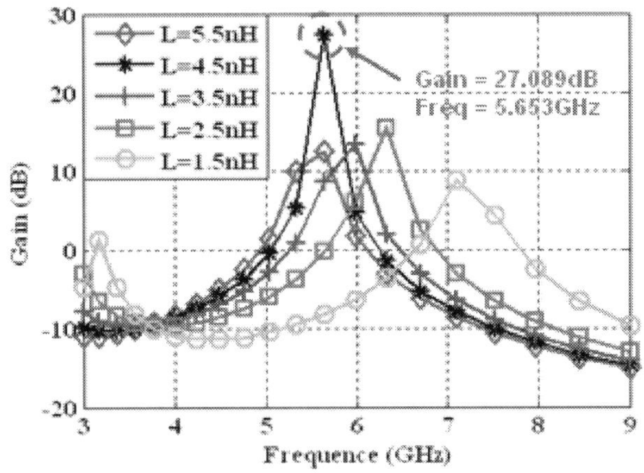

Fig. 9. Dynamic Feedback LNA Conversion Gain with (C = 0.4pF).

Regarding the noise figure issue, and according to the Friis equation for cascaded stages, the overall noise figure is mainly determined by the first amplification stage, provided that it has sufficient gain. You achieve low noise performance by carefully selecting the low noise transistor, DC biasing point, and noise-matching at the input, and the noise performance is characterized by NF value, defined as the ratio between the input signal-to-noise ratio and the output signal-to-noise ratio (9).

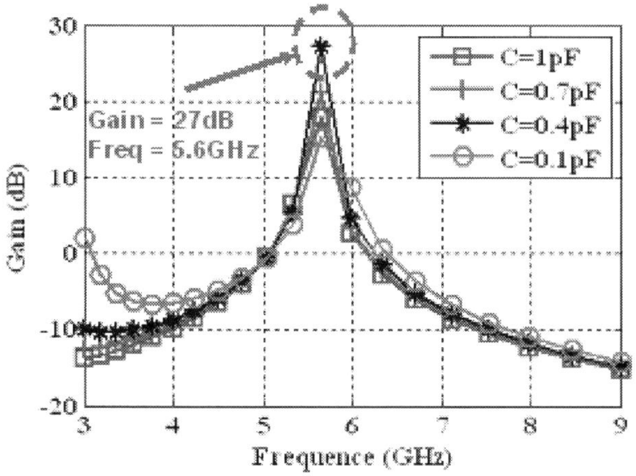

Fig. 10. Dynamic Feedback LNA Gain Conversion Optimization.

$$NF = \frac{\left(\frac{S}{N}\right)_{Out}}{\left(\frac{S}{N}\right)_{In}} = \frac{N_{Out}}{G_A . N_{In}} \tag{9}$$

Thus, one other advantage when considering the multi-block LNA design methodology, as depicted in (Fig. 2), is the fact that the trade-off between the conversion gain and the noise figure is no longer needed, since, as detailed earlier, the conversion gain could be optimised by properly shaping the over all LNA circuit transfer function. Consequently, the multi-block design LNA circuit noise figure, can be lowered by means of proper input stage circuit, and feedback circuit biasing, considering only the power consumption limitations. Concretely, by introducing a dynamic feedback, with a distinct biasing for the input stage circuit, we actually de-correlate between the available noise power from the source (N_{in}), and the available noise power to the load (N_{out}), and hence, one can be able to reduce the global NF value. Effectively, the figure 11 shows that, the dynamic feedback LNA noise figure values, vary now from 3.86dB down to 2.78dB in the 5-6GHz frequency range, when considering the inductance optimum value (L=4.5nH), depicted in black curve. As expected, this presents a 0.78 dB average gain with respect to the 4.1dB LNA minimum noise figure, developed by the common-gate made device in (Cusmai & Brandolini, 2006), even when biased at 5mA. However, the dynamic feedback LNA input stage where biased at 3.8mA, with marginal power consumption for its ultra low-power feedback circuit.

In terms of linearity, compared to the LNA circuit proposed by (Cusmai & Brandolini, 2006), the dynamic feedback significant narrow-band conversion gain improvement, was produced at the cost of slight linearity reduction, with a 1dB compression and desensitizing point falling at +1,-2 dBm respectively (Fig. 12), as depicted in (Tab. 1), witch reports the proposed LNA related performances, in comparison with a various recently published UWB

LNAs, including common-source degenerated devices. We also note that, the common-source input stage LNA (Park, et al., 2005), show a poor linearity performance, even with an ultra low-power made devices (Shameli & Heydari, 2006), suggesting that the trade-off between, conversion gain, noise figure, linearity, and power consumption could be relaxed, only when considering a multi-block design methodology, with distinct biasing circuits.

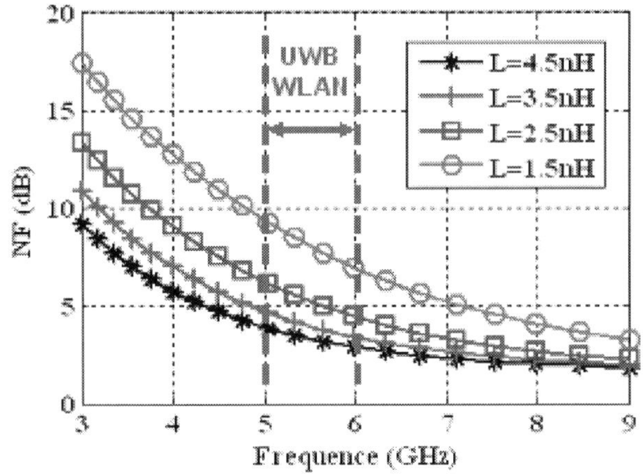

Fig. 11. Dynamic Feedback LNA Noise Figure.

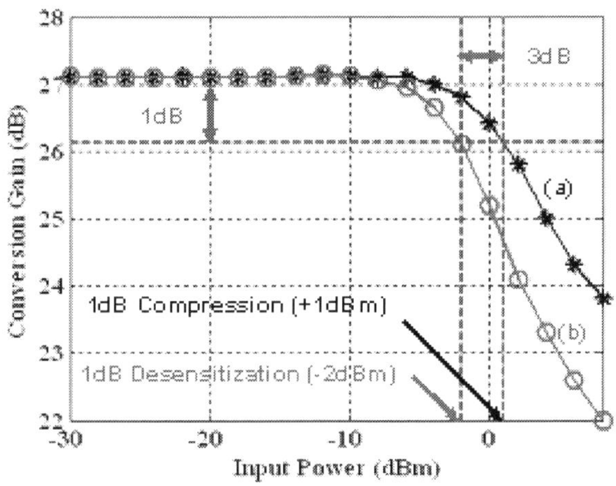

Fig. 12. Dynamic Feedback LNA Linearity Simulations. (a) Gain versus Signal Power. (b) Small Signal Gain versus the Closest Interferer Signal Power (7GHz, Group#3 Signal Power).

	Tech.	CG [dB]	NF [dB]	1dB C.P. [dBm]	1dB Desensitization [dBm]	Power [mW]
(Cusmai & Brandolini, 2006)	0.18 μm CMOS	16	4.1	+1.5	-1.5	9
(Park, et al., 2005)	0.18 μm SiGe	13	3.3	-17	-	9.6
(Shameli & Heydari, 2006)	0.18 μm CMOS	16.8	3.9	-21	-	0.1
LNA Fig. 2.	0.18 μm CMOS	27	3.3	+1	-2	7

Table 1. Comparison With Previously Published UWB LNA.

3.2 Downconversion Mixer

The choice of a single-balanced mixer instead of its double-balanced alternative is due to the converter would be required after the LNA witch increase the power, and the higher noise introduced by the double-balanced solution.

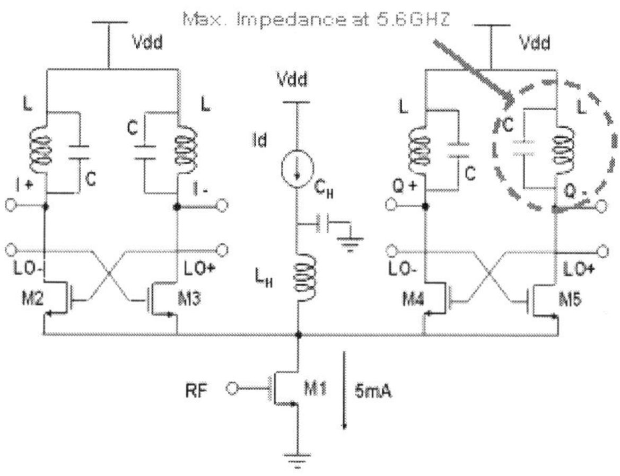

Fig. 13. Quadrature Mixer Schematic.

The mixer schematic is shown in (Fig. 13). A single common-source g_m-transistor (M_1) injects the RF signal in two single-balanced quadrature commutating pairs. When compared to the conventional solution adopting two separate transconductors, this choice allows a higher switching pair current gain (Sjoland & Karimi-Sanjaani, 2003).

A current source is used to set transconductor and switching stage current independently, in order to lower to DC current in the switching stage, witch leads to a lower noise (Darabi & Abidi, 2003). The inductor L_H extend the commutation bandwidth with benefits to conversion gain, noise and linearity (Razavi, 2007). The bias current of the g_m-transistor (M_1)

should be higher enough (~5mA) to achieve the desired conversion gain, noise figure and IIP3. The Vgs of the LO switches is set near the V_t to achieve a low bias current, and at the same time ensure that the required LO amplitude remains at a reasonable level (300mVpp) for complete current commutation. The LC circuit present a high impedance at 5.6GHz, such that the output AC current of (M_1) will flow into the LO switches. The quadrature mixer achieves 5.8dB CG, 8.8 dB and +1.68 dBm IIP3 at 5.6GHz (Fig. 14).
The DC offset in mixers is a critical parameter for direct conversion receivers, since most of the gain occurs after the downconversion of the input signal and the receiver can be saturated if the offset is too large, but the direct-conversion architecture lends itself to UWB receivers, because static and time varying DC offsets can be easily removed in the adopted OFDM modulation where the subcarrier falling at DC is not used (Batra et al., 2004), and because of the wide bandwidth makes the (1/f) noise less critical.

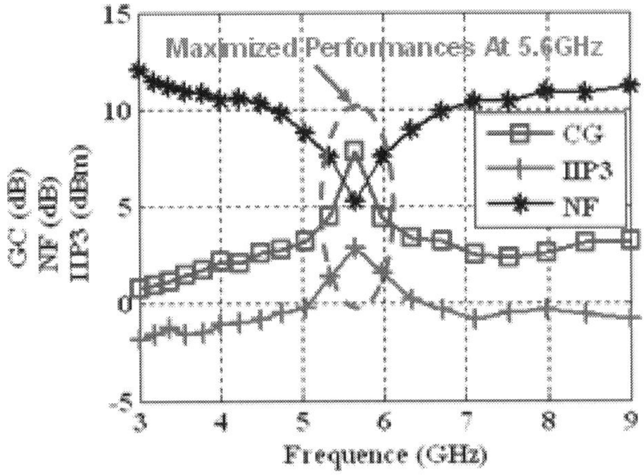

Fig. 14. Quadrature Mixer Frequency Response of CG, NF and IIP3.

3.3. Baseband Filter

An SK filter (Razavi, 2006) is designed in conjunction with the above mixer. The core amplifier is a simple low-gain circuit to obtain flat-band behaviour across 300MHz. Consequently, the voltage swings reduction removes the compression bottleneck at the mixer output; however, the loop gain does not force a virtual ground at these nodes. The baseband filter is therefore designed with a 2dB limited loop gain, this is mainly due to the substantial narrow-band conversion gain produced by the downconversion mixer at the 5-6Ghz frequency band, therefore, the later is likely to experience a compression at it's output. Finaly, table 2 reports the proposed selective, time-domain front-end performances, in comparison with the selective UWB front-end presented in (Cusmai & Brandolini, 2006). One can note that, the high interferer rejection developed by the multi-block LNA design methodology; very useful to overcome the UWB transform-domain receiver problem, has been achieved with an excellent front-end linearity, noise figure, and even power consumption performances. Therefore, the front-end subsequent stages design requirements, were greatly relaxed, when the multi-block LNA design methodology has been introduced.

	0.18 μm CMOS Selective UWB WLAN Front-end	**0.18 μm CMOS Selective Front-end in (Cusmai & Brandolini, 2006)**
Max. Voltage Gain [dB]	34.8	22.8
Min. NF [dB]	6.42	5.2
Min. IIP3 [dBm]	-4.35	-3.5
Current [mA]	10.9	10
Voltage Supply [V]	1.8	1.8
Interferer Rejection [dBc]	-35	-
1 dB Desensitization [dBm]	-8	-6.5

Table 2. Time-Domain Front-End Performances Summary Comparison

4. Conclusion

In this work, a very robust quadrature time-domain CMOS front-end for transform-domain UWB WLAN receiver has been presented, showing 1-dB desensitization point as high as -2dBm, with 27dB narrow-band conversion gain, and 35dBc interferer rejection, witch helps minimizing the loss of orthogonality effect, introduced by the short windowing, in the analog basis expansion of the input signal. The introduced multi-block design LNA, based on highly linear voltage-voltage dynamic feedback topology, filter out the UWB interferers in group #1 and #3, while amplifying the UWB WLAN signal, and shows a better trade-off between linearity, conversion gain, and power consumption.

The downconversion mixer is single-balanced, with the two quadrature pairs sharing the same input transconductor. Further research, will focusing on the implementation of the frequency-domain part of the transform-domain UWB WLAN receiver, where the receiver expands the signal over a basis set, and then operates on the basis coefficients, in order to better use the time-domain front-end performances .

5. Referring

Hoyos, S.; Sadler, B. M. (2006). UWB Mixed-Signal Transform-Domain Direct-Sequence Receiver, IEEE Transactions On Wireless Communications, vol. 6, No.8, (August 2006) (3038-3046), ISSN: 10.1109/TWC.2007.051069.

Hoyos, S. et al., (2004). High-Speed A/D conversion for Ultra-Wideband signals based on signal projection over basis functions, Proc. (ICASSP '04), pp. 537-540, ISBN: 10.1109/ICASSP.2004.1326882, International Conference on Acoustics Speech and Signal Processing, May 2004, Montreal, Canada.

Prakasam, P. K. et al., (2008). Applications of Multipath Transform-Domain Charge-Sampling Wide-Band Receivers, IEEE Transactions on Circuits and Systems -II, vol. 55, No. 4, (April 2008) (309-313), ISSN: 10.1109/TCSII.2008.919480.

Razavi, B. (1997). Design Considerations for Direct-Conversion Receivers, IEEE Transactions On Circuits and Systems-II: Analog and Digital Signal Processing, Vol. 44, No. 6, (June 1997) (428-435), ISSN: 10.1109/82.592569.

Federal Communications Commission, (2002). Revision of Part 15 of the Commission's Rules Regarding Ultra Wide-band Transmission Systems. [Online].Available: http://www.fcc.gov/Document_Indexes/Engineering_Technology/2002__index_OET_Order_html.

Blazquez, R. et al., (2005). Direct Conversion Plused UWB Transceiver Architecture, IEEE Proc. (DATE '05), pp. 94-95, ISBN: 10.1109/DATE.2005.122, the Design, Automation and Test in Europe Conference and Exhibition, 2005.

Chen, P. & Chiueh, T. (2006). Design of A Low Power Mixed-Signal Rake Receiver, IEEE Proc. (ISCAS '06), pp. 2796, ISBN: 10.1109/ISCAS.2006.1693204, International Symposium on Circuits and Systems, May 2006, Island of Kos, Greece.

Park, Y. et al., (2005). A Very Low Power SiGe LNA for UWB Application, ISBN: 10.1109/MWSYM.2005.1516847, IEEE MTT-S International Microwave Symposium Digest, June 2005, Long Beach, CA, USA.

Yu, Y-H. et al., (2007). A 0.6-V Low Power CMOS LNA, IEEE Microwave and Wireless Components letters, Vol. 17, No. 3, (March 2007) (229-239), ISSN: 10.1109/LMWC.2006.890502.

Shameli, A. & Heydari P. (2006). A Novel Ultra-Low Power (ULP) Low Noise Amplifier Using Differential Inductor Feedback, Proc. (ESSCIRC '06), ISBN: 10.1109/ESSCIR.2006.307603, 32nd European Solid-States Circuits Conference , Sept. 2006, Montreux, France.

Yo, S-S. & Yoo, H-J. (2007). A Low Power Current-reused CMOS RF Front-end with Stacked LNA and Mixer, ISBN: 10.1109/SMIC.2007.322780, Topical Meeting on Silicon Monolithic Integrated Circuits in RF Systems, Jan. 2007, Valence, France.

Cusmai, et al., (2006). A 0.18 μm CMOS Selective Receiver Front-End for UWB Applications, IEEE Journal Of Solid-State Circuits. Vol. 41, No. 8, (August 2006) (1764-1771), ISSN: 10.1109/ISCAS.2007.377992.

Razavi, B. (2001). Design of analog CMOS integrated Circuits, Boston, MA; Toronto: McGraw-Hill, c2001.

Razavi, B. (2006). Fundamentals of Microelectronics, B.John & Wiley Sons, Inc. April 2006.

Sjoland, et al., (2003). A merged CMOS LNA and mixer for a WCDMA Receiver, IEEE J. Solid-Sate Circuits, Vol. 38, No. 6, (Jun. 2003), (1045-1050), ISSN: 10.1109/JSSC.2003.811952.

Darabi, H. A. & Abidi, A. (2000). Noise in RF-CMOS mixers: a simple physical model, IEEE J. Solid-Sate Circuits, Vol. 35, No. 1, (Jan. 2000), (15-25), ISSN: 10.1109/4.818916.

Razavi, B. (2007). Design Considerations for Future RF Circuits, Proc. (ISCAS '07), ISBN: 10.1109/ISCAS.2007.377992. IEEE International Symposium on Circuits and Systems May 2007, New Orleans, USA.

Batra, A. et al., (2004). Multi-Band OFDM Physical Layer Proposal for IEEE 802.15 Task Group 3a. Mar. 2004 [Online]. Available : https://www.multibandofdm.org .

5

Flexible Power Amplifier Architectures for Spectrum Efficient Wireless Applications

Alessandro Cidronali, Iacopo Magrini and Gianfranco Manes
Department of Electronics and Telecommunications, University of Firenze, Italy

1. Introduction

The wireless systems evolution known as "beyond the 3rd generation" (B3G) will make use of dynamic spectrum access techniques to provide wide bandwidth to mobile users via heterogeneous wireless networks. A consistent step toward this scenario is represented by the outcome of the last World Radiocommunication Conference [1] which established new primary frequency bands allocation spanning from the UHF band to low microwaves and thus reflecting the increasing demands for broadband mobile and cellular systems.

We have become used to the doubling of processing power of chips based on Moore's law, but the progress in radio interface technologies still poses significant challenges.

High spectrum efficiency performance becomes therefore another major requirement of the design, along with the more consolidated ones: energy efficiency, integration, cost and reliability.

While the IMT-advanced roadmap foresees a 100 Mbps data rate for mobile users and a peak of 1 Gbps is expected for nomadic users, the available spectrum for legacy wireless communications is fragmented and reaches the amount of 750 MHz in the S-C band. A radio technology that is expected to interact with a multi-services network should be able to change between different operating bands and adapt its features according with the different available standard and requirements. Most of the research efforts performed during the last years dealt with issues related to the physical layer of the communication stack [2]; however, despite the growing interest in multi-standard operation, less attention has been devoted to the radio-frequency front-end, which therefore remains one of the most challenging parts of a multi-band radio. One main reason for the delay in effectively implementing multi-standard transceivers can be attributed to the implementation of the RF transmit power amplifier (PA). Today, dedicated, single standard PAs achieve very good power added efficiency (PAE) and, in this way, long battery lifetime. Any multi-standard PA, needed for the support of different, not always predefined, communication systems, should compete with such dedicated solutions. A conceptual framework to this is provided by the so-called software-defined radio (SDR), i.e. a radio communication system, using software for the reconfiguration of the digital and analog parts in order to perform the modulation and demodulation of radio signals, [3]. In practice, however, due to the difficulty of implementing the fast signal processing implied in the SDR approach, most of the systems on the market, based on more traditional approaches, are still supporting only a very limited number of standards (e.g. 4 GSM frequencies, UMTS and, possibly, Bluetooth). In the near future, further standards will have to be supported, and more could have to be added during the handset lifetime, hopefully without hardware reconfiguration. This will determine the need of multiband PAs capable to transmit efficiently more than one service with variable radio access schemes.

Example of realizations in different technologies are provided in this Chapter as demonstrators of the discussed multiband design methods. The flexibility of the operative frequency is thus introduced by analyzing new PA architectures and design methodologies which consider the inclusion of tunable and switching components to enable a change in the operative frequency. A review of the most promising circuit topologies suitable to design reconfigurable matching networks is given in this Chapter. Varactor diodes based and MOS switched based topologies are compared, highlighting their point of strength and weakness. It is shown as a concurrent dual-band PA implemented by the proper combination of frequency-dedicated PAs, each of them optimized to work in a given bandwidth would be an easy approach, it becomes unsuited due to the complexity of the power combiners. For this reason the true concurrent dual band PA presented in this chapter is to be considered as an enabler components for high efficiency multiband systems.

Spectrum efficiency is just one of the challenges a wireless system designer faces, further come from linearity and energy efficiency resulting from the use of multicarrier and complex envelope modulation schemes. As the spectrum efficiency increases a more demanding requirement in term of PA linearity faces to the designers. Energy efficiency and linearity are conventionally traded-off considering that increasing the power back-off increases the linearity at the expenses of lower energy efficiency. To maintain signal integrity, the resulting waveforms in turn require linear transmission paths for their successful deployment. A way to match signal integrity and energy efficiency consists in the use of digital predistortion algorithm applied at base-band and implemented in the digital section of the transmitter. In spite of their large development in frequency dedicated PA architectures, the development of a technique suitable for multi-band applications is not yet completely available. In this Chapter a comprehensive treatment a novel technique for Dual Band Digital Predistortion (DB-DP) is discussed. The DB-PD is based on the simultaneous predistortion of both channels at intermediate frequency (IF), it uses a single band memory polynomial DP for linearization, while the feedback path is based on a subsampling receiver. The memory polynomial DB-DP system is presented by simulation with Matlab-Simulink® for a deep understanding of performance.

2. A possible applicative scenario for multi-band transmitters

Extending the scenario to already experienced 3G voice/data systems, users may be moving while simultaneously operating in a broadband data access or multimedia streaming session. The need to support low latency and low packet loss handovers of data streams as users transition from one access point to another may require the concurrent use of more than one frequency band at the time. For full-mobile data services, no user interaction will be required to adapt their service expectations because of environmental limitations that are technically challenging but not directly relevant to the user (such as being stationary or moving). The enabling front-end of future mobile unit thus will accommodate more than one system in a effective and efficient way to make possible the connectivity capabilities depicted in Fig. 1.

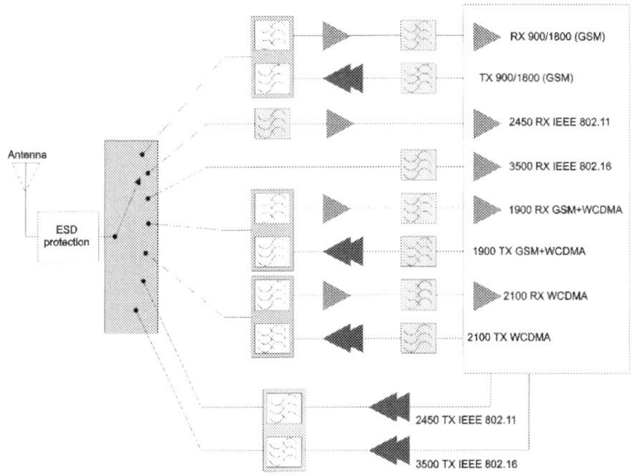

Fig. 1. Concept of a multi-band transmitter

The Wireless Local Area Network (WLAN) industry has become one of the fastest growing segments of the communications industry. This growth is due, in large part, to the introduction of standards-based WLAN products, regulated by the IEEE 802.11. The expectation of the WLAN's continuing growth stems from the promise of new standardized WLAN technologies, from improved cost/performance of WLAN systems, and from the growing availability of WLAN solutions that consolidate voice, data, and mobility functions. This, combined with market forecasts reporting that WLAN will experience a continuous growth in the next years, show that WLAN technologies will play a significant role in the future and will have a significant impact on our business and personal life styles. The WiMAX is an alternative and complementing standard for high data rate transmission, which will transform the world of mobile broadband by enabling the cost-effective deployment of metropolitan area networks based on the IEEE 802.16 standard to support notebook PC and mobile users on move. There are many advantages of systems based on 802.16, e.g. the ability to provide service even in areas that are difficult for wired infrastructure to reach and the ability to overcome the physical limitations of traditional wired infrastructure. The standard will offer wireless connectivity of up to 30 miles. The major capabilities of the standard are its widespread reach, which can be used to set up a metropolitan area network, and its data capacity of 75 Mbps. This high-speed wireless broadband technology promises to open new, economically viable market opportunities for operators, wireless Internet service providers and equipment manufacturers. The flexibility of wireless technology, combined with high throughput, scalability and long-range features of the IEEE 802.16 standard helps to fill the broadband coverage gaps and reach millions of new residential and business customers worldwide.

With WLAN 802.11 and now WiMAX 802.16, there has been a growing interest in technologies that allow delivery of higher data rates over large geographical areas. The IEEE 802.16 family of standards (802.16-2004 and 802.16e) are intended to provide high bandwidth wireless voice and data for residential and enterprise use. The modulation used to achieve these high data rates is orthogonal frequency-division multiplexing (OFDM). WiMAX OFDM features a minimum of 256 subcarriers up to 2048 subcarriers, each modulated with either BPSK, QPSK, 16 QAM or 64 QAM modulation. Having these carriers orthogonal to each other minimizes self-interference. This standard also supports different signal bandwidths, from 1.25 MHz to 20 MHz to facilitate transmission over longer ranges and to accommodate different multipath environments. This represents a significant increase in system profile complexity as compared to the 802.11 standard, mostly to guarantee a wider, more efficient, more robust network. More subcarriers and variable-length guard intervals contribute to this enhancement.

The ability to develop and manufacture a single reconfigurable terminal, which can be configured at the final stage of manufacture to tailor it to a particular market, clearly presents immense benefits to equipment manufacturers. With the design, components used, and hardware manufacturing processes all being identical for all terminals worldwide, the economy of scale would be huge. This has the potential to offset the additional hardware costs which would be inevitable in the realisation of such a generic device.

Based on this, the scenario adopted reflects in the request for transceiver architectures capable to support cellular phone, WLAN and WiMAX in an 'always and everywhere connected' solution. The transceiver performance in this multi-standard operation, however, comes at the expense of RF specifications that are more difficult to achieve. Furthermore, the choice and definition of the proper transceiver architecture becomes a difficult task, since several parameters - as now imposed by two standards - must be taken into account.

3. Suitable architectures for multiband-multimode transmitters

The concept of a multiband or general coverage terminal is, strictly speaking, an extension of the basic SDR concept into that of a broadband flexible architecture radio, since the basic reconfigurability and adaptability aspects of operation do not depend upon multiband coverage. It would be possible, for example, to construct a useful SDR which operated in the 800-900 MHz area of spectrum and which could adapt between AMPS, GSM, DAMPS, PDC, and CDMA. It is now normal, however, for a handset to have multi-frequency operation and hence the extension of this principle to a SDR is a natural one. The international business traveler market is still seen as both large and lucrative, particularly in terms of call charges, hence making this type of handset attractive to both manufacturers and network providers. An ideal SDR is shown in Fig. 2; note that the A/D converter is assumed to have a built-in anti-alias filter and that the D/A is assumed to have a built-in reconstruction filter.

The ideal software defined radio has the following features [4]:

- The radio access scheme (i.e. modulation scheme, channelization, coding) and equalization for transmitter and receiver are all determined in software within the digital processing subsystem. This is shown containing a DSP in Fig. 2
- The ideal circulator is used to separate the transmit and receive path signals, without the usual frequency restrictions placed upon this function when using filter-based solutions (e.g., a conventional diplexer). This component relies on ideal matching between itself and the antenna and power amplifier impedances and so is unrealistic in practice over a broad frequency band. Since the primary

alternative, a diplexer, is very much a frequency-dedicated component, its elimination is a key element in a multiband or even multimode transceiver.

- The linear, or linearised, PA ensures an ideal transfer of the RF modulation from the DAC to a high-power signal suitable for transmission, with ideally no adjacent channel emissions. Note that this function could also be provided by an RF synthesis technique, in which case the DAC and power amplifier functions would effectively be combined into a single high-power RF synthesis block.
- Anti-alias and reconstruction filtering is clearly required in this architecture (not shown in Fig. 2
- . It should, however, be relatively straightforward to implement, assuming that the ADC and DAC have sampling rates of many gigahertz. Current transmit, receive, and duplex filtering can achieve excellent roll-off rates in both handportable and (especially) base-station designs. The main change would be in transforming them from bandpass (where relevant) to lowpass designs.

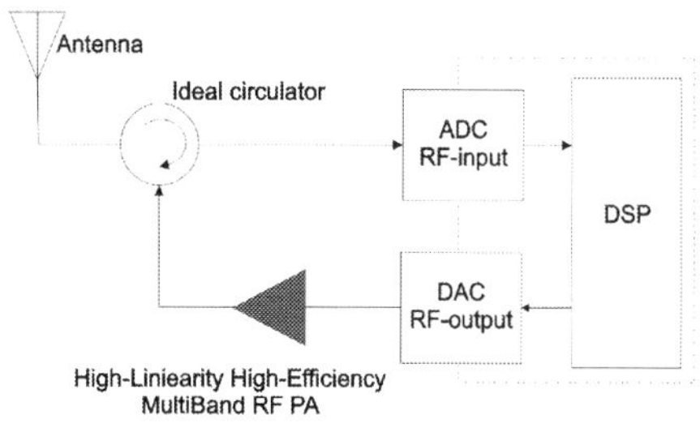

Fig. 2. Ideal software defined radio architecture

Possibly the most important element of any SDR system, whether in a base station or handset, is the linear or linearised multiband transmitter. Receiver systems have always required a high degree of linearity, as they must possess a good signal handling capability, in addition to good low-noise performance. In the case of transmitters, however, a high degree of linearity is a relatively recent requirement, arising predominantly from the widespread adoption of multi symbols envelope-varying digital modulations.

This follows from the fact that most modern modulation formats incorporate some degree of envelope variation, the only significant exception at present being GSM and its derivatives (DCS and PCS). The basic architecture of a SDR transmitter revolves around the creation of a baseband version of the desired RF spectrum, followed by a linear path translating that spectrum to a high-power RF signal.

Nevertheless the implementation of a true SDR poses a further very critical issues, i.e. the power consumption of the analogue-digital converter. Let's consider for instance the use of a flash converter, largely available in the market with a maximum number of bit about 18 preceded by a sample and hold circuit. Carrying out a simplified calculation, given the converter dynamic range, Dc, the power consumption of this systems is:

$$P_{dc}\frac{kT}{t_s}10^{Dc/10}, \quad (1)$$

where the k is the Boltzmann's constant = 1.38×10^{-23} J/K, T is the device temperature and t_s the sampling time. Furthermore the dynamic range of the converter is given by:

$$D_c = 6.02N + 1.76 - PAR + 10\log_{10}\left[2OSR\right], \qquad (2)$$

where number of bit, N, a peak-average ratio for the signal, PAR, and an oversampling ratio, OSR. From this easy calculation we can straightforwardly estimate the AD power consumption Pdc in a significant scenario for SDR. Assuming to digitize a frequency band from 800 MHz to 5.5 GHz with a 11GS/s ADC and assuming that the receiver dynamic range is from -20 dBm to -120 dBm, with a SNR of 12 dB at minimum sensitivity, the average PAR of 4, the required N is 20; it results that a such ADC consumes hundred of watt, thus preventing the use of the ideal architecture in Fig. 2. in practical implementation.

4. Reconfigurable Matching Networks

The multiband-multimode demands of today's wireless market, is fulfilled by implementations based on parallel line-ups completed by antenna diplexers and switches to meet the specific requirements of each communication standard, (c.f. Fig. 1). Utilizing only one adaptive transmit path to replace the parallel path concept is conceptually simple, but practical design considerations place severe design constraints and technology. Major challenges consists in creating the tunable filters and PAs [5]. Addressing these challenges means to develop flexible PAs capable to maintain the power-added efficiency (PAE) and linearity while moving among different operating frequencies. In conventional PA implementations, the linearity requirement typically results in the use of class-AB operation for the output, which provides a workable compromise between linearity and efficiency. When considering linearity, the class-AB output stage must be dimensioned in such a way that it can provide its peak output power without saturation. As a result, for a given peak output power and battery voltage, the load impedance for a class-AB stage at the fundamental frequency is fixed to $R_L \approx 0.5 \cdot V_{cc}^2 / P_{Peak}$ .Unfortunately, class-AB operation provides its highest efficiency only under maximum drive conditions. When operated at the required back-off level, due to linearity reasons for a given communication standard, a rather dramatic loss in efficiency occurs. For these reasons improving amplifier efficiency, while maintaining linearity, is currently a major research topic in wireless communications. In linearity-focused researches, the circuit is designed so that the resulting overall linearity performance of the PA module is improved. In this way, the active device can be operated closer to its peak-power capabilities and still be able to meet the linearity requirements.
Techniques that address the efficiency in the back-off mode are dynamic biasing or regulation of the supply voltage of the output stage

[6]. Dynamic biasing provides only modest improvements in efficiency, and supply voltage regulation requires an efficient DC-to-DC conversion, increasing system cost and complexity and operative bandwidth. Nevertheless this techniques appear very promising for future transmitter architectures. An alternative for improved class-AB efficiency is load-line adjustment as a function of output power using an adaptive or reconfigurable output matching network.
An ideal Reconfigurable Matching Network has to provide:

- Low Loss
- High linearity
- High Tuning Speed
- Sufficient impedance coverage
- Low complexity
- Low area usage

Power handling of matching networks is a critical issue in PA applications. To reduce the losses in a matching network, the use of a limited number of reactive elements is mandatory, beside the choice of high Q tunable components. Typically, such a network is based on varactor diodes, PIN-diodes or FET switching of matching elements like inductors, transmission-lines or capacitors, also involving micro electromechanical systems to improve the power handling capability [7].

We can conclude that these integrated adaptive networks will play an important role for the realization of the next generation of adaptive transceivers and this paragraph is aimed to describe the ongoing basic researches on this subject.

4.1 Varactor based switching matching network

Varactor diodes, although characterized by a relatively low Q factor at microwave frequencies, can be a choice for enabling RF tuning. Unfortunately, because of their inherently non linear behavior, their use with modern communication standards (characterized by high peak-to-average power ratios), has to be carefully analyzed according to the specific case considered. In Fig. 3 are shown varactor diode based circuit topologies [5] suited to provide matching tuning overcoming the issue related to the linearity of the electron devices.

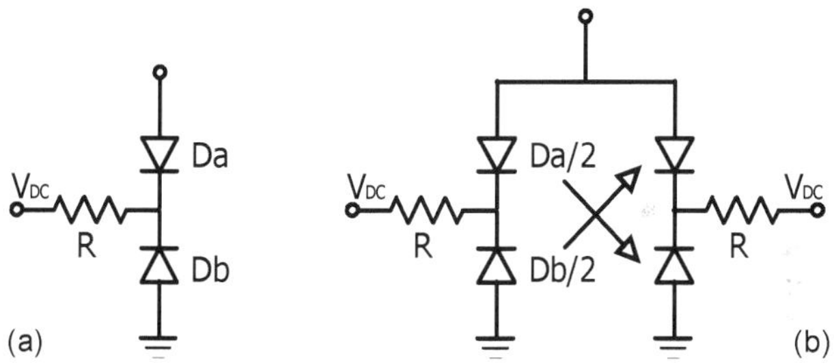

Fig. 3 - (a) Anti-series connection of varactor diodes to minimize third-order distortion, (b) Anti-series/anti-parallel connection of varactor diodes to minimize second and third-order distortion.

Basically, the capacitance of a single varactor diode can usually be expressed as:

$$C(V) = \frac{K}{(\varphi + V)^n} \quad (3)$$

where φ is the built-in potential of the diode, V is the applied voltage, n is the power law exponent of the diode capacitance, and K is the capacitance constant. The power law exponent can exhibit wide variation in different situations, from a value of n≈0.3 for an

implanted junction to n≈0.5 for a uniformly doped junction to n≈1.5/2 for a hyper-abrupt junction.
If the applied dc voltage is V_{DC}, then the incremental capacitance of a single varactor diode as a function of the incremental voltage v can be expressed as

$$C(v) = C_0 + C_1 v + C_2 v^2 + ... \tag{4}$$

where the term C_1 gives rise to second-order distortion and the C_2 term gives rise to third-order distortion. The diode configuration in Fig. 3a can be employed to realize a voltage variable capacitor with theoretically no distortion. Indeed, (referring to the circuit in Fig. 3a) evaluating the expressions for the linear and nonlinear terms of the capacitance, and setting to s the ratio of the diode areas Db/Da, it follows that:

$$\begin{aligned} C_0 &= \frac{sK}{(1+s)(\varphi + V_{DC})^n} \\ C_1 &= \frac{(1-s)nC_0}{(1+s)(\varphi + V_{DC})} \\ C_2 &= \frac{C_0\left[(s^2+1)(n+1) - s(4n+1)\right]}{2(\varphi + V_{DC})^2 (1+s)^2} \end{aligned} \tag{5}$$

It can be noted that for n ≥ 0.5, C_2 can be made equal to zero, resulting in zero third-order distortion, by setting

$$s = \frac{4n + 1 + \sqrt{12n^2 - 3}}{2(n+1)} \tag{6}$$

It can be observed from eq. (6), that the particular case of constant doping profile in the diode (the abrupt junction case where n=0.5) results in a value of s=1. This case is particularly attractive because this set of conditions (n=0.5, s=1), sets both C_2 and C_1 equal to zero. A more elaborate analysis shows that all higher order distortion terms also vanish, yielding (in theory) a "distortion-free" operation for this unique case. When dealing with process technologies where n>0.5, eq. (6) provides a direct means of calculating the required diode area ratio to minimize C_2.
For example, in the case where n=1, the required area ratio is exactly two. In the case of n=2, which corresponds to the ideal hyper-abrupt junction, the required area ratio is 2.6.
Although this approach can minimize C_2, it is clear from eq. (5), that a value of s≠1 will result in a finite value of C_1. In this case, a relatively high third-order distortion product will unfortunately still arise, resulting from the secondary mixing of the fundamental with the second-order non-linearity C_1. Fortunately, this distortion contribution can be eliminated, by placing an identical varactor stack in anti-parallel configuration (see Fig. 3b).
The linear capacitance of the circuits of Fig. 3a and b are identical, but the circuit of Fig. 3b has C_1=C_2=0 when the proper area ratio is set. It must be underlined that, in this configuration (Fig. 3b), all the even-order coefficients are zero (C_1, C_3, C_5, …) in this topology, but the higher coefficients that create odd-order distortion (C_4, C_6, C_8, …) are not

zero, although the IM3 contributions due to the 5th and higher order nonlinearities are very small.

The implications of this analysis can be summarized as follows:

- The classical configuration of Fig. 3a provides theoretically a "distortion-free" varactor stack when n=0.5, corresponding to a uniform doping profile of the varactors.
- The more generalized configuration of Fig. 3b provides an ultra-low distortion varactor stack for any value n>0.5, by setting the proper ratio of the diode areas, which sets C_1 and C_2 to zero, providing more freedom for use in different process technologies.

It must be underlined that each of the circuits in Fig. 3 requires a very high tap impedance (R) for proper operation. A high tap impedance, limits the impact that forward biasing of one of the diodes by RF signal has on linearity.

An effective way to implement the high impedance while keeping the RC time limited for the control signal is the anti-parallel diode pair depicted in Fig. 4.

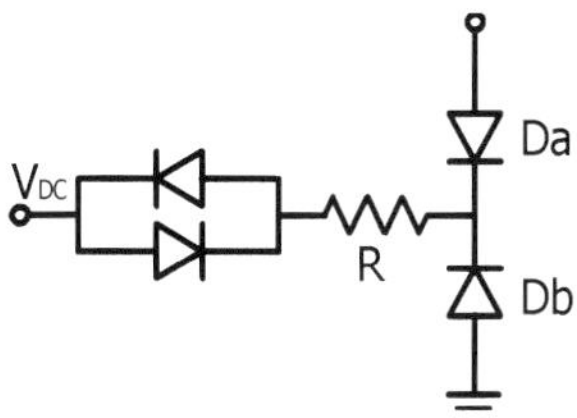

Fig. 4.- Anti-series connection of varactor diodes with modified center tap.

Analyzing, by a two tone test the described matching and comparing them with a single varactor diode we can observe the linearity of the different topologies. The IM2 and IM3 distortion of the circuit in Fig. 3a are comparably low, while the IM3 of the circuits in Fig. 3b is limited by fifth-order distortion due to complete cancellation of the third-order products. In this case indeed, a 1:5 slope dependence for the IM3 components can been found [8], confirming the elimination of the C_1 and C_2 contributions.

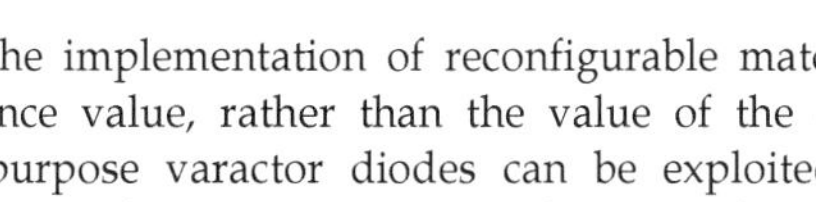

Fig. 5 - (a) Equivalent circuit of the "tunable inductor", (b) Simplified circuit of a varactor-based "tunable inductor".

A different option for the implementation of reconfigurable matching network is to act directly on the inductance value, rather than the value of the capacitors as discussed hereinabove. For this purpose varactor diodes can be exploited to create a "tunable inductor" [9]. By considering the circuits in Fig. 5, the equivalent impedance Z of the LC circuits can be written as:

$$Z = \frac{1 - \left(\omega/\omega_c\right)^2}{j\omega C} \tag{7}$$

where $\omega_c = \frac{1}{\sqrt{LC}}$ is the resonant frequency of the circuit. When the condition $\omega > \omega_c$ is achieved, the impedance Z represents an equivalent inductor L_{eq}, that, for the simplified circuits in Fig. 5, is analytically expressed as:

$$L_{eq} = \frac{\left(\frac{\omega}{\omega_c}\right)^2 - 1}{\omega C} \tag{8}$$

The resulting equivalent inductor is a "tunable inductor" whose value is related to the varactor capacitance C (and also ω_c). In Fig. 5b, the varactor parasitic elements have not been represented, however they have to be taken into account during the design of the "tunable inductor".

4.2 Switching matching networks

If the output power of the PA is so high to make the use of varactor diodes prohibitive, a possible solution could be to replace the varactor diodes with fixed value capacitors controlled by switching PIN diodes. By observing to the examples in Fig. 6, if the PIN diode connected to the capacitor is in the ON state, the capacitor adds its own capacity to the global circuit. On the other hand, if the PIN diode is in the OFF state, the capacitor does not affect the global circuit.

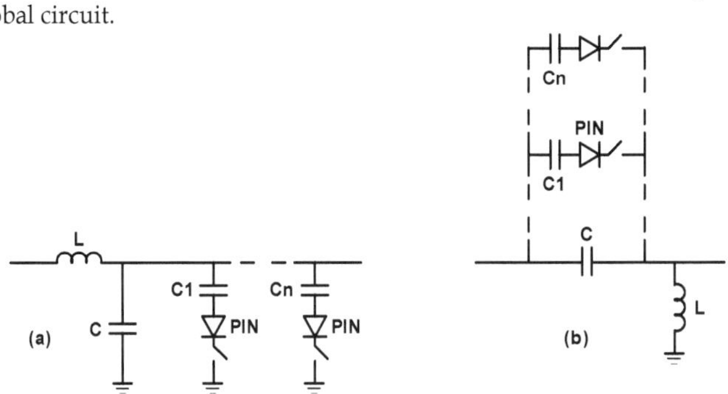

Fig. 6.- (a) Low-Pass Matching Network example, (b) High-Pass Matching Network Example

By using a set of digital signals that change the state of the PIN diodes to active or inverse condition, it is possible to generate $N=2^n$ impedances (where n is the number of PIN diodes used). Unfortunately, such a technique permits only a discrete set of tuning possibilities, resulting inappropriate where fine tuning is requested, unless a considerable number of components is used, thus increasing losses, area usage and costs. A different option of switched matching network is based on MOS devices . The use of MOS devices rather than PIN diodes makes them more suitable for IC designs. The operative principle of the network is discussed by a following example.

Let's consider a Si-Ge HBT device, the optimum fundamental loads at one dB compression point are 15 Ω at 2.45 GHz and 5 Ω and 3.5 GHz respectively, the reactive parts are negligible. These resistance values have been determinate accounting for the output power level to be supplied by the power amplifier in the two different bands. The aim of the output network consists in synthesizing these loads using a MOS switching network topology, schematically depicted in Fig. 7 and investigating the general features.

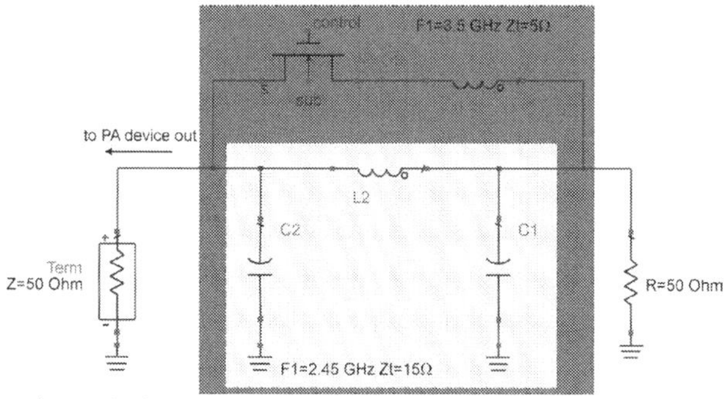

Fig. 7. MOS based switched matching network π topology

The basic simple network topology is based on a π-structure, with an additional branch composed by a NMOS device acting as a switch, with in series an inductor to change the network impedance when the switch is closed. The behavior of the network depends on the switch condition:

1. when the switch is ON (i.e. its R_{ds} value is low, zero in the ideal case) the network should present at its input an impedance value equal to 15 Ω that is the optimum value for the device at 2.45 GHz.
2. When the switch is OFF (i.e. its R_{ds} value is high, infinite in the ideal case) of course, the frequency behavior of the network change. In particular the equivalent value of the inductor between the two shunt capacitors becomes equal to the parallel of L_1 and L_2 and the input impedance of the network decreases to 5 Ω at 3.5 GHz.

Moving on the actual schematic, for the switch it has been necessary to introduce the biasing network in order to guarantee the right switching functionalities. The simulations have been performed biasing the NMOS device, with dc voltages on drain and source set to V_D=3V and V_S=3V respectively, to ensure that the NMOS device is properly biased in the origin of its output characteristics. The gate control voltage is raised to V_G=5V, when the NMOS switch has to realize a short circuit condition, and dropped to V_G=2V when the NMOS switch has to realize an open circuit condition. The feed lines for all three terminals have been realized using 4.8 KΩ resistor for each lines to guarantee the request isolation. Fig. 8 shows the small signal parameters S11 and S21 of the networks, as a function of frequency from 1 GHz to 4 GHz, respectively when the switch is OFF, e.g. when the network has to synthesize the load at 3.5 GHz, and when the switch is ON, e.g. when has to be synthesized the load at 2.45 GHz.

In particular in Fig. 8 left, the blue line in the right part of the figure represents the real part of the input impedance, which can be note is roughly equal to 15 ohm. Unfortunately, the input impedance shows a residual imaginary part due to the non ideal behavior of the inductor and capacitor elements present in the network. Similarly, Fig. 8 right reports the same features when the switch is OFF, e.g. when the network has to realize the load required at 3.5GHz. Also in this case, the network exhibits the requested real part of the input impedance, with a residual imaginary part of the input impedance.

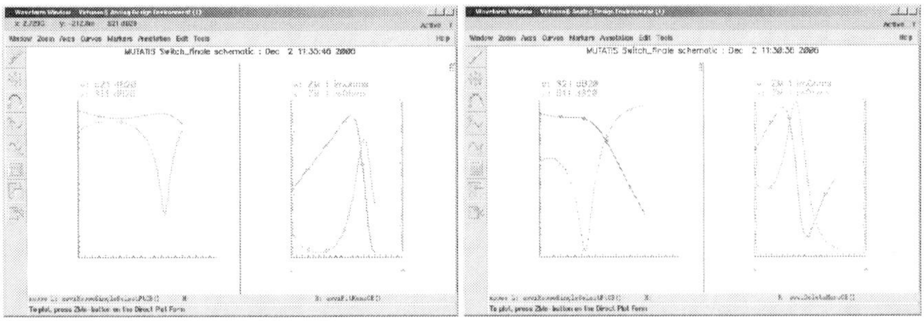

Fig. 8. S11 and S21 frequency behavior together with the real and imaginary parts of the input impedance when the switch is ON (left) and OFF (right).

5. Dual-Band reconfigurable SiGe HBT amplifier design

The above described matching network can been adopted to implement the two matching networks for a class-AB amplifier. In the following example of such design approach is presented. The design of the reconfigurable PA is based on a power device composed of 17x8 elementary SiGe HBT with and emitter area of 8.49 μm^2. The bias circuitry of the power section is designed to provide other than the required base bias current, a circuit-level linearization. Let's start introducing this latter part, referencing to the Fig. 9. The size of the devices used for the bias circuitry and the values of passive components scale accordingly, so that accurate (i.e. matched) current mirroring can take place. In addition, resistor R4 introduces base ballasting and its value is selected to be 350 Ω per emitter. It helps in reducing the risks due to thermal runaway. A larger value would have better effect, however, that could result in high DC voltage drop at high power drive, and thus early gain compression of T4, the power transistor, would take place.

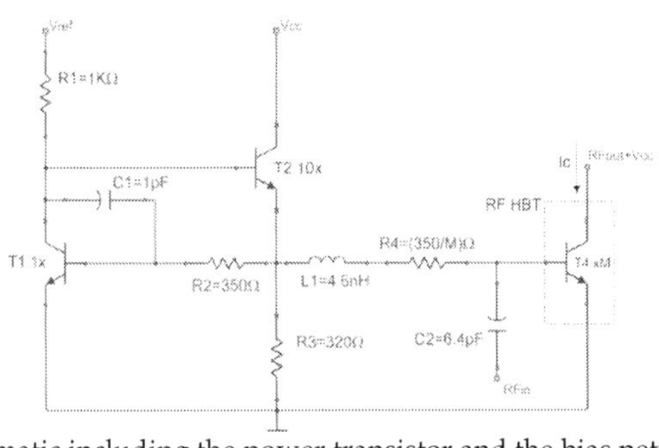

Fig. 9. PA core schematic including the power transistor and the bias network

The power device, which is shown in Fig. 9 within the dotted outline, is biased through the current mirror. By inspection of this figure the current through the reference transistor T1 is:

$$I_{REF} = \left(V_{REF} - 2V_{be}\right) \Big/ R1 \,, \tag{9}$$

with Vbe ~ 0.8 V. If the ratio between the reference device and the RF device is M, in this case M = 17x8, then the current that will flow through the RF transistor is M*I_{ref}. in addition, the current mirror behaves like an ideal voltage source, since its output impedance is

$$R_{out} \approx \frac{1}{g_m A_{loop}} \,, \tag{10}$$

where g_m is the transconductance of the emitter follower device, and A_{loop} is the loop gain of the loop formed by the reference device and the emitter follower (buffer) transistor. At high frequencies, R_{out} becomes inductive and its value should be kept low throughout the frequency band of interest and could be further reduced if resistor R_3 was further reduced in value, at the expenses of an higher current trough this resistor.
At 1 dB compression point, the power transistor should be biased at a collector current which has a value near the optimum value that guarantees that the maximum transition frequency (f_t) of the transistor is achieved. Due to self bias effect, which forces the DC average collector current to increase with increasing input power, the value of the quiescent bias current is chosen to be much lower compared to the value that it reaches under full power drive conditions. So, when the transistor T4 is self biased it starts to draw more current. Since the current flowing through R_2 is constant, as the base-emitter voltage of T_1 is constant, this additional current is supplied by transistor T_2. The capacitor C_1 in the bias network helps to stabilize the loop formed by the reference transistor, T_1, and the emitter follower transistor, T_2. The stability analysis of the loop formed by T_1 and T_2 shown that for unity loop gain the phase margin is approximately 78 degrees, which guarantees the loop unconditional stability. Fig. 10 reports the DC current flowing in the collector of the power device, T4, as a function of the input power at the frequency of 2.45 GHz. From the curve is seen that the quiescent current increases according with the description given above and reaches the value related to the peak f_T, at the 108 mA for an input power of 6 dBm. The value of the bias current for low input power is defined according with the eq. (9), adjusting the value of either R_1 or V_{ref}, in this case to the R_1 = 1 kΩ, corresponds a V_{ref} = 3.3 V.

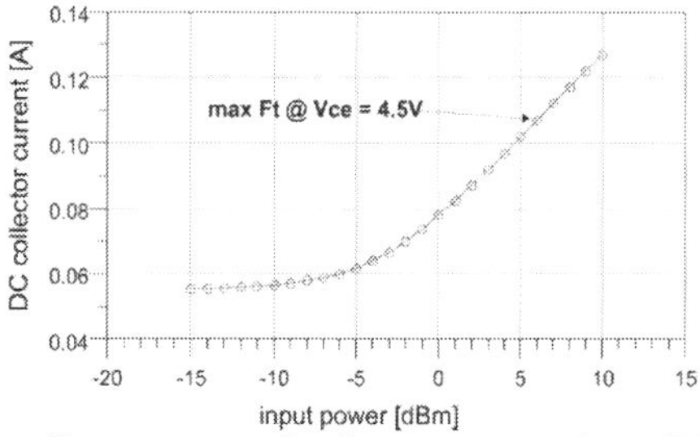

Fig. 10. Behavior of the collector current for the power transistor, T4 in Fig. 9, as a function of the input power at 2.45 GHz.

The layout of the reconfigurable PA is illustrated in Fig. 11 (left), where are clearly visible the inductors adopted for the matching and the area for the active part of circuit. The total size of the layout with the bonding pads is 1x1mm. While in operation, the voltage in the node between the inductors swings between positive and negative values. This causes the control MOS drain-bulk np junction to be forwarded biased, which degrades switch performance and may results in latchup. In order to overcome the specific drawback, we set an offset voltage (pin Vofs) which shifts the voltage in the abovementioned node in positive values.

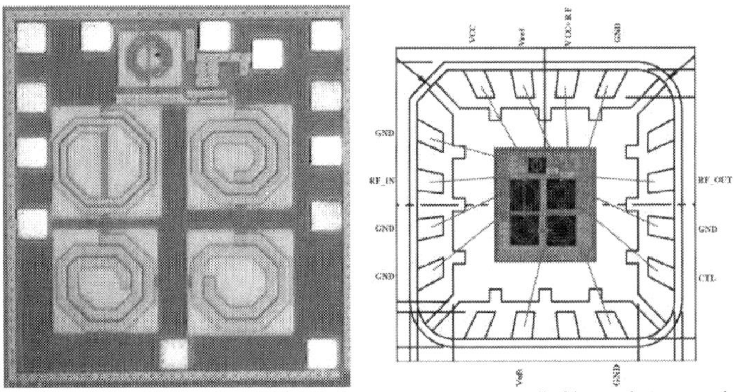

Fig. 11. Layout of the reconfigurable SiGe-PA prototype (left) and its packed wiring diagram (right)

For the design of the inductive part of the matching networks a planar electromagnetic simulator should be used, this permits to calculate any mutual inductances, between the inductors that are in proximity along with the losses in the silicon. Due to the large number of controls and dc-supply for the operation of the components a package is normally required and has to be taken into account during the design. The selected solution consists in the QFN package which allows up to 16 leads. The bonding diagram for this component is reported in the Fig. 11 (right), along with the pin description. At the time of this report editing, the packaged components were not yet available, for the reason the measurement results that will follow consider only the on-wafer device.

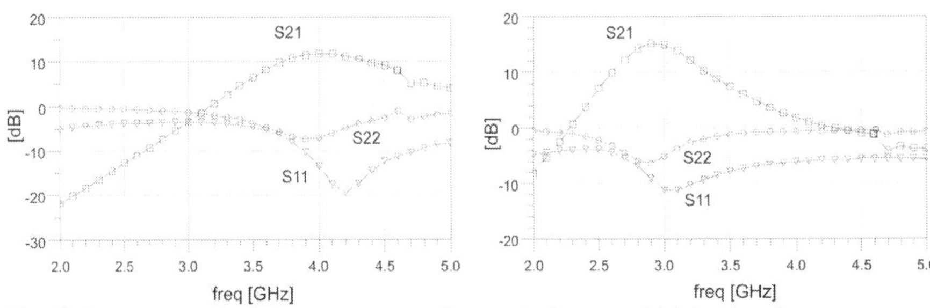

Fig. 12. S-parameters in the state corresponding to the lower and higher bands.

The first test of measured data considers s-parameter in the range 2 to 5 GHz. The S21-S11 and S22 in dB for the state corresponding to the lower and higher frequency band are reported in Fig. 12. From the figures is clearly observed the shift in frequency due to the above mentioned reasons, which is estimated in the range of 500 MHz. While the small-signal gain maximum is 2.5 dB lower than the simulated in the lower frequency band and 1 dB less in the higher band. This reduced matching have to be attributed to the matching which, being lower than estimated for the discussed reasons, introduces and matching loss consistent with the reduced gain observed during the characterization phase. We exclude problems related to the measurement set and calibration, although a problem related to measurements, the ringing in the 4.7GHz, was observed.
The characterization of the sample in the large-signal regime is reported in the below figures. It is carried out at the frequencies where the device exhibits the maximum gain and matching in the two states that are respectively at 2.9GHz and 3.9GHz, which correspond to 500MHz frequency shift from the design target as discussed above. The data related to the large signal gain up to the compression are reported in the Fig. 13, respectively for the state corresponding to the lower and higher frequency. In this figure the CW single tone signal is applied in the two states and the Pout-Pin curves are recorded. The value of the gain is consistent with the small-signal gain while the compression point is estimated to be about 16 dBm for the lower band and 15 dBm for the higher frequency. These figures are 1 dB lower that estimated during the simulations. All this data are consistent with the simulation and again supports the hypothesis that introducing the additional inductive parts due to the packaging the proper frequency behavior can be reached.
An intermodulation product characterization considering a two-tone signal with a center frequency at the selected frequencies of 2.9GHz and 3.9GHz and 1MHz offset, was applied to the device. The input power was swept from -10dBm to 13dBm. The results of the characterization, in terms of the higher IM3 are reported in the Fig. 13. It is observed that any consistent change in slope is observed in the traces. This allows concluding that the MOS involved in the switched matching networks doesn't introduce any additional nonlinearities. In fact, if the additional nonlinearity required changing the state of the device, was effectively excited we would have observed an addition component in the IM3. It is also worth to observe that the input power dealt by the device is below the threshold for which this effects become evident. This threshold from simulation is estimated in the range of 28-30 dBm.

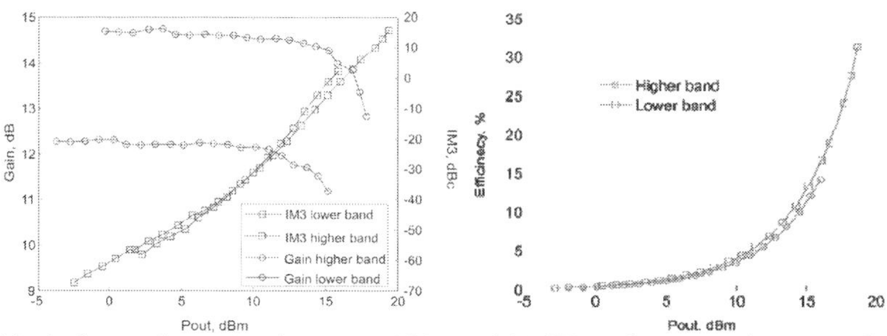

Fig. 13. single tone large-signal gain at 2.9GHz and 3.9GHz and two-tone large-signal inter-modulation offset frequency 1MHz, data reported for the higher IM3 (left), PAE

6. Dual-band power amplifier architectures

The main objective of this paragraph is about the consistent and quantitative evaluation of a two possible architectures of dual-band PA both suitable for their involvement in the concurrent dual-band systems, [12]. The first is based on two dedicated PAs combined by a frequency diplexer while the second is specifically designed to be operated in dual-band state. For the sake of the comparison the operative frequency are defined as 1.98 and 3.42 GHz respectively suitable for WCDMA and OFDM radio access technologies.
The two dual-band architectures considered in this paragraph are based on the schematic representations reported in Fig. 14. In the first (Fig. 14, left) the PA is implemented by making use of two dedicated PAs combined by a frequency diplexer. This latter device has to be designed to combine the two PAs introducing band-pass and band-stop behavior in each of the two branches. This is implement by the most innovative technique and technology and it still represents a very critical part of the entire PA structure. Indeed, this component must guarantee an almost lossless behavior in the two transmission paths and as much as possible isolation between them; without sacrificing the matching. In particular the transmission loss characteristic is required to preserve the combined efficiency of the entire structure, while the isolation is a required feature to avoid the cross-modulation between the two dedicated PAs. The constraints on the diplexer become more critical in the case of closer operative spectrum bands. During the two dedicated PAs design, the eventual combination with the diplexer implies a specific additional PA design consideration related to an accurate evaluation of the out-of band termination, which might degrade the output power and efficiency of the two units. The treatment of the harmonic termination of the due to the diplexer is out of the scope for the present treatment.

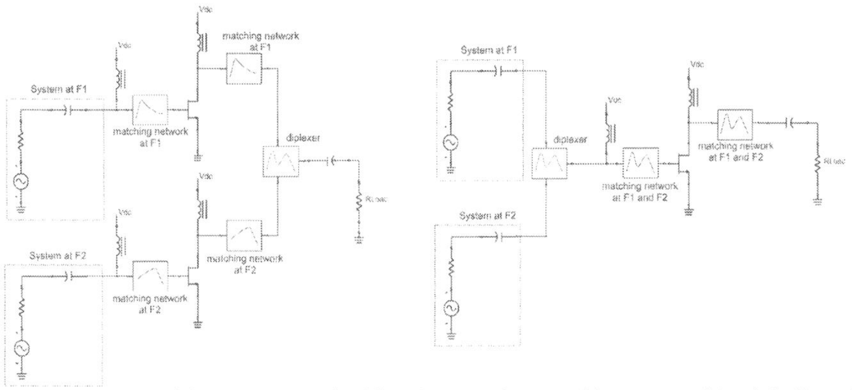

Fig. 14. Schematic of the concurrent dual-band PA implemented by two combined dedicated PAs (left), and by a dual-band PA (right).

Either the dedicated frequency PAs and concurrent PA topology consist of conventional class-AB PA designs, where the tuned matching networks can be synthesized by either passive and\or distributed elements properly dimensioned, without external tuning controls. For the concurrent PA (Fig. 14, right), the two signal sources are combined prior to

be applied to the PA input. The design method of concurrent dual-band PA based on multi-tuned networks composed of lumped elements is discussed in [10].
The investigation carried out in this chapter relies on prototypes designed and fabricated using low cost off the shelf active devices along with discrete SMD passive components assembled on FR4 0.8 mm thick evaluation printed circuit board designed with microstrip technology. For this specific test several electron device technology can be considered, eider bipolar or FET fabricated using several different material, spanning from Si to GaAs and possibly GaN. In the present case we choose as active device a GaAs FET, namely the ATF50189 from AVAGO Technologies, a medium power enhanced mode p-HEMT with a cut off frequency of 6 GHz and a 1-dB compression point of 29 dBm at 2 GHz. Optimum bias point for efficiency, linearity and gain can be fund either from manufacturer specifications or CAD simulations, on the basis of nonlinear model analysis. For the specific device possible bias point is found to be 4.5V drain supply voltage with a corresponding quiescent current of 200 mA. The chosen bias point drives the ATF50189 transistor in the AB class operation. The design was based on load and source pull simulations carried out at the two fundamental frequencies of 1.98 and 3.42 GHz, adopting a nonlinear device model which included the package parasitic. Simulations provided saturated output power of 28 dBm and 26 dBm respectively in the lower and higher frequency bands with a power added efficiency of approximately 40% and 35% at the 1-dB gain compression point. The resulting load and source constant power contours are shown respectively in Fig. 15.

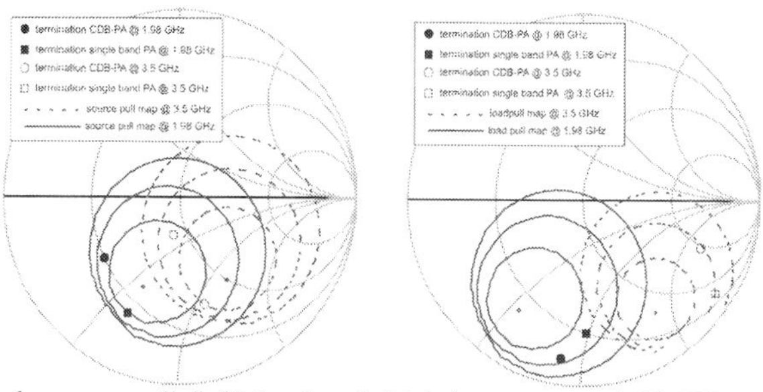

Fig. 15. Simulated source-pull (left) load-pull (right)contours at 1.98 GHz and 3.42GHz, 1 dB steps, and terminations at fundamentals for the single band and dual-band prototypes

The implementation of the source and load terminations defined by the source- and load-pull analysis was obtained by using lumped elements matching networks. This technique, by employing a different approach with respect to standard microstrip technology, enabled the achievement of highly compact prototypes. All the designed PAs adopt the same general topology for the input and output matching networks. Different nominal values and the absence of some of the components determine the difference between the prototypes. In addition, the input network accommodates a stabilizing network which has bee implemented by all three prototypes. The presence of shunt capacitors at both the gate and drain terminals, provide a short circuit to the second harmonic.

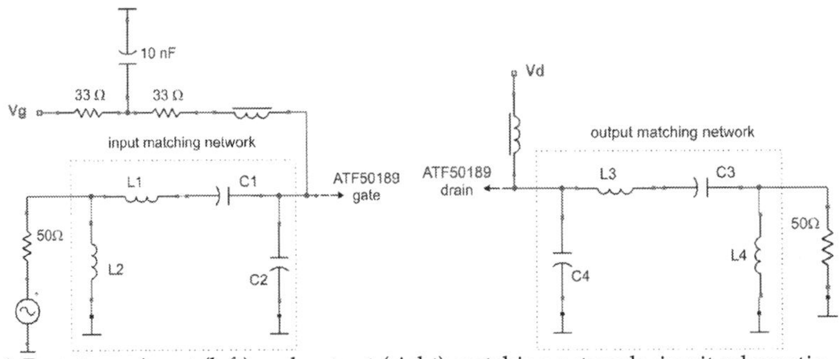

Fig. 16. Prototypes input (left) and output (right) matching network circuit schematic

The selected matching networks pi-topology, exhibits high out of band frequency roll off and a null in the transfer characteristic between the two fundamental frequency bands at 1.98 GHz and 3.42 GHz so enhancing isolation between frequency bands. In order to properly define the networks the additional conditions for maximum efficiency and 1-dB compression output power under large signal excitations were taken into account. To simplify the description we can consider that the three prototypes adopt the same general topology for the input and output matching networks, whose schematics are represented in Fig. 16. As far as the networks dimensioning is of concern, it is possible to show that the 4 unknown (2 inductors and 2 capacitors for each network) are calculated solving a nonlinear system of four equations, where the number of equations results directly equating the two real parts and the two imaginary parts of the equivalent impedance, in symbolic format, with the required optimum impedances. Different nominal values and the absence of some of the components will determine the difference between the prototypes. In addition to matching purpose the input network accommodates a stabilizing network which is common for all the three prototypes. The resulting matching networks result very compact and capable to satisfy the conditions for the optimization of both the output power and gain. The presence of shunt capacitors at both the gate and drain terminals, C2 and C4 in the figures, provide a short circuits at the second harmonics. The resulting best values for the SMD capacitances and inductances are those indicated in Table 1 and Table 1. Input matching network L-C values

respectively for the input and output matching networks. The achieved impedance are reported in the Fig. 16 where the mismatch between the actual values and the optimum impedances for power level take into account for the additional condition of gain and commercial availability of the nominal values.

	C1	L1	C2	L2
concurrent dual band	1.7 nH	0.33 pF	7.6 nH	0.3 pF
single band at 1.98GHz	0.6 nH	1.47 pF	n.a.	n.a.
single band 3.42GHz	1.6 nH	0.26 pF	n.a.	n.a.

Table 1. Input matching network L-C values

	C4	L4	C3	L3
concurrent dual band	0.6 pF	3.2 nH	0.77 pF	2.88 nH
single band at 1.98GHz	n.a.	n.a.	1.7 nH	1.9 pF
single band 3.42GHz	n.a.	n.a.	1.1 nH	0.5 pF

Table 2. Output matching network L-C values

The three PA modules were fabricated using FR4 PCB technology and then adopted to implement the two dual-band PA configurations, namely the combined PAs and the dual-band PA. The diplexer used in the large-signal test benches it is realized in microstrip technology and provides an insertion loss of 0.6 dB and 0.8 dB respectively at 1.9 GHz and 3.4 GHz, and isolation between the two channels better than 30 dB and a return loss higher than 20 dB.

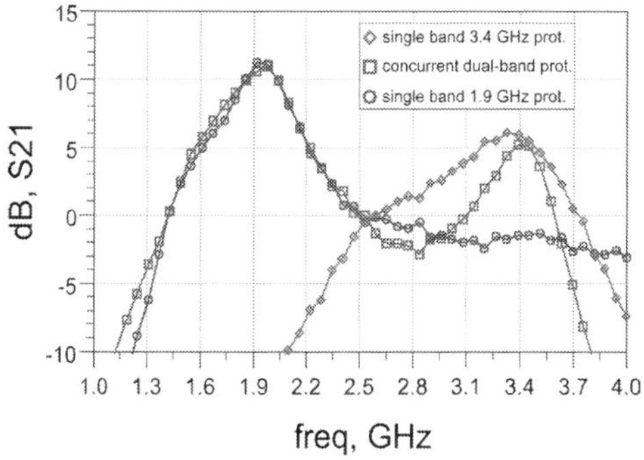

Fig. 17. Measured small-signal gain for the concurrent dual-band and the two single band prototypes

The preliminary test was performed in small signal regime to assess the prototypes performance and to verify the consistency of the comparison. The measured small signal gains associated with the concurrent dual-band PA and with the two single-band PAs prototypes are compared in Fig. 17. The figure indicates that at 1.98 GHz the maximum linear gain is approximately 11dB for both single band and dual band circuits. In the 3.4 GHz band the PA prototypes exhibit a maximum linear gain of 6 dB at 3.42 GHz, and present a 0.5dB gain bandwidth of approximately 60 MHz. Input and output return losses are not reported but are below -15 dB in the respective frequency bands for all the PAs. Small signal characterizations have indicated a very close correspondence between the single band circuits and the concurrent dual band PA in terms of both input/output return loss and gain. This results show that the design of a concurrent dual-band PA using compact lumped elements is feasible without loss of performance at small-signal and makes the characterization and comparison with large and modulated signals meaningful.

The first set of large signal measurements test bench is deployed to fully characterise the PA prototypes with CW large signal excitations. The data are useful to compare the maximum linear power, the gain and the efficiency of the two architectures. In particular the comparison between the power gains as a function of the two CW signals at 1.98 GHz (namely F1) and 3.4 GHz (namely F2), reported in Fig. 18, shows that the combined PA architecture is capable to maintain the two channels mostly separated producing a gain compression which is insensible from the concurrent signal at the side band. In fact the contour plots show a linear behaviour which is independent of the power level in the other channel. At the contrary for the dual-band PA, the mutual interaction between signals at the two frequencies is evidenced by bended constant gain loci, see Fig. 19. Nevertheless, although the effect of the side band is quite evident, the effect of the output diplexer is such that the maximum output power was slightly higher. For example, we can notice that at 1.98 GHz for input power equal to 16 dBm the gain is 8.7 dB for the combined PAs while 9.5 dB for the dual band. This latter condition is maintained only when the input signal at 3.4 GHz is lower than 10 dBm. Similar behaviour is observed in respect to the power gain calculated at 3.4 GHz. From the contour plots it is observed that the loci corresponding to 8.5 dB and 4 dB correspond to the linear gain that the dual-band PA can provide and, consequently, the input ranges which guarantee the linear operation of the PA. Differently, in the case of the combined PAs, the linearity in one frequency band does not depend of the side band.

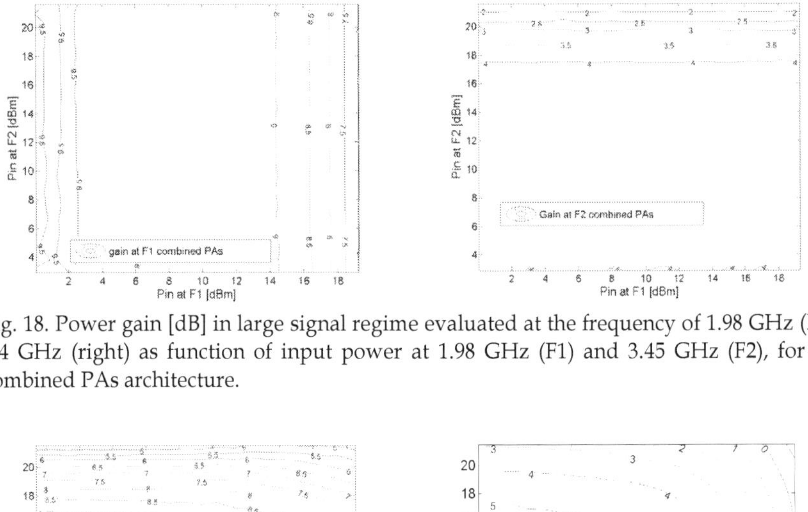

Fig. 18. Power gain [dB] in large signal regime evaluated at the frequency of 1.98 GHz (left) 3.4 GHz (right) as function of input power at 1.98 GHz (F1) and 3.45 GHz (F2), for the combined PAs architecture.

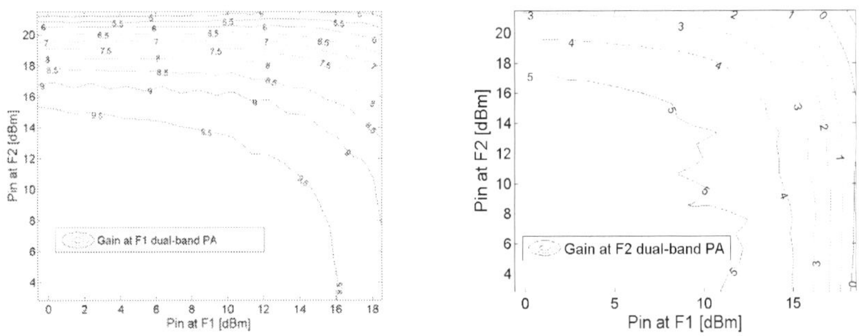

Fig. 19. Power gain [dB] in large signal regime evaluated at the frequency of 1.98 GHz (left) 3.4 GHz (right) and as function of input power at 1.98 GHz (F1) and 3.45 GHz (F2), for the dual-band PA architecture.

From the contour plots it is observed in the case of the combined PAs and in the case of the dual band PA, the presence of the diplexer and the mutual interaction between carriers respectively, determine the output power in correspondence of a 1-dB compressed power gain reported in Table 3. This latter consideration leads to the conclusion that from the point of view of the CW output power there are not differences between the two architectures. This latter consideration leads to the conclusion that from the maximum output power in CW condition there are not differences between the two possible architectures.

architecture	1.98 GHz	3.4 GHz
Combined PAs	25 dBm	22 dBm
Dual-band PA	24.5 dBm	23 dBm

Table 3. Simultaneous maximum linear output power (at 1 dB gain compression)

A further very significant figure is represented by the Power Added Efficiency (PAE) for the two PA architectures. In the case of dual-band concurrent PA the PAE is calculated by:

$$PAE = \frac{\left(P_{load}^{F1} - P_{av}^{F1}\right) + \left(P_{load}^{F2} - P_{av}^{F2}\right)}{P_{dc}}, \tag{11}$$

The above equation admits that the two signal are uncorrelated and where P_{dc} takes into account for the total current drawn by the PA modules. The PAE as a function of the input power at the two carrier frequencies, respectively for the combined PAs and the dual band PA architectures are reported in Fig. 20.

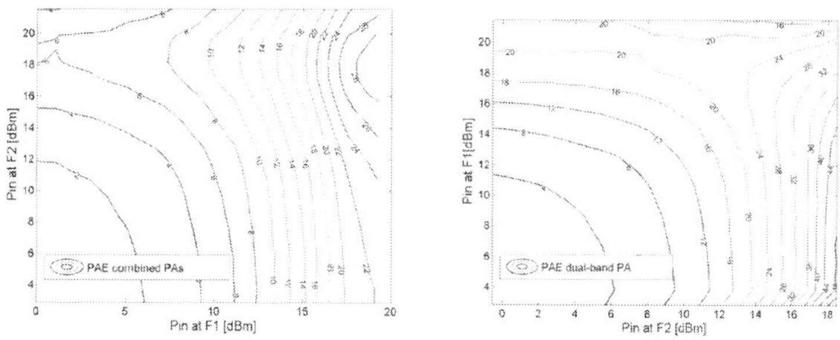

Fig. 20. Power added efficiency [%] in large signal regime as function of input power at 1.98 GHz (F1) and 3.45 GHz (F2), for the combined PAs architecture, (left), and the dual-band PA architecture, (right).

From these experimental verification it is confirmed the intuition that the dual-band PA PAE takes advantage from the current reuse which is inherent in the use of a single power device, when compared with the case of the combined PAs which need twice of the DC power to bias the two PAs. This determine an almost factor 2 in the PAE for the dual-band PA for almost the entire range of evaluation. In particular at 1-dB gain compression the PAE achieved with the combined PAs architecture is in the range of 20%, as evidenced in Fig. 20 left, while in the case of the dual-band PA it reaches 32 %, see Fig. 20 right. The maxima are 28 % and 44% respectively for the combined PAs and the dual band PA architectures. By this figure we can observe a significant improvement of the dual-band PA with respect to the combined PAs architecture. The absolute maximum power for the two PAs are reported in the Fig. 21, calculated by summing the power level at the two carrier frequencies, assumed uncorrelated. Form the contour plots is observed that, regardless linearity concerns, the total power provided by the two systems are slightly the same.

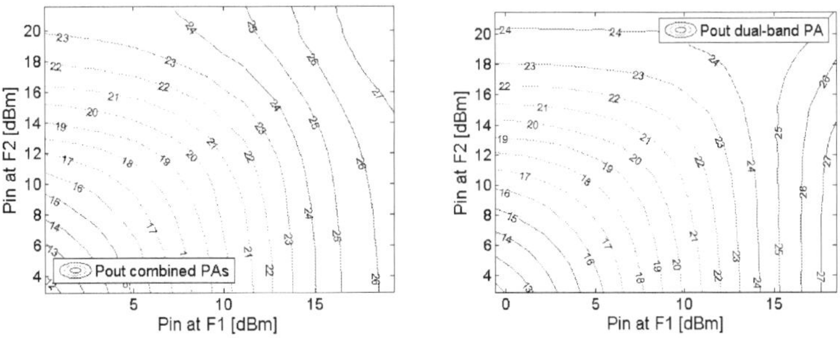

Fig. 21. Total output power [dBm] in large signal regime as function of input power at 1.98 GHz (F1) and 3.45 GHz (F2), for the combined PAs architecture (left) and for the dual-band PA architecture, (right).

The next test concerns about the capability of the dual-band PA architecture to deal with modulated signals and its performance are compared with the combined single band PAs; henceforth the combined PAs architecture is not longer considered. In this case, the base-band signals were down-loaded in the arbitrary signal generators (Agilent ESG 4438C) by using the tools available in the Agilent ADS2006A systems. Two different digitally modulated signals were employed to evaluate PA performance: a 3GPP up-link W-CDMA 3.84 MHz chip rate signal at 1.98 GHz and a 5MHz OFDM 16-QAM signal at 3.42GHz corresponding to one of the WiMAX modes. The output of the PA under test was connected to the VSA (Agilent N9020, 26MHz bandwidth) which was synchronized with the two arbitrary signal generators. The first set of data refers to the large signal gain plotted against the output power for the three PA modules; the comparisons between several operating conditions are shown in Fig. 22, left and right, for the lower and higher frequency bands respectively, which include also CW for the sake of a better comparison.

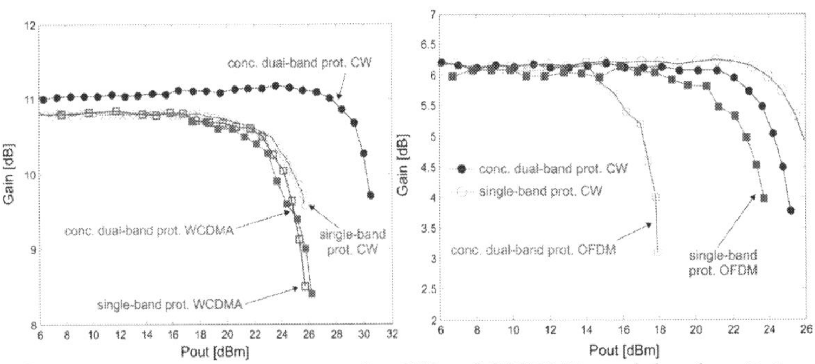

Fig. 22. Gain curve versus output power for CW and WCDMA modulated excitations, both with carrier at 1.98 GHz (left) and 3.42 GHz (right).

It is observed that when the amplifiers are driven by a single modulated signal peak power at 1dB gain compression point decreases: this effect is explained by the fact that gain compression in PAs driven by digitally modulated signals occurs at lower power levels than for 1-tone CW signals. In addition, load pull CAD analysis and successive design were performed based on a CW test signal, while experimental results show that the optimum load impedance for maximum linear output power as well as peak efficiency varies depending on the characteristics of the input signal, i.e. pulsed, modulated or CW. Concurrent mode was then operated by simultaneously feeding the dual-band PA with OFDM and WCDMA signals at the two center band frequencies. Reduction of peak output power with respect to single-channel excitations is mainly due to the simultaneous presence of two modulated signals in the same device which cause cross-modulation between the two time varying envelopes. A resulting 4 dB and a 4.5 dB peak power reduction at 1.98 GHz and at 3.42 GHz respectively were measured with respect to the single channel cases.

Moving on to system level figures, the 5.6% EVM WiMax standard limit and a minimum ACPR of 33dBc for a WCDMA signal as settled by the 3GPP specifications have been taken as a reference for power and efficiency values. The goal of the large signal characterisation has so being focussed on the evaluation of the peak output powers and the resulting PAE levels achievable in the two frequency bands with both concurrent dual-band and single-band excitations so as to satisfy EVM and ACPR constrains.

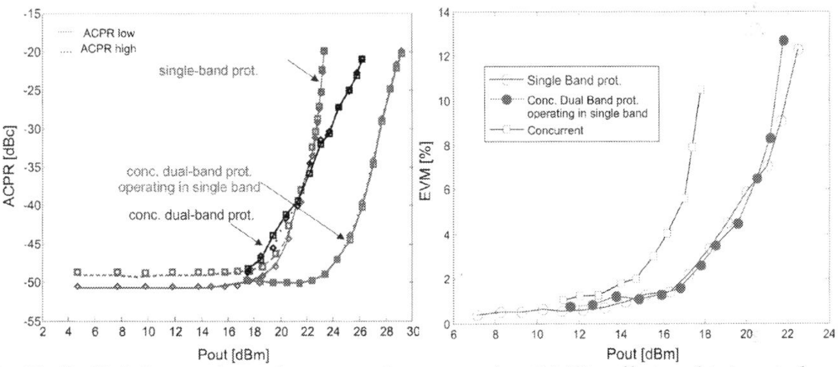

Fig. 23. (Left) Adjacent channel power ratio measured at 5 MHz offset and integrated over the bandwidth, for the single band and the dual-band prototypes with the WCDMA signal at 1.98 GHz; (Right) Error vector magnitude measured for the single band and the dual-band prototypes with the OFDM signal at 3.42 GHz.

From Fig. 23 it is observed that at 1.98 GHz the maximum achievable output power, due to ACPR constrains, is 27.5 dBm when the dual-band PA is working in single-channel mode, while for the concurrent dual-band case this limit decreases to 23 dBm. Data in Fig. 23 show EVM versus the output power results, for single channel operation and dual band mode at 3.42 GHz: a maximum output power of 20 dBm is achieved in the first case while when the dual-band PA is working in concurrent mode, maximum output power settles to 17 dBm. The above data indicate that a significant change in performance arises when the PA is driven in the concurrent dual band mode, specifically resulting in a peak power back off of about 4.5 dB and 3dB respectively for the lower and higher frequency bands due to meet the

EVM and ACPR restrictions. Envelope cross-modulation and inter-modulation explain the EVM and ACPR increased growth with input power when compared with single channel mode. Experimental data showed that a 2 dB back-off is necessary with concurrent operation to maintain the EVM at 4.1%. It can be concluded that the proposed solution is capable to provide the same system level performance of more conventional solutions while increasing the overall PAE and allowing a significant reduction of the system complexity.
A further implementation of dual band PA in GaN technology can be found in [13].

7. Dual-band power amplifier digital linearization

As discussed in the above paragraphs, the application of digitally modulated signals to a PA, which is considered hereinafter as an nonlinear (NL) dynamic system, causes in-band distortion and spectrum spreading and finally determine a degradation of the signal quality. The most effective broadband linearization systems have usually been based upon the feedforward technique [14]. However, RF and baseband predistortion linearization techniques have become an attractive solution owing to their reduced cost and complexity.
For multicarrier PA applications, an effort has been placed to increase the bandwidth of predistortion linearization to combat fast memory effects. Baseband Digital Predistortion (DP) seems to be the most promising one. It works by the introduction, in the digital baseband, of an opposite NL of the PA's one, allowing for greater efficiency through a significant power backoff reduction. The most of the available DP techniques deal with single-band operation, although recently an approach to deal with multi-carrier and potentially for multiband systems was presented in [15]. In this technique the modulation bandwidth in several bands and then a DP algorithm is applied selectively the inband and interband third-order intermodulation distortion (IMD3) As the approach relies on third order Volterra model the accuracy of the DP depends upon the identification procedure and the frequency band spacing.
Here we discuss a novel method of Dual Band DP (DB-DP), based on the simultaneous predistortion of both channels at intermediate frequency (IF), [16]. The proposed method uses a single band memory polynomial DP for linearization. As a feedback path we propose a subsampling receiver.

7.1 Basic principles of digital linearization

Let's start reviewing the basic concepts of the DP system. It consists basically in the introduction, at the baseband, of a subsystems which has a transfer function which is opposite to the one of the PA, as in Fig. 24.

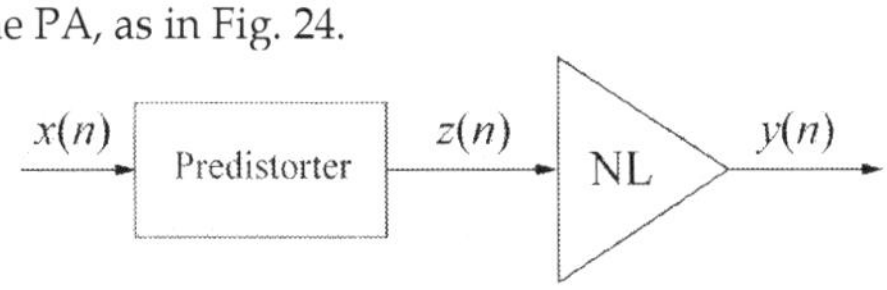

Fig. 24. Predistorsion Principle applied to a nonlinear system

The predistorter can be Look-Up Table (LUT) based or polynomial-based: in the first case, a LUT indexed by the input power is filled with complex coefficients, and the input $x(n)$ is multiplied with the corrisponding one; in the second, the complex coefficients of a k-order polynomial approximating the inverse of the PA's characteristics are found, and the DP output $z(n)$ is given by:

$$z(n)=a_1x(n)+a_2x(n)+a_3x(n)+\cdots \tag{12}$$

The coefficients vector a can be found through a recursive algorithm based on the Indirect Learning architecture shown in Fig. 25. The name derives from the fact that the polynomial coefficients are found without passing by the determination of the PA's characteristics.

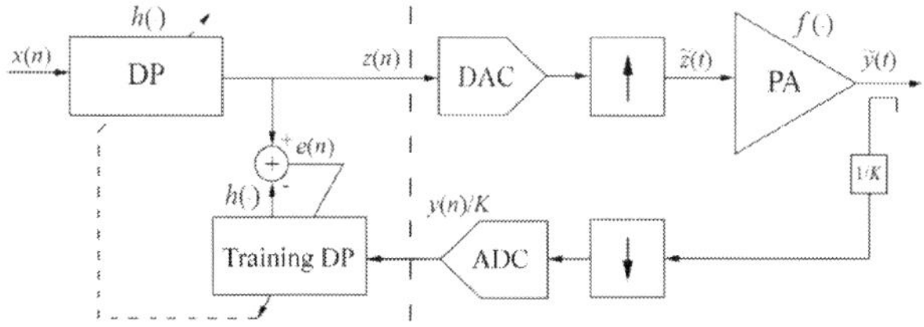

Fig. 25. Indirect Learning schematic principle for a DPD system

The indirect learning techniques works by two identical predistorters, the first - the actual one - in the transmission path and the second - the training one - in the feedback path. The outputs of both are compared to produce an error signal:

$$e(n)=\hat{z}(n)-z(n) \tag{13}$$

where $\hat{z}(n)$ is the output of the training DP. It can be demonstrated that when the error energy is zero the PA's baseband output $y(t)$ is linear with the baseband input $x(n)$, that is the cascade of predistorter and amplifier becomes linear. If the PA's memory length is comparable to the envelope variations of the signal, the baseband model - which we call memory polynomial - can be adopted:

$$y(n)=\sum_{\substack{k=1,\\ k\ \text{odd}}}^{K}\sum_{l=0}^{L-1} b_{k,l}\, z(n-l)\left|z(n-l)\right|^{k-1} \tag{14}$$

where K represent the order of the DP, while L is the number of memory samples; the baseband equivalent input $z(n)$, the output $y(n)$ and the coefficients $b_{k,l}$ of the model, are all complex valued in general. The predistorter's output can be written the same way:

$$z(n)=\sum_{\substack{k=1,\\ k\ \text{odd}}}^{K}\sum_{l=0}^{L-1} a_{k,l}\, x(n-l)\left|x(n-l)\right|^{k-1} \tag{15}$$

It has to be equal to the Training DP's output $\hat{z}(n)$ to minimize the error energy, that is:

$$z(n)=\sum_{\substack{k=1,\\ k\ \text{odd}}}^{K}\sum_{l=0}^{L-1} a_{k,l}\, \frac{y(n-l)}{K}\left|\frac{y(n-l)}{K}\right|^{k-1} \tag{16}$$

The objective consists in finding the parameters $a_{k,l}$ that define the predistorter. Since $z(n)$ is linear in the $a_{k,l}$, the latter can be estimated by a simple least-squares method. By defining a new sequence:

$$u_{k,l}(n) = \frac{y(n-l)}{K}\left|\frac{y(n-l)}{K}\right|^{k-1} \quad (17)$$

we can rewrite $z(n)$ in matrix form as:

$$\mathbf{z} = \mathbf{Ua} \quad (18)$$

where $\mathbf{z}=[z(0),\cdots,z(n-1)]^T$, $\mathbf{U}=[\mathbf{U}_0,\cdots,\mathbf{U}_L]$, $\mathbf{U}_l=[\mathbf{u}_{1l},\cdots,\mathbf{u}_{Kl}]$, $\mathbf{u}_{kl}=[u_{kl}(0),\cdots,u_{kl}(N-1)]^T$ and $\mathbf{a}=[a_{10},\cdots,a_{K0},\cdots,a_{1Q},\cdots,a_{KQ}]^T$. The least-squares solution for $\mathbf{a}$ is given by:

$$\mathbf{a} = [\mathbf{U}^H\mathbf{U}]^{-1}\mathbf{U}^H\mathbf{z} \quad (19)$$

where $(\cdot)^H$ denotes complex conjugate transpose. A direct implementation of the polynomial predistorter is difficult, because it requires several sample-per-sample multiplications and power raisings. However, an efficient implementation is possible by observing that (15) is equivalent to:

$$z(n) = \left[\sum_{\substack{k=1,\\k\text{ odd}}}^{K} a_{k,0}\,|x(n)|^{k-1}\right]x(n) + \left[\sum_{\substack{k=1,\\k\text{ odd}}}^{K} a_{k,1}\,|x(n-1)|^{k-1}\right]x(n-1) + \cdots + \left[\sum_{\substack{k=1,\\k\text{ odd}}}^{K} a_{k,L-1}\,|x(n-L+1)|^{k-1}\right]x(n-L+1) \quad (20)$$

The nonlinear polynomial can be implemented with a LUT indexed by the input magnitude, $|x(n-l)|$ [1]. This way, only L complex multiplications per sample are needed. LUT coefficients calculation is performed once the $a_{k,l}$ are found. The performance of the memory polynomial-LUT predistorter depends on the number of quantization points, on the memory length L and on the order of the polynomial, K.

7.2 Sub-sampling receiver

A key component for the DB-DP is the sub-sampling receiver, it operates on the principle of the band-pass sampling theorem, and it is used as feedback path of the DP system. If RF signals have a narrow bandwidth *B*, they can be sampled with a frequency:

$$f_s \geq 2B \quad (21)$$

As a result of the sampling process, spectrum aliases are generated around all the multiples of f_s as in Fig. 26. The image that falls in $[0; f_s/2]$ (first Nyquist zone) is the exact representation of the input signal, unless a potential phase inversion, and can be digitized. The same principle can also be used to convert two (or more) band-pass signals s_1 and s_2, located at different carrier frequencies f_{c1} and f_{c2}, with band-widths B_1 and B_2. With a proper sampling frequency there will be replicas of the two signals located side-by-side in the first Nyquist zone with no overlap, as shown in Fig. 27. The proper sampling frequency respect the condition:

$$f_s \geq 2(B_1 + B_2) \tag{22}$$

That is, a Nyquist Zone must be wider than the sum of the two bands.

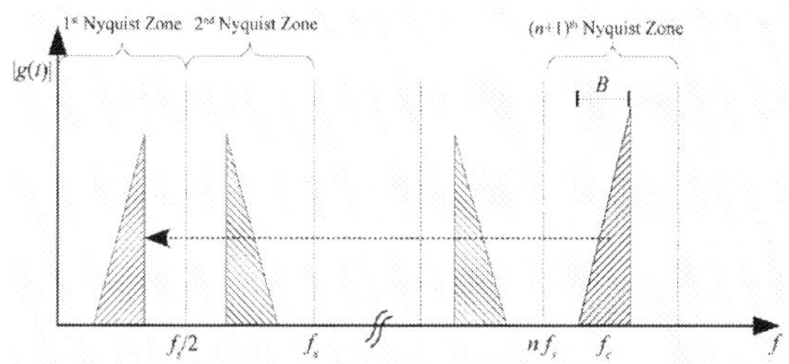

Fig. 26. Single band band-pass sub-sampling principle

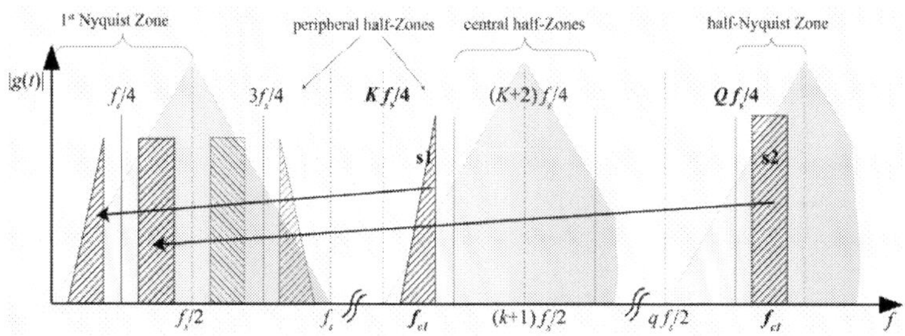

Fig. 27. Dual band sub-sampling principle

The condition of no overlap consists of the both signals to be comprised in a single half-Nyquist zone, i.e. $[nf_s / 4; (n+1)f_s / 4]$, where n is integer. If we define:

$$K = floor\left(\frac{f_{c1} - B_1 / 2}{B_1}\right) \quad Q = floor\left(\frac{f_{c2} - B_2 / 2}{B_2}\right) \tag{23}$$

where $floor()$ is the operation of rounding to the lower integer, the conditions of no overlap are first given by:

$$\begin{cases} kf_s / 4 + B_1 / 2 \leq f_{c1} \leq (k+1)f_s / 4 - B_1 / 2 \\ qf_s / 4 + B_2 / 2 \leq f_{c2} \leq (q+1)f_s / 4 - B_2 / 2 \\ k \leq K \\ q \leq Q \end{cases} \tag{24}$$

where k and q are integers identifying the order of the half-Zone in which the first and the second signals stand, respectively. The other condition, i.e. standing in central vs. peripheral half-zones, are given by:

$$\begin{cases} K = 4n & \Leftrightarrow \quad Q \neq 4n, 4n-1 \\ K = 4n-1 & \Leftrightarrow \quad Q \neq 4n, 4n-1 \\ K = 4n+1 & \Leftrightarrow \quad Q \neq 4n+2, 4n+1 \\ K = 4n+2 & \Leftrightarrow \quad Q \neq 4n+2, 4n+1 \end{cases} \tag{25}$$

These conditions lead to a not closed form formulation which require an iterative approach for the solution. Once the suitable sampling frequency is found, the two signals replicas in the first Nyquist zone are located at the frequencies f_{bb1} and f_{bb2} which are given by:

$$\begin{aligned} f_{bb1} &= \begin{cases} f_{c1} - floor(k/4)\cdot f_s & k = 4n, k = 4n+1 \\ \left(floor(k/4)+1\right)\cdot f_s - f_{c1} & k = 4n+2, k = 4n-1 \end{cases} \\ f_{bb2} &= \begin{cases} f_{c2} - floor(q/4)\cdot f_s & q = 4n, q = 4n+1 \\ \left(floor(q/4)+1\right)\cdot f_s - f_{c2} & q = 4n+2, q = 4n-1 \end{cases} \end{aligned} \tag{26}$$

The distortion introduced by a sub-sampling receiver is due in large part to the transfer function of the sampling device. In general, a T/H is preferred over a S/H, because of the lower distortion and higher sampling frequency reachable. The transfer function of a T/H is:

$$G_s(f) = \sum_{n=-\infty}^{\infty} G\left(f - \frac{n}{T_s}\right)\left[\frac{\tau}{T_s}\operatorname{sinc}\left(\tau f\right)e^{-j\pi\tau f} + \frac{T_s - \tau}{T_s}\operatorname{sinc}\left(\frac{n(T_s - \tau)}{T_s}\right)e^{-j\pi(T_s+\tau)\frac{n}{T_s}}\right], \tag{27}$$

where T_s is the sampling period and τ is the length of the hold period. Due to the sinc() in order to avoid an amplitude distortion, τ should be as low as possible to move at high frequency the first null. Also, the baseband aliases should be as near as possible to the zero. As regards the phase, different replicas have a different offset depending on the order n and the frequency of the alias. Replicas falling into the first Nyquist zone have a phase offset depending on k and f_{BB1}, or q and f_{BB2}. This offset must be compensated if a synchronism between the two signals is necessary, as in our proposed Dual Band DP method.

This approach exhibits some critical points, [17]. The first ones to be considered are noise aliasing and aperture jitter; then out-of-bands signals and wideband noise must be filtered out before the sampler. That noise would otherwise, after sampling, translate and accumulate into the first Nyquist zone. Besides, as even a perfect filter would reject the noise introduced by downstream circuits, low noise components have to be chosen. However, noise aliasing reduces with sampling frequency increase. Aperture jitter can be treated as a white noise if the jitter is low, and it doesn't depend on the sampling frequency. When designing a sub-sampling receiver, another important parameter to take care of is the analog bandwidth of the sampler, that must be greater than the highest frequency of the RF signals.

7.3 Dual Band Digital Predistortion Architecture

The DP-DP is achieved by a RF-level predistortion: a signal predistorter (as opposed to a data predistorter) is able to treat any kind of signal, that is it doesn't depend either on the bandwidth or the center frequency. Let's consider an input signal made of the superposition of two signals at different center frequencies, that is $x(n) = x_1(n) + x_2(n)$. The input is predistorted ($z(n)$), converted into analog ($\tilde{z}(t)$) and amplified ($\tilde{y}(t)$). A portion of $\tilde{y}(t)$ is drawn to have a feedback signal and to train the DP. A scheme is shown in Fig. 28. The main problem with this setup is the lack of sufficiently fast D/A and A/D converters, that will remain so in the foreseeable future because ADC dynamic range and conversion are known to progress at a rate much slower than Moore's law. Also, a RF predistortion is not possible at the moment, because it must be performed sample-per-sample and the sample rate is at least twice the maximum RF frequency (baseband sampling theorem).

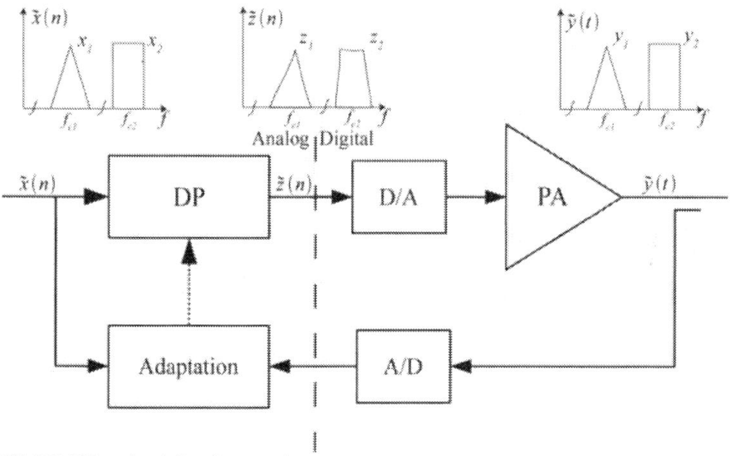

Fig. 28. RF DB-DP, principle of operation

Actually, the converters related problem can be easily overcome. The RF DAC can be replaced by two baseband DAC preceded by a proper digital filtering and digital frequency conversion system. In a similar way, the RF ADC can be replaced by two frequency converters and two baseband ADCs.

There remains the sample rate problem. The last limit can be overcome by introducing a new architecture which is capable to lowering the sample rate, that is predistorting at intermediate frequency (IF). In this case the baseband digital signals $x_1(n)$ and $x_2(n)$ are shifted to f_{IF1} and f_{IF2} then summed, creating $\tilde{x}'(n)$. This IF signal is predistorted ($\tilde{z}'(n)$), and the two bands are separated and shifted to the baseband to be analog converted. The analog PA's input $\tilde{z}'(t)$ is built by those baseband signals, shifted to the RF frequencies f_{c1} and f_{c2}. It is amplified ($\tilde{y}'(t)$) and a portion of it is drawn to create the feedback signals. As a feedback path we propose a subsampling receiver: the two bands composing $\tilde{y}'(t)$ are aliased side-by-side in the baseband, then digitized by a single ADC. In the digital domain, the bands are separated and shifted to IF, composing the signal $\tilde{y}''(n)$ that will be compared to $\tilde{x}'(n)$.

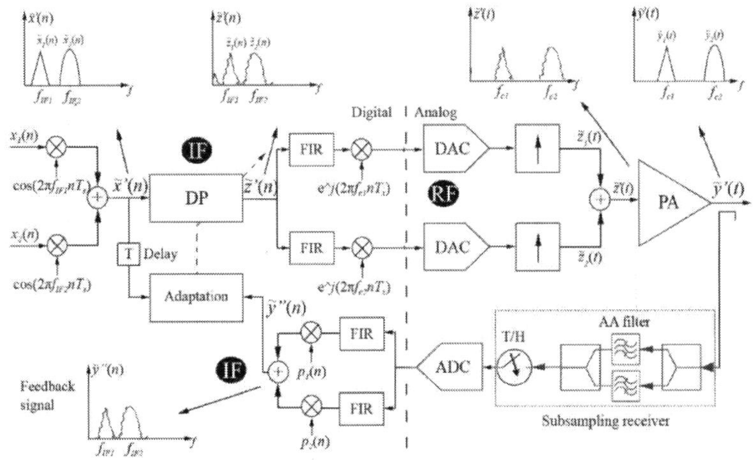

Fig. 29. DB-DP system with IF predistortion and subsampling feedback
The block diagram of the whole system is shown in

When using a subsampling receiver, it is necessary to compensate the different phase offset applied to both bands. This may be done in the digital domain. If a T/H is used, the right phase shift can be calculated through eq. (27). Anti-aliasing filters must be carefully designed with in general out of band rejection. The IFs setting is a crucial point of the system design. They have to be far enough to leave room for out-of-band distortion and to simplify filtering; on the other side, they should be as low as possible to reduce computational constraints. As a rule, for the proposed DB-DP you may consider a sample rate at least four times higher than in a SB-DP system.
The DB-DP was simulated by Matlab/Simulink®. We considered two 16 QAM signals, with amplitudes $P = -10\text{dBm}$ and centre frequencies $f_{c1} = 2.1\,\text{GHz}$ and $f_{c2} = 3.5$ GHz; the sampling frequency was set to $f_s = 146.5\,\text{MHz}$. The PA was modeled with the Wiener-Hammerstein model. LTI blocks preceeding and following the memoryless NL were set to have the following transfer functions:

$$H(z) = \frac{1 + 0.5z^{-2}}{1 - 0.4z^{-1}}, \quad G(z) = \frac{1 - 0.3z^{-2}}{1 - 0.4z^{-1}} \tag{28}$$

It was chosen a tanh-shaped AM/AM NL, that has G=20dB, IP3=38dB and whose AM/PM is linear, with 5°/dB slope.

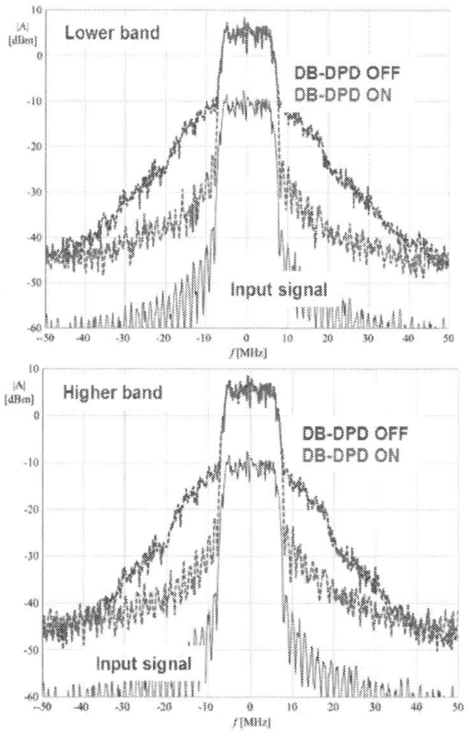

Fig. 30. Spectra comparison for lower and higher channels, between transmitted signal and input signal, with DB-DP OFF and DB-DP ON (left.

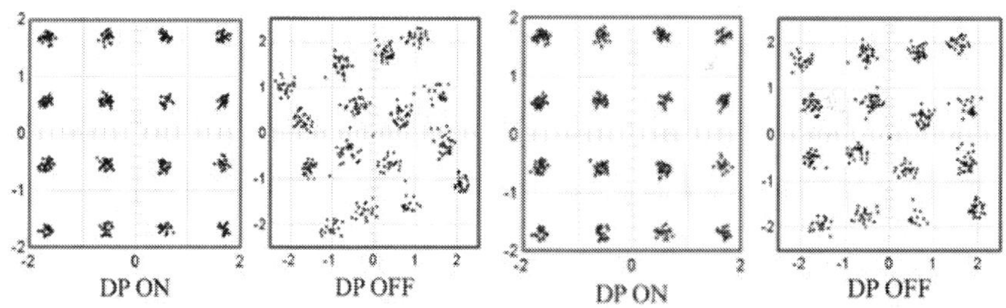

Fig. 31. Constellations comparison for lower (left) and higher (higher) channel, between transmitted signal and input signal, with DB-DP OFF and DB-DP ON

For the implementation of the DB-DPD we used a memory polynomial DP, with a memory length of 4 taps, a polynomial order K=9 and a LUT predistorter with a size of 512. Polynomial coefficients were estimated on a basis of 8192 samples. Simulation results for both channels are shown in Fig. 30 and Fig. 31, where an ACPR and EVM significant reduction is observed. The method proved to be able to correct most NLs, but it is not as good as a SB-DP. While in that case we obtained a Normalized Mean Square Error (NMSE) of 3e-4, in the DB-DP case we obtained an NMSE of 1e-3.

8. Concluding Remarks

The design of flexible PAs and multiband transmitter architectures is at a crucial stage; the number of research teams and projects that approached this field increased over the recent years. The number of special sessions and workshops in the main international conferences confirmed this interest. Some commercial products appeared recently, although they remain mainly based on very simple arrangements of frequency dedicated PAs with limited tuning control. Some technological and methodological problem have to be solved. The first set are related to the device technologies for both the RF power devices and the control devices. Indeed, the energy efficiency and peak power have to be maintained for wideband operation, making the device technology more challenging. Design approach have to take into account for multiband driving which reduce sensibly the power handling capability of the power device. Control devices, like switches and tuning elements have to cope with high peak power increasing the demand of linearity and efficiency, in this field MEMS appears a promising technology. An additional consideration is due for the architectures of multiband-multiband transmitters. Other than flexibility they have to provide excellent signal quality, which is much more threated by simultaneous concurrent signals. Polar transmitters versus Cartesian architectures are investigated as the two mainstreams for future transmitter architectures.

9. Acknowledgement

The contents of this chapter are mainly based on the results of the research activities performed in the context of the project TARGET- "Top Amplifier Research Groups in a European Team" supported by the Information Society Technologies Programme of the EU under contract IST-1-507893-NOE, www.target-net.org.

10. References

[1] Hashimoto, A.; Yoshino, H.; Atarashi, H., "Roadmap of IMT-advanced development," Microwave Magazine, IEEE , vol.9, no.4, pp.80-88, Aug. 2008
[2] F. K. Jondral, "Software-Defined Radio Basics and Evolution to Cognitive Radio", Journal on Wireless Communications and Networking, 2005, vol. 3, 275-283
[3] A. A. Abidi, "The Path to the Software-Defined Radio Receiver", IEEE Journal of Solid-State Circuits, Vol. 42, no. 5, May 2007, pp. 954-966
[4] P. B. Kennington, RF and Baseband Techniques for Software Defined. Radio. Norwell, MA: Artech House, 2005.
[5] J. Laskar, R. Mukhopadhyay, Y. Hur, C. -H. Lee, and K. Lim, "Reconfigurable RFICs and modules for cognitive radio", Digest of Topical Meeting on Silicon Monolithic Integrated Circuits in RF Systems, 2006. Jan. 2006 pp. 18-20

[6] F. Wang, D. F. Kimball, J. D. Popp, A. H. Yang, D. Y. Lie, P. M. Asbeck, L. E. Larson, "An Improved Power-Added Efficiency 19-dBm Hybrid Envelope Elimination and Restoration Power Amplifier for 802.11g WLAN Applications," Trans. On Microwave Theory and Techniques, Vol. 54, Dec. 2006, pp. 4086-4099

[7] Q. Shen, N. S. Barker "Distributed MEMS tunable matching network using minimal-contact RF-MEMS varactors," *Microwave Theory and Techniques, IEEE Transactions on* , vol.54, no.6, pp.2646-2658, June 2006

[8]K. Buisman, L.C.N. de Vreede, L.E. Larson, M. Spirito, A. Akhnoukh, T.L.M. Scholtes, L.K. Nanver "Distortion-free varactor diode topologies for RF adaptivity", Microwave Symposium Digest, 2005 IEEE MTT-S International,12-17 June 2005 pp. 157-160

[9]A. Jrad, A.-L. Perrier, R. Bourtoutian, J.-M. Duchamp, P. Ferrari, "Design of an ultra compact electronically tunable microwave impedance transformer", Electronics Letters, Volume 41, Issue 12, 9 June 2005 pp. 707 - 709

[10]P. Colantonio, F. Giannini, R. Giofrè, L. Piazzon, "Simultaneous Dual-Band High Efficiency Harmonic Tuned Power Amplifier in GaN Technology", European Microwave Conference Digest, Munich Oct., 2007

[11] W.C.E. Neo, Yu Lin, Xiao-dong Liu, L.C.N. de Vreede, L.E. Larson, M. Spirito, M.J. Pelk, K. Buisman, A. Akhnoukh, A. de Graauw, L.K. Nanver, "Adaptive Multi-Band Multi-Mode Power Amplifier Using Integrated Varactor-Based Tunable Matching Networks," *Solid-State Circuits, IEEE Journal of* , vol.41, no.9, pp.2166-2176, Sept. 2006

[12] A. Cidronali, I. Magrini, N. Giovannelli, M. Mercanti, G. Manes "Experimental system level analysis of a concurrent dual-band power amplifier for WiMAX and WCDMA applications"; International Journal of Microwave and Wireless Technologies, Cambridge University Press and the European Microwave Association, Vol.1 Special Issue 02, April 2009 pp 99-107

[13]P. Colantonio, F. Giannini, R. Giofre, L. Piazzon, "Simultaneous dual-band high efficiency harmonic tuned power amplifier in GaN technology", European Microwave Integrated Circuit Conference, 8-10 Oct. 2007 pp.127 - 130

[14]R. Meyer, R. Eschenback, and W. Edgerley, Jr., "A wideband feedforward amplifier," IEEE J. Solid-State Circuits, vol. SCC-9, no. 6, pp. 422–448, Jun. 1974.

[15]P. Roblin, S. K. Myoung, D. Chaillot, Y. Gi Kim, A. Fathimulla, J. Strahler, S. Bibyk"Frequency-Selective Predistortion Linearization of RF Power Amplifiers" IEEE Trans on Microwave Theory and Tech., Vol. 56, Jan. 2008, pp 65-76

[16]A. Cidronali, I. Magrini, R. Fagotti, G. Manes, "A new approach for concurrent Dual-Band IF Digital PreDistortion: System design and analysis," Workshop on Integrated Nonlinear Microwave and Millimetre-Wave Circuits, 2008. INMMIC 2008, pp.127-130, 24-25 Nov. 2008

[17]G. Avitabile, A. Cidronali, G. Manes, 'A S-band digital down converter for radar applications based on GaAs MMIC fast sample&hold', *IEE Proceedings-Circuits, Device and Systems, Vol.143, No.6 Dec. 1996 pp.337-342*

6

The Doherty Power Amplifier

Paolo Colantonio, Franco Giannini, Rocco Giofrè and Luca Piazzon
University of Roma Tor Vergata
Italy

1. Introduction

The Doherty Power Amplifier (DPA) was invented in the far 1936 by W. H. Doherty, at the Bell Telephone Laboratories of Whippany, New Jersey (Doherty, 1936). It was the results of research activities devoted to find a solution to increase the efficiency of the first broadcasting transmitters, based on vacuum tubes. The latter, as it happens in current transistors, deliver maximum efficiency when they achieve their saturation, i.e. when the maximum voltage swing is achieved at their output terminals. Therefore, when the signal to be transmitted is amplitude modulated, the typical single ended power amplifiers achieve their saturation only during modulation peaks, keeping their average efficiency very low. The solution to this issue, proposed by Doherty, was to devise a technique able to increase the output power, while increasing the input power envelope, by simultaneously maintaining a constant saturation level of the tube, and thus a high efficiency. The first DPA realization was based on two tube amplifiers, both biased in Class B and able to deliver tens of kilowatts.

Nowadays, wireless systems are based on solid state technologies and also the required power level, as well as the adopted modulation schemes, are completely different with respect to the first broadcasting transmitters. However, in spite of more than 70th years from its introduction, the DPA actually seems to be the best candidate to realize power amplifier (PA) stage for current and future generations of wireless systems. In fact, the increasing complexity of modulation schemes, used to achieve higher and higher data rate transfer, is requiring PAs able to manage signals with a large time-varying envelope. The resulting peak-to-average power ratio (PAPR) of the involved signals critically affects the achievable average efficiency with traditional PAs. For instance, in the European UMTS standard with W-CDMA modulation, a PAPR of 5-10 dB is typical registered. As schematically reported in Fig. 1, such high values of PAPR imply a great back-off operating condition, dramatically reducing the average efficiency levels attained by using traditional PA solutions.

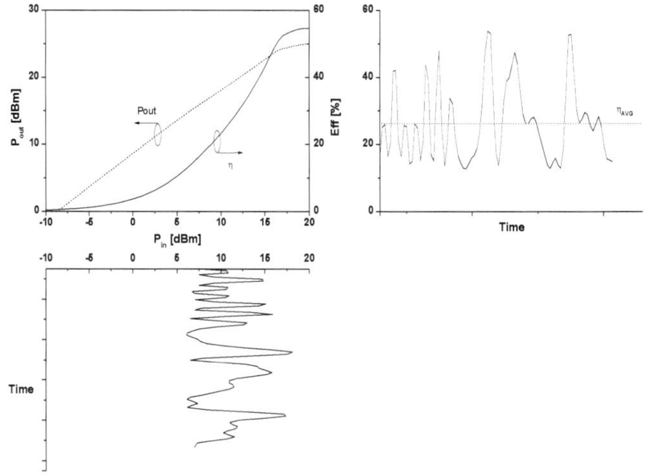

Fig. 1. Average efficiency using traditional PA.

To stress this effect, it is helpful to refer to an ideal Class B PA, which delivers an efficiency of 78.6% at its maximum output power, whereas it becomes only 25% at 10dB back-off. Therefore, when dealing with amplitude modulation signal, it is more useful to refer to the average efficiency, which is defined as the ratio of the average output power ($P_{out,AVG}$) to the average supply DC power ($P_{DC,AVG}$) (Raab, 1986):

$$\eta_{AVG} = \frac{P_{out,AVG}}{P_{DC,AVG}} \tag{1}$$

Clearly, the average efficiency depends on both the PA instantaneous efficiency and the probability density function (*PDF*), i.e. the relative amount of time spent by the input signal envelope at different amplitudes. Therefore, to obtain high average efficiency when time-varying envelope signals are used, the PA should work at the highest efficiency level in a wide range of its output (i.e. input) power. This requirement represents the main feature of the DPA architecture, as shown in Fig. 2, where its theoretical efficiency behavior is reported.

The region with almost constant efficiency identifies the DPA Output Back-Off (OBO) range, and it is fixed according to the PAPR of the signal to be amplified. As will be later detailed, the OBO value represents the first parameter to be chose in the design process.

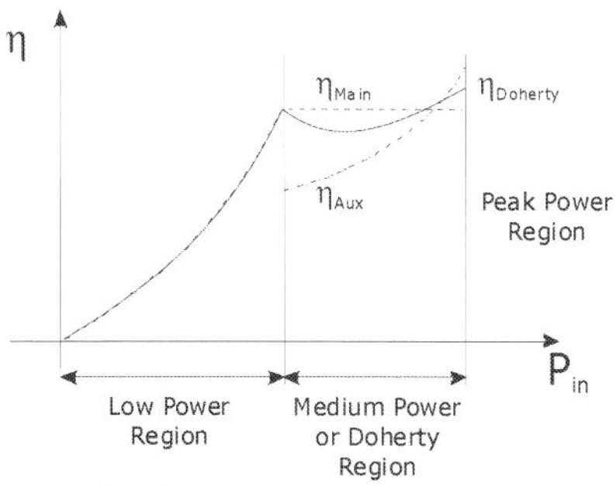

Fig. 2. Typical DPA efficiency behavior versus input power.

Due to this attractive characteristic and the relative simple implementation scheme, the DPA is being the preferred architecture for new communication systems.
The Doherty technique is usually adopted to design PA for wireless systems and, in particular, in base stations, working in L-S-C Band with time-varying envelope signals such as WiMax, WLAN, Cellular network etc. In this field, a lot of experimental results have been published using different active device technologies such as Si LDMOS, GaN HEMT, GaAs PHEMT and GaAs HBT. Typically, these DPAs are realised in hybrid form and they work around 2.14 GHz with W-CDMA input signals. Drain efficiencies up to 70% have been demonstrated for output powers between 5W and 10W (Kim et al., 2008 - Lee et al., 2008 - Markos et al., 2007 - Kim et al., 2005), whereas 50% of drain efficiency has been demonstrated for 250W output power (Steinbeiser et al., 2008). Also for high frequency applications the DPA has been successfully implemented using GaAs MMIC technologies (McCarroll et al., 2000 - Campbell, 1999 - Tsai & Huang, 2007). For instance, in (Tsai & Huang, 2007) it has been reported a fully integrated DPA at millimeter-wave frequency band with 22dBm and 25% of output power and efficiency peak, respectively. Also DPA realizations based on CMOS technology was proposed (Kang et al., 2006 - Elmala et al., 2006 - Wongkomet et al., 2006). However, in this case, due to the high losses related to the realization of required transmission lines, the achieved performances are quite low (peak efficiency lower than 15%).
In this chapter the theory and the design guidelines of the DPA will be reviewed in deep detail with the aim to show to the reader the proper way to design a DPA.

2. The Doherty operating principle

The DPA operating principle is based on the idea to modulate the load of the active device, namely Main (or Carrier) typically biased in Class AB, exploiting the active load pull concept (Cripps, 2002), by using a second active device, namely Auxiliary (or Peaking), usually biased in Class C.
In order to understand the active load-pull concept, it is possible to consider the schematic reported in Fig. 3, where two current sources are shunt connected to an impedance Z_L.

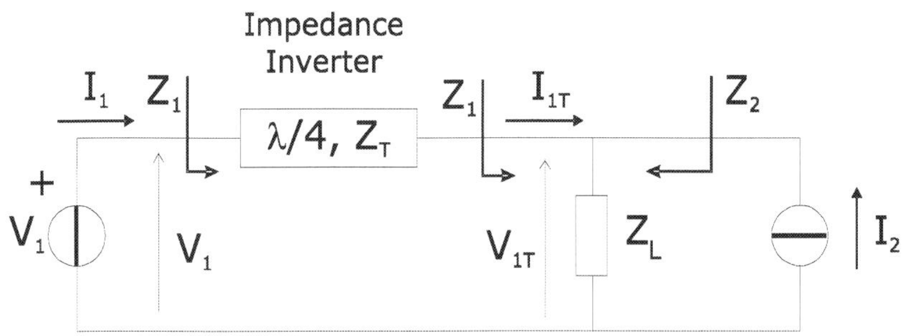

Fig. 3. Schematic of the active load-pull.

Appling Kirchhoff law, the voltage across the generic loading impedance Z_L is given by:

$$V_L = Z_L(I_1 + I_2) \tag{2}$$

Where I_1 and I_2 are the currents supplied by source 1 and 2, respectively. Therefore, if both currents are different from zero, the load seen by each current source is given by:

$$Z_1 = Z_L \cdot \left(1 + \frac{I_2}{I_1}\right) \tag{3}$$

$$Z_2 = Z_L \cdot \left(1 + \frac{I_1}{I_2}\right) \tag{4}$$

Thus, the actual impedance seen by one current source is dependent from the current supplied by the other one.
In particular, if I_2 is in phase with I_1, Z_L will be transformed in a higher impedance Z_1 at the source 1 terminals. Conversely, if I_2 is opposite in phase with I_1, Z_L will be transformed in a lower impedance Z_1. However, in both cases also the voltage across Z_L changes becoming higher in the former and lower in the latter situation.
Replacing the current sources with two equivalent transconductance sources, representing two separate RF transistors (Main and Auxiliary respectively), it is easy to understand that to maximize the efficiency of one device (i.e. Main) while its output load is changing (by the current supplied by the Auxiliary device), the voltage swing across it has to be maintained constant. In order to guarantee such constrain, it is necessary to interpose an Impedance Inverting Network (IIN) between the load (Z_L) and the Main source, as reported in Fig. 4.
In this way, the constant voltage value V_1 at the Main terminals will be transformed in a constant current value I_{1T} at the other IIN terminals, independently from the value of Z_L.

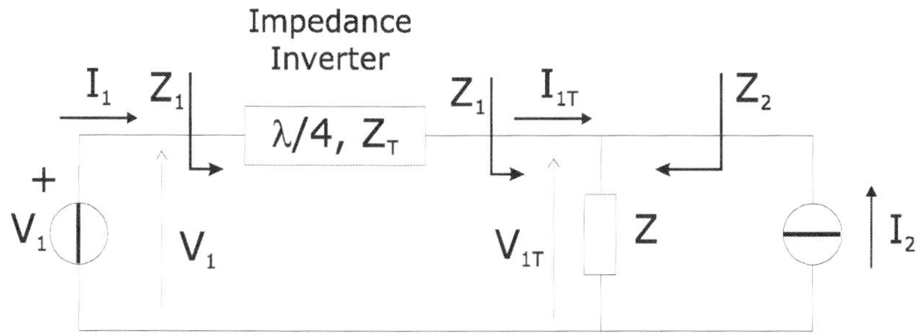

Fig. 4. Simplified schema of the DPA.

For the IIN implementations, several design solutions could be adopted (Cripps, 2002). The most typical implementation is through a lambda quarter transmission line (λ/4 TL), which ABCD matrix is given by:

$$\begin{pmatrix} V_1 \\ I_1 \end{pmatrix} = \begin{pmatrix} 0 & j \cdot Z_0 \\ \frac{j}{Z_0} & 0 \end{pmatrix} \cdot \begin{pmatrix} V_2 \\ I_2 \end{pmatrix} \tag{5}$$

being Z_0 the characteristic impedance of the line.
From (5) it is evident that the voltage at one side (V_1) is dependent only on the current at the other side (I_2) through Z_0, but it is independent from the output load (Z_L) in which the current I_2 is flowing.
Thus, actual DPAs are implemented following the scheme reported in Fig. 5, which is composed by two active devices, one IIN connected at the output of the Main branch, one Phase Compensation Network (PCN) connected at the input of the Auxiliary device and by an input power splitter besides the output load (R_L). The role of the PCN is to allow the in phase sum on R_L of the signals arising from the two active devices, while the splitter is required to divide in a proper way the input signal to the device gates.

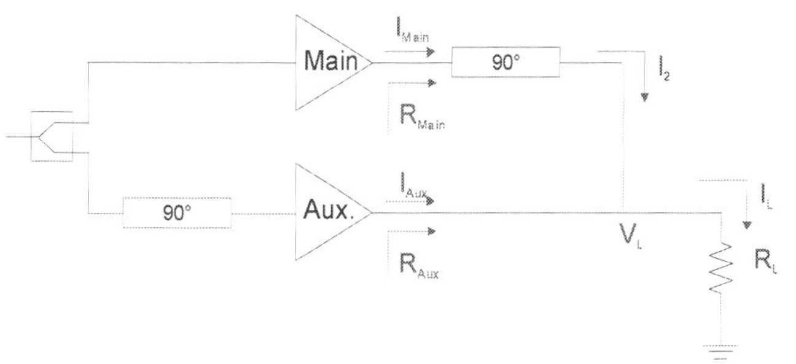

Fig. 5. Typical DPA structure.

In order to easy understand the DPA behavior, the following operating regions can be recognized (Raab, 1987).

For low input power level (i.e. Low Power Region, see Fig. 2), the DPA acts as a typical PA, since the Main device is conducting while the Auxiliary is OFF due to its Class C bias condition.

Increasing the input power level, the current supplied by the Main device to R_L increases reaching the device saturation ($I_{critical}$), thus the maximum efficiency condition. The corresponding input power level reaches a "break point" condition, while the expected load curve of both active devices are indicated in Fig. 6 with the letter A. For higher input power level ($P_{in\_DPA}>P_{in\_DPA(break\ point)}$), the Auxiliary device will automatically turned on, injecting current into the output load R_L. Consequently, the impedance (Z_1) seen by the Main device is modulated and, thanks to the λ/4 TL, its value becomes lower with respect to the one at the break point (load curve "A" in the Fig. 6). In this way, the efficiency of the Main device remains constant, due to the constant level of saturation, while the efficiency of the Auxiliary device starts to increase (see Fig. 2). As a result, the overall DPA efficiency shows the typical behavior reported in Fig. 2.

At the end of the DPA dynamic, i.e. for the peak envelope value, both devices achieve their saturation corresponding to the load curves "C" in Fig. 6.

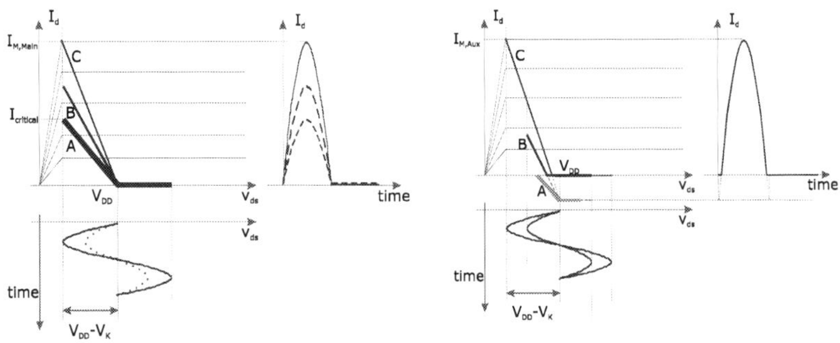

Fig. 6. Evolution of the load curves for both DPA active devices: Main (left) and Auxiliary (right) amplifiers.

3. The Doherty design guidelines

In order to infer useful design relationships and guidelines, simplified models are assumed for the elements which are included in the DPA architecture. In particular, the passive components (λ/4 TLs and power splitting) are assumed to be ideally lossless, while for the active device (in the following assumed as a FET device) an equivalent linearised model is assumed, as shown in Fig. 7. It is represented by a voltage-controlled current source, while for simplicity any parasitic feedback elements are neglected and all the other ones are embedded in the matching networks.

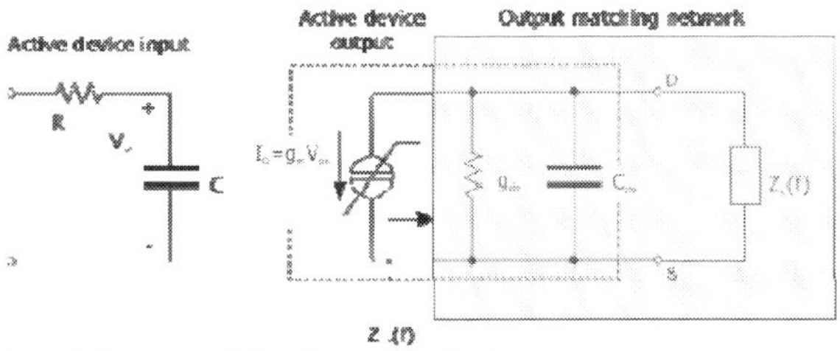

Fig. 7. Simplified model assumed for the active device.

The device output current source is described by a constant transconductance (g_m) in the saturation region, while a constant ON resistance (R_{ON}) is assumed for the ohmic region, resulting in the output I-V linearised characteristics depicted in Fig. 8.

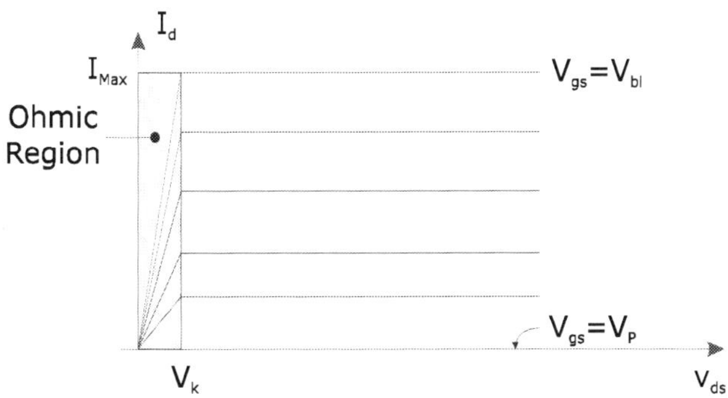

Fig. 8 I-V output characteristics of the simplified model assumed for the active device.

The main parameter taken into account to represent the simplified I-V characteristics are the maximum achievable output current (I_{Max}), the constant knee voltage (V_k) and the pinch-off voltage (V_p).

As it commonly happens in the amplifiers design, some parameters are assumed as starting requirements, thus imposed by the designer, while other ones are consequently derived. Obviously, the following guidelines outline only one of the possible design flows.

The design starts by fixing the OBO level, required to the DPA, accounting for the peculiar PAPR of the application which the DPA is oriented for. The OBO can be defined by the following equation:

$$OBO = \frac{P_{out,DPA(x=x_{break})}}{P_{out,DPA(x=1)}} = \frac{P_{out,Main(x=x_{break})}}{P_{out,Main(x=1)} + P_{out,Aux(x=1)}} \tag{6}$$

where the subscripts are used to refer to the entire DPA or to the single amplifiers (Main and Auxiliary respectively). Moreover a parameter x ($0 \le x \le 1$) is used to identify the dynamic point in which those quantities are considered. In particular $x=0$ identifies the quiescent state, i.e. when no RF signal is applied to the input, while $x=1$ identifies the saturation condition, i.e. when the DPA reaches its maximum output power level. Similarly, $x=x_{break}$ identifies the break point condition, i.e. when the Auxiliary amplifier is turned on.

Clearly, eqn. (6) is based on the assumption that only the Main amplifier delivers output power until the break point condition is reached, and the output network is assumed lossless.

In order to understand how the selected OBO affects the design, it is useful to investigate the expected DLLs of the Main and Auxiliary amplifiers for $x=x_{break}$ (load curves "A" in Fig. 6) and $x=1$ (load curves "C" in Fig. 6). It is to remark that the shape of the DLLs is due, for sake of simplicity, to the assumption of a Tuned Load configuration (Colantonio et al., 2002) both for Main and Auxiliary amplifiers.

Assuming a bias voltage V_{DD}, the drain voltage amplitude of the Main device is equal to V_{DD}-V_k both for $x=x_{break}$ and $x=1$

The same amplitude value is reached by the drain voltage of the Auxiliary device for $x=1$, as shown by the load curve "C" in Fig. 6.

Consequently the output powers delivered by the Main and Auxiliary amplifiers in such peculiar conditions become:

$$P_{out,Main(x=x_{break})} = \frac{1}{2}\cdot\left(V_{DD} - V_k\right)\cdot I_{1,Main(x=x_{break})} \tag{7}$$

$$P_{out,Main(x=1)} = \frac{1}{2}\cdot\left(V_{DD} - V_k\right)\cdot I_{1,Main(x=1)} \tag{8}$$

$$P_{out,Aux(x=1)} = \frac{1}{2}\cdot\left(V_{DD} - V_k\right)\cdot I_{1,Aux(x=1)} \tag{9}$$

where the subscript "1" is added to the current in order to refer to its fundamental component.

Referring to Fig. 5, the power balance at the two ports of the λ/4 both for $x=x_{break}$ and $x=1$ is given by:

$$\frac{1}{2}\cdot\left(V_{DD} - V_k\right)\cdot I_{1,Main(x=x_{break})} = \frac{1}{2}\cdot V_{L(x=x_{break})}\cdot I_{2(x=x_{break})} \tag{10}$$

$$\frac{1}{2}\cdot\left(V_{DD} - V_k\right)\cdot I_{1,Main(x=1)} = \frac{1}{2}\cdot\left(V_{DD} - V_k\right)\cdot I_{2(x=1)} \tag{11}$$

being I_2 the current flowing into the load R_L from the Main branch.

From (11) it follows:

$$I_{1,Main(x=1)} = I_{2(x=1)} \tag{12}$$

Moreover, remembering that the current of one side of the λ/4 is function only of the voltage of the other side, it is possible to write

$$I_{2(x=x_{break})} = I_{2(x=1)} \tag{13}$$

since the voltage at the other side is assumed constant to V_{DD}–V_k in all medium power region, i.e. both for $x=x_{break}$ and $x=1$.

Consequently, taking into account (11), the output voltage for $x=x_{break}$ is given by:

$$V_{L(x=x_{break})} = (V_{DD} - V_k) \cdot \frac{I_{1,Main(x=x_{break})}}{I_{1,Main(x=1)}} = \alpha \cdot (V_{DD} - V_k) \tag{14}$$

where α defines the ratio between the currents of the Main amplifier at $x=x_{break}$ and $x=1$:

$$\alpha = \frac{I_{1,Main(x=x_{break})}}{I_{1,Main(x=1)}} \tag{15}$$

Regarding the output resistance (R_L), its value has to satisfy two conditions, imposed by the voltage and current ratios at $x=x_{break}$ and $x=1$ respectively:

$$R_L = \frac{V_{L(x=x_{break})}}{I_{2(x=x_{break})}} = \frac{\alpha \cdot (V_{DD} - V_k)}{I_{1,Main(x=1)}} \tag{16}$$

$$R_L = \frac{V_{L(x=1)}}{I_{2(x=1)} + I_{1,Aux(x=1)}} = \frac{(V_{DD} - V_k)}{I_{1,Main(x=1)} + I_{1,Aux(x=1)}} \tag{17}$$

Therefore, from the previous equations it follows:

$$I_{1,Aux(x=1)} = \frac{1-\alpha}{\alpha} \cdot I_{1,Main(x=1)} \tag{18}$$

Consequently, substituting (7)-(9) (9) in (6)and taking into account for (18), the following relationship can be derived:

$$OBO = \alpha^2 \tag{19}$$

which demonstrates that, selecting the desired OBO, the ratio between the Main amplifier currents for $x=x_{break}$ and $x=1$ is fixed also.

Since the maximum output power value is usually fixed by the application requirement, it represents another constraints to be selected by the designer. Such maximum output power is reached for $x=1$ and it can be estimated by the following relationship:

$$P_{out,DPA(x=1)} = P_{out,Main(x=1)} + P_{out,Aux(x=1)} = \frac{1}{\alpha} \cdot \frac{1}{2} \cdot (V_{DD} - V_k) \cdot I_{1,Main(x=1)} \tag{20}$$

which can be used to derive the maximum value of fundamental current of Main amplifier ($I_{1,Main(x=1)}$), once its drain bias voltage (V_{DD}) and the device knee voltage (V_k) are selected. Knowing the maximum current at fundamental, it is possible to compute the values of R_L by (16)(16) and the required characteristic impedance of the output λ/4 TL (Z_0) by using:

$$Z_0 = \frac{(V_{DD} - V_k)}{I_{1,Main(x=1)}} \tag{21}$$

which is derived assuming that the output voltage (V_L) reaches the value V_{DD}-V_k for $x=1$. Clearly the maximum value $I_{1,Main(x=1)}$ depends on the Main device maximum allowable output current I_{Max} and its selected bias point.

Referring to Fig. 9, where it is reported for clearness a simplified current waveform, assuming a generic Class AB bias condition, the bias condition can be easily identified defining the following parameter

$$\xi = \frac{I_{DC,Main}}{I_{Max,Main}} \tag{22}$$

being $I_{DC,Main}$ the quiescent (i.e. bias) current of the Main device.
Consequently, ξ=0.5 and ξ=0 refer to a Class A and Class B bias conditions respectively, while 0<ξ<0.5 identifies Class AB bias condition.

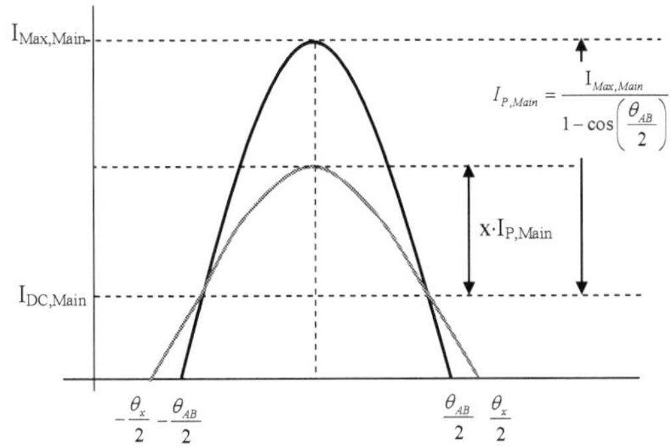

Fig. 9. Current waveform in time domain of the Main amplifier.

The current waveform of Fig. 9 can be analytically described by the following expression:

$$i_{D,Main} = I_{DC,Main} + x \cdot \frac{I_{Max,Main}}{1-\cos\left(\frac{\theta_{AB}}{2}\right)} \cdot \cos(\theta) \tag{23}$$

whose fundamental component can be written as following:

$$I_{1,Main(x=1)} = \frac{I_{Max,Main}}{2\pi} \cdot \frac{\theta_{AB} - \sin(\theta_{AB})}{1-\cos\left(\frac{\theta_{AB}}{2}\right)} \tag{24}$$

being θ_{AB} the current conduction angle (CCA) of the Main output current, achieved for $x=1$.
The bias point ξ and the CCA θ_{AB} can be easily related by the following relationship:

$$\theta_{AB} = 2\pi - 2\arccos\left(\frac{\xi}{1-\xi}\right) \tag{25}$$

Manipulating (24), the value of $I_{Max,Main}$, required to reach the desired maximum power, can be estimated, once the bias point ξ of the Main amplifier has been selected (the last parameter should be fixed by the designer).
As made with Main amplifier, the value of the Auxiliary maximum current can be obtained by using the equation of the first order coefficient of the Furier series, since the value of $I_{1,Aux,(x=1)}$ should fulfill (18).
Consequently, it follows:

$$I_{1,Aux(x=1)} = \frac{I_{Max,Aux}}{2\pi} \cdot \frac{\theta_C - \sin(\theta_C)}{1 - \cos\left(\frac{\theta_C}{2}\right)} \tag{26}$$

being θ_C the CCA of the Auxiliary device output current for $x=1$.
Referring to Fig 10, where it is reported the current waveform of the Auxiliary amplifier, assuming a virtual negative bias point, the Auxiliary device current can be written similarly to (23), thus:

$$i_{D,Aux} = I_{DC,Aux} + x \cdot \frac{I_{Max,Aux}}{1 - \cos\left(\frac{\theta_C}{2}\right)} \cdot \cos(\theta) \tag{27}$$

Moreover, for a proper behavior of the Auxiliary amplifier, the peak of the current has to reach zero for $x=x_{break}$, as highlighted in Fig10. Consequently the following condition has to be taken into account.

$$x_{break} \cdot \frac{I_{Max,Aux}}{1 - \cos\left(\frac{\theta_C}{2}\right)} = -I_{DC,Aux} \tag{28}$$

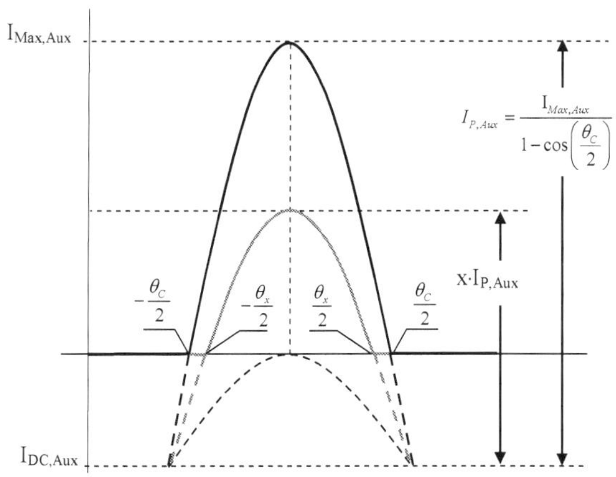

Fig. 10. Current waveform in time domain of the Auxiliary amplifier for $x=x_{break}$ and $x=1$.

Substituting (28) in (27), it is possible to refer the value of θ_C directly to x_{break}:

$$\theta_C = 2 \cdot \arccos(x_{break}) \tag{29}$$

Now, from (15) and replacing the respective Fourier expressions, it follows:

$$x_{break} \cdot \left[\theta_{Main(x=x_{break})} - \sin\left(\theta_{Main(x=x_{break})}\right) \right] = \alpha \cdot \left(\theta_{AB} - \sin\left(\theta_{AB}\right)\right) \tag{30}$$

where from (23) it can be inferred:

$$\theta_{Main(x=x_{break})} = 2\pi - 2\arccos\left(\frac{\xi}{x_{break} \cdot (1-\xi)} \right) \tag{31}$$

The value of x_{break} has to be numerically obtained solving (30), having fixed the OBO (i.e. α) and the Main device bias point (i.e. ξ).
Once the value of $I_{Max,Aux}$ is obtained, the one of $I_{DC,Aux}$ is immediately estimable manipulating (28):

$$I_{DC,Aux} = -I_{Max,Aux} \cdot \frac{x_{break}}{1 - x_{break}} \tag{32}$$

At this point, an interesting consideration can be done about the ratio between the maximum currents required by the devices. Fig. 11 reports this ratio as function of OBO and ξ. As it is possible to note, the dependence on ξ can be practically neglected, while the one by the OBO is very high. Moreover, the same amount of maximum current is required from both devices in case of nearly 5dB as OBO, while an higher current has to be provided by the Auxiliary device for greater OBO.
From the designer point of view, the maximum currents ratio can be used as an useful information to choice the proper device periphery. In fact, supposing for the used technology a linear relationship between maximum current and drain periphery, Fig. 11 gives the possibility to directly derive the drain periphery of the Auxiliary device, once the Main one has been selected in order to respect the maximum output power constraint.

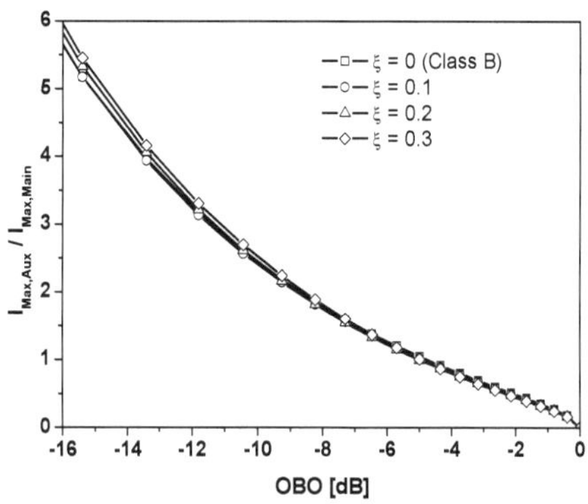

Fig. 11. Ratio between Auxiliary and Main maximum currents as function of OBO and ξ.

3.1. Power splitter dimensioning

In this subsection the dimensioning of the input power splitter is discussed, highlighting its critical role in the DPA architecture.

Following the simplified analysis based on an active device with constant transconductance (g_m), the amplitude of the gate voltage for $x=1$, for Main and Auxiliary devices respectively, can be written as

$$V_{gs,Main(x=1)} = \frac{I_{Max,Main} - I_{DC,Main}}{g_{m,Aux}} = \frac{I_{Max,Main} \cdot (1-\xi)}{g_{m,Main}} \tag{33}$$

$$V_{gs,Aux(x=1)} = \frac{I_{Max,Aux} - I_{DC,Aux}}{g_{m,Aux}} = \frac{I_{Max,Aux}}{g_{m,Aux}} \cdot \frac{1}{1-x_{break}} \tag{34}$$

Using the previous equations, it is possible to derive the powers at the input of the devices by using the following relationships:

$$P_{in,Main(x=1)} = \frac{1}{2} \cdot \frac{\left(V_{gs,Main(x=1)}\right)^2}{R_{in,Main}} = \frac{1}{2} \cdot \frac{\left(I_{Max,Main} \cdot (1-\xi)\right)^2}{R_{in,Main} \cdot \left(g_{m,Main}\right)^2} \tag{35}$$

$$P_{in,Aux(x=1)} = \frac{1}{2} \cdot \frac{\left(V_{gs,Aux(x=1)}\right)^2}{R_{in,Aux}} = \frac{1}{2} \cdot \frac{\left(I_{Max,Aux}\right)^2}{R_{in,Aux} \cdot \left(g_{m,Aux} \cdot (1-x_{break})\right)^2} \tag{36}$$

where $R_{in,Main}$ and $R_{in,Aux}$ are the input resistances respectively of Main and Auxiliary devices.

Therefore, it is possible to compute the power splitting factor, i.e. the amount of power delivered to the Auxiliary device with respect to the total input power, by using:

$$\Lambda_{Aux} = \frac{P_{in,Aux(x=1)}}{P_{in,Main(x=1)} + P_{in,Aux(x=1)}} = \frac{1}{\left(\frac{I_{Max,Main}}{I_{Max,Aux}} \cdot \frac{1-\xi}{1-x_{break}} \cdot \frac{g_{m,Aux}}{g_{m,Main}}\right)^2 \cdot \frac{R_{in,Aux}}{R_{in,Main}} + 1} \tag{37}$$

and consequently for the Main device:

$$\Lambda_{Main} = 1 - \Lambda_{Aux} \tag{38}$$

In Fig. 12 is reported the computed values for Λ_{Aux}, as function of OBO and ξ parameters, assuming for both devices the same values for g_m and R_{in}.

Fig. 12 highlights that large amount of input power has to be sent to the Auxiliary device, requiring an uneven power splitting. For example, considering a DPA with 6dB as OBO and a Class B bias condition (i.e $\xi=0$) for the Main amplifier, 87% of input power has to be provided to Auxiliary device, while only the remaining 13% is used to drive the Main amplifier. This aspect dramatically affects in a detrimental way the overall gain of the DPA, which becomes 5-6 dB lower if compared to the gain achievable by using a single amplifier only.

Nevertheless, it has to remark that this largely unbalanced splitting factor has been inferred assuming a constant transconductance (g_m) for both devices. Such approximation is sufficiently accurate in the saturation region ($x=1$), while becomes unsatisfactory for low power operation. In this case, the actual transconductance behavior can be very different depending on the technology and bias point of the selected active device. In general, it is possible to state that the transconductance value of actual devices, in low power region, is lower than the average one, when the chosen bias point is close to the Class B. Thus, if the bias point of Main amplifier ξ is selected roughly lower than 0.2, the predicted gain in low power region is higher than the experimentally resulting one, being the former affected by the higher value assumed for the transconductance in the theoretical analysis.

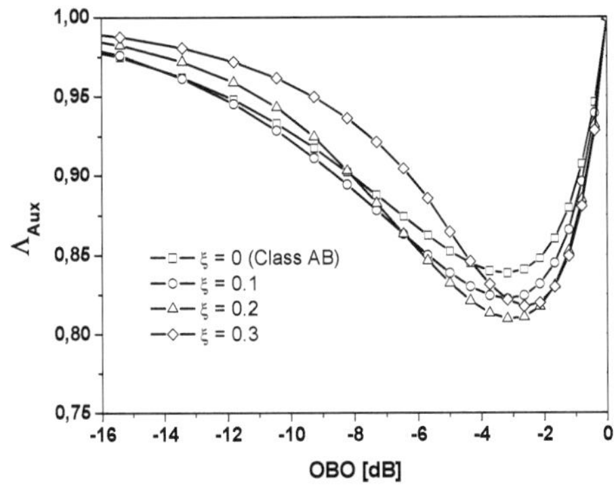

Fig. 12. Λ_{Aux} behavior as a function of OBO and ξ, assuming for both devices the same values for g_m and R_{in}.

From a practical point of view, if the theoretical splitting factor is assumed in actual design, usually the Auxiliary amplifier turns on before the Main amplifier reaches its saturation (i.e. its maximum of efficiency). Consequently a reduction of the unbalancing in the power splitter is usually required in actual DPA design with respect to the theoretical value, in order to compensate the non constant transconductance behavior and, thus, to switch on the Auxiliary amplifier at the proper dynamic point.

3.2. Performance behavior

Once the DPA design parameters have been dimensioned, closed form equations for the estimation of the achievable performances can be obtained. Since the approach is based on the electronic basic laws, it will be here neglected, in order to avoid that this chapter dull reading and to focus the attention on the analysis of the performance behavior in terms of output power, gain, efficiency and AM/AM distortion. The complete relationships can be found in (Colantonio et al., 2009 - a).

The theoretical performance of a DPA designed to fulfill 7dB of OBO and 6W as maximum output power, are shown in Fig. 12. Moreover, the same physical parameters have been assumed for both Main and and Auxiliary devices: $V_k=0V$, $g_m=0.22S$ and $R_{in}=50\Omega$. Finally the drain bias voltage and the Main amplifier quiescent point have been fixed to $V_{DD}=10V$ and $\xi=0.1$ respectively.

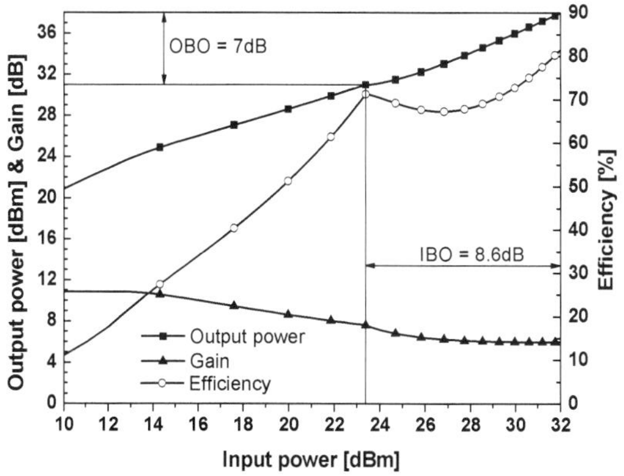

Fig. 12. Theoretical performances of a DPA with 7dB OBO and 6W as maximum output power.

As it appears looking at Fig. 13, the efficiency value at the saturation is higher than the one at the break point. The latter, in fact, is the one of the Main device, which is a Class AB amplifier. The efficiency at the saturation, instead, is increased by the one of the Auxiliary device, which has a Class C bias point, with a consequent greater efficiency value.

It is possible to note as the gain behaves linearly until 13dBm of input power, while becomes a monotonic decreasing function up to about 23.5dBm. Along this dynamic region, the Main amplifier only is working and the variation of the gain behavior is due to the pinch-off limitation in the output current.

In particular, until 13dBm, the Main device operates as a Class A amplifier, since its DLL did not reach yet the pinch-off physical limitation. Then, the Main device becomes a Class AB amplifier, coming up to the near Class B increasing the input power, with a consequent decreasing of the gain. However this evident effect of class (and gain) changing is due to the assumption of a constant transconductance model for the active device. In actual devices, in fact, the value of the transconductance is lower than the average one, when the selected bias point is close to the Class B, as it has been discussed in section 3.1. Consequently, in practical DPA design, the gain, for small input power levels, is lower than the theoretical one estimated by the average g_m value, thus reducing the effect highlighted in Fig. 12.

In the Doherty region, from 23.5dBm up to 32dBm of input power, the gain changes its behavior again. The latter change is due to the combination of the gain decreasing of the Main amplifier, whose output resistance is diminishing, and the gain increasing of the Auxiliary amplifier, which passes from the switched off condition to the proper operative Class C.

The non constant gain behavior is further highlighted in Fig. 12 by the difference between the resulting OBO and input back-off (IBO), resulting in an AM/AM distortion in the overall DPA. In order to deeply analyze this effect, Fig. 13 reports the difference between OBO and IBO for several values of ξ.

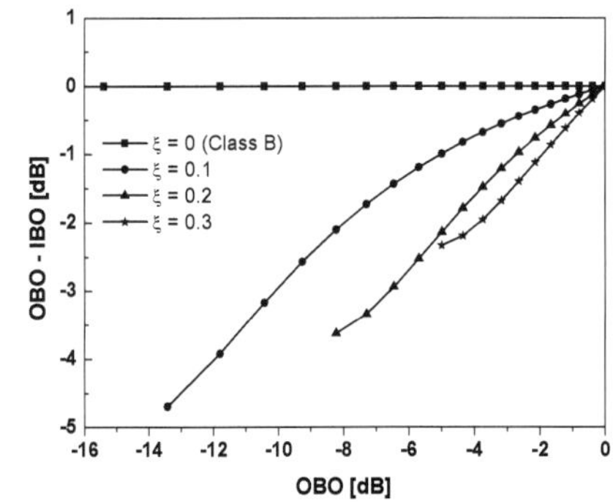

Fig. 13. Theoretical difference between OBO and IBO for several values of ξ.

In order to proper select the Main device bias point ξ to reduce AM/AM distortion, it is useful to introduce another parameter, the Linear Factor (LF), defined as:

$$LF = \frac{1}{1 - x_{break}} \cdot \int_{x_{break}}^{1} \left[P_{out,DPA}(x) - \left(x^2 \cdot P_{out,DPA(x=1)} \right) \right] dx \qquad (39)$$

The Linear Factor represents the variation in the Doherty region of the DPA output power, with respect to a linear PA having the same maximum output power and represented in (39)(39) by $x^2 \cdot P_{out,DPA(x=1)}$. Thus LF gives the simplified estimation of the average AM/AM distortion in the Doherty region.

Consequently, the optimum bias condition should be assumed to assure LF=0. Obviously this condition, if it exists, can be obtained only for one ξ, once the OBO has been selected.

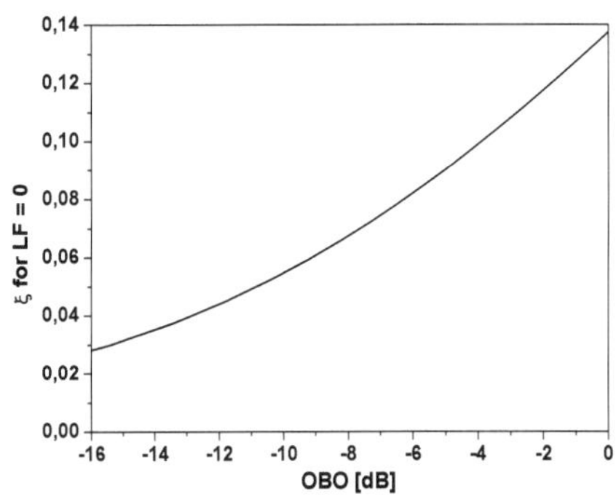

Fig. 14. Values of ξ assuring LF=0, as function of the OBO.

Fig. 14 shows the values of ξ, which theoretically assures LF=0, as function of the selected OBO. This design chart provides a guideline to select the proper bias point of the Main amplifier (ξ), having fixed the desired OBO of the DPA.
In order to further clarify the DPA behavior, Fig. 15 shows the fundamental drain currents and voltages for both Main and Auxiliary devices. These behaviors can be used in the design flow to verify if the DPA operates in a proper way. In particular, the attention has to be focused on the Main voltage, which has to reach, at the break point (x_{break}), the maximum achievable amplitude (10V in this example) in order to maximize the efficiency. Moreover the Auxiliary current can be used to verify that the device is turned on in the proper dynamic instant. Finally, the designer has to pay attention if the Auxiliary current reaches the expected value at the saturation ($x=1$), in order to perform the desired modulation of the Main resistance. This aspect can be evaluated also observing the behavior of Main and Auxiliary resistances, as reported in Fig. 17.

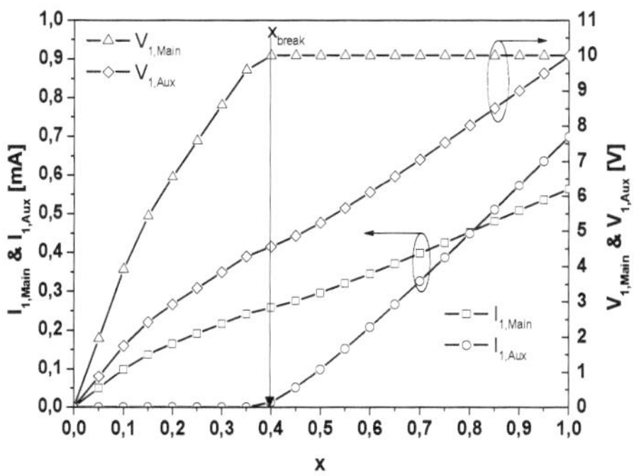

Fig. 15. Fundamental current and voltage components of Main and Auxiliary amplifiers, as function of the dynamic variable x.

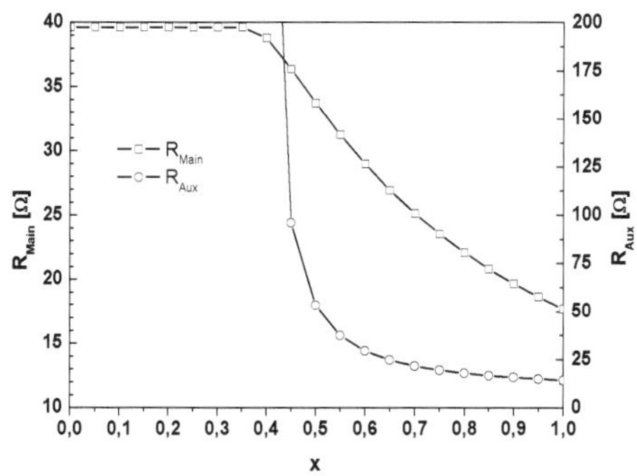

Fig. 17. Drain resistance at fundamental frequency of Main and Auxiliary amplifiers, as function of the dynamic variable x.

4. Advanced DPA Design

In the previous paragraphs the classical Doherty scheme based on Tuned Load configuration for both Main and Auxiliary amplifiers has been analyzed. Obviously, other solutions are available, still based on the load modulation principle, but developed with the aim to further improve the features of the DPA, by using additional some free design parameters.

4.1. DPA Design by using different bias voltage

For instance, the adoption of different drain bias voltage for the two amplifiers (Main and Auxiliary) could be useful to increase the gain of the overall DPA. In fact, in the DPA topology the voltage at the output common node, V_L in Fig. 5, at saturation is imposed by the Auxiliary drain bias voltage ($V_{DD,Aux}$) in order to fulfill the condition $V_L = V_{DD,Aux}- V_{k,Aux}$. Thus, assuming a different bias, i.e. $V_{DD,Main}$ and $V_{DD,Aux}$ for the Main and Auxiliary devices respectively, and defining the parameter

$$\beta = \frac{V_{DD,Main} - V_{k,Main}}{V_{DD,Aux} - V_{k,Aux}} \tag{40}$$

then the design relationships previously inferred have to be tailored accounting for such different supplying voltages.
Therefore, the DPA elements R_L and Z_0 becomes:

$$R_L = \frac{\alpha^2}{\beta^2} \cdot R_{Main}\left(x_{break}\right) \tag{41}$$

$$Z_0 = \frac{V_{DD,Aux} - V_k}{I_{1,Main}\left(\theta_{AB}\right)} \tag{42}$$

where

$$R_{Main}\left(x_{break}\right) = 2\frac{V_{DD,Main} - V_{k,Main}}{I_{M,Main}} \frac{\pi}{\alpha} \cdot \frac{1-\cos\left(\frac{\theta_{AB}}{2}\right)}{\theta_{AB} - \sin\left(\theta_{AB}\right)} \tag{43}$$

Moreover, the Auxiliary and Main devices maximum output currents are now related through the following relationship:

$$I_{M,Aux} = \beta \cdot I_{M,Main} \cdot \frac{1-\alpha}{\alpha} \cdot \frac{1-\cos\left(\frac{\theta_C}{2}\right)}{\theta_C - \sin\left(\theta_C\right)} \cdot \frac{\theta_{AB} - \sin\left(\theta_{AB}\right)}{1-\cos\left(\frac{\theta_{AB}}{2}\right)} \tag{44}$$

Which, clearly, highlights that a suitable selection of the Auxiliary device supply voltage, i.e. $\beta<1$, could imply a lower maximum output current required from the Auxiliary device.
Conversely, the saturated output power of the Doherty (for $x=1$) is still related to both the Main device supply voltage and its maximum output current, i.e.:

$$P_{out,DPA(x=1)} = \frac{1}{2}\left(V_{DD,Main} - V_k\right) \cdot \frac{I_{1,Main}\left(\theta_{AB}\right)}{\alpha} = \frac{1}{\alpha} \cdot P_{out,Main,Max} = \frac{1}{\alpha^2} \cdot P_{out,Main,break} \tag{45}$$

Thus being not affected by the different drain supply voltage adopted for the Auxiliary device.

4.2. DPA Design by using Harmonic Tuning strategies

To further improve the overall efficiency in a Doherty configuration, high efficiency design strategies can be adopted in the synthesis of both Main and Auxiliary amplifiers. For this purpose, harmonic tuning strategies have been proposed (Colantonio et al., 2002).

However, due to the Class C bias condition for the Auxiliary device, thus implying a wrong phase relationships between current (and voltage) harmonic components, the optimum solution for such amplifier is the classical Tuned Load one.

Conversely, for the Main amplifier, which is normally operating in a Class AB bias, the efficiency can be improved by using for instance Class F strategy (Raab, 2001). Such design strategy implies that the second harmonic current component I_2 should be short circuited, while the fundamental (I_1) and the third one (I_3) should be terminated on impedance R_1 and R_3, respectively, to obtain a proper voltage harmonic component ratio (Colantonio et al., 2002):

$$k_3 = \frac{V_3}{V_1} = \frac{R_3 \cdot I_3}{R_1 \cdot I_1} = 0.167 \tag{46}$$

Thus, in a Class F Doherty amplifier (i.e. with Main amplifier in Class F configuration), the proper output harmonic loading conditions have to be fulfilled across the Main device, accounting for the load modulation effect in the medium power region.

In particular, it is possible to compute the theoretical load modulation required for the third harmonic, in order to fulfill (46) accounting for the modulation of R_1. In Fig. 16 is reported the ratio between the values required for R_3 at the end of the low power region ($x=x_{break}$) and at the end of the Medium (or Doherty) power region, i.e. at saturation ($x=1$), as a function of the Main device bias point (ξ) and the selected OBO.

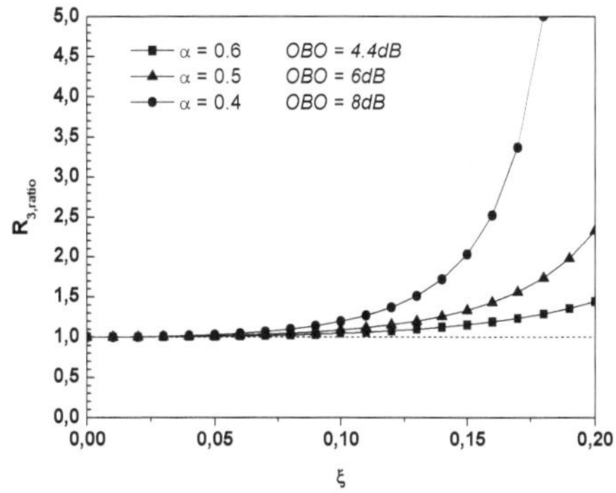

Fig. 16. $R_{3,ratio}$ as function of ξ for different OBO (α) values.

As it can be noted, the $R_{3,ratio}$ (i.e. the degree of modulation required for the third harmonic loading condition) increases with the bias point (ξ) and OBO values (α).
Nevertheless, the modulation of R_3 through the output $\lambda/4$ line and the Auxiliary current, critically complicate the design and can be usually neglected if the Main device bias point is chosen nearly Class B condition, i.e. $\xi<0.1$, being $R_{3,ratio}\approx 1$.
Under such assumption, it is possible to compute the Class F DPA design parameter as compared to the Tuned Load case.
It can be inferred, referring to Fig. 3, that the output load R_L and the characteristic impedance of the output $\lambda/4$ TL become

$$R_{L,F} \simeq \frac{R_L}{1.15} \qquad Z_{0,F} = Z_{0,TL} \tag{47}$$

being R_L given equivalently by (16) or (17) and Z_0 by (21).
Finally, regarding the power splitter dimensioning, it is required a different power splitting ratio, resulting in:

$$\Lambda_{Aux,F} = \frac{1.322 \cdot \Lambda_{Aux}}{1+0.322 \cdot \Lambda_{Aux}} > \Lambda_{Aux} \quad \Lambda_{Main,F} = 1 - \Lambda_{Aux,F} < \Lambda_{AB} \tag{48}$$

The expected behavior of the output current and voltage fundamental components for the Main and the Auxiliary devices are reported in Fig. 19, assuming ξ=0.082 and OBO=6dB (i.e. α=0.5).
Similarly, in Fig. 17 are reported the comparisons in terms of output power and efficiency of Class F DPA with respect to Tuned Load DPA, normalized as functions of the input signal x.

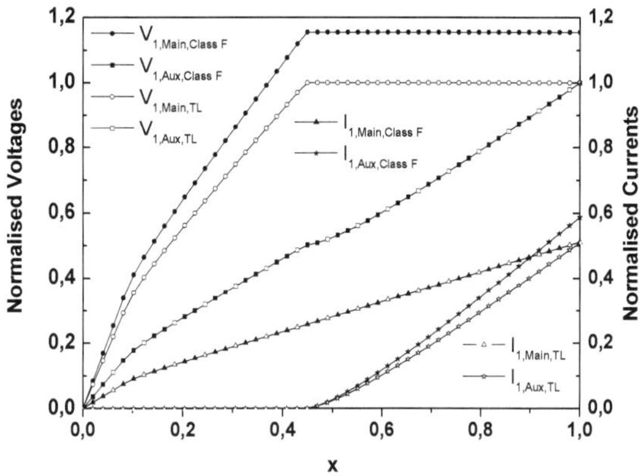

Fig. 19. Theoretical behavior of drain current and voltage fundamental components for Main and Auxiliary devices, assuming Class F or Tuned Load design strategies for Main amplifier.

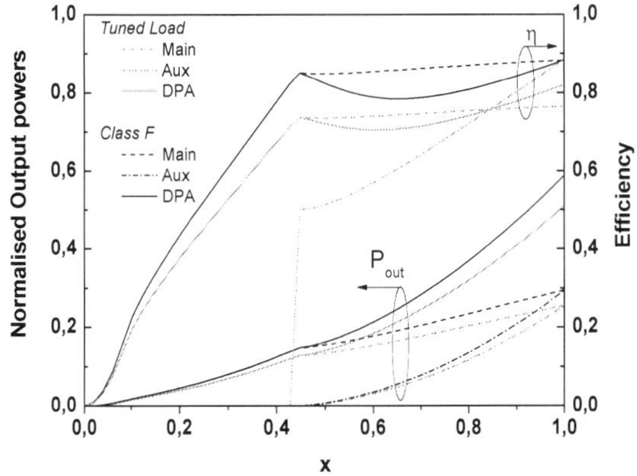

Fig. 17. Expected drain efficiency and output power behaviors for Class F and Tuned Load Doherty amplifiers.

4.3. Multi-way Doherty amplifiers

In order to reduce the Auxiliary device size, while still providing the required current for the load modulation, a different solution is based on the so called Multi-Way Doherty configuration, usually referred as N-Way Doherty amplifier also (Yang et al., 2003 - Kim et al., 2006 - Cho et al., 2007). It is realized by paralleling one Main amplifier and N-1 Auxiliary amplifiers, aimed to acquire an N-1 times larger-sized Auxiliary amplifier, as schematically shown in Fig. 18.

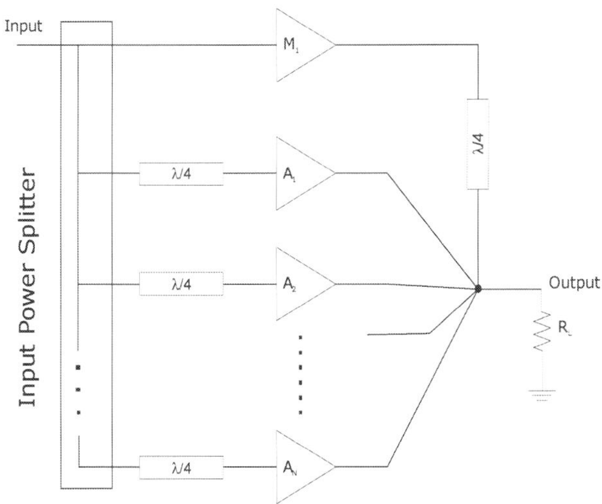

Fig. 18. Schematic diagram of the N-way Doherty amplifier.

With the proposed device combination, it becomes possible to implement larger OBO using smaller devices, resulting in the theoretical efficiency performance shown in Fig. 19.

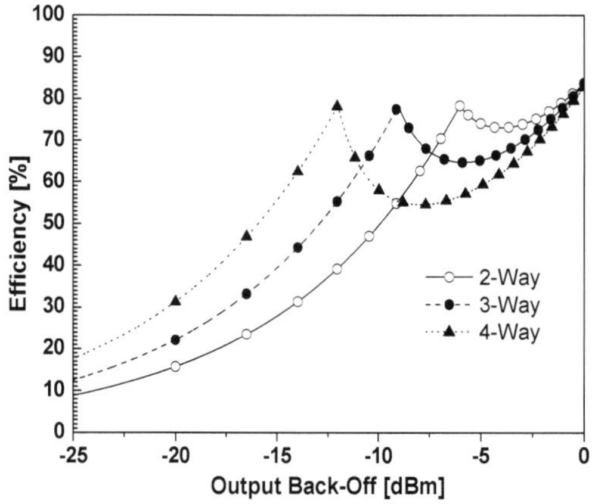

Fig. 19. Theoretical efficiency of the N-Way Doherty amplifier.

4.4. Multi-Stage Doherty amplifiers

The Multi-Stage Doherty amplifier is conceptually different from the Multi-Way configuration, since it is based on a subsequent turning on condition of several Auxiliary devices, with the aim to assure a multiple Doherty region in a cascade configuration, overcoming the reduction of the average value due to the increased drop-down phenomenon in efficiency, especially when larger OBO are required (Neo et al., 2007 - Pelk et al., 2008 - Srirattana et al., 2005).

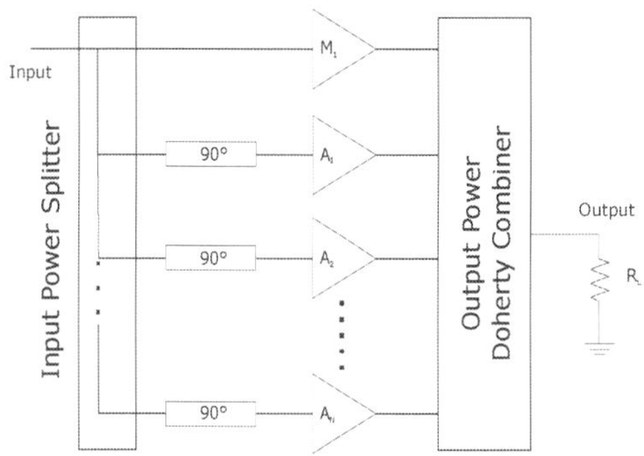

Fig. 20. Theoretical diagram of a Multi-Stage Doherty amplifier.

For this purpose, referring to the theoretical diagram shown in Fig. 23, amplifiers M_1 and A_1 have to be designed to act as Main and Auxiliary amplifiers in a standard Doherty configuration. Then, when both amplifiers are approaching their saturation, amplifier A_2 is turned on operating as another Auxiliary amplifier, thus modulating the load seen by the previous M_1-A_1 pair, that must be considered, from now onward, as a single amplifier. Such concept is then iterated inserting N Auxiliary amplifiers, each introducing a new break-point, resulting in a theoretical efficiency behavior as depicted in Fig. 21.

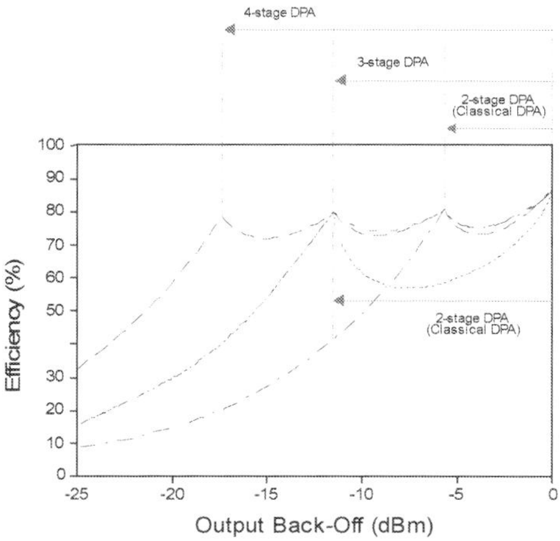

Fig. 21. Theoretical behavior of the efficiency for a Multi-Stage Doherty amplifier.

From the design issues, it is easy to note that the most critical one resides in the practical implementation of the output power combining network, required to properly exploit the load modulation concept for all the cascaded stages.

A proposed solution is reported in (Neo et al., 2007 - Pelk et al., 2008), based on the scheme depicted in Fig. 22, where the relationships to design the output $\lambda/4$ transmission lines adopted are given by

$$Z_{0,i} = R_L \cdot \prod_{j=1}^{i} \frac{1}{\alpha_j}$$
$$\prod_{j=k}^{\frac{i+k}{2}} \alpha_{2j-k} = 10^{-\frac{OBO_i}{20}} \tag{49}$$

where *i=1,2,...N*, *k=1* (for odd *i*) or *k=2* (for even *i*), *N* is the total number of Auxiliary amplifiers, OBO_i is the back-off level from the maximum output power of the system at which the efficiency will peak (i.e. the turning on condition of the Auxiliary A_i).

The R_L value is determined by the optimum loading condition of the last Auxiliary stage, according to the following relationship:

$$R_L = (1-\alpha_1) \cdot R_{opt,Aux_N} \tag{50}$$

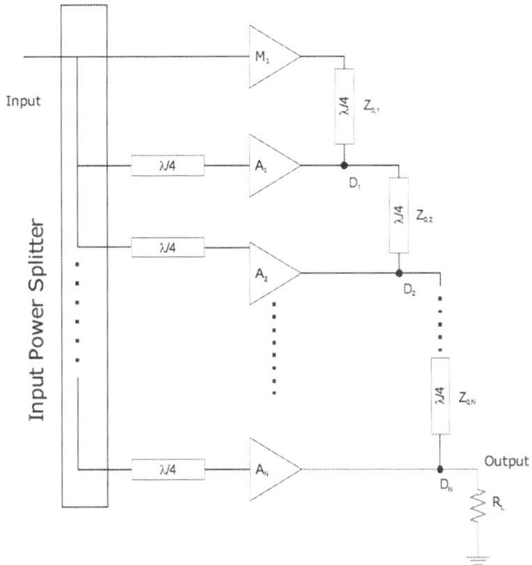

Fig. 22. Proposed schematic diagram for a multi-stage Doherty amplifier.

However, some practical drawbacks arise from the scheme depicted in Fig. 22. In fact, the Auxiliary device A_1 is turned on to increase the load at D_1 node and consequently, due to the $\lambda/4$ line impedance $Z_{0,1}$, to properly decrease the load seen by M_1. However, when A_2 is turned on, its output current contributes to increase the load impedance seen at D_2 node. Such increase, while it is reflected in a suitable decreasing load condition for A_1 (at D_1 node), it also results in an unwanted increased load condition for M_1, still due to the $\lambda/4$ line transformer. As a consequence, such device results to be overdriven, therefore saturating the overall amplifier and introducing a strong non linearity phenomenon in such device. To overcome such a drawback, it is mandatory to change the operating conditions, by turning on, for instance, the corresponding Auxiliary device before the Main device M_1 has reached its maximum efficiency, or similarly, changing the input signal amplitudes to each device (Pelk et al., 2008).

Different solutions could be adopted for the output power combiner in order to properly exploit the Doherty idea and perform the correct load modulation, and a optimized solution has been identified as the one in (Colantonio et al., 2009 - a).

5. References

Campbell, C. F. (1999). A Fully Integrated Ku-Band Doherty Amplifier MMIC, IEEE Microwave and Guided Wave Letters, Vol. 9, No. 3, March 1999, pp. 114-116.

Cho, K. J.; Kim, W. J.; Stapleton, S. P.; Kim, J. H.; Lee, B.; Choi, J. J.; Kim, J. Y. & Lee, J. C. (2007). Design of N-way distributed Doherty amplifier for WCDMA and OFDM applications, Electronics Letters, Vol. 43, No. 10, May 2007, pp. 577-578.

Colantonio, P.; Giannini, F.; Leuzzi, F. & Limiti, E. (2002). Harmonic tuned PAs design criteria, *IEEE MTT-S International Microwave Symposium Digest*, Vol. 3, June 2002, pp. 1639-1642.

Colantonio, P.; Giannini, F.; Giofrè, R. & Piazzon, L. (2009 - a). AMPLIFICATORE DI TIPO DOHERTY, Italian Patent, No. RM2008A000480, 2009.

Colantonio, P.; Giannini, F.; Giofrè, R. & Piazzon, L. (2009 - b). The AB-C Doherty power amplifier. Part I: Theory, *International Journal of RF and Microwave Computer-Aided Engineering*, Vol. 19, Is. 3, May 2009, pp. 293–306.

Cripps, S. C. (2002). *Advanced Techniques in RF Power Amplifiers Design*, Artech House, Norwood (Massachusetts).

Doherty, W. H. (1936). A New High Efficiency Power Amplifier for Modulated Waves, *Proceedings of Institute of Radio Engineers*, pp. 1163-1182, September 1936.

Elmala, M.; Paramesh, J. & Soumyanath, K. (2006). A 90-nm CMOS Doherty power amplifier with minimum AM-PM distortion, *IEEE Journal of Solid-State Circuits*, Vol. 41, No. 6, June 2006, pp. 1323–1332.

Kang, J.; Yu, D.; Min, K. & Kim, B. (2006). A Ultra-High PAE Doherty Amplifier Based on 0.13-μm CMOS Process, *IEEE Microwave and Wireless Components Letters*, Vol. 16, No. 9, September 2006, pp. 505–507.

Kim, J.; Cha, J.; Kim, I. & Kim, B. (2005). Optimum Operation of Asymmetrical-Cells-Based Linear Doherty Power Amplifier-Uneven Power Drive and Power Matching, *IEEE Transaction on Microwaves Theory and Techniques*, Vol. 53, No. 5, May 2005, pp. 1802-1809.

Kim, I.; Cha, J.; Hong, S.; Kim, J.; Woo, Y. Y.; Park, C. S. & Kim, B. (2006). Highly Linear Three-Way Doherty Amplifier With Uneven Power Drive for Repeater System, *IEEE Microwave and Wireless Components Letters*, Vol. 16, No. 4, April 2006, pp. 176-178.

Kim, J.; Moon, J.; Woo, Y. Y.; Hong, S.; Kim, I.; Kim, J. & Kim, B. (2008). Analysis of a Fully Matched Saturated Doherty Amplifier With Excellent Efficiency, *IEEE Transaction on Microwaves Theory and Techniques*, Vol. 56, No. 2, February 2008, pp. 328-338.

Lee, Y.; Lee, M. & Jeong, Y. (2008). Unequal-Cells-Based GaN HEMT Doherty Amplifier With an Extended Efficiency Range, *IEEE Microwave and Wireless Components Letters*, Vol. 18, No. 8, August 2008, pp. 536–538.

Markos, Z.; Colantonio, P.; Giannini, F.; Giofrè, R.; Imbimbo, M. & Kompa, G. (2007). A 6W Uneven Doherty Power Amplifier in GaN Technology, *Proceedings of 37th European Microwave Conference*, pp. 1097-1100, Germany, October 2007, IEEE, Munich.

McCarroll, C.P.; Alley, G.D.; Yates, S. & Matreci, R. (2000). A 20 GHz Doherty power amplifier MMIC with high efficiency and low distortion designed for broad band digital communication systems, *IEEE MTT-S International Microwave Symposium Digest*, Vol. 1, June 2000, pp. 537–540.

Neo, W. C. E.; Qureshi, J.; Pelk, M. J.; Gajadharsing, J. R. & de Vreede, L. C. N. (2007). A Mixed-Signal Approach Towards Linear and Efficient N-Way Doherty Amplifiers, *IEEE Transaction on Microwaves Theory and Techniques*, Vol. 55, No. 5, May 2007, pp. 866-879.

Pelk, M. J.; Neo, W. C. E.; Gajadharsing, J. R.; Pengelly, R. S. & de Vreede, L. C. N. (2008). A High-Efficiency 100-W GaN Three-Way Doherty Amplifier for Base-Station Applications, *IEEE Transaction on Microwaves Theory and Techniques*, Vol. 56, No. 7, July 2008, pp. 1582-1591.

Raab, F. H. (1987). Efficiency of Doherty RF power-amplifier systems, *IEEE Transaction on Broadcasting*, Vol. BC-33, No. 3, September 1987, pp. 77–83.

Raab, F. H. (2001). Class-E, Class-C and Class-F power amplifiers based upon a finite number of harmonics, *IEEE Transaction on Microwaves Theory and Techniques*, Vol. 49, No. 8, August 2001, pp. 1462-1468.

Srirattana, N.; Raghavan, A.; Heo, D.; Allen, P. E. & Laskar, J. (2005). Analysis and Design of a High-Efficiency Multistage Doherty Power Amplifier for Wireless Communications, *IEEE Transaction on Microwaves Theory and Techniques*, Vol. 53, No. 3, March 2005, pp. 852-860.

Steinbeiser, C.; Landon, T.; Suckling, C.; Nelson, J.; Delaney, J.; Hitt, J.; Witkowski, L.; Burgin, G.; Hajji, R. & Krutko, O. (2008). 250W HVHBT Doherty With 57% WCDMA Efficiency Linearized to -55 dBc for 2c11 6.5 dB PAR, *IEEE Journal of Solid-State Circuits*, Vol. 43, No. 10, October 2008, pp. 2218–2228.

Tsai, J. & Huang, T. (2007). A 38–46 GHz MMIC Doherty Power Amplifier Using Post-Distortion Linearization, *IEEE Microwave and Wireless Components Letters*, Vol. 17, No. 5, May 2007, pp. 388–390.

Wongkomet, N.; Tee, L. & Gray, P. R. (2006). A 31.5 dBm CMOS RF Doherty Power Amplifier for Wireless Communications, *IEEE Journal of Solid-State Circuits*, Vol. 41, No. 12, December 2006, pp. 2852–2859.

Yang, Y.; Cha, J.; Shin, B. & Kim, B. (2003). A Fully Matched N-Way Doherty Amplifier With Optimized Linearity, *IEEE Transaction on Microwaves Theory and Techniques*, Vol. 51, No. 3, March 2003, pp. 986-993.

7

Distortion in RF Power Amplifiers and Adaptive Digital Base-Band Predistortion

Mazen Abi Hussein, Yide Wang and Bruno Feuvrie
Polytechnique school of the university of Nantes (IREENA)
France

1. Introduction

Future communications systems will have to integrate multiple radio technologies in order to support a large number of wireless standards (cellular technology, WMAN, WLAN and WPAN). The concept of Software-Defined Radio (SDR) in which a common hardware platform is used to cover any communication channel in a wide frequency spectrum with any modulation and bandwidth, supports this trend (Mitola, 1995; Haykin, 2005). In light of the functional flexibility and platform programmability requirements, one of the major challenges of SDR clearly resides in energy consumption. Besides the energy constraints, spectrum is becoming also a major resource bottleneck because of the steady increase in the number of wireless users and applications. The accelerated deployment of broadband personal communication and the continuously increasing demand for higher data rates will crow the limited and precious frequency spectrum.

The power amplifier (PA) is one of the basic building blocks that imposes both spectral and power constraints while designing SDR transmitters, mainly due to its nonlinear characteristics. In fact, SDR systems should operate with third-generation (3G) air interface standards, where non constant envelope modulation formats (e.g, M-QAM, OFDM) are required as part of their functionalities. These signals are particularly vulnerable to nonlinearity and, in this case, the PA must be backed off from its normal operating point, as a simple solution to avoid nonlinear spectral distortion. Efficiency of this system is likely to be poor and size and cost may well be greater than would otherwise be required to deliver the same output power. Therefore, linearization and efficiency enhancement techniques (Kenington, 2000; Cripps, 2006) prove essential in the realization of efficient and cost-effective PAs and/or transmitters (Kenington, 2002). Such techniques, and particularly those that are digital input in nature, are a key enabling technology without which SDR cannot succeed.

This chapter makes the following two contributions. First, it provides new elements for modeling nonlinear distortion of RF PAs by reconsidering power series analysis. This analysis has been traditionally used as an intuitive introduction to understand and quantify nonlinear phenomena. However, in order to have a tractable analysis, it would be necessary to ignore higher order terms, which affects modeling accuracy. Our approach allows us to extend this analysis to an arbitrary high order in some particular cases. More interestingly, it allows to take advantage from the static power measurements and/or the parameters found in a PA's data sheet to identify the coefficients of a relatively high order RF polynomial model. Second, we experimentally evaluate one of the most promising linearization techniques for SDR transmitters, the adaptive digital baseband predistortion technique. A measurement testbed which is designed to be fully automatic has been used. This testbed is configured so that the computer plays virtually the role of the digital predistorter in the transmission chain.

The chapter is organized as follows. Section 2 discusses nonlinear distortion. In section 3, we present advanced analytical development validated by simulations in Matlab. Section 4 describes the measurements test bench and experimentally evaluates the performance of the adaptive digital baseband predistortion technique. Section 5 concludes and discusses some perspectives.

2. Nonlinear Distortion

In order to analyze a given system, it is relevant to represent it by a mathematical model. A linear system (Fig. 1) may be described either by its impulse response $g(t)$ or by its frequency response $G(f)$, which is the Fourier transform of $g(t)$. The output signal $y(t)$ is related to the input signal $x(t)$ and the impulse response $g(t)$ by the convolution integral, $y(t) = \int_{-\infty}^{+\infty} x(\tau)g(t-\tau)d\tau$. Equivalently, the output of the system, expressed in the frequency domain is, $Y(f) = X(f)G(f)$. Thus, a linear system modifies the spectral components of its input signal according to the special form of its frequency response $G(f)$. Modifications on the amplitude spectrum of the input signal results from a simple product ($|Y(f)| = |X(f)||G(f)|$), and those on the phase spectrum from a simple addition ($\angle Y(f) = \angle X(f) + \angle G(f)$). In order to preserve the particular form of the information-bearing signal, the output signal $y(t)$ could differ from the input signal only by a constant factor a and a constant delay τ, $y(t) = ax(t-\tau)$. Every linear system whose amplitude response is not constant or whose phase response is not linear, may incur distortion to the input signal. This is what we call linear amplitude and linear phase distortion respectively.

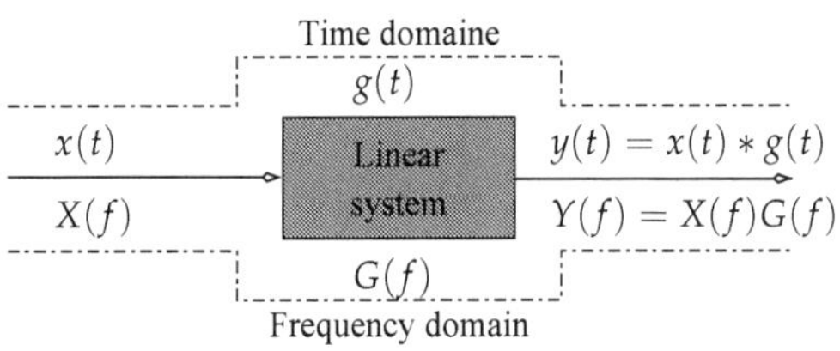

Fig. 1. Linear system

In nonlinear (NL) systems, however, a different type of distortion is encountered, which is not a simple modification of the spectral content of the input signal. Output signals are enriched with new spectral components that may be totally absent from the input signal, and they are often interpreted as an additional source of distortion (they could be exploited in some circumstances). Here, it is worthy to note that the linearity assumption on which relies the analysis of complex electronic systems is only an approximation. In reality, all circuits present some non linear effects due to design imperfections. Thus, in order to understand real phenomena encountered in electronic systems, the analysis of nonlinearity and its effects is crucial. This will be the subject of the next section.

3. Power Series Analysis

Nonlinear distortions in the transmitter are typically introduced by RF Power Amplifiers (PAs). The nonlinear behavior of this device may seriously change the properties of the transmitted signals. One important and, perhaps, the simplest way for understanding and quantifying nonlinear effects of PAs, is to describe their behavior via a power series under special types of excitation signals. In this case, the power amplifier is implicitly considered as a memoryless *"weakly"* NL system which causes only amplitude distortion (Bosch & Gatti, 1989). In fact, when the system introduces phase distortion, it certainly possesses a certain amount of memory which cannot be taken into account by power series. Thus, we confine first our attention to nonlinear amplitude distortion. Nonlinear phase distortion will be discussed later in this chapter, through an appropriate baseband modeling of the PA.

Under the memoryless assumption, the RF output signal, denoted $y_a(t)$ (where the subscript a stands for amplitude distortion), is related to the RF input signal, $x(t)$, by the following polynomial model:

$$y_a(t) = \sum_{k=1}^{K_a} a_k x^k(t) \tag{1}$$

where the coefficients a_k are real constants that can be determined experimentally, and K_a is the polynomial (nonlinearity) order.

While traditional analysis is established in time-domain, which limits the development to lower order degrees for mathematical tractability, we propose here a novel development in the frequency domain, and give closed form expressions of the output spectrum. Such a development allows us to quantify the effect of an arbitrary high order nonlinearity on special types of test signals. Moreover, it allows to take advantage from the static power measurements and/or the parameters found in the PA's data sheet to identify the coefficients a_k of a relatively high order RF polynomial model. The equivalent frequency-domain expression for equation (1) is

$$Y_a(f) = \sum_{k=1}^{K_a} a_k \coprod_{*}^{k} X(f). \tag{2}$$

where $\coprod_*^k X(f) = X(f) * X(f) * \ldots * X(f)$ is the *k-conv* of $X(f)$, i.e. k-times the convolution product of $X(f)$.

According to the equation above, we can see that the spectrum of the output signal $Y_a(f)$ contains one term proportional to the spectrum of the input signal $X(f)$, and a set of new terms each of which is proportional to a *k-conv* of $X(f)$, $k = 2, 3, \ldots, K_a$. Thus, one of the most obvious properties of a NL system is its generation of new frequencies absent from the input signal. If the spectrum of the input signal contains one or more frequency components, or it covers a limited bandwidth, the convolution products in (2) will multiply the number of frequency components in the first case, and spread the signal bandwidth in the second one. The formulation (2) will be analyzed and detailed in the case of special types of excitation signals: one- and two-tone signals.

We precise that, for illustration purposes, the coefficients a_k in (2) have been determined using the parameters found in the data sheet of a High power amplifier (HPA) from Mini-Circuit, the ZHL-100W-52 – (US patent 7,348,854). A 9th order polynomial model, with odd degree terms only, has been identified and used in all the simulations presented below. The model identification method will be explained later in this section.

3.1 One-Tone Signal

One simple test to describe the behavior of the PA is the one-tone test. It consists of generating a one-tone signal at the input and sweeping the power over a finite range. Analyzing the spectral distortion of the output signal gives some important information which allow to quantify the degree of nonlinearity of the PA. Let us consider a one-tone input signal, $x(t) = A\cos(2\pi f_0 t)$. The Fourier transform of $x(t)$, denoted $X_{f_0}(f)$, is

$$X_{f_0}(f) = \frac{A}{2}\left(\delta(f - f_0) + \delta(f + f_0)\right). \quad (3)$$

Given the distributive property of the convolution with respect to addition, the *n-conv* (i.e. $\coprod_*^n(\cdot)$) of X_{f_0} may be written as

$$\coprod_*^n X_{f_0}(f) = \left(\frac{A}{2}\right)^n \sum_{i=0}^{n} \binom{n}{i} \delta\left(f + (n-i)f_0 - if_0\right). \quad (4)$$

Hence, using Eq. (4), and substituting (3) in (2), the output spectrum of the PA may be written in the form

$$Y_a(f) = \sum_{k=1}^{K_a} a_k \left(\frac{A}{2}\right)^k \sum_{i=0}^{k} \binom{k}{i} \delta(f + (k-2i)f_0). \quad (5)$$

We can notice the generation of a remarkable number of new components at frequencies $(k - 2i)f_0$ in Eq. 5. Each successive term generates more new frequencies than the previous one. Besides, odd (even) degree terms generate only odd- (even-) order frequency components since the parity of $k - 2i$ is equal to the parity of k. For instance, the term corresponding to $k = 3$ generates 4 frequencies at $(\pm 3f_0, \pm f_0)$, while the term corresponding to $k = 2$ generates 3 frequencies at $(\pm 2f_0, 0)$. Such new frequency components, which appear at a multiple of the original frequency f_0, are called the *harmonics*.

3.1.1 AM/AM Characteristic

In one-tone tests, the amplitude nonlinearity which is the nonlinear relationship between the input power and the output power, could be quantified by a characteristic curve, called the *AM-AM* (Amplitude Modulation) characteristic. It is also called the *AM-AM conversion*, since it is the conversion introduced by the PA between the amplitude modulation present on the input signal, and the modified amplitude modulation present on the output signal.

In practice, this curve is obtained by measuring the output power at the fundamental frequency f_0 for different values of the average input power. Theoretically, the average power of the signal is proportional to the sum of the squares of its spectral components (Parseval's identity). Thus, given the symmetry of the spectrum around the origin $f = 0$, the average power of the one-tone signal, measured in dBm, is equal to

$$P_{avg}^{in} = 10\log_{10} \frac{2\left|X_{f_0}(+f_0)\right|^2}{R10^{-3}} \quad (6)$$

where R is the resistance for terminal circuits and measurement instruments interface. It is of common use in telecommunications to choose a reference power of $1mW$ (the factor 10^{-3} in (6))[1]. For the output signal, every odd-order term in Eq. (5) generates one spectral component

[1] dBm is an abbreviation for the power ratio in decibel (dB) of the measured power referenced to one milliwatt (mW).

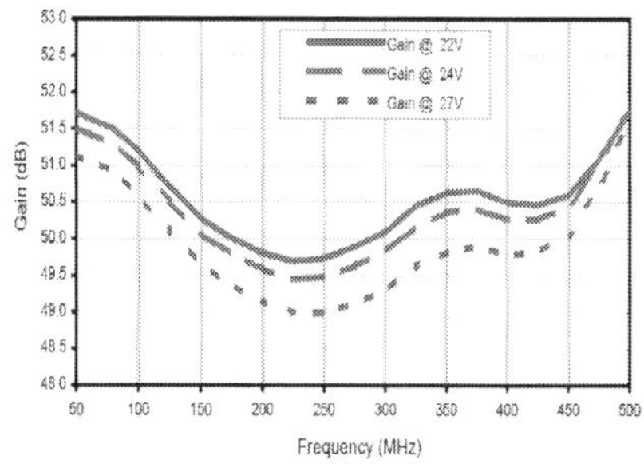

Fig. 2. Gain flatness of the ZHL-100W-52 in its operation bandwidth (data sheet, US patent 7,348,854)

at f_0. Once again, given the symmetry around the origin $f = 0$, the power $P_{f_0}^{out}$ of the output signal at the frequency f_0, is given by

$$P_{f_0}^{out} = 10\log_{10}\frac{2\left|Y_a(+f_0)\right|^2}{R10^{-3}} \tag{7}$$

where $Y_a(+f_0)$ is the amplitude of the sum of all the components at $+f_0$ in Eq. (5)

$$Y_a(+f_0) = \sum_{k=1}^{K} a_{2k-1}\left(\frac{A}{2}\right)^{2k-1}\binom{2k-1}{k}. \tag{8}$$

Substituting $Y_a(+f_0)$ in Eq. (7) by its value in Eq. (8), the former can be expressed in function of the average input power or the input power $P_{f_0}^{in}$ at f_0

$$P_{f_0}^{out} = P_{f_0}^{in} + G + Gc_{f_0} \tag{9}$$

where

$$G = 10\log_{10}(a_1^2), \tag{10}$$

$$Gc_{f_0} = 10\log_{10}(1+S)^2, \tag{11}$$

and $S = \sum_{k=2}^{K}\frac{a_{2k-1}}{a_1}C_{2k-1}^{k}(A/2)^{2(k-1)}$. For low input power values, the Gc_{f_0} factor (11) is negligible, and the output power $P_{f_0}^{out}$ at f_0 is a linear function of the input power. In this range of power, the PA is considered to be linear. The factor G is the gain of the amplifier expressed in dB. This gain, which does not depend on the input power, varies often in function of frequency in the operating bandwidth of the PA. For instance, the typical gain of the ZHL-100W-52 is equal to 50dB (± 1.2, typical variation) and figure 2, extracted from its data sheet, shows the gain variation in its operating bandwidth ($50 - 500$ MHz).

For higher input power levels, the summation S in (11) takes negative values between -1 and 0, $-1 < S < 0$. Consequently, the factor Gc_{f_0} takes negative values incurring a gain decrease w.r.t the gain G in the linear region. Hence, the PA operates in a nonlinear mode and the notation Gc_{f_0} is chosen to indicate the amount of gain compression in the nonlinear region. The amplifier reaches finally a region of saturation, where the output power does not increase anymore with the input power.

The AM-AM characteristic of the ZHL-100W-52, modeled by the 9th-order polynomial model, is presented in figure 3. As we can see the polynomial model is unable to model the saturation region. This is one of the limitations of polynomial models which are restricted to describe the behavior of weakly NL systems. Note that the fundamental frequency has been chosen in the operating bandwidth of the PA, $f_0 = 250$ MHz. But since the polynomial model does not take into account memory effects, the choice of the fundamental frequency does not change the model behavior. In order to study NL systems with memory effects, more complicated models are required, such as the Volterra series and neural networks (Schetzen, 2006; Morgan et al., 2006; Ibnkahla, 2000; Liu et al., 2004; Wood et al., 2004; Isaksson et al., 2005).

3.1.2 Compression and Interception Points

For an ideal PA, Gc_{f_0} in (9) is equal to zero (Fig. 3) on the whole operating power range. In order to quantify the gain compression phenomenon, the deviation of the AM/AM characteristic of the real PA from the ideal one is determined for different values of the input power (or output power) in the nonlinear operating region. The *compression points* at 1 and 3 dB are often determined and indicated in the data sheet of PAs. For instance, 1 and 3 dB compression points of our case study PA, the ZHL-100W-52, are specified in its data sheet by typical output power values of 47 and 48 respectively, as shown in Fig. 3.

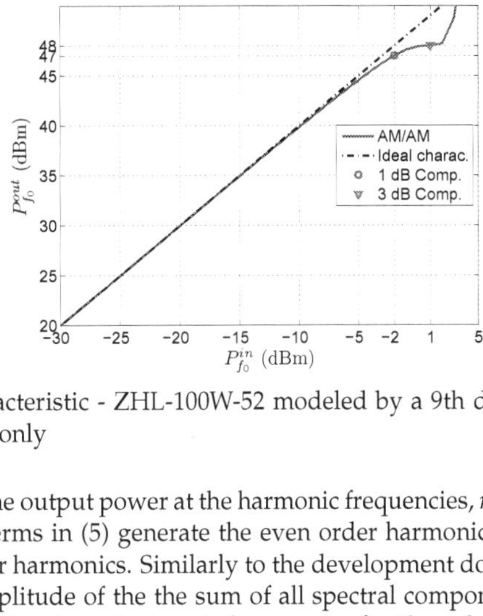

Fig. 3. AM/AM characteristic - ZHL-100W-52 modeled by a 9th degree polynomial model, with odd-order terms only

Now, let us consider the output power at the harmonic frequencies, nf_0 ($n > 1$). As mentioned before, even degree terms in (5) generate the even order harmonics, while odd order terms generate the odd order harmonics. Similarly to the development done in subsection 3.1.1, we can determine the amplitude of the the sum of all spectral components generated by all the terms of the power series at any harmonic frequency nf_0. The value of $Y_a(f)$ (5) for positive even order harmonics $+2mf_0$, $m = 1, 2, \ldots$, can be written as

$$Y_a(+2mf_0) = \sum_{k=m}^{K} a_{2k} \left(\frac{A}{2}\right)^{2k} \binom{2k}{k-m} \tag{12}$$

and for odd order harmonics, $Y_a(+(2m-1)f_0)$ can be written as

$$Y_a(+(2m-1)f_0) = \sum_{k=m}^{K} a_{2k-1} \left(\frac{A}{2}\right)^{2k-1} \binom{2k-1}{k-m}. \tag{13}$$

Then, to illustrate, the output power at any odd-order harmonic frequency $(2m-1)f_0$ ($m = 1, 2, \ldots$) can be written in a form similar to (9)

$$P^{out}_{(2m-1)f_0} = (2m-1)P^{in}_{f_0} + G_{h_{2m-1}} + Gc_{h_{2m-1}} \tag{14}$$

where

$$G_{h_{2m-1}} = 10\log_{10}(a^2_{2m-1}) - 32(m-1) \tag{15}$$

and

$$Cg_{h_{2m-1}} = 10\log_{10}(1+S)^2. \tag{16}$$

and $S = \sum_{k=m+1}^{K} \frac{a_{2k-1}}{a_{2m-1}} C^{k-m}_{2k-1} (A/2)^{2(k-m)}$. Note that, in the particular case where $m = 1$, Eq. (14) is equivalent to Eq. (9). And as for Eq. (9), the factor $Gc_{h_{2m-1}}$ is negligible for low input power levels. In this region, the output power $P^{out}_{(2m-1)f_0}$ is a linear function of the input power $P^{in}_{f_0}$. Furthermore, according to (14), we observe that the output power $P^{out}_{(2m-1)f_0}$ increases $2m-1$ times more rapidly than the output power $P^{out}_{f_0}$ at the fundamental frequency in function of the input power $P^{in}_{f_0}$. At first glance, one could expect a point for which the fundamental component and the $(2m-1)^{th}$ harmonic have the same output power, for a given input level. However, the gain compression phenomenon on the fundamental component (9) as well as on the harmonic $(2m-1)f_0$ (14), occurs before reaching this point. This point is called the $(2m-1)^{th}$ *interception point* (Kenington, 2000), and is determined either at the input and, in this case, we will denote it by IP^{in}_{2m-1}, or at the output and it will be denoted IP^{out}_{2m-1}. In practice, this level can never be reached, because of the gain compression of the PA. Note that an equation equivalent to (14) can be determined for even order harmonics by repeating a similar development starting from Eq. (12). In almost all PA data sheets, only third order interception point is given. We will see below that another, more commonly used definition for the interception point can be determined from the two-tone test.

3.2 Two-Tone Signal

When the input signal is composed of more than one frequency component, that is a multi-tone signal, a new type of distortion will occur. In this case, a new set of spectral components will be generated, in addition to those generated by the nonlinearity on each component considered separately. Such components are called the mixing products or more commonly *intermodulation products*. Power series analysis helps in this case in understanding the nonlinear effects of PAs on pass-band communications signals used in real applications. One of the simplest multi-tone tests is the two-tone test, which consists of driving the PA by a two-tone excitation signal, the sum of two sinusoids

$$x(t) = A_1\cos(2\pi f_1 t + \varphi_1) + A_2\cos(2\pi f_2 t + \varphi_2). \tag{17}$$

As in section 3.1, the analysis in this paragraph is established in the frequency domain. For simplicity, we assume that the two sinusoids are of equal amplitude ($A_1 = A_2 = A$) and null phases ($\varphi_1 = \varphi_2 = 0$). The Fourier transform of (17) may be written as

$$X(f) = X_{f_1}(f) + X_{f_2}(f) \tag{18}$$

where $X_{f_1}(f)$ and $X_{f_2}(f)$ are the Fourier transforms of the first and second sinusoids in (17), respectively. Substituting (18) in (2), the Fourier transform of the output signal may be written as

$$Y_a(f) = \sum_{k=1}^{K_a} a_k \sum_{i=0}^{k} \binom{k}{i} \coprod_{*}^{k-i} X_{f_1}(f) * \coprod_{*}^{i} X_{f_2}(f). \tag{19}$$

In this summation, for $i = 0$ and $i = k$ (for all values of k) we obtain the responses of the PA for one-tone excitations at frequencies f_1 and f_2, respectively. In order to compute the term $\coprod_*^{k-i} X_{f_1}(f) * \coprod_*^{i} X_{f_2}(f)$, the following relationship can be used

$$\coprod_{*}^{n_1} X_{f_1}(f) * \coprod_{*}^{n_2} X_{f_2}(f) = \left(\frac{A}{2}\right)^{n_1+n_2} \sum_{i_1=0}^{n_1} \sum_{i_2=0}^{n_2} \binom{n_1}{i_1} \binom{n_2}{i_2} \delta\left(f + (n_1 - 2i_1)f_1 + (n_2 - 2i_2)f_2\right) \tag{20}$$

where $\coprod_*^0 X_{f_i}(f) = \delta(f)$ and $n_1, n_2 \in N$. Hence, equation (19) can be rewritten in the form

$$\begin{aligned} Y_a(f) &= Y_{f_1}(f) + Y_{f_2}(f) + \sum_{k=2}^{K_a} a_k \left(\frac{A}{2}\right)^k \sum_{i=1}^{k-1} \sum_{i_1=0}^{k-i} \sum_{i_2=0}^{i} \binom{k}{i} \binom{k-i}{i_1} \binom{i}{i_2} \\ &\qquad \delta(f + (k - i - 2i_1)f_1 + (i - 2i_2)f_2) \end{aligned} \tag{21}$$

where $Y_{f_1}(f)$ ($Y_{f_2}(f)$, respectively) is the Fourier transform of the output signal of the PA when excited by a one-tone signal at the frequency f_1 (f_2, respectively), that is components at f_1, f_2, and their harmonics. However, the third term on the right hand side of Eq. (21) contains a new set of components in addition to those generated by one-tone excitation. These components are the Intermodulation Products (IMPs), and are located around f_1, f_2 and their harmonics. To illustrate, Fig. 4 shows the output spectrum of the ZHL-100W-52 PA, modeled by an odd order power series, and excited by a two-tone signal. We observe that IMPs appear only around odd order harmonics. In fact, even order harmonics and IMPs around them do not appear because even order terms have been neglected in the adopted polynomial model. This choice does not affect our analysis because, in communication systems, we are generally concerned with IMPs around the fundamental frequencies, as they cannot be easily rejected by filters. Even order IMPs usually occur at frequencies well above or below the fundamentals, and thus are often of little concern.

The two-tone signal is in fact constituted by two amplitude modulated quadrature carriers. Eq.(17) may be written in the form

$$\begin{aligned} x(t) &= \{A_1 \cos(2\pi f_m t - \varphi_1) + A_2 \cos(2\pi f_m t + \varphi_2)\} \cos(2\pi f_c t) \\ &\quad - \{A_1 \sin(2\pi f_m t - \varphi_1) - A_2 \sin(2\pi f_m t + \varphi_2)\} \sin(2\pi f_c t) \end{aligned} \tag{22}$$

where $f_m = \frac{1}{2}(f_2 - f_1)$ ($f_2 > f_1$) and the *carrier frequency* $f_c = \frac{1}{2}(f_1 + f_2)$. In the particular case where $A_1 = A_2 = A$ and $\varphi_1 = \varphi_2 = 0$, Equation (22) is written

$$x(t) = 2A \cos(2\pi f_m t) \cos(2\pi f_c t). \tag{23}$$

The two-tone signal used is, hence, an amplitude modulated carrier.

The components in (21), $\delta(f + (k - i - 2i_1)f_1 + (i - 2i_2)f_2)$, can be expressed in function of f_c and f_m, $\delta(f + (k - 2(i_1 + i_2))f_c - (k - 2i - 2i_1 + 2i_2)f_m)$. So, as mentioned before, only odd-degree terms generate components in the vicinity of f_c (i.e. the vicinity of f_1 and f_2), since the

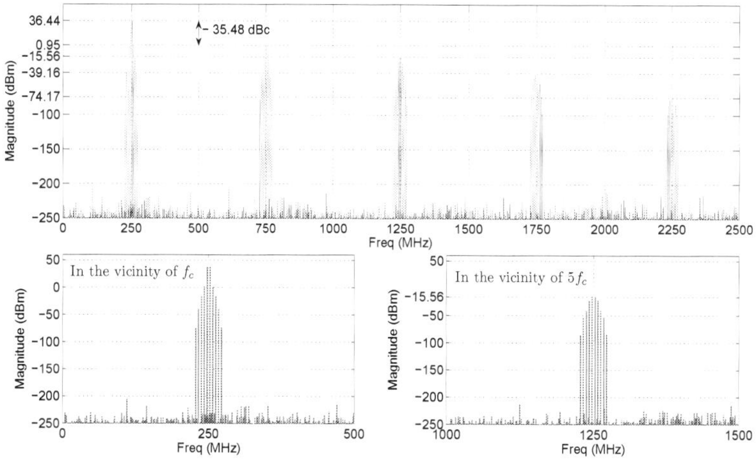

Fig. 4. Output spectrum of the ZHL-100W-52 PA, modeled by a 9th-order polynomial model with odd degree terms only, and excited by a two-tone signal ($P_{avg}^{in} = -7\,\text{dBm}$)

condition $k - 2(i_1 + i_2) = \pm 1$ must be verified. The set of components in the vicinity of $+f_c$ can be extracted from (21) to form the following equation

$$Y_a(f)\Big|_{+f_c} = \sum_{k=1}^{K} a_{2k-1} \left(\frac{A}{2}\right)^{2k-1} \sum_{i=0}^{2k-1} \sum_{\substack{i_1=k-i \\ 2k-i-1 \geq i_1 \geq 0}}^{k} \binom{2k-1}{i} \binom{2k-i-1}{i_1} \binom{i}{k-i_1}$$
$$\delta(f - f_c - (4k - 2i - 4i_1 - 1)f_m). \tag{24}$$

Note that the components at frequencies $f_1 = f_c - f_m$ and $f_2 = f_c + f_m$ are included in (24). Similarly to the one tone test, we are interested here in analyzing the nonlinearity on all the components close to f_c when sweeping the input power level on a finite range of power. In the following, we discuss the nonlinearity on the fundamental components and provide a new definition of interception points.

3.2.1 AM/AM Characteristic

In a two-tone test the AM/AM characteristic is often not determined, and no related parameters are given in the data sheet of the power amplifier. However, it is often advantageous to determine the amount of nonlinearity on the two fundamental components w.r.t the nonlinearity on the unique fundamental component in the one-tone test. Therefore, the AM/AM characteristic of the two-tone test will be determined by measuring the output power at the two fundamentals, f_1 and f_2, for different values of the average input power (i.e., the input power at the two fundamental components). Every term in (24), generates one or more frequency components at $+f_1$ and $+f_2$. To illustrate, let us take the sum of all frequency

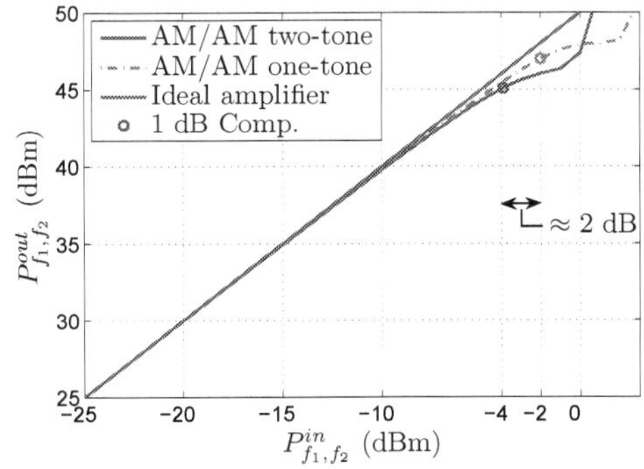

Fig. 5. AM-AM characteristics: one- and two-tone excitation signals

components at $+f_2 = f_c + f_m$

$$Y_a(+f_2) \quad = \quad \sum_{k=1}^{K} a_{2k-1} \left(\frac{A}{2}\right)^{2k-1} \sum_{i=1}^{k} \binom{2k-1}{2i-1} \binom{2(k-i)}{k-i} \binom{2i-1}{i-1}. \tag{25}$$

The output power, $P^{out}_{f_1,f_2}$, at the two frequencies f_1 and f_2, can be determined from (25), if we exploit the symmetry of the spectrum around the origin $f = 0$ and the symmetry of IMPs close to f_1 and f_2 around the carrier frequency f_c

$$P^{out}_{f_1,f_2} = 10\log_{10} \frac{4\,|Y_a(+f_2)|^2}{R10^{-3}} \tag{26}$$

Similarly to (9), this power can be expressed in the form

$$P^{out}_{f_1,f_2} = P^{in}_{f_1,f_2} + G + Gc_{f_1,f_2} \tag{27}$$

where

$$Gc_{f_1,f_2} = 10\log_{10}\left(1+S\right)^2, \tag{28}$$

$S = \sum_{k=2}^{K} \frac{a_{2k-1}}{a_1} (A/2)^{2(k-1)} \sum_{i=1}^{k} C^{2i-1}_{2k-1} C^{k-i}_{2(k-i)} C^{i-1}_{2i-1}$ and G is the gain of the PA (10). The input power at f_1 and f_2, $P^{in}_{f_1,f_2}$ is equal to $10\log_{10}\left(\frac{4|X(+f_2)|^2}{R10^{-3}}\right)$, which is equal to the average input power P^{in}_{avg}. For values of i equal to k in the summation of equation (28), we get the summation S of (11). Denoting it by S_1, Eq.(28) can be expressed in the form

$$Gc_{f_1,f_2} = 10\log_{10}(1+S_1+S_2)^2 \tag{29}$$

where $S_2 = \sum_{k=2}^{K} \frac{a_{2k-1}}{a_1} (A/2)^{2(k-1)} \sum_{i=1}^{k-1} C^{2i-1}_{2k-1} C^{k-i}_{2(k-i)} C^{i-1}_{2i-1}$.

According to Eq. (29), the gain compression on the two fundamental components of the two-tone signal is faster than the gain compression on the unique component of the one-tone signal, as shown in figure 5. The compression region in this case is shifted down and consequently the 1dB compression point is backed off.

In order to interpret this phenomenon, a number of parameters are needed. As already mentioned, a two-tone signal is a varying envelope signal (23). Two parameters are often used to characterize this type of signals (Cripps, 2006):

- The *PEP* (Peak Envelope Power), P_{peak}, often measured in dBm and defined by

$$P_{peak} = 10\log_{10}\frac{(A_{peak}/\sqrt{2})^2}{R10^{-3}}(\text{dBm})$$

  where A_{peak} is the peak amplitude of the signal (equal to $2A$ in the case of a two-tone signal).

- The *PAPR* (Peak-to-Average Power Ratio), which is the ratio between the PEP and the average power of the signal. It is therefore the difference in dB between the PEP and the average power P_{avg} expressed both in dBm, $\text{PAPR} = P_{peak} - P_{avg}(\text{dB})$

In the case of a two-tone signal, a deterministic signal, the PAPR is equal to 3 dB ($P_{peak} = P_{avg} + 10\log_{10}2$).
Other parameters are also used in order to locate the input/output signal w.r.t the AM/AM characteristic of the PA[2].

- The IPBO (Input Peak Backoff), which is the difference in dB between the input saturation power of the PA P^{in}_{sat} and the peak power P^{in}_{peak} of the input signal, $\text{IPBO} = P^{in}_{sat} - P^{in}_{peak}(\text{dB})$. Similarly, the Output Peak Backoff (OPBO) is given by $\text{OPBO} = P^{out}_{sat} - P^{out}_{peak}$, where P^{out}_{sat} is the output saturation power of the PA and P^{out}_{peak} is the peak power of the output signal.
- The IBO (Input Backoff), which is the difference in dB between P^{in}_{sat} and the average input power P^{in}_{avg}, $\text{IBO} = P^{in}_{sat} - P^{in}_{avg}(\text{dB})$. Simlarly, the Output Backoff (OBO) is given by $\text{OBO} = P^{out}_{sat} - P^{out}_{avg}$.

To this we add a last definition, which is the instantaneous power of a signal $x(t)$, $P_{inst}(t)$ expressed in dBm

$$P_{inst}(t) = 10\log_{10}\frac{\left(x(t)/\sqrt{2}\right)^2}{R10^{-3}} \tag{30}$$

Let us consider, now, the particular case when the gain compression on the fundamental components of the two-tone signal is equal to 1 dB. Referring to Fig. 5, the average input power P^{in}_{avg} causing this amount of gain compression is approximately equal to -4 dBm. For this same P^{in}_{avg} value, we observe that the compression on the fundamental component of the one-tone signal is less than 1 dB. Thus, we can conclude that the gain compression is related to the instantaneous input power rather than the average input power.
If we interpret this by considering a discrete representation of the signals, we can conclude that some samples of the two-tone signal can undergo gain compression bigger than 1 dB since its P^{in}_{peak} exceeds the 1 dB compression point. On the other hand, the samples of the one-tone signal, which is a non-varying envelope signal, take always values less than its average power, which is equal to its P^{in}_{peak}. This interpretation is better illustrated in Fig. 6, where the

[2] Note that when we talk about the AM/AM characteristic without specifying the signal used to extract it, we designate the one corresponding of a one-tone test.

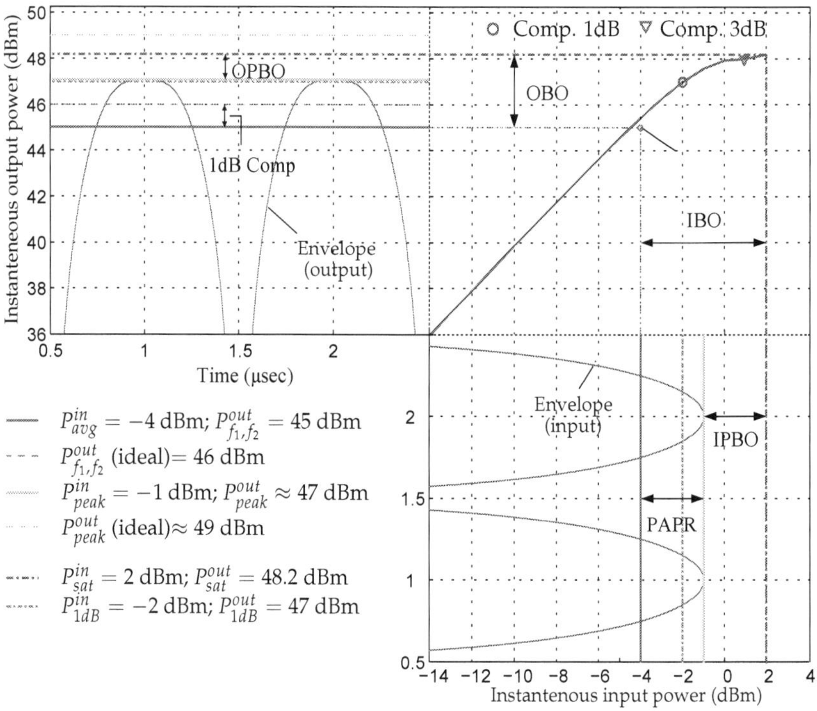

Fig. 6. Input/output envelope variation of a two tone signal with respect to AM/AM characteristic

instantaneous power of the input/output two-tone envelope signals are presented in comparison with the AM/AM characteristic. All the parameters already defined are also shown on this figure (Fig. 6), for this particular case. Therefore, we conclude that the envelope variation of the two-tone signal makes it more vulnerable to nonlinearity.

Besides, it is important to note that for varying envelope signals, it is very difficult to predict or to quantify the amount of gain compression at the output of the PA (the OBO and OPBO values), even in the case of deterministic signals like the two-tone signal. This is due to the fact that various parameters and nonlinear relations are involved.

In general, reducing the nonlinear distortion is at the expense of power efficiency. In fact, the power efficiency of the PA is likely to increase when its operating point approaches the saturation. But, for all the reasons already mentioned, and in order to preserve the form of the envelope of the input signal, large power backoffs are classically imposed. For an ideal amplification, the IPBO must be sufficiently important to prevent the envelope from penetrating the compression region. Doing so the power efficiency of the PA decreases considerably, especially if the input signal has a high PAPR value. In the second part of this chapter, we will present an important technique used to resolve this problem.

3.2.2 Interception Points

In the one-tone test, the nth interception point was determined as the input (output) power level for which the fundamental frequency component and the nth harmonic have the same output power level. In a two-tone test we are interested to the power levels of the odd-order IMPs close to the fundamental frequencies (f_1 and f_2). Due to the symmetry of these IMPs around f_c[3], we will consider below the products at the right side of $+f_c$ only, that is at the frequencies $f_c + (2m-1)f_m$ where $m = 1, 2, \ldots$. Every term in the summation (24) having an odd-degree equal or greater than $2m-1$, generates a spectral component at the frequency $f_c + (2m-1)f_m$. Depending on the value of m, the sum of all the components can be expressed by one of the two following forms. Equation (31) corresponds to even values, while equation (32) corresponds to odd values

$$Y_a(f_c + (2m-1)f_m) = \sum_{k=m}^{K} a_{2k-1}\left(\frac{A}{2}\right)^{2k-1}\sum_{i=\frac{m}{2}}^{k-\frac{m}{2}}\binom{2k-1}{2i}\binom{2k-1-2i}{k-i+\frac{m}{2}-1}\binom{2i}{i-\frac{m}{2}} \quad (31)$$

$$Y_a(f_c + (2m-1)f_m) = \sum_{k=m}^{K} a_{2k-1}\left(\frac{A}{2}\right)^{2k-1}\sum_{i=\frac{m+1}{2}}^{k-\frac{m-1}{2}}\binom{2k-1}{2i-1}\binom{2(k-i)}{k-i+\frac{m-1}{2}}\binom{2i-1}{i-\frac{m+1}{2}}. \quad (32)$$

Note that for $m = 1$, Eq. (32) is equivalent to Eq. (25). The output power of any of those IMPs can be determined from equations (31) and (32). For even values of m, the power of the $(2m-1)th$ IMP, may be expressed (similarly to (14)) in function of the average input power $P_{avg}^{in} = P_{f_1,f_2}^{in}$, or in function of the power of one of the two fundamental components, $P_{f_1}^{in}$ or $P_{f_2}^{in}$. The latter relation is the more commonly used and we will adopted hereafter,

$$P_{IMP_{2m-1}}^{out} = (2m-1)P_{f_1}^{in} + G_{IMP_{2m-1}} + Gc_{IMP_{2m-1}} \quad (33)$$

where

$$G_{IMP_{2m-1}} = 10\log_{10}(C_{2m-1}^{m}a_{2m-1})^2 - 32(m-1), \quad (34)$$

$$Gc_{IMP_{2m-1}} = 10\log_{10}(1+S)^2 \quad (35)$$

and $S = \sum_{k=m+1}^{K}\frac{a_{2k-1}}{a_{2m-1}}\left(\frac{A}{2}\right)^{2(k-m)}\sum_{i=\frac{m}{2}}^{k-\frac{m}{2}} C_{2k-1}^{2i} C_{2k-1-2i}^{k-i+\frac{m}{2}-1} C_{2i}^{i-\frac{m}{2}}$. For odd values of m, only the summation in equation (35) will change, and becomes in this case, $S = \sum_{k=m+1}^{K}\frac{a_{2k-1}}{a_{2m-1}}\left(\frac{A}{2}\right)^{2(k-m)}\sum_{i=\frac{m+1}{2}}^{k-\frac{m-1}{2}} C_{2k-1}^{2i-1} C_{2(k-i)}^{k-i+\frac{m-1}{2}} C_{2i-1}^{i-\frac{m+1}{2}}$. Note that in (33), $P_{f_1}^{in}$ is equal to $P_{avg}^{in} - 3\,\text{dB}$.

Hence, the IMPs, like the fundamental components, undergo gain compression. For low input power levels, the gain compression is negligible. In this range of power, the output power of the $(2m-1)^{th}$ product, $P_{IMP_{2m-1}}^{out}$, $m = 1, 2, \ldots$, increase linearly in function of $P_{f_1}^{in}$ (33). However, this evolution is $2m-1$ times faster than the power at one of the fundamental frequencies, if taken separately (one-tone test, Eq. (9)). Therefore, if the gain compression phenomenon does not occur, one could expect an input (output) power level, for which, the $(2m-1)^{th}$ IMP and one of the fundamental components will have the same power level. In this case, The input (output) power level is called the $(2m-1)^{th}$ input (output) interception

[3] The symmetry of IMPs around f_c is related to the memoryless assumption.

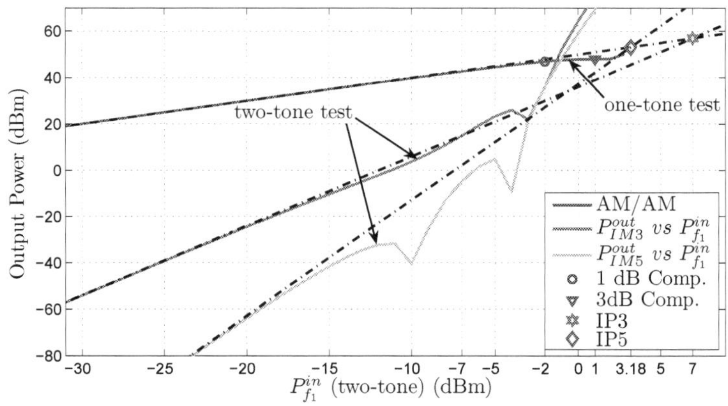

Fig. 7. 1 and 3dB compression points, and third and fifth interception points

point, denoted IP_{2m-1}^{in} (IP_{2m-1}^{out}). This point is thus the interception point of the linear extrapolations of power evolution curves (9) and (33). This new definition of the interception point is often preferred to the first one (Section 3.1), since it gives an indication on the amount of spectral regrowth in real applications, when the PA is excited by a band-pass signal (Sec. 3.3). Figure 7 shows the 3rd and 5th interception points, as well as the AM-AM characteristic, and its corresponding compression points for our case-study PA, the ZHL-52-100W. Here, we observe that the power series model is able to describe the nonlinearity on IMPs, only over a limited power range.

3.2.3 Model Identification

Some parameters presented in the preceding sections appear in almost every RF PA data sheet. They are adopted to give a first indication on the nonlinearity of the PA. In this section, we present how such parameters could be used to determine the coefficients a_k of the ZHL-100W-52 $9th$-order polynomial model, adopted all along our simulations. As mentioned before, even-order terms are neglected, and thus only odd-order coefficients will be identified. Recall that the identified model takes into account nonlinear amplitude distortion only.
The first coefficient of the power series a_1 can be determined simply from the gain of the PA (10), $a_1 = 10^{G/20}$. For the ZHL-100W-52, the gain is equal to 50 dB, and, hence $a_1 = 316.23$.
On the other hand, referring to equations (9), (33) and (34), we can write

$$IP_{2m-1}^{in} + G = (2m-1)IP_{2m-1}^{in} + 10\log_{10}((C_{2m-1}^{m}a_{2m-1})^2) - 32(m-1). \quad (36)$$

Thus, if the $(2m-1)^{th}$ input interception point is known, IP_{2m-1}^{in}, we can determine the coefficient of the same order, a_{2m-1} of the power series

$$a_{2m-1} = \frac{10^{\frac{32(m-1)+G-2(m-1)IP_{2m-1}^{in}}{20}}}{C_{2m-1}^{m}}. \quad (37)$$

In our case-study PA, the third interception point at the output is specified, $IP_3^{out} = 57\,\text{dBm}$, allowing thus to determine the third coefficient a_3. Note that, given IP_{2m-1}^{out}, IP_{2m-1}^{in} could be

simply obtained by setting Gc_{f_0} to zero in Eq. (9), $IP^{out}_{2m-1} = IP^{in}_{2m-1} + G$, since the interception point is always on the ideal PA characteristic. Thus, a_3 is equal to -837.3.
Moreover, based on Eq. (36), we could express a_{2m-1} in function of the amplitude corresponding to the $(2m-1)^{th}$ interception point at the input, $A_{IP_{2m-1}}$, and the coefficient a_1

$$a_{2m-1} = \frac{a_1}{C^m_{2m-1}(A_{IP_{2m-1}}/2)^{2(m-1)}}. \tag{38}$$

In traditional power series analysis, the order of the used model is usually limited to 3. Thus, we can determine a relation between A_{1dB} and A_{IP3}, which correspond to the 1 dB compression point (one-tone test), and the third interception point (two-tone test), respectively. Here, if we set Gc_{f_0} in 11, we find the following relation

$$A^2_{1dB} = \frac{4a_1(10^{-1/20} - 1)}{3a_3}. \tag{39}$$

Note that, the coefficient a_3 should have, in this case, a negative value in order to model the gain compression of the PA. Now, substituting a_3 from (38), in Eq. (39), we obtain a new relation between A_{1dB} and A_{IP3}

$$\left(\frac{A_{1dB}}{A_{IP3}}\right)^2 = 1 - 10^{-1/20}. \tag{40}$$

Eq. (40) and Eq. (38) for $m = 2$, could be found in almost all classical studies on modeling the PA via power series (e.g. chap. 9, (Cripps, 2006)).
Now, thanks to Eq. (11) obtained from our development, every point of the AM/AM characteristic can be used to determine a new higher-order coefficient. For example, two compression points are given in the data sheet of the ZHL-100W-52 PA, and a third point near saturation could be deduced since the maximum input power is specified (Maximum Input power no damage) and is equal to 3 dBm. Hence, we can choose for example an additional point at 2 dBm input power with a compression greater than 3 dB, let us say 3.8 dB and we suppose that this point is the saturation of the PA. We have added this point to reinforce the modeling capacity of the power series model by extending its power range validity (in other words, delaying its divergence point) and to define a saturation point useful for the development in this chapter (Fig. (6)). Finally, setting Gc_{f_0} to -1, -3, and -3.8 successively, a linear system of three equations with three unknowns can be established. In a matrix notation, this system can be written

$$\mathbf{C}\boldsymbol{x} = \boldsymbol{b} \tag{41}$$

where $\mathbf{C}^T = [\boldsymbol{v}_{1dB}\ \boldsymbol{v}_{3dB}\ \boldsymbol{v}_{3.8dB}]$, $\boldsymbol{v}^T_{pdB} = \left[C^3_5(A_{pdB}/2)^4\ C^4_7(A_{pdB}/2)^6\ C^5_9(A_{pdB}/2)^8\right]$, $\boldsymbol{x}^T = [a_5\ a_7\ a_9]$ et $\boldsymbol{b}^T = [u_{1dB}\ u_{3dB}\ u_{3.8dB}]$ where $u_{pdB} = a_1\left(10^{-p/20} - 1\right) - a_3C^2_3(A_{pdB}/2)^2$, $(\cdot)^T$ being the transpose operator. The solution of this linear system (41) can be simply written

$$\boldsymbol{x} = \mathbf{C}^{-1}\boldsymbol{b}. \tag{42}$$

If the matrix $\mathbf{C}$ is not singular, we can find an exact unique solution of (42), and otherwise the least square method can be used. The founded values of a_5, a_7 et a_9 for the ZHL-100W-52, are respectively 11525.2, -224770 and 952803.3. Note that, we can exploit all the power points of the measured AM-AM characteristic to identify more coefficients and to improve the least square method accuracy.

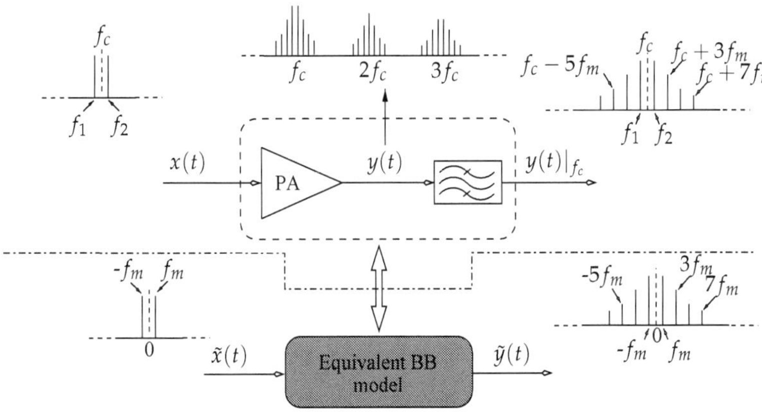

Fig. 8. Equivalent baseband modeling of the PA

3.3 Band-pass signals and baseband equivalent Modeling

So far, we have discussed the nonlinearity on one- and two-tone signals. However, in real modern communications systems, more complex signals are used to transmit digital information by some type of carrier modulation. Besides, due to bandwidth constraints, narrowband *band-pass* signals are generated in most applications. Signals are termed narrowband band-pass signals or, shortly, band-pass signals, when they satisfy the condition that their bandwidth is much smaller than the carrier frequency. Such a signal can be expressed by

$$\begin{aligned} x(t) &= \Re\left\{\tilde{x}(t)e^{j2\pi f_c t}\right\} \\ &= \frac{1}{2}\left(\tilde{x}(t)e^{j2\pi f_c t} + \tilde{x}^*(t)e^{-j2\pi f_c t}\right) \end{aligned} \tag{43}$$

where f_c is the carrier frequency and $\tilde{x}(t)$ is the complex envelope of the signal or the *baseband* signal. Substituting (43) into Eq. (1), and using the binomial theorem, the output signal of the PA modeled by a power series model can be written

$$\begin{aligned} y_a(t) &= \sum_{k=1}^{K_a} a_k \frac{1}{2^k}\left(\tilde{x}(t)e^{j2\pi f_c t} + \tilde{x}^*(t)e^{-j2\pi f_c t}\right)^k \\ &= \sum_{k=1}^{K_a} a_k \frac{1}{2^k} \sum_{i=0}^{k} \binom{k}{i} \tilde{x}^i(t)\tilde{x}^{*(k-i)}(t)e^{j2\pi(2i-k)f_c t}. \end{aligned} \tag{44}$$

In the above equation, only odd-degree terms generate frequency components close to f_c, since the condition $i = (k \pm 1)/2$ must be verified. The sum of all these components, denoted $y_a(t)\Big|_{f_c}$, can be extracted from (44) to form the following equation

$$\begin{aligned} y_a(t)\Big|_{f_c} &= \sum_{k=1}^{K} a_{2k-1} \frac{1}{2^{2k-1}} C_{2k-1}^{k} \left|\tilde{x}(t)\right|^{2(k-1)} \left(\tilde{x}^*(t)e^{-j2\pi f_c t} + \tilde{x}(t)e^{j2\pi f_c t}\right) \\ &= \sum_{k=1}^{K} a_{2k-1} \frac{1}{2^{2(k-1)}} C_{2k-1}^{k} \left|\tilde{x}(t)\right|^{2(k-1)} x(t). \end{aligned} \tag{45}$$

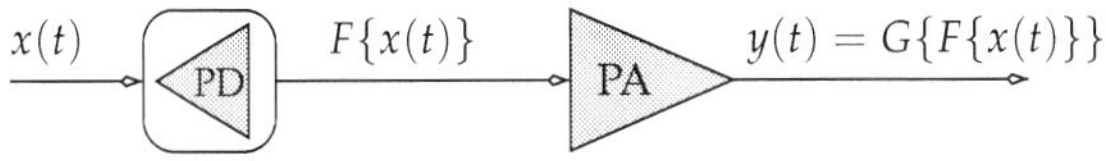

Fig. 9. Predistortion technique

Since we are interested only by the frequency content near f_c, this result (45) suggests that it is sufficient to study the nonlinearity of the PA on the complex envelope of the input signal. Denoting by $\tilde{y}(t)$ the complex envelope of the output signal (44) filtered by a band-pass filter centered on f_c, Eq. (45) can be written

$$\tilde{y}(t) = \sum_{k=1}^{K} a'_{2k-1} \left| \tilde{x}(t) \right|^{2(k-1)} \tilde{x}(t) \tag{46}$$

where $a'_{2k-1} = a_{2k-1} \frac{1}{2^{2(k-1)}} C^k_{2k-1}$ (Benedetto & Biglieri, 1999). Eq. (46) constitutes a baseband equivalent model of the RF power series model (1) used before. The baseband model is in fact valid in all cases where band-pass signals are used. It is particularly interesting for digital simulators since baseband signals require relatively low sampling rate w.r.t the carrier frequency. In addition to its capacity of representing simply power amplifiers, this model is often used in baseband predistortion techniques, when the PA does not represent strong memory effects. To illustrate, the baseband signal of a two-tone signal, which can be considered as a band-pass signal, is a sinusoidal signal $\tilde{x}(t) = 2A\cos(2\pi f_m t)$ (Eq. (23)), and the equivalent baseband system is illustrated in Fig. 8.

As mentioned before, a memoryless nonlinear system can induce amplitude distortion only, but never phase distortion. However, PAs with weak memory effects can be considered as quasi-memoryless systems (Bosch & Gatti, 1989), where nonlinear amplitude and phase distortion at instant t depend only on the amplitude of the input envelope signal at the same instant. Hence, the output complex envelope can be expressed in the general form

$$\begin{aligned} \tilde{y}(t) &= G\left(\left|\tilde{x}(t)\right|\right)\tilde{x}(t) \\ &= G_a\left(\left|\tilde{x}(t)\right|\right)\exp\left\{j\Phi\left(\left|\tilde{x}(t)\right|\right)\right\}\tilde{x}(t) \end{aligned} \tag{47}$$

where $G_a(\cdot)$ and $\Phi(\cdot)$ are nonlinear functions of the amplitude, $|\tilde{x}(t)|$, of the input complex envelope. The equivalent baseband power series model (46) can be used to describe the behavior of a quasi-memoryless system, and it is often called the quasi-memoryless polynomial (QMP) model. In this case, the coefficients a'_{2k-1} are complex valued, and from (46), the complex gain G of the PA (47) is equal to $\sum_{k=1}^{K} a'_{2k-1} \left|\tilde{x}(t)\right|^{2(k-1)}$. In this modeling approach, the nonlinear functions $G_a(\cdot)$ and $\Phi(\cdot)$, which are the module and the phase of the complex gain of the PA respectively, represent the AM/AM et AM/PM conversions of the PA.

In the next part of this chapter, a digitally modulated signal (a band-pass signal) is used while evaluating a linearization technique. Thus, we will observe the nonlinear effects that can be incurred by the PA on such a type of signals.

4. Adaptive digital baseband predistortion technique

Linearization techniques aim to linearize the behavior of the PA in its nonlinear region, or to extend the linear behavior over its operating power range. Generally speaking, this can be done by acting on the input and/or output signals without changing the internal design of the PA. Two linearization techniques were first applied to PAs, both invented by H. S. Black (Black, 1928; 1937): the Feedback (FB) and Feedforward (FF) techniques. Different implementation approaches have been proposed in the literature but the main idea behind these techniques is to generate a corrective signal by comparing the distorted output signal to the input signal, and to combine it either to the input (FB) or output signal (FF). The FB technique, as any feedback system, suffers from instability problems which limit its deployment to narrow-band applications. On the other hand, FF technique is inherently an open-loop process and, thus, it can be applied to wide-band applications but it has many disadvantages, mainly due to signals combination at the output of the PA. More recently, a new technique, called the predistortion technique, has been proposed and widely used. This technique consists in inserting a nonlinear circuit, the *predistorter* (PD), prior to the RF PA such that the combined transfer characteristic of both is linear (Fig. 9). Denoting by G and F the transfer characteristic of the PA and the PD respectively, the output signal $y(t)$ of the cascade of the two circuits, PA and PD, may be written

$$y(t) = G\{F\{x(t)\}\} = Kx(t) \tag{48}$$

where K is a positive constant representing the gain of the linearized PA, and $x(t)$ is the input signal. Different approaches, relying on analog, digital or hybrid circuits, could be employed while designing the PD. In the following, however, we will be interested in Adaptive Digital Predistortion (ADPD), which is a promising and cost-effective technique for SDR transmitters. Given the considerable processing power now available from Digital Signal Processing (DSP) devices, the digital implementation offers high precision and flexibility.

4.1 ADPD: An overview

The digital predistortion technique is basically relying on the equivalent baseband modeling of the PA and/or its inverse. For digital signal processing convenience, it is very desirable to implement the PD in baseband. To this end, we resort to an equivalent low-pass or baseband representation of the band-pass system. Thus, the cascade of the equivalent baseband behavioral model of the PD and the PA should form a global linear system, as shown in figure 10. Hereafter, F and G will represent the transfer characteristic at baseband of the PA and the PD respectively. To illustrate, we will assume that the PD and the PA are quasi-memoryless systems, and thus G and F are nonlinear functions of the amplitude of their input signals. Hence, the output $x_p(t)$ of the PD can be written in function of the input signal $x_i(t)$ as follows

$$x_p(t) = F(|x_i(t)|)x_i(t) \tag{49}$$

Accordingly, the output of the PA is written as

$$\begin{aligned} y(t) &= G(|x_p(t)|)x_p(t) \\ &= G(|F(|x_i(t)|)x_i(t)|)F(|x_i(t)|)x_i(t) \\ &= G_{lin}(|x_i(t)|)x_i(t) \end{aligned} \tag{50}$$

where $y(t)$ is the output baseband signal and $G_{lin}(.)$ the characteristic function of the linearized PA, LPA (i.e. cascade of the PA and the PD). In an ideal scenario, the module and phase of this function must be constant for the whole amplitude range up to saturation. Thus, according to Eq. (50), a linear behavior can be obtained if the following condition is fulfilled

$$|G(|x_p(t)|)F(|x_i(t)|)| = K \tag{51}$$

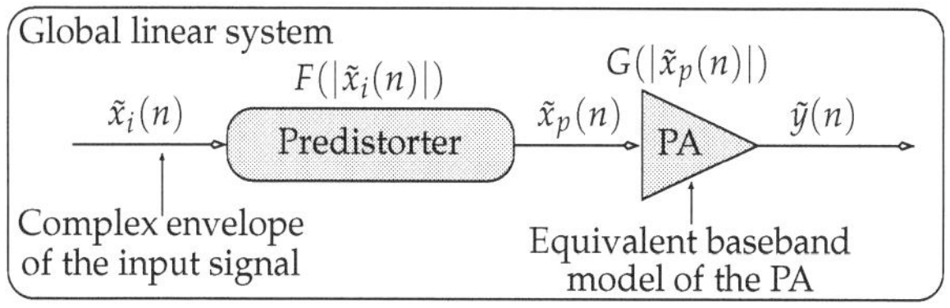

Fig. 10. Baseband predistortion

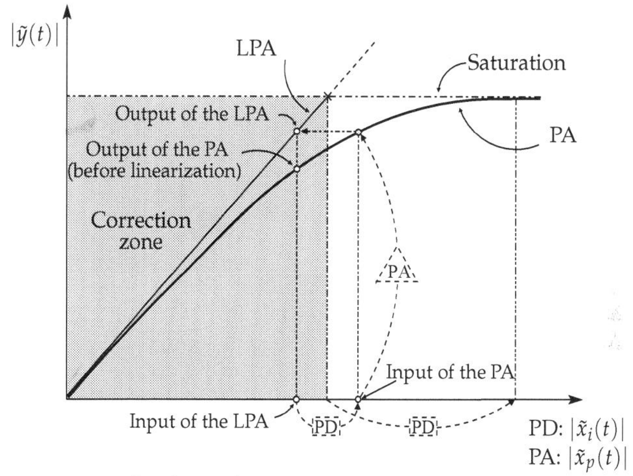

Fig. 11. Instanteneous amplitude predistortion

where K, a positive constant, is the global gain of the LPA. For further illustration, Fig. 11 shows the instantaneous predistortion mechanism in the simple case where the PA introduces amplitude distortion only. The insertion of the PD makes linear the amplitude response of the PA over a large amplitude range, covering part of the compression zone, before reaching saturation. A phase predistortion should be also performed since the phase distortion of the PA has considerable effects on the output signal.

There are different configurations of the digital baseband predistortion system. However, all these configurations have the same principle presented in Fig. 12. The transmitted RF signal at the output of the PA is converted to baseband, and its quadrature components are digitized by an analog to digital converter. The samples in baseband are then treated by a digital signal processor (DSP) with an algorithm that compares them to the corresponding samples of the reference input signal. The PD's parameters are identified while trying to minimize the error between the input and the output, or another appropriate cost function. After a short time of convergence which characterizes the identification algorithm, the PD could perform as the exact pre-inverse of the equivalent baseband model of the PA.
In modern SDR transmitters, most of the components must be reconfigurable in order to switch, ideally on the fly, from one standard to another. In such systems, the digital predistortion technique seems to be the unique applicable linearization technique. In this case,

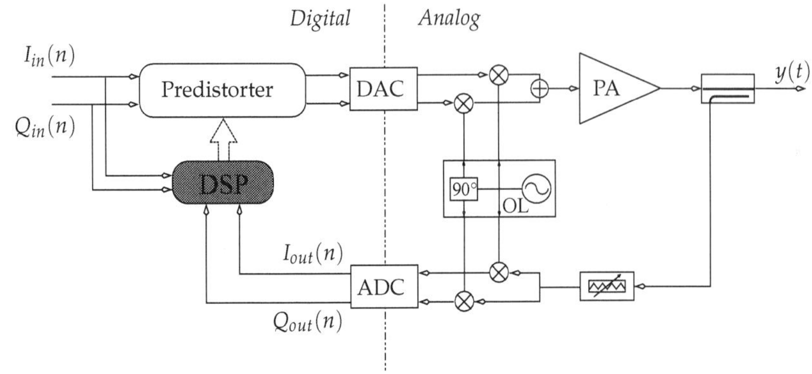

Fig. 12. Adaptive digital baseband predistortion

the PD must be updated on a continuous or quasi-continuous basis in order to keep a good linearization performance, and thus the lowest energy dissipation.

4.2 Performance evaluation

In this section, we first describe the test bench designed for our experiments. Then, we evaluate the performance of the digital baseband predistortion technique, using a medium power PA from Mini-Circuits, the ZFL-2500 , driven by 16-QAM modulated signal. To this end, we first identify a model of this PA from the input/output signals acquired using the test bench. This model is used in simulations to determine the best expected performance of the digital baseband predistortion technique, in the ideal scenario (without measurement noise). Second, we present the experimental results, and compare them to the theoretical ones. The performance of the PD has been evaluated by measuring two important parameters, the Adjacent Channel Power Ratio (ACPR) and Error Vector Magnitude (EVM), for different backoff values. However, due to space limitation, we will only present the results obtained for the ACPR parameter.

4.2.1 Test bench description

The measurement testbed consists of a vector signal generator (VSG) and a digital oscilloscope (DO) (Fig. 13). This testbed was designed to be fully automatic using the instrument toolbox of Matlab. The measurement technique concept consists in generating data in Matlab to send out to the VSG and then to read data into Matlab for analysis. The VSG (Rhode & Schwartz SMU 200A) receives the complex envelope data via an Ethernet cable (TCP/IP) from a personal computer (PC) and using a direct up-conversion from baseband to RF, produces virtually any signal within its bandwidth limits. Note that, once the data have been sent to the VSG, the latter will send the corresponding modulated signal repeatedly to the PA. A marker can be activated to trigger the DO every time the sequence is regenerated. The microwave input and output signals of the PA are then sampled simultaneously in the real time oscilloscope (LeCroy, 4 channels Wave master 8600, 6GHz bandwidth, 20 GS/sec), transferred via an Ethernet cable to the PC, and recorded in the workspace of Matlab. The acquisition time in the DO is fixed to be equal to the duration of the baseband signal generated by Matlab. In this way, the acquired RF signals correspond exactly to the original signal of Matlab. After

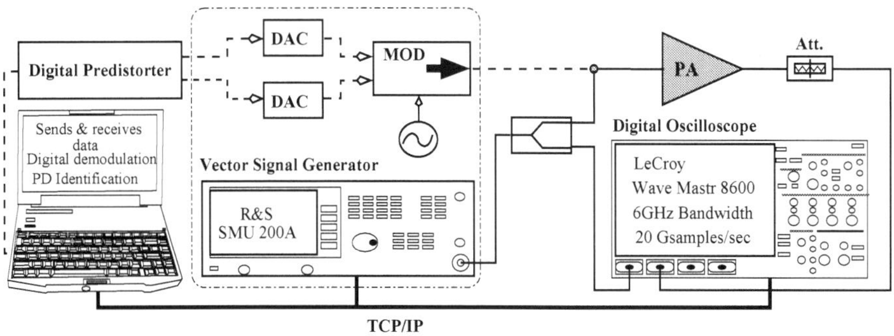

Fig. 13. Measurements setup

that, the two sequences are digitally demodulated in Matlab, adjusted using a subsampling synchronization algorithm (Isaksson et al., 2006), and processed in order to identify the parameters of the PD. The baseband signal is then processed by the predistortion function and loaded again to the VSG. Finally, the output of the linearized PA is digitized in the DO and sent back to the PC to evaluate the performance of the particular PD scheme. This evaluation can be done by comparing the output spectra (ACPR) and constellation distortion (EVM) of the PA with and without linearization, for different back-off values. The time of this entire test is several minutes since this test bench is fully automatic. In other words, the transmission and the signals acquisition, identification and performances evaluation can be implemented in a single program in Matlab which run without interruption. Note that, for signals acquisition, the spectrum analyzer "Agilent E4440A" has been also used as an alternative method for precision, comparison and verification. In this case, the signal analysis software provided with this device can be used to demodulate and acquire the input and output signals separately. The signals can then be synchronized by correlating them with the original signal of Matlab.

4.2.2 Experimental results

Measurements have been carried out on a PA from the market, the ZFL 2500 from Mini-circuits. This wide-band (500-2500 MHz) PA is used in several types of applications, typically in GPS and cellular base stations. According to its data sheet, it has a typical output power of 15 dBm at 1 dB gain compression, and a small signal gain of 28 dB (± 1.5). The modulation adopted through the measurements is 16 QAM. The pulse shaping filters are raised cosine filters with a roll-off factor of 0.35 extending 4 symbols on either side of the center tap and 20 times oversampled. The carrier frequency is 1.8 GHz and the bandwidth 4 MHz. In order to acquire a sufficient number of samples for an accurate PD identification, 5 sequences of 100 symbols ($2k$ samples) each, have been generated and sent to the VSG successively, i.e. a total number of $10k$ samples have been used for identification and evaluation.

Static power measurements

In order to validate the study presented in Sec. 3, we have performed the one- and two-tone tests on this PA. The defined parameters, namely, compression and interception points and the output saturation power, are also very useful for the experimental evaluation of the DPD technique. Figure 14 shows the AM/AM characteristic of the PA under test, its compression points and the corresponding power series model identified using the development presented in sec-

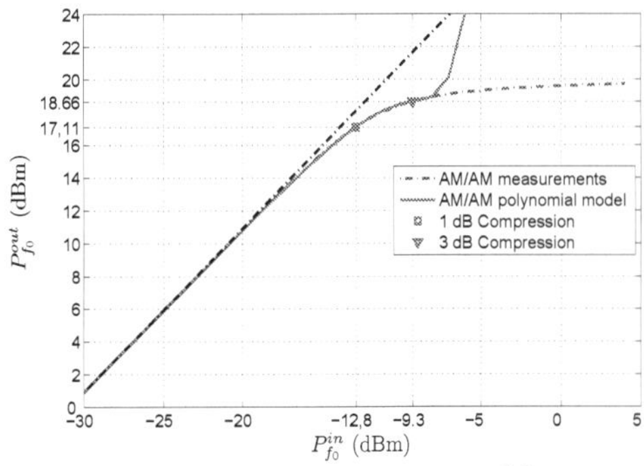

Fig. 14. RF polynomial model of the ZFL-2500 PA extracted from static power measurements (compression and interception points)

P_{1dB}^{out}	17.11 dBm
P_{3dB}^{out}	18.66 dBm
P_{sat}^{out}	19.73 dBm
IP_3^{out}	29.46 dBm

Table 1. Parameters from the static power measurements

tion 3. As we can see from this figure, the power series model fits well the measured AM/AM characteristic up to approximately the 4 dB compression point, after which it diverges. Table 1 shows the different parameters measured from the one- and two-tone tests at the carrier frequency of 1.8 GHz.

Nonlinearity on modulated signals

In the first part of this chapter, we have analyzed amplitude nonlinear distortion of PAs on special excitation signals, the one- and two-tone signals. We have observed that, in the case of a two-tone excitation, some frequency components, the intermodulation products (IMPs), appear very close to the fundamental frequencies and consequently cannot be rejected by filtering. If the number of tones increases in the excitation signal, approaching thus real communications bandpass signals (Sec. 3.3), the number of IMPs increases drastically. Here, a simple quantification of the nonlinearity at one IMP becomes no more sufficient to appropriately represent the real distortion incurred on such a signal. In fact, the IMPs fall inside or very close to, the bandwidth of the original signal, causing in band and out of band distortions.
Fig. 15 shows the input/output spectra of the ZFL-2500 PA, and the constellation of its output signal for an average output power equal to 16.52 dBm. As shown in Fig. 15a, the out of band distortion appears as spurious components in the frequency domain in the vicinity of the original signal bandwidth, which is often referred by *spectral regrowth*. In real communications, this out of band distortion may result in unacceptable levels of interference to other users, which is often quantified by the ACPR parameter. On the other hand, the in band distortion appears on the warped constellation of the output signal, as shown in Fig. 15b, where the constellation points are no more located on a rectangular grid. This may increase the bit error rate (BER) in the system, and is measured by the EVM parameter.

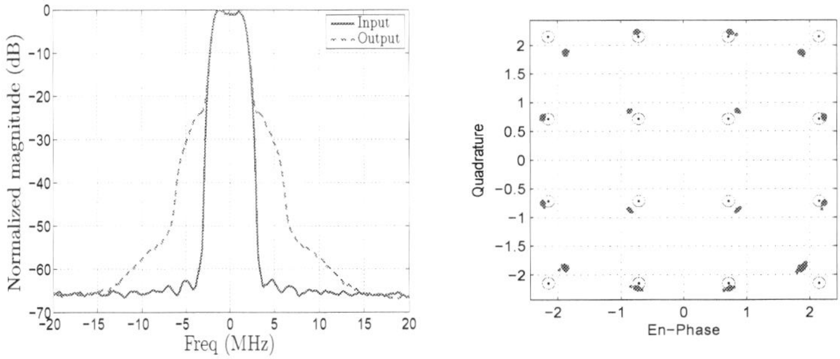

(a) Input and output spectra of the ZFL-2500 PA, ACPR ≈ -30 dB

(b) Constellation at the output of the ZFL-2500 PA, EVM $\approx 10.32\%$

Fig. 15. Nonlinear distortion on modulated signals, $P_{avg}^{out} \approx 16.52$ dBm

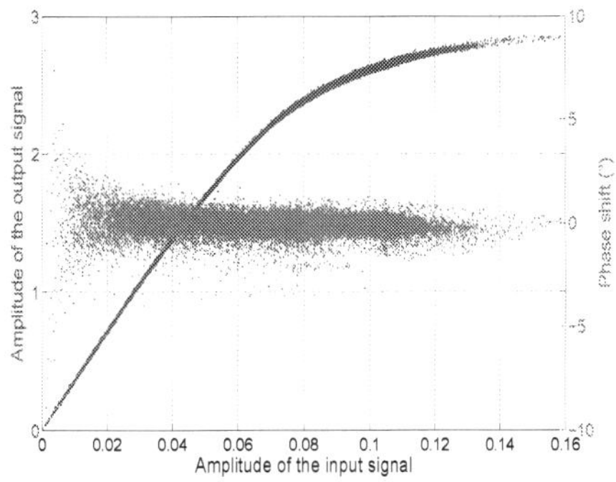

Fig. 16. Dynamic AM/AM et AM/PM characteristics of the ZFL-2500 PA

Rapp model

The dynamic AM/AM and AM/PM characteristics of the ZFL-2500 PA are shown in Fig. 16. They are defined as being, respectively, the instantaneous amplitude variation of the output signal $|\tilde{y}(n)|$, and the instantaneous phase shift $\varphi(n) = \angle\tilde{y}(n) - \angle\tilde{x}_i(n)$, in function of the instantaneous amplitude of the input signal $|\tilde{x}_i(n)|$. Although a relatively high dispersion appears on the AM/PM characteristic, the nonlinear phase distortion can be considered as negligible on the whole input amplitude range. It is a typical characteristic of low power Solid State PAs (SSPAs), which generally do not present strong memory effects. The quasi-memoryless Rapp model (Rapp, 1991) is often used in this case to model such PAs. Assuming that the phase distortion is negligible, the output signal may be expressed as follows

$$\tilde{y}(n) = G(|\tilde{x}_i(n)|)\tilde{x}_i(n) \tag{52}$$

where

$$G(|\tilde{x}_i(n)|) = \frac{K_r}{(1 + (\frac{K|\tilde{x}_i(n)|}{A_{sat}})^{2p})^{1/2p}} \tag{53}$$

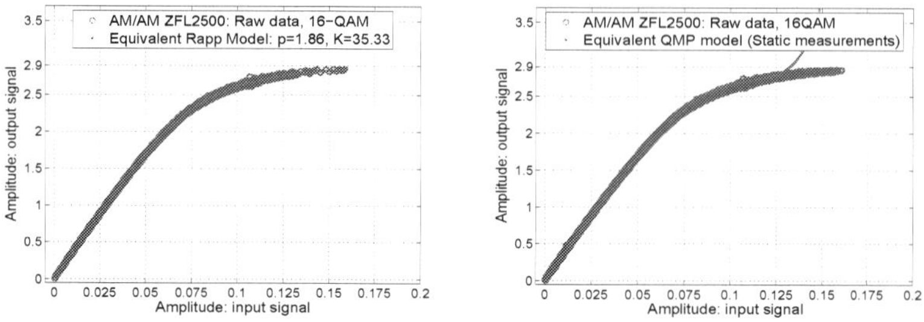

(a) Rapp Model (b) QMP model (static measurements)

Fig. 17. ZFL-2500 models: Rapp identified from the acquired samples of the 16-QAM modulated signal, and the QMP model extracted from the measured compression and interception points (Sec. 3)

is the gain function of the PA, K_r the small signal gain, A_{sat} the saturation amplitude at the output, and $p > 0$ a parameter to control the transition form of the AM-AM curve between the linear region and saturation. The Rapp model corresponding to the ZFL-2500 PA has been identified from the acquired input/output samples, with a 16QAM excitation signal. In Fig. 17 we show the dynamic AM-AM characteristics of the ZFL-2500 and its corresponding Rapp model (Fig. 17a). For comparison, we present also on Fig. 17b the AM-AM characteristic of the quasi-memoryless polynomial (QMP) model. The latter is identified from the static measurements (one- and two-tone tests) and relying on the theoretical development presented in the first part of this chapter. One could obviously notice that the Rapp model fits better the measured dynamic AM/AM characteristic than the QMP model. However, we should not forget that the QMP model is identified from a completely different excitation signals. When the signals acquisition, i.e. input/output samples, are not available, the QMP model could be useful for a first description of the behavior of the PA. Unlike the polynomial model, the Rapp model has the desirable property of being able to model the PA behavior close to saturation, that is, strong nonlinearities. While evaluating the DPD technique we are particularly interested in its linearity performance near saturation where the PA reaches its highest power efficiency. For this reason we will adopt the Rapp model, as mentioned before, for a first evaluation via simulations.

Predistorter Performance

For simplicity, the characteristic function of the PD, $F(\cdot)$, has been implemented using a constant gain Look-Up-Table (LUT) (Cavers, 1990) in simulations and measurements. Figure 18 shows the ACPR performance over a varying output power values in simulations (Fig. 18a) and in measurements (Fig. 18b). In both cases, the maximum correction is achieved at an output power close to 12 dBm. Simulations were conducted with a very high precision, using $80k$ samples and a sweep power step equal to 0.1 dB. We can conclude first that measurements and simulations results are of high agreement. While a correction of 19 dB could be achieved in simulations, a close improvement has been reached in measurements of 17.5 dB. The small disagreement between simulations and measurements is due to unavoidable noise effects.

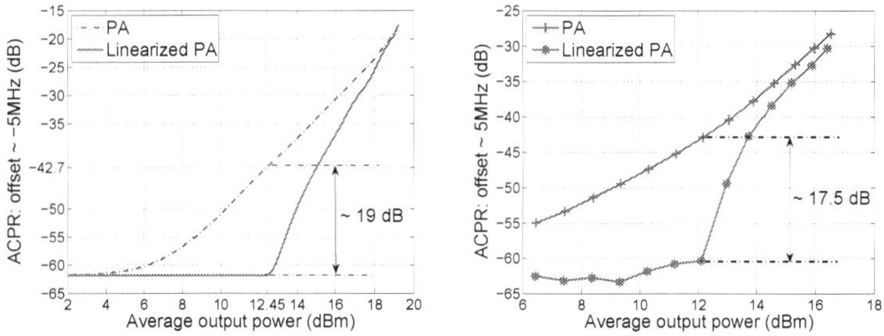

(a) Simulations (b) Measurements

Fig. 18. ACPR performance vs output power of the PA without and with linearization

We can notice from figure 18 the rapid deterioration in the performance of the PD for an output power greater than 12.45 dBm. In fact, from the knowledge of the output saturation power of the PA, we can determine the maximum theoretical output power of the linearized power amplifier (LPA), denoted P^{lin}_{max}. This power corresponds to the minimum backoff value, OBO^{lin}_{min} for an ideal amplification of the cascade PD and PA. In fact, knowing the saturation power at the output of the PA P^{out}_{sat} and the PAPR of the input signal, it is easy to show that $P^{lin}_{max} = P^{out}_{sat} - \mathrm{PAPR}$. In our case, the PAPR of the 16QAM modulated signal, filtered by a raised cosine pulse shaping filter, is equal to 7.25 dB (20 times averaging, 500 ksymbs). The output saturation power has been found equal to 19.7 dBm (Tab. 1). Thus, $P^{lin}_{max} = 12.45\,\mathrm{dBm}$ and $\mathrm{OBO}^{lin}_{min} = \mathrm{PAPR} = 7.25\,\mathrm{dB}$. If the output power exceeds P^{lin}_{max}, the signal will reach the saturation of the PA, which is a very strong nonlinearity and will deteriorate rapidly the performance of the PD. We can deduce that by reducing the PAPR of the input signal, i.e. its envelope variation, smaller values of backoff could be used, and hence, approaching the maximum power efficiency of the PA. Most of the linearization systems today, combine special techniques to reduce the PAPR of the modulated signals to linearization techniques.

Finally, from the above results, we can say that the DPD technique could have linearization performances very close to ideal, if the system is provided with sufficiently digital power processing.

5. Conclusion

PA nonlinearity is a major concern in the realization of modern communications systems. In this chapter, we have provided some of the basic knowledge on power amplifier nonlinearity and digital baseband predistortion technique. In the first part the traditional power series analysis was repeated with a new interesting development in frequency domain. This analysis was validated in simulations under Matlab and through measurements on a real PA. The second part of this chapter was dedicated to a brief overview on the adaptive digital baseband predistortion technique and an experimental evaluation of this technique. A fully automatic test bench was used.

The most interesting perspective of this study is make further generalization of the power analysis when more complicated signals are used. For the digital predistortion techniques, there remain a lot of efforts to deploy, especially on fast adaptation algorithms, and nonlinear memory effects modeling accuracy.

6. References

Benedetto, S. & Biglieri, E. (1999). *Principles of Digital Transmission: With Wireless Applications*, Kluwer Academic Publishers, Norwell, MA, USA.

Black, H. (1928). Translating system, US Patent, number 1,686,792.

Black, H. (1937). Wave translation system, US Patent, number 2,102,671.

Bosch, W. & Gatti, G. (1989). Measurement and simulation of memory effects in predistortion linearizers, *Microwave Theory and Techniques, IEEE Transactions on* **37**(12): 1885–1890.

Cavers, J. (1990). Amplifier linearization using a digital predistorter with fast adaptation and low memory requirements, *Vehicular Technology, IEEE Transactions on* **39**(4): 374–382.

Cripps, S. C. (2006). *RF Power Amplifiers for Wireless Communications, Second Edition (Artech House Microwave Library (Hardcover))*, Artech House, Inc., Norwood, MA, USA.

Haykin, S. (2005). Cognitive radio: brain-empowered wireless communications, *Selected Areas in Communications, IEEE Journal on* **23**(2): 201–220.

Ibnkahla, M. (2000). Neural network predistortion technique for digital satellite communications, *Acoustics, Speech, and Signal Processing, 2000. ICASSP '00. Proceedings. 2000 IEEE International Conference on* **6**: 3506–3509 vol.6.

Isaksson, M., Wisell, D. & Ronnow, D. (2005). Wide-band dynamic modeling of power amplifiers using radial-basis function neural networks, *Microwave Theory and Techniques, IEEE Transactions on* **53**(11): 3422–3428.

Isaksson, M., Wisell, D. & Ronnow, D. (2006). A comparative analysis of behavioral models for rf power amplifiers, *Microwave Theory and Techniques, IEEE Transactions on* **54**(1): 348–359.

Kenington, P. (2002). Linearized transmitters: an enabling technology for software defined radio, *Communications Magazine, IEEE* **40**(2): 156–162.

Kenington, P. B. (2000). *High Linearity RF Amplifier Design*, Artech House, Inc., Norwood, MA, USA.

Liu, T., Boumaiza, S. & Ghannouchi, F. (2004). Dynamic behavioral modeling of 3g power amplifiers using real-valued time-delay neural networks, *Microwave Theory and Techniques, IEEE Transactions on* **52**(3): 1025–1033.

Mitola, J. (1995). The software radio architecture, *Communications Magazine, IEEE* **33**(5): 26–38.

Morgan, D., Ma, Z., Kim, J., Zierdt, M. & Pastalan, J. (2006). A generalized memory polynomial model for digital predistortion of rf power amplifiers, *Signal Processing, IEEE Transactions on* **54**(10): 3852–3860.

Rapp, C. (1991). Effects of HPA-nonlinearity on a 4-dpsk/ofdm-signal for a digital sound broadcasting system, *Proc. 2nd Eur. Conf. Satellite Communications, Liege, Belgium* pp. 179–184.

Schetzen, M. (2006). *The Volterra and Wiener Theories of Nonlinear Systems*, Krieger Publishing Co., Inc., Melbourne, FL, USA.

Wood, J., Root, D. & Tufillaro, N. (2004). A behavioral modeling approach to nonlinear model-order reduction for rf/microwave ics and systems, *Microwave Theory and Techniques, IEEE Transactions on* **52**(9): 2274–2284.

8

Spatial power combining techniques for semiconductor power amplifiers

Zenon R. Szczepaniak
Przemysłowy Instytut Telekomunikacji S.A.
Poland

1. Introduction

Growing demand on special signal modulation schemes in novel radars and ability to transmit relatively long pulses cause the Travelling Wave Tubes (TWT) to be constantly replaced by new concepts of power amplifiers. Solid-state power amplifiers appear to be a good candidate, however, the output power from a single transistor module is still relatively low. The only available solution is that of combining output power from a number of semiconductor amplifiers. To accomplish this, one can use, classical and well-known, two-way power combiners (like Willkinson type) or specially-designed new type of multi-input combiners. Current requirements for radar working modes imply using active antenna arrays, thereby providing multifunction ability. The active antenna concept assumes the use of transmit-receive modules (T/R), each comprising a power transistor. The overall transmitted power is then a function of the sum of the output powers from each T/R module, and the power summing operation is performed in free space.

On the other hand, in some radar applications (or generally, where a power amplifier is needed, be it electronic warfare or jamming), the central power transmitter is still desired. The older applications are based on TWTs, and although they give enough power, they carry a number of disadvantages. The main are as follows:

- TWTs generally offer low duty factor (although some of them are approaching up to 100%),
- they need special power supplies, which are dangerous due to tube working voltages in the range of kVolts
- reliability is limited due to erosion of inner electrodes inside the TWT
- reliability system is two-state, a tube works or does not; any failure results in a complete malfunction of the radar.

Additionally, in higher bands there are no solid-state power sources with enough power.

The conclusion and current trends are that there is a constant need for combining power from a number of sources.

2. General combining techniques

2.1 Types

2.1.1. Multilevel combining

Combining a number of sources with the use of basic two-input power combiners implies the necessity of using a number of them. As a result, the overall power combiner is formed as a tree-like structure. The number N of power sources (transistors) has to be a power of 2. For N amplifiers, the resulting combiner structure contains $p = \log_2 N$ levels (Fig. 1).

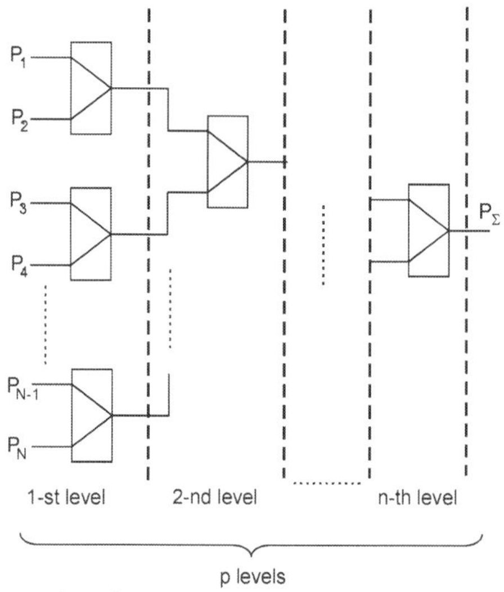

Fig. 1. Combining structure based on two-input power combiners

For a cascaded network combining N input signals the number of N-1 basic two-input combiners has to be used. The multilevel combining scheme is easy to implement. The two-input power combiners are well-known and their design is well-developed. Depending on the chosen power transmitter structure, the multilevel structure may be fabricated on one big PCB, forming a packet-like power module, or each of the two-input combiners may be assembled and packed separately. Due to the fact that they form p levels, insertion losses of the final structure are p times insertion losses of the basic structure. Therefore for each of the input power path there is insertion loss p times higher than that of one basic two-input structure. Another serious drawback of cascaded devices is possible accumulation of phase mismatches introduced by each of the basic structures.

2.1.2. Spatial combining

The term "spatial combining" means combining a number of input power sources with the use of simultaneous addition of input signals in a kind of special structure with multi-couplings or multi-excitations. Input signal sources are distributed in space and excite their own signal waves inside a specially designed space intended for power addition. The

structure of spatial power combiner may have a number of input ports and one output port, whereas the combining takes place inside the structure.
To complete a power amplifier system two of such structures are needed. The first one acts as power splitter, connected to a number of amplifying submodules, and the second collects output power from these submodules.
The second available solution is when the combiner has got only one input and one output port. The amplifying modules, or simply transistors, are incorporated inside the combining structure, most frequently a hollow metal waveguide-like structure, which contains a set of specially designed probes/antennas inside, each one connected to a power transistor, and the same set at the transistor outputs. The input set of probes reads EM field distribution, it is then amplified, and finally the output set recovers field distribution with amplified magnitude. This structure may be regarded as a section of an active waveguide.

2.2 Theory

For the basic two-input structure the relationship between input power and output power is given by:

$$P_{\Sigma}\big|_2 = T^2\left(\sqrt{P_1} + \sqrt{P_2}\right)^2 \tag{1}$$

where the combiner is characterised by the scattering matrix [S] (Srivastava & Gupta, 2006):

$$[S] = \begin{bmatrix} R & T & T \\ T & R & I \\ T & I & R \end{bmatrix} \tag{2}$$

For purposes of simplification, isolation (I) is assumed to be equal to 0 and the combiner is perfectly matched at all its ports (reflection R=0).
In order to design a power combining network one needs to be familiar with the influence of the combining structure on final output power. This has to cover the influence of individual characteristics of combining sub-structures and the number of levels, as well as, the output power degradation as a function of failed input amplifiers. Such knowledge allows to calculate and predict a drop of radar cover range in case the amplifying modules fail.
For higher value of N (and number of levels p), when the equivalent insertion losses become higher, a specialised spatial combiner is worth considering. In reality, it may turn out that insertion losses of a specially designed multi-input combiner (with, for example, eight-input port) may be comparable to those of a two-input structure. That means that usually it exhibits lower insertion losses than the equivalent cascaded network.
Assuming approach shown in Fig. 2 the combined output power is associated with the normalized wave b_n in case the input ports from 1 to n-1 are excited by input powers P_1 to P_N (i.e. $N = n-1$).

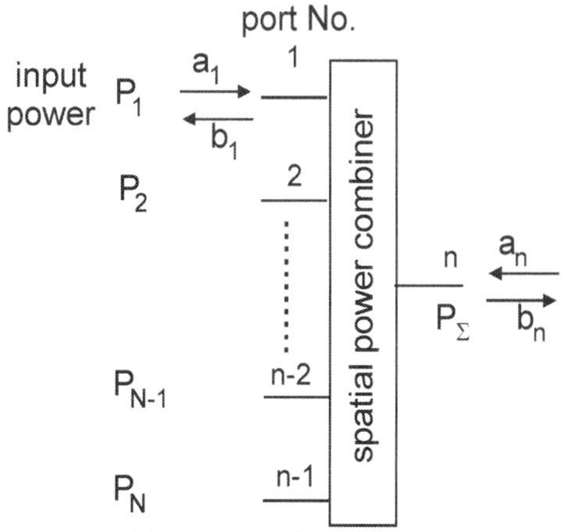

Fig. 2. General spatial power combiner - excitation of the ports

$$\begin{bmatrix} b_1 \\ \vdots \\ b_{n-1} \\ b_n \end{bmatrix} = \begin{bmatrix} S_{11} & \cdots & \cdots & S_{1n} \\ \vdots & \ddots & & \vdots \\ \vdots & & \ddots & \vdots \\ S_{n1} & \cdots & \cdots & S_{nn} \end{bmatrix} \cdot \begin{bmatrix} a_1 \\ \vdots \\ a_{n-1} \\ 0 \end{bmatrix} \tag{3}$$

In ideal case (neglecting the insertion losses, and assuming ideal matching and isolations) the general formula for power combining is as follows:

$$P_\Sigma\big|_N = \frac{1}{N}\left(\sum_{i=1}^{N}\sqrt{P_i}\right)^2 \tag{4}$$

where P_Σ is the transmitter output power (summed) and subscript N denote the quantity of power sources.

2.3 Benefits

The use of power combining techniques allows, first of all, to replace a TWT transmitter and not to suffer from its disadvantages. The main advantage is the reliability. A transmitter with many power sources will still emit some power, when a number of them are damaged. The detailed analysis of this effect is presented in Chapter 3 (also Rutledge et al, 1999).
The structure often used consists of power submodules, each containing power transistors, an input power splitter and an output power combiner. It may be configured in distributed amplifier concept, with power submodules placed along the waveguide. The solutions with separate power submodules, exhibit several substantial advantages. Due to their extended metal construction they have an excellent heat transfer capability, which makes cooling easy to perform. Furthermore, they provide an easy access to amplifying units in case they are damaged and need replacing. Finally, once the structure is made, it can be easily upgraded to a higher power by replacing the amplifying units with new ones with a higher output

power. Another way is to stack several transmitters with the use of standard waveguide tree-port junctions.

However, the disadvantage of waveguide distributed amplifiers concerns the frequency band limitation due to spatial, wavelength-related periodicity. The working bandwidth decreases when the number of coupled amplifying units is increased. Hence, there is a power-bandwidth trade-off.

The process of summing the output power from a number of power amplifiers has its inherent advantage. As far as multi-transistor amplifier is concerned, there is always the question to the designer whether to use lower number of higher power amplifiers (transistors) or higher number of lower power amplifiers. Intuitively, one is inclined to use the newest available transistors with maximal available power.

However, taking into consideration that every active element generates its own residual phase noise, the phase noise at the output of combiner is a function of the number of elements. Assuming, that the residual phase noise contributions from all the amplifiers are uncorrelated, then the increase of the number of single amplifiers causes the improvement of output signal to noise ratio (DeLisio & York, 2002).

For a fixed value of output power in a spatial power combining system the increase of the number of single amplifiers gives the increase of intercept point IP3 and spurious-free dynamic range SFDR.

The real advantage of using a spatial combiner is that the combining efficiency is approximately independent of the number of inputs. Then, for given insertion losses of a basic two-input combiner there is a number of power sources (input ports for multilevel combiner) where a spatial power combiner (naturally, with its own insertion losses) becomes more efficient.

In real cases, efficiency of any combiner is limited by channel-to-channel uniformity. Gain and phase variations, which arise from transistor non-uniformities and manufacturing tolerances, can lead to imperfect summation of power and a reduction in combining efficiency. However, considering that the variations of gain have statistical behavior the use of higher number of inputs enables one to average and then minimize the influence.

On the other hand, the variations of phase shift between summing channels have a crucial influence on the output power of the combiner. In the case of multilevel combiner, the phase variations of individual two-input combiners may accumulate and therefore degrade power summing efficiency. Moreover, taking into account that the amplifying modules may have their own phase variations, introducing individual tuning for two-input combiners, becomes extremely difficult for real manufactured systems. In the case of spatial combiner, it is possible to introduce individual correcting tuning for each summing channel (arm). For a higher number of channels, the tuning becomes demanding, yet still possible to be made. It is worth developing an automatic tuning system, involving a computer with tuning algorithm and electronically driven tuners, for example screw tuners moved by electric step motors.

3. Power degradation

3.1 Combined power dependence in case of input sources failures

The output power degradation mechanism in a tree structure is the same as in the spatial one. It may be derived from S-matrix calculations for various numbers of active ports.

Assuming equal input power P_{in} on each of the input ports the relationship for the output power vs number m of non-active ports is expressed as:

$$P_\Sigma\big|_{N-m} = \frac{P_{in}}{N}(N-m)^2 \tag{5}$$

where m equals from 0 to N.
It may be derived from the analysis of dependence of output wave b_n versus varying number of input waves (a_1 to a_{n-1}) equal to zero.
It means that for a two-input basic network a failure of one of input power sources P_{in} will result in $0.5P_{in}$ output power. Compared to power of $2P_{in}$, available when there is no failure, the penalty equals 6 dB.

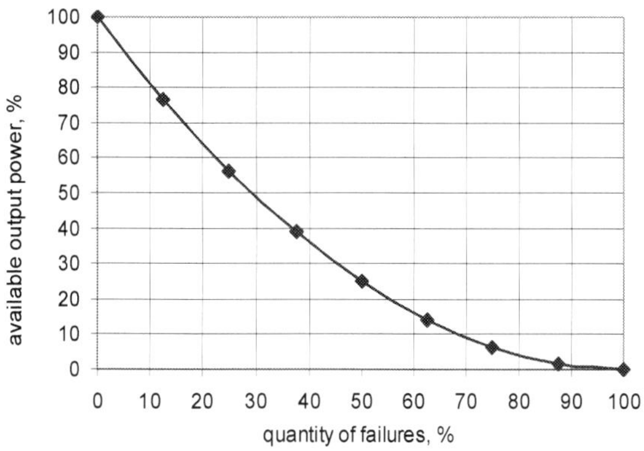

Fig. 3. Combined output power vs number of failed input sources

The power degradation is calculated as the ratio of max output power without failures (when m=0) to power expressed as a function of m for different number of sources N.

$$\frac{P_\Sigma\big|_{N-m}}{P_\Sigma\big|_N} = \left(1-\frac{m}{N}\right)^2 \tag{6}$$

where P_Σ is the transmitter output power (summed) and subscripts N-m and N denote the quantity of working modules.
A graphical illustration of Eq (6) is shown in Fig. 3, where the quantity of failures is defined as m/N and expressed in percentages.

3.2 Influence of power degradation on radar cover range

Information presented here is necessary to predict radar range suppression as the function of failures in its solid-state transmitter. The transmitter output power degradation vs number of damaged power modules is given by Eq (6).

Considering the radar range equation and assuming that the received power is constant, in order to achieve proper detection for the same target, the suppression of the range R may be expressed as:

$$\frac{R|_{N-m}}{R|_N} = \left(\frac{P_\Sigma|_{N-m}}{P_\Sigma|_N}\right)^{1/4} = \left(1-\frac{m}{N}\right)^{1/2} \tag{7}$$

Here R is the radar cover range and subscripts N-m and N denote quantity of working modules.
It may be seen that for 50% of modules failed, radar coverage decreases to 70% of its maximal value (Fig. 4).

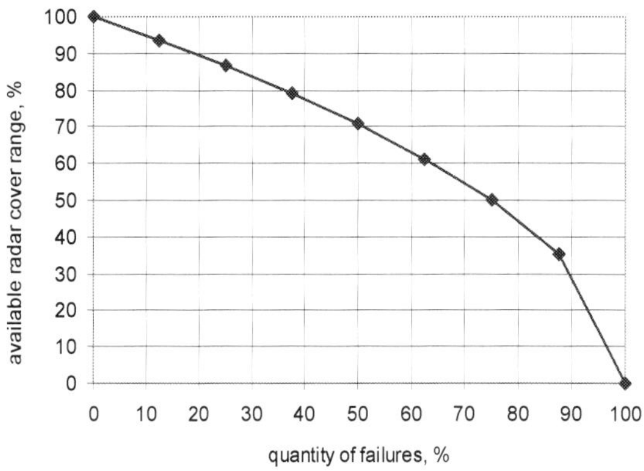

Fig. 4. Radar cover range vs. number of failed input sources

All these considerations assume perfect matching and isolations between channels in spatial power combiner. In real case, the isolations are not ideal and a failure of power transistor might result in different output impedance thereof, from open to even short circuit. Therefore, the real output combined power may differ from the ideal one.

4. Examples of multi-input splitters/combiners

The need for replacing TWT high power amplifiers in higher frequency bands contributes to the invention of new methods of power combining from many single semiconductor amplifiers. Those already known that involve planar dividers/combiners based on Wilkinson or Gysel types offer noticeable power losses in higher bands (X, K) especially when used as complex tree-structure for combining power from many basic amplifying units. The methods of spatial power combining may be divided into two main ideas. The first method is to place a two-dimensional matrix of amplifier chips with micro-antennas inside a waveguide. The second comprises the use of separate multiport input splitter and output combiner networks. It employs the use of specially designed structures (Bialkowski

& Waris, 1996; Fathy et al, 2006; Nantista & Tantawi, 2000; Szczepaniak, 2007; Szczepaniak & Arvaniti, 2008) or a concept of distributed wave amplifier, where amplifying units are coupled with input and output waveguides by means of a set of probes inserted into the waveguides.

4.1 Waveguide built-in 2D array of amplifiers

The solution presented here may be regarded as a technique of so called quasi-optical power combining. Quasi-optical method of power combining assumes multidimensional diffraction and interference of incoming and outgoing waves at input and output of a power combining system. The most typical example of such a solution is two-dimensional matrix of amplifiers, each with mini-antennas at their inputs and outputs (Fig. 5). The 2-D amplifying matrix may be inserted into a waveguide (sometimes oversized) or illuminated by means of a horn antenna, additionally with dielectric lenses. The second horn antenna collects output power from all the transistors. There are many technical examples of the amplifying grid construction and splitting/combining structures (Belaid & Wu, 2003; Cheng et al, 1999-a; ; Cheng et al, 1999-b; Zhang et al, 2007).

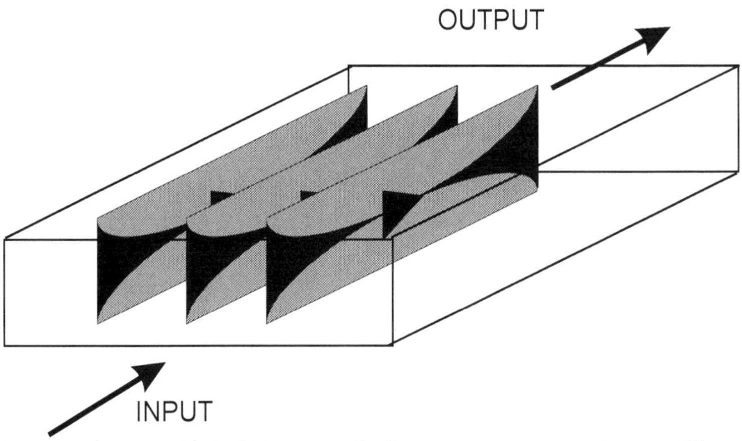

Fig. 5. Concept of waveguide built-in array of microantennas connected to amplifiers

A grid of amplifiers may contain even several hundred of active devices. In the case of insertion 2-D amplifiers set into a waveguide, the input antennas matrix probes the E-M field distribution inside the waveguide. After amplifying, the output antennas matrix restores field distribution and excites a wave going towards the waveguide output. The whole structure may be regarded as a section of an "active" waveguide. The main development is being done in the concept and structure of a transistors array. The transistors may be placed on the plane (in real case dielectric substrate) perpendicular to the waveguide longitudal axis (called grid amplifiers), or they may be stacked in sandwich-like structure, where layers are parallel to the waveguide longitudal axis (called active array amplifiers).

The main advantage of waveguide built-in concepts is their compact structure and wide frequency bandwidth of operation. However, there are some disadvantages, for example, difficult heat transfer, especially when high power is desired, the necessity of special simulation and design, and inconvenient repairing.

4.2 Distributed waveguide splitter/combiner

The most frequently used structure of distributed splitter/combiner scheme assumes the use of hollow waveguide, e.g. rectangular one working with H_{10} mode, with a number of probes inserted into the waveguide and periodically distributed along its longitudal axis. The period equals half-wavelength of guided wave $\lambda_w/2$ at the center frequency. The waveguide is ended with a short, which is at quarter-wavelength distance from the last probe. The structures of the splitter and the combiner are identical. The differences between subsequent solutions are in the concept of EM field probes (Bashirullah & Mortazawi, 2000; Becker & Oudghiri, 2005; Jiang et al, 2003; Jiang et al, 2004; Sanada et al, 1995).

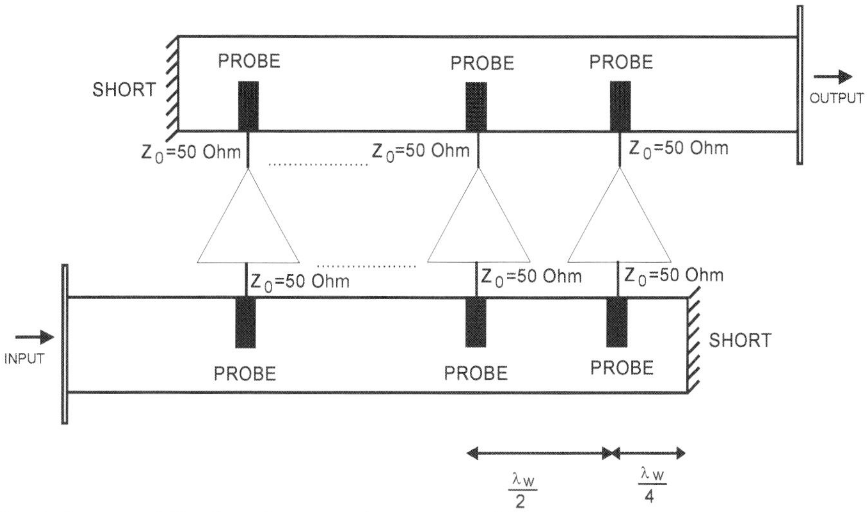

Fig. 6. Concept of distributed waveguide power amplifier

For the centre frequency the short-ended section of the waveguide is transformed into the open-circuit and the half-wavelength sections of waveguide transforms adjacent probes impedance with no changes. Therefore the equivalent circuit of the splitter contains N probes impedances in parallel connected to the input waveguide impedance. Each probe transforms the 50 Ohm impedance of the amplifier into the value required to obtain equal power splitting ratio from circuit input port to each of the output. Spatial distribution of the probes along the waveguide causes frequency dependence of power transmission to each probe. As the result for increasing number of outputs (probes) the frequency band of splitter/combiner operation becomes narrower.

The simplest solution is a coax-based probe inserted through a hole in the wider waveguide wall. The length of the probe, its diameter and distance from the narrow waveguide sidewall results from design optimization for minimal insertion losses and equal transmission coefficient for each channel.

There are many solutions for the probe design, for example, it may be a microstrip line with a piece of substrate laminate inserted into the waveguide, or specially designed slot in the wider waveguide wall, coupled with a planar circuit on the laminate with ground metallization removed.
An example of application of four amplifying modules connected to four-probe waveguide splitter and combiner is shown in Fig. 7 (Szczepaniak et al, 2009).

Fig. 7. Example of power amplifier using five-port distributed waveguide power splitter/combiner

4.3 Four-input microwave rectangular waveguide combiner

The four-input waveguide splitter, according to this concept, offers a very high working bandwidth and low insertion losses, providing a good reason for the design of solid-state high power modules. The concept of construction may be applied to any rectangular waveguide, in frequency band related to its dimensions.
The example structure shown below is assumed to work in X-band (Szczepaniak, 2007). It is based on a standard X-band rectangular waveguide R-100 with a short at one end. The input port for the splitter is the waveguide and the four output ports are of 50 Ohm coaxial type. The cross-sectional diagrams of the discussed structure are shown in Fig 8.

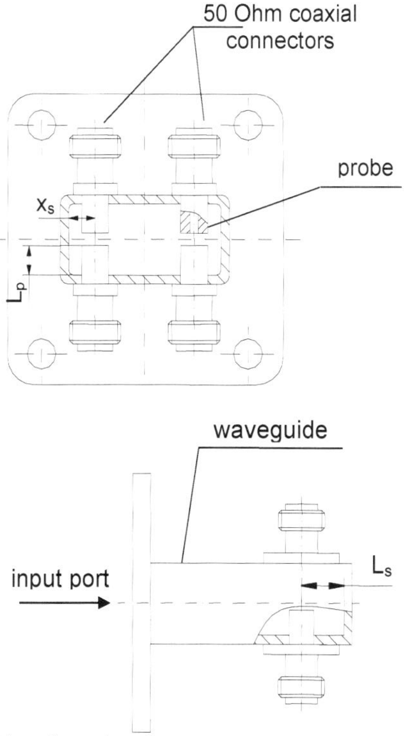

Fig. 8. Five-port waveguide splitter/combiner

The coupling is performed by means of four coaxial probes, made from 50 Ohm coaxial line with the outer conductor removed. The probes are inserted into the waveguide in the plane, at a certain distance L_S from the shorted end.

In order to achieve four-way equal power dividing and avoid reflections, the equivalent impedance in the plane of the probes insertion must equal wave impedance Z_w. As the four probes are connected parallelly, each one should be regarded as the impedance four times higher than that of the wave. The probes are connected to further microwave devices, e.g. amplifying modules, which have input/output reference impedance Z_0=50 Ohm. Therefore, each probe must work as an impedance transformer from value Z_0 to 4*Z_w (Eq. 8)

$$Z_{probe} = 4 \cdot Z_w = n^2 \cdot Z_0 \tag{8}$$

where n is the equivalent transforming ratio, which fulfills the condition (8). This is a simplified case and it can be assumed for the ideal transforming probes and a single frequency. In such case, the presence of a quarter-wavelength section of the waveguide and its frequency dependence can be temporarily neglected. The situation is shown in Fig. 9

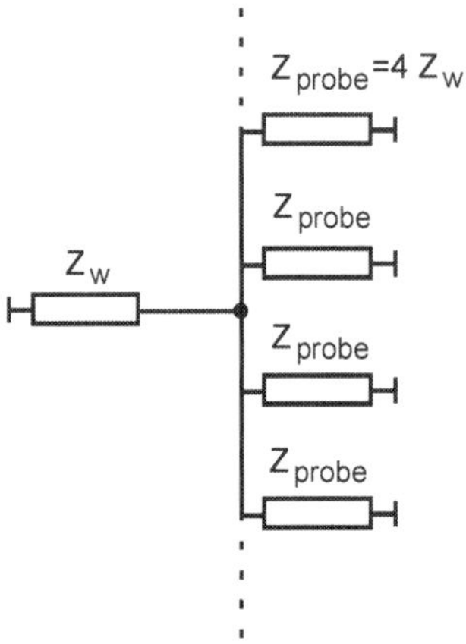

Fig. 9. Simplified equivalent circuit of four-probe splitter at centre frequency

The length of the probe L_p and the distance x_s (Fig. 8) from the lower waveguide wall are the parameters to be optimized in order to obtain a wide frequency bandwidth where the input reflection coefficient is as close as possible to the desired value. Furthermore, one can begin the design of the power splitter from the design of one probe which fulfils the condition (8). In such case, the starting point of the design is a section of the waveguide with length equal to the odd number (2m+1) of quarter guided wavelength λ_w, as shown in Fig.10.

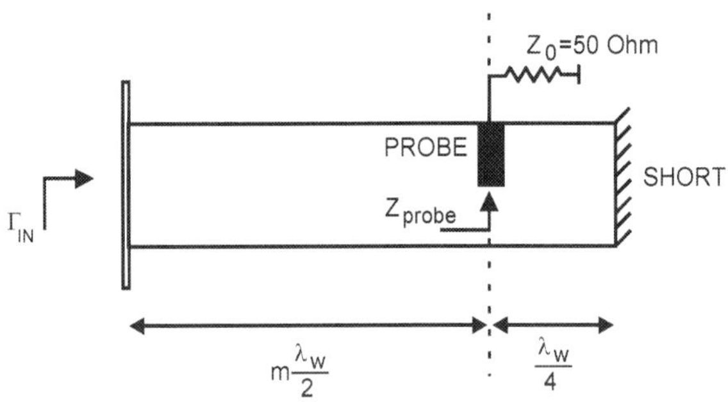

Fig. 10. Configuration of one probe inside a waveguide

After one probe has been optimized fully, a four-probe circuit is to be simulated. The pre-optimized probes are inserted symmetrically with respect to the main longitudal axis of the waveguide. Next, the second issue must be considered. Inserting the probes into the waveguide causes disturbance of the field distribution, excitation of higher order modes, and, therefore, creates additional parasitic susceptances which add to the admittance seen via the probe. As a result, the final stage of design concerns simulation of four pre-designed probes with the shorted section of the waveguide. The optimization of the probe's parameters (the same as before) together with the length of the shorted waveguide section gives the final matching of susceptances in the plane of probes insertion. This way it is possible to obtain broadband matching, which gives a wide frequency range of a very low reflection coefficient for full power splitter and flat transmission to each of four outputs approaching to ideal value of -6dB.

The splitter structure proposed here has an interesting additional feature. The transmission from the waveguide port to the coaxial ports placed on one of the wide waveguide walls differs in phase from the transmission to the ports on the other wide wall. It is because the probes are inserted in parallel to the lines of the electric field of H_{10} mode. The phase difference equals π. Due to the fact that all the probes are inserted close to each other without any additional shielding, which may be additionally considered, the values of the isolation between them are not high. This is about -3dB for opposite probes (between wide walls) and about -10dB for adjacent probes (on the same wide wall).

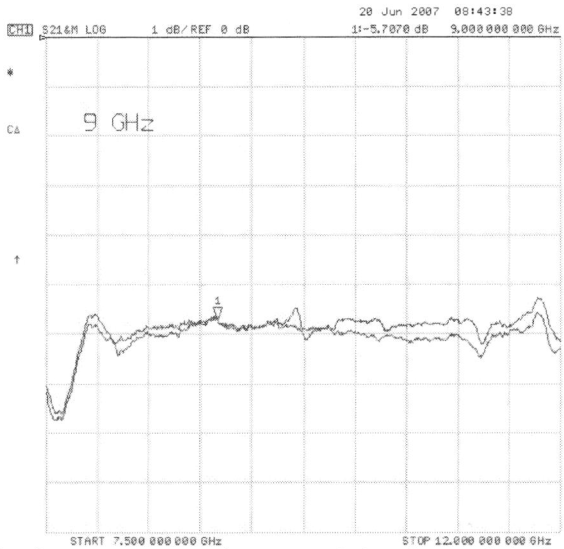

Fig. 11. Example of measurement results - transmission characteristics from waveguide to coax ports for two coax outputs

The structure of the four-way power splitter shown here offers a frequency bandwidth many times wider than an equivalent, four-way distributed waveguide structure. It may be used together with standard waveguide T-junctions in order to achieve simple eight-way power splitter by connecting two of them. Such a structure will still have wider bandwidth than a distributed wave one. The insertion losses are sufficiently small, about 0.2÷0.3 dB, and the input reflection coefficient is low enough to make this splitter an attractive alternative to distributed waveguide structures. There is a phase difference, equal to π between transmissions from waveguide port to outputs placed on the opposite wider walls of the waveguide, which may be useful in some measurement applications. The proposed structure is also very simple. And finally, although it has a relatively low isolation between output coaxial ports, in the case of symmetrical power combining, when the amplifiers are designed to have good matching to 50 Ohm, this power splitter/combiner works properly.

An example of application is shown in Fig. 12 (Szczepaniak et al, 2009). Here four amplifying modules are connected to two identical splitting/combining structures.

Fig. 12. Example of high power X-band amplifier using four-input rectangular waveguide power splitter and combiner.

4.4 Eight-input microwave circular waveguide combiner

The following splitter structure comprises a section of cylindrical microwave waveguide and nine coax-based probes (Szczepaniak & Arvaniti, 2008). The waveguide has two circular walls which transform the waveguide into a resonator. The input probe is inserted in the center of one of the circular walls and the remaining eight probes are inserted into the second circular wall. The probe insertion points form a circle, whose center corresponds to the center of the second wall.
The cross-section and 3D view of the splitter structure are shown in Fig. 13 and 14.

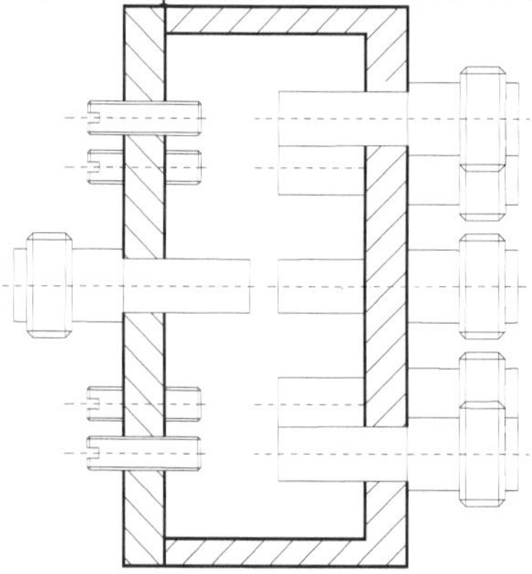

Fig. 13. Nine-port circular waveguide power splitter/combiner

In order to provide tuning possibility for all the output probes eight screw tuners are inserted in the wall containing the input probe. The tuners are placed according to the positions of the output probes but on the opposite wall.

All the probes are made as sections of 50 Ohm coaxial line with outer conductor removed from the part inserted inside the cavity. In this case special coaxial jacks from Radiall containing special Teflon-covered pin with diameters corresponding to a 50 Ohm line have been used.

The inner cavity dimensions and probes placement are optimized to obtain minimal input reflection coefficient and uniform power division. Each of the probes transforms 50 Ohm line characteristic impedance to a value loading the cavity. Symmetrical probe insertion gives symmetrical field disturbance and distribution. Careful design gives optimal power transfer from center probe to eight output probes and vice versa.

The number of output probes may be different. It depends on a designer's needs. The key factor is to control the field distribution inside the cavity during the design process. For each desired number of inputs the optimization procedure gives the dimensions and positions of the probes and the dimensions of the cavity.

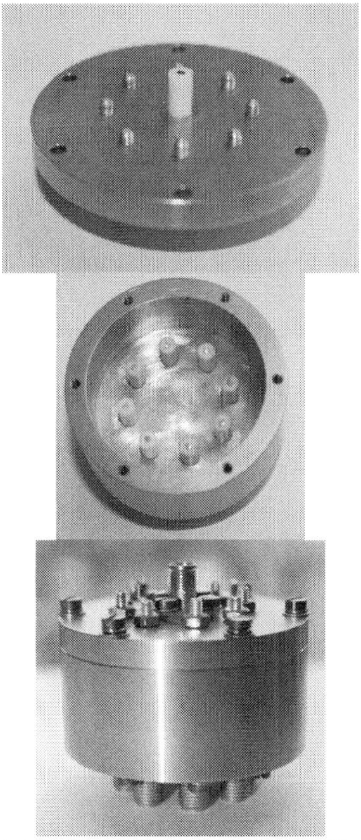

Fig. 14. Manufactured model structure of nine-port power splitter/combiner

This example structure is designed to work in X-band. Assuming that the working bandwidth is defined by 0.5 dB drop of transmission coefficient, the obtained bandwidth is equal to about 8.3-10.7 GHz. Within the working bandwidth, all the measured characteristics fall within the range -9 dB +/- 0.5 dB. Depending on the application, the useful working bandwidth may be defined differently, for example on the basis of 1dB-drop of the transmission.

For purposes of power combining from microwave amplifiers the combining losses should be as low as possible. The test structure presented here does not have silver or gold plating inside the cavity, therefore, the insertion losses may be decreased further. The input reflection coefficient has an acceptable value lower than -10dB within the working band.

An example of measurements results for the test structure of the splitter is shown in Fig. 15.

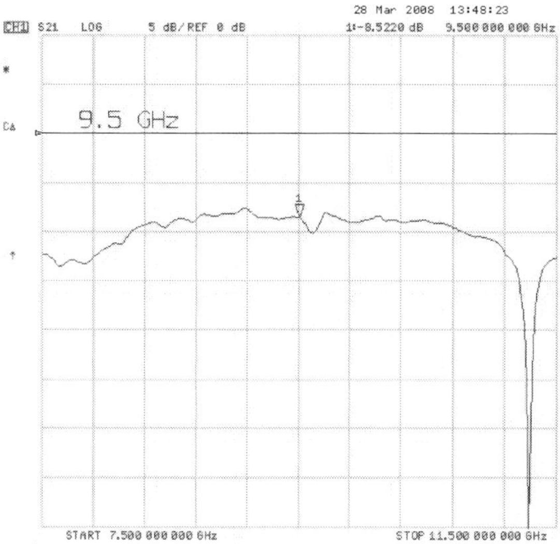

Fig. 15. Example of measurement results - transmission characteristics from centre coax input to one of coax output port

5. Conclusion

Solid-state power sources based on spatial power combining may successfully replace TWT central transmitters. This method of power combining offers several advantages compared to the use of multi-level three-port based approach. In high power transmitters it is important to reduce the combining losses to as low as possible. Spatial combining does not suffer from additive accumulation of insertion losses and phase mismatches of individual devices as in the tree-structure of cascaded two-input combiners, which is the reason why it is very promising.

In case of failures of power transistors, solid-state transmitter exhibits soft output power degradation. The radar coverage, which may be calculated for a given number of working modules, reduces softly while failures proceed. It, therefore, gives additional reliability to radar systems using power sources based on spatial combining.

According to most recent developments, in the case of single transistor/semiconductor amplifiers, we are approaching the limits of power density and combining efficiency. On the other hand, combining large numbers transistors on-chip eventually becomes impractical. It results in most of the semiconductor area being occupied by the passive matching and combining circuitry. Furthermore, losses in the semiconductor transmission lines are relatively high. These factors limit combining efficiency. In order to realize solid-state components with higher power and efficiency, new kinds of combining techniques have to be used. They should integrate large numbers of devices with minimal signal splitting and combining losses. Additionally, the desired amplitude and phase relationships between summing channels should be maintained. Spatial or quasi-optical techniques provide a possible solution. Additionally they give promising phase noise degradation for power transmitter.

The future challenges are as follows: critical power in one combiner (to avoid discharge or damage of a probe), effective cooling and heat transfer from individual power transistors, automatic failure detection and current temperature sensing, easy access to repair, or finally application of automated tuning procedures and circuits for testing and output power optimization.

6. References

Bashirullah, R., Mortazawi, A. (2000). A Slotted-Waveguide Power Amplifier for Spatial Power-Combining Applications. *IEEE Transactions on Microwave Theory and Techniques*, Vol. 48, No. 7, July 2000, pp. 1142-1147, 10.1109/22.848497.

Becker, J., and Oudghiri, A. (2005). A Planar Probe Double Ladder Waveguide Power Divider, *IEEE Transactions on Microwave Theory and Techniques*, Vol. 15, No. 3, March 2005, pp.168-170, 10.1109/LMWC.2005.844214.

Belaid, M., and Wu, K. (2003). Spatial Power Amplifier Using a Passive and Active TEM Waveguide Concept. *IEEE Transactions on Microwave Theory and Techniques*, Vol. 51, No. 3, March 2003, pp. 684-689, 10.1109/TMTT.2003.808698.

Bialkowski, M., and Waris, V. (1996). Analysis of an N-Way Radial Cavity Divider with a Coaxial Central Port and Waveguide Output Ports, *IEEE Transactions on Microwave Theory and Techniques,*, Vol. 44, No. 11, November 1996, pp.2010-2016, 10.1109/22.543956.

Cheng, N., Fukui, K., Alexanian, A., Case, M.G., Rensch, D.B., and York, R. A. (1999-a). 40-W CW Broad-Band Spatial Power Combiner Using Dense Finline Arrays. *IEEE Transactions on Microwave Theory and Techniques*, Vol. 47, No. 7, July 1999, pp. 1070-1076, 10.1109/22.775438.

Cheng, N., Jia, P., Rensch, D. B., and York, R.A. (1999-b). A 120-W -Band Spatially Combined Solid-State Amplifier. *IEEE Transactions on Microwave Theory and Techniques*, Vol. 47, No. 12, December 1999, pp. 2557-2561, S 0018-9480(99)08455-0.

DeLisio, M.P., and York, R.A. (2002). Quasi-Optical and Spatial Power Combining. *IEEE Transactions on Microwave Theory and Techniques*, Vol. 50, No. 3, March 2002, pp. 929-936, S 0018-9480(02)01959-2.

Fathy, A.E., Lee, S., and Kalokitis, D. (2006). A Simplified Design Approach for Radial Power Combiners. *IEEE Transactions on Microwave Theory and Techniques*, Vol. 54, No. 1, January 2006, pp. 247-255, 10.1109/TMTT.2005.860302

Jiang, X., Liu, L., Ortiz, S.C., Bashirullah, R., and Mortazawi, A. (2003). A Ka-Band Power Amplifier Based on a Low-Profile Slotted-Waveguide Power-Combining/Dividing Circuit, *IEEE Transactions on Microwave Theory and Techniques*, Vol. 51, No. 1, January 2003, pp. 144-147. 10.1109/TMTT.2002.806927.

Jiang, X., Ortiz, S., and Mortazawi, A. (2004). A Ka-Band Power Amplifier Based on the Traveling-Wave Power-Dividing/Combining Slotted-Waveguide Circuit. *IEEE Transactions on Microwave Theory and Techniques*, Vol. 52, No. 2, February 2004, pp.633-639, 10.1109/TMTT.2003.822026.

Nantista, C. and Tantawi, S. (2000). A Compact, Planar, Eight-Port Waveguide Power Divider/Combiner: The Cross Potent Superhybrid. *IEEE Microwave and Guided Wave Letters*, Vol. 10, No. 12, December 2000, pp.520-522, 10.1109/75.895089.

Rutledge, D.B., Cheng, N., York, R.A., Weikle II, R.M., and De Lisio, M.P. (1999). Failures in Power-Combining Arrays. *IEEE Transactions on Microwave Theory and Techniques*, Vol. 47, No. 7, July 1999, pp. 1077-1082, S 0018-9480(99)05305-3.

Sanada, A., Fukui, K., Nogi, S., and Sanagi, M. (1995). Traveling-Wave Microwave Power Divider Composed of Reflectionless Dividing Units. *IEEE Transactions on Microwave Theory and Techniques,*, vol. 43, No. 1, January 1995, pp. 14-20, 10.1109/22.363014.

Srivastava, G.P, and Gupta, V.L. (2006). *Microwave devices and circuit design* . Prentice-Hall of India, New Delhi, ISBN 81-203-2195-2.

Szczepaniak, Z. (2007). Broadband Waveguide Power Splitter for X-band Solid-state Power Amplifiers. *Proceedings of Asia-Pacific Microwave Conference APMC 2007*, Volume 4, pp. 2587-2590, Bangkok, Thailand, December 11-14, 2007.

Szczepaniak, Z. and Arvaniti, A. (2008). Eight-way microwave power splitter. *Proceedings of IASTED Circuits and Systems CS2008* , pp. 134-137, Kailua-Kona, USA, August 18-20, 2008.

Szczepaniak, Z., Arvaniti, A., Popkowski, J., and Orzel-Tatarczuk, E. (2009). X-band power transmitting module based on waveguide spatial power combining. *Proceedings of 10th Wireless and Microwave Technology WAMICON 2009*, April 20-21, 2009, Clearwater, Florida, USA.

Zhang Y., Kishk, A.A., Yakovlev, A.B., and Glisson, A.W. (2007). Analysis of Wideband Dielectric Resonator Antenna Arrays for Waveguide-Based Spatial Power Combining. *IEEE Transactions on Microwave Theory and Techniques,*, Vol. 55, No. 6, June 2007, pp. 1332-1340, 10.1109/TMTT.2007.896777

9

Field Plate Devices for RF Power Applications

Alessandro Chini
Department of Information Engineering
University of Modena and Reggio Emilia
Italy

1. Introduction

Microwave power transistor play a key role in today's communications system and they are a necessary component for all major aspect of human activities for entertainment, business and military applications. Recent developments in wireless communications have drastically increased the need for high-power, high efficiency, linear, low-cost, monolithic solid-state amplifiers in the 1-30 GHz frequency range. Because of these needs, there has been a significant investment in the development of high performance microwave transistors and amplifiers based on Si/SiGe, GaAs, SiC and GaN.
Improving device performance by improving the semiconductor physical properties is one of the method that can be followed in order to fabricate better devices. As proposed by Johnson (Johnson, 1965) the power - frequency product depends from the carrier saturation velocity and the semiconductor critical electric field. This means that once a semiconductor material is chosen the device performance will not improve behind certain values, unless material properties improves. On the other hand, it has been shown in the literature that device performance can be greatly enhanced by adopting dedicated device structure and fabrication methods without changing the semiconductor material. One of these structures is the so called field plate structure. This structure has been successfully implemented in RF GaAs- and GaN-based devices (Asano et al., 1998; Ando et al., 2003; Chini et al., 2004; Chini et al., 2008; Wu et al., 2004; Wu et al. 2006) boosting device power performance by 2-4 times compared to conventional ones. The origin of this improvement has been associated by many authors to at least two reasons. The first one is related to the observed increase in device breakdown voltage. Increasing the device breakdown voltage means that the device can operate at higher voltages and thus, keeping constant the device current, higher output power levels. The second one is instead related to a reduction of a parasitic effect which is called DC-to-RF dispersion or drain current-collapse (Asano et al., 1998, Ando et al.,2003; Chini et al., 2004; Chini et al., 2008). When the device is affected by this phenomenon, drain current levels reached during RF operation are lower than those recorded during DC measurements. As a consequence, the device output power during RF operation decreases and device performance are lower than expected. Several authors have experimentally observed a reduction in current-collapse for device fabricated with a field plate structure

pointing out that beside increasing the device operating voltage the field plate structure helps also in preventing drain current-collapse resulting in improved large signal RF performance compared to device without field plate.
The aim of this chapter is to provide to the reader insights into field plate operation and its geometrical optimization. After giving some basic definitions concerning the operation of an RF-power device, which will be used in order to quantify the performance of the devices studied, the optimization of a gate-connected single field-plate GaAs-based pHEMT will be presented. Field plate geometrical parameters will be varied in order to show how they can affect device properties such as breakdown voltage, maximum output power and small signal performances. It will be thus possible to quantify the maximum improvement that can be achieved by using a gate connected single field plate. Finally, some advanced field plate structure will be discussed and compared in order to point out their advantages with respect to the gate connected single field plate structure.

2. Simulated device structure and simulation parameters

For the evaluation of the field plate benefits this author has decided to focus on a typical GaAs-based pHEMT device structure for power applications. All the numerical simulations that will be presented have been carried out by means of the commercial DESSIS-ISE (Synopsis Inc.) simulator. The device structure used for numerical simulations in this work is depicted in figure 1 and is composed as follows, starting from the bottom: a semi-insulating GaAs substrate, an undoped 50nm thick AlGaAs back-barrier, an undoped 15nm thick InGaAs channel, a 5nm thick AlGaAs spacer which is n-doped with a $2x10^{17}(cm^{-3})$ concentration, a delta-doped layer with a concentration of $2x10^{12}(cm^{-2})$, a 25nm thick AlGaAs barrier which is n-doped with a $2x10^{17}(cm^{-3})$ concentration, a 20nm thick GaAs cap layer which is $2x10^{17}(cm^{-3})$ n-doped. Although not necessary for the simulation process a brief description of a possible process for the realization of the simulated device is also provided in the following. The fabrication of pHEMT devices typically starts with the deposition of the source and drain ohmic contacts on the cap-layer followed by device isolation carried out either by ion-implantation or mesa isolation. A this point a SiN passivation layer is deposited, and its thickness (t_{SIN}) will be one of the parameter that will be varied in order to evaluate field plate operation. After that SiN layer has been deposited a window is defined trough the SiN layer and the GaAs cap-layer is wet etched. In our case the defined window is 0.5μm long which corresponds to the gate length of the simulated device. At this point, after a realignment lithographic step, the gate metal is evaporated forming both the gate contact and a field-plate which is formed by covering with the gate metal a portion of the SiN layer from the gate-edge toward the drain contact. The extension of the field plate (L_{FP}) is the second parameter that will be analyzed in order to evaluate the effects of adding a gate connected single field plate structure. There are however others methods that can be used in order to fabricate field plated devices, although the resulting device behaves similarly to the one chosen here for carrying out numerical simulations. As proposed by (Chini et al., 2004) the field plate terminal can be formed on a passivated device by evaporating a second gate on top of the passivation layer and by forming an electrical connection between the gate and field plate terminal by using the common path of gate-pad and gate-feeder in the extrinsic device region.

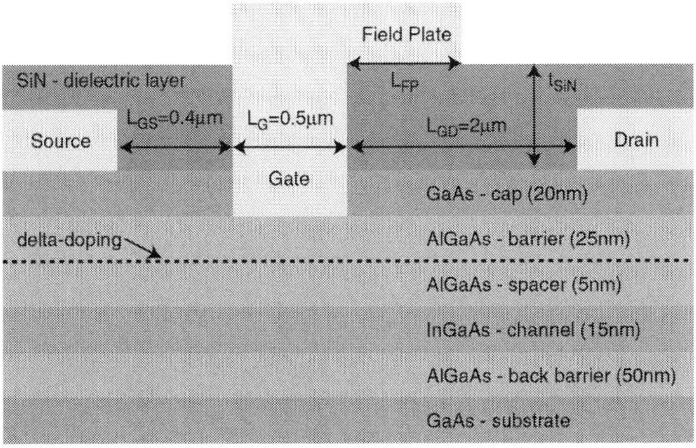

Fig. 1. Cross section of the gate connected single field plate device that will be used for the numerical simulations.

As previously stated, the device structure that will be used for numerical simulations represents a typical GaAs-based pHEMT device. This device has been chosen for the following reasons. First of all GaAs-based pHEMT are already commercially available and widely used while other devices (such as GaN HEMTs) have not reached yet a full commercialization stage. Secondly, the GaAs, AlGaAs and InGaAs material have been widely studied in the past and the physical parameters of these materials are better known than those of Nitride based ones. Since this chapter will deal with a simulated device , semiconductor parameters such as impact ionization coefficient are easier to find for GaAs-based devices, so this author decided to focus on a GaAs pHEMT device.
Concerning the physical parameters and the numerical simulations, the device structure in figure 1 has been simulated by means of hydrodynamic simulation by taking into account both gate tunnelling effects from the gate terminal and impact ionization phenomena in the InGaAs, GaAs and AlGaAs region of the device. Particularly, impact ionization coefficient used for simulation are those reported in (Robbins et al., 1988) for GaAs and AlGaAs and (Bhattacharya et al., 1986) for the InGaAs. Finally, during simulation a donor trap located at the SiN/GaAs interface with a $8x10^{12}$cm-2 density has been taken into account. The $8x10^{12}cm^{-2}$ density represent a comparable value with those reported in (Sung et al.,1994; Chini et al., 2006).
After having described the device structure let us move now on the device parameter that will be simulated in order to evaluate the effects of the field plate geometry on device performance. Since we are dealing with an RF power device and since we are interested in evaluating the improvement in its performance due to the adoption of a field plate structure it is mandatory to summarize some concepts and parameter extraction methods before that this analysis can be presented. One of the most interesting parameter for a device is its maximum output power density, typically measured in W/mm, which corresponds to the maximum output power that a 1mm wide device can deliver to a load. However, before any prediction of device performance is carried out we have to firstly define how the expected

maximum output power can be extracted from the output I-V characteristics of said device. It can be shown (Cripps, 1999) that if the device drives a maximum current which is represented by I_{MAX}, has a knee-voltage given by V_{KNEE} and that the maximum applicable voltage is given by the breakdown voltage V_{BREAK} the maximum linear power that can be obtained from the device when used as a class A linear amplifier is given by:

$$P_{OUT,LIN}=I_{MAX} * (V_{BREAK}-V_{KNEE}) / 8 \quad (1)$$

If the maximum drain current I_{MAX} is expressed in terms of A/mm equation 1 yields the maximum linear output power density. Another parameter that can be extracted, and usually easier to measure experimentally, is the saturated output power density. It can be demonstrated (Cripps, 1999) that the saturated output power is 2.1dB higher than the output linear power, or equivalently that:

$$P_{OUT,SAT}=1.61*I_{MAX} * (V_{BREAK}-V_{KNEE}) / 8 \quad (2)$$

Thus, in order to predict the maximum output power that a device can deliver to a load with respect to the two field plate parameters (L_{FP} and t_{SIN}) simulations concerning the open-channel condition, i.e. high drain currents low drain voltages, and simulations aimed at the extraction of the breakdown voltage need to be performed. In order to extract the I_{MAX} and V_{KNEE} parameters the device has thus been simulated by applying a positive gate-source voltage of 0.8V and by increasing the drain voltage up to 2V. As can be seen in figure 2 the

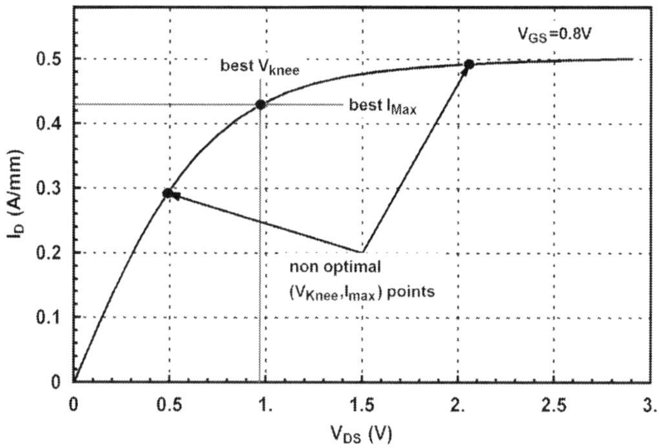

Fig. 2. Simulated output I-V characteristics for V_{GS}=0.8V. The choice of the best V_{KNEE},I_{MAX} point of the characteristics is illustrated.

drain current linearly increases until it reaches the saturation region for drain voltages higher than 1V. At this point it should be stressed that the device knee voltage and the maximum drain current have to be chosen as a point of the simulated I-V characteristics. If

the knee voltage is chosen in the linear region the device current will be lower and thus output power will be lower, as predicted by equation 1. If the knee voltage value is chosen in saturation the term (V_{BREAK}-V_{KNEE}) in equation 1 will decrease inducing a decrease in the device output power. For this reason, for each of the simulated structure, the optimum current-voltage point of the I-V characteristics have been selected for the estimation of the maximum output power.
After describing the simulation procedure used for extracting I_{MAX} and V_{KNEE} parameters lets move now to the simulation used in order to extract the device breakdown voltage. Experimentally the device breakdown voltage can be measured by adopting the method proposed by (Bahl et al., 1993). For the device studied in this chapter the experimental measurement was emulated by means of numerical simulations. With the source terminal grounded, a constant drain current level of 1mA/mm was forced into the device while the gate voltage was swept from 0V to -1.5V. By monitoring the drain voltage it has been possible to obtain the experimental data depicted in figure 3, which qualitatively corresponds to the data that can typically be obtained on real devices (Bahl et al., 1993). As described in (Bahl et al., 1993) the drain-source breakdown voltage is given by the highest value reached from the V_{DS} characteristic during the gate voltage sweep.

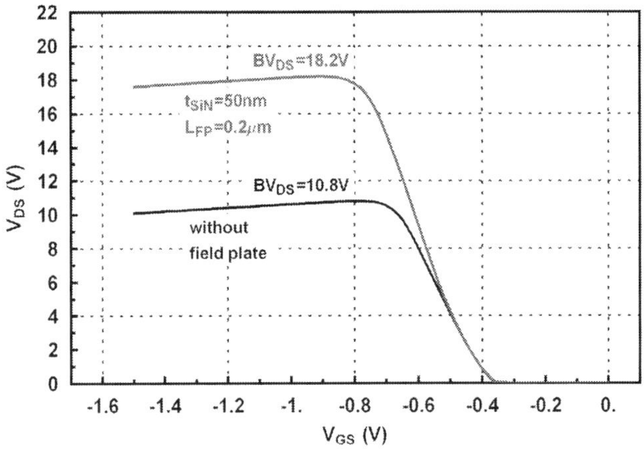

Fig. 3. Simulated off state breakdown measurements at a drain current level of 1mA/mm for a device without field plate and a device with LFP=0.2mm and tSiN=50nm. The highest drain voltage reached during the measurement (BV_{DS}) represents the maximum drain-source voltage that can be applied before reaching breakdown condition.

After defining the equation used for the evaluation of the device maximum output power, and the simulation methods used for extracting the device breakdown, knee-voltage and maximum drain current we can move to the next stage of this section that is represented by the analysis of the dependence of breakdown voltage and output power from the field plate parameters L_{FP} and t_{SiN}.

3. Breakdown dependence from field plate geometry

After describing the device used for the simulation and the parameter used, it is now possible to start analyzing the effects of the field plate geometry on device breakdown. As previously stated, field plate geometry has been varied by acting on two parameters: the field plate length L_{FP} and the silicon nitride dielectric layer thickness (t_{SiN}). Particularly , values of 0.2,0.4,0.6,0.9,1.2 and 1.6μm have been taken into account for L_{FP}, while thicknesses ranging from 30 to 90nm have been used for t_{SiN}. Various simulation have been carried out in order to simulate all the devices and the results in terms of breakdown voltage are summarized in figure 4. At a first glance it is possible to notice that, except for the case

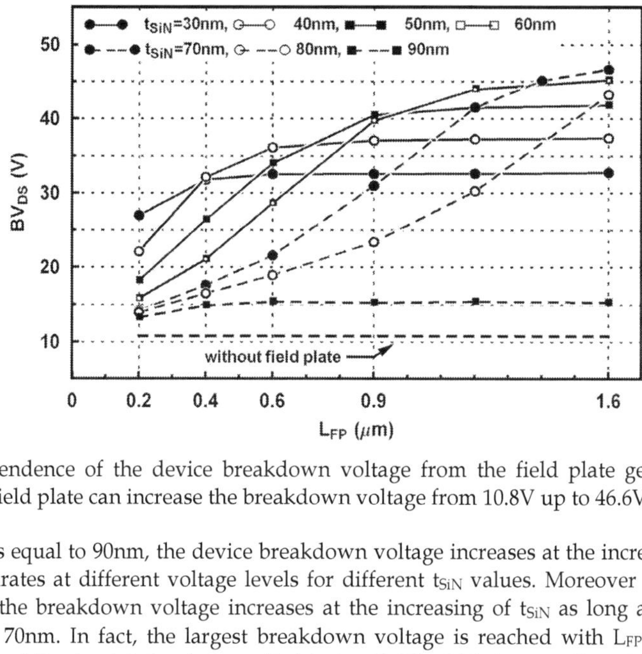

Fig. 4. Dependence of the device breakdown voltage from the field plate geometry. An optimized field plate can increase the breakdown voltage from 10.8V up to 46.6V.

where t_{SiN} is equal to 90nm, the device breakdown voltage increases at the increasing of L_{FP} until it saturates at different voltage levels for different t_{SiN} values. Moreover we can also notice that the breakdown voltage increases at the increasing of t_{SiN} as long as t_{SiN} is not larger than 70nm. In fact, the largest breakdown voltage is reached with L_{FP}=1.6μm and t_{SiN}=70nm and its simulated value resulted to be 46.6V which is more than 4 times larger than the breakdown of the device without field plate which resulted to be 10.8V (V_{DG} at breakdown is approximately 11.5V), see figure 3. By increasing the t_{SiN} value over 70nm the breakdown voltage start to decrease quite rapidly reaching a 15.3V value when t_{SiN} is equal to 90nm.

Running all the simulation with different geometrical parameters brings us to the following conclusions, that of course will be explained in the following:

1) increasing L_{FP} initially increases the breakdown voltage
2) increasing L_{FP} after a certain value does not give any further increase in device breakdown voltage
3) there is an optimum SiN thickness that maximize the device breakdown voltage

Now, in order to better understand the field plate "action" it is necessary to look at the electric field profile at breakdown condition for the various geometry tested. First of all, a comparison between the device without field plate and a device with field-plate can explain where the increase in breakdown voltage comes from. As can be seen in figure 5 the electric field profile of the device without field plate presents a single peak located at the

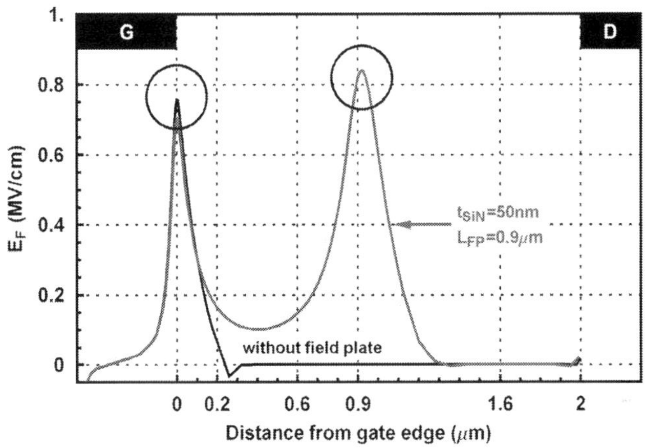

Fig. 5. Electric field profiles at breakdown in the device InGaAs channel for a device without field plate and for a device with field plate. When a field plate is added the electric field profile shows two peaks, one located at the gate contact edge, the other located at the field plate contact edge.

drain edge of the gate contact. This high electric field gives raise to at least two mechanisms that contribute to drive the device into breakdown. The high electric field at the edge of the gate contact enhances electron tunnelling from the gate to the device channel increasing, in absolute value, the total gate current (Meneghesso et al., 2003). The other mechanisms that take places are instead impact ionization phenomena which gives raise to the formation of electrons and holes pairs. The electrons are collected from the drain contact while holes are collected from the gate and the source terminal (Meneghesso et al., 2003). Since holes are coming out from the gate terminal their current has the same sign as the electrons one. As a consequence gate current becomes more negative when impact ionization phenomena are taking places. Since both of this mechanisms are triggered by high electric fields, it is clear that one way to increase the device breakdown is to lower electric field values in the gate-drain device region while increasing the area of the electric field profile. In fact this is what happens if we observe the electric field profile at breakdown for a device with a field plate. First of all two electric field peaks are present in the gate-drain device region, and secondly the electric field profile has a largest area which corresponds to an higher breakdown voltage. So the ability of the field plate structure in increasing the breakdown voltage is related to the splitting of the electric field peaks and its distribution across the gate-drain region.

3.1 Dependence of Breakdown from the dielectric layer thickness

Let us move now to some electric field profile obtained for t_{SiN}=30,70 and 90nm with a constant L_{FP} of 1.6 μm. As can be seen in figure 6 the electric field profile at breakdown for t_{SiN}=30nm presents two electric field peaks but the one at the gate edge is smaller than that at the field plate edge. On the other hand the electric field profile at breakdown for t_{SiN}=70nm presents two balanced electric field peaks while the electric field profile at

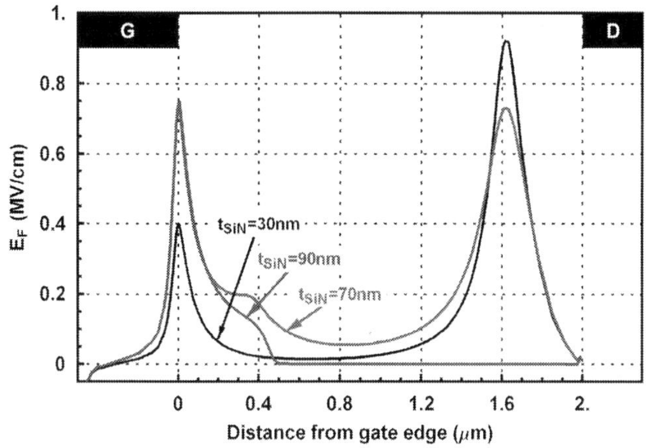

Fig. 6. Electric field profiles at breakdown in the device InGaAs channel for field plated devices with different dielectric layer thicknesses.

breakdown for t_{SiN}=90nm shows only one electric field peak located at the gate edge. From figure 6 it is also straightforward to notice that the electric field profile with the largest area is the one with t_{SiN}=70nm which actually corresponds to the field plate geometry that yields the highest breakdown voltage. In order to better understand the mechanism relating the device breakdown voltage with the thickness of the SiN layer it is now useful to consider the pinch-off voltage of the MIS structure formed by the field plate terminal, the SiN layer and the active layers of the pHEMT. Numerical simulations carried out on the structure depicted in figure 7 by applying a small drain to source voltage of 0.1V and by sweeping the field plate voltage towards negative values show that the pinch-off voltage of this structures increases at the increasing of the SiN thickness. As can be seen in figure 8, the pinch-off voltage for a 30nm SiN thick MIS structure is about -8V, while it increases up to -27V for a 90nm SiN thick MIS structure.

Since for t_{SiN}=30nm the field-plate terminal will deplete the InGaAs and GaAs layers located below it once a total reverse gate-drain voltage of 8V is applied, the electric field peak at the gate edge will be frozen at the value reached for V_{DG}=8V and when the V_{DG} voltage will be increased another electric peak will form at the field plate edge. Since the pinch-off voltage for t_{SiN}=30nm is much smaller than the V_{GD} voltage at which the device without field plate reaches breakdown condition, see figure 3, the electric field peak value will be lower at the gate edge (about 0.4MV/cm) with respect to the value reached at breakdown for the device without field plate (about 0.75MV/cm, see figure 3). For V_{DG} voltages larger than 8V the

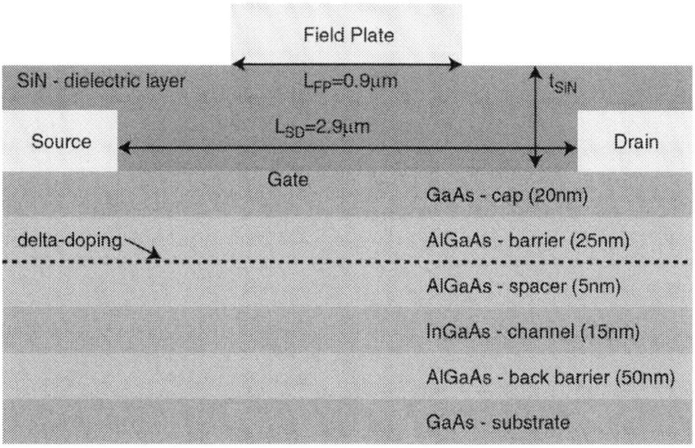

Fig. 7. Cross section of the simulated MISpHEMT structure.

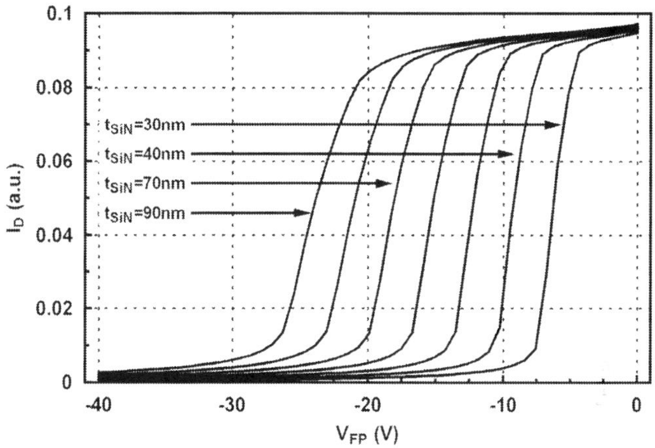

Fig. 8. Simulated pinch-off voltages for the MISpHEMT structures for different values of the dielectric thickness. For t_{SiN}=30nm the pinch-off voltage is approximately -8V while it increases, in absolute value, up to -28V when t_{SiN} is equal to 90nm.

device with t_{SiN} experiences thus the formation of a second electric field peak at the field plate edge which eventually reaches a level of 0.9MV/cm when the device breakdown condition occurs. Thus, for small values of tSiN the electric field profile shows two peaks the smaller one located at the gate edge.
An opposite behaviour can be observed instead for t_{SiN}=90nm. Since the pinch-off voltage of the MIS structure is larger (about -27V) the field plate is not able to deplete the gate drain access region before breakdown condition at the gate edge occurs. The electric field peak is thus located at the gate edge and since the second peak does not form at the field plate edge

the increase in the electric field profile area is very low. As a consequence, the improvement in terms of breakdown voltage is very low. Finally, when analyzing the electric field profile for t_{SiN}=70nm it is straightforward to notice that the electric field peaks at the gate and field plate edges are both approximately 0.7MV/cm. This means that the field plate has started to deplete the gate-drain access region just before the device was reaching breakdown at the gate junction. This is the best solution in order to achieve high breakdown voltages, since once the field plate depletes the gate-drain region any other increase in the V_{DG} reverse voltage will give rise to an increase in the depletion region at the field plate edge while the electric field at the gate edge will remain almost unchanged. Although the pinch-off voltage of the MIS structure for tSiN=70nm is about -21V the reader might ask why the field plate is able to increase the breakdown voltage that should happen for a V_{GD} of approximately 12V. It should be noted that the field-plate starts to deplete the gate drain access region at V_{GD} voltages of about 11-12V. Even if small, any decrease in the charge concentration of the gate drain access region will help in improving the breakdown voltage and it is this small modulation that prevents the device for reaching breakdown before the field plate fully depletes the gate drain access region. This is the reason for which the field operation it is still possible also for t_{SiN}=70nm although the pinch-off voltage of the MIS structure is slightly larger.

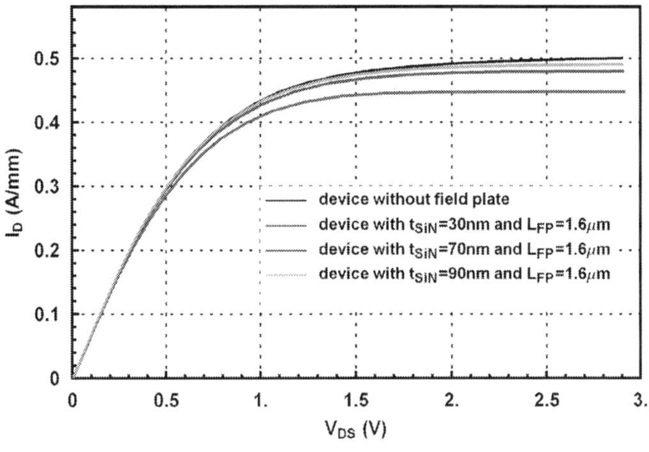

Fig. 9. Simulated output I-V characteristics with V_{GS}=0.8V for different values of the dielectric layer thickness. Thin dielectric layers help in lowering the electric field peak at the gate edge, thus reducing the device output conductance when the device operates in the saturation region.

Concerning the dependence from t_{SiN} of the field plate operation it is thus possible to conclude that if the dielectric layer is too thin the field plate will give some advantages in terms of device breakdown but they might not be the best one achievable. Increasing t_{SiN} will bring to the best result which correspond in having two balanced peaks at the gate and field plate edge in the electric field profile. Finally, if t_{SiN} is too thick there might not be any field plate operation at all since the device will reach breakdown condition before that the field plate can actually start to deplete the gate drain access region. It is interesting to notice also that the dependence of the electric field peak at the gate edge from the thickness of the SiN layer can be seen when comparing the output I-V characteristic in the saturation region for the simulated devices. As can be seen in figure 9 the device with the thinnest silicon nitride layer has a lower output conductance which increases at the increasing of the t_{SiN} parameter. This is a consequence of the lowering of the electric field peak value at the gate edge compared to the value reached for the device without field plate.

3.2 Dependence of Breakdown from the field plate extension

After gaining some insights in the dependence from t_{SiN} of the breakdown voltage it is possible now to analyze its dependence from the field plate length parameter by keeping the SiN thickness constant to a value of 50nm. As can be seen in figure 10 the electric field profile for L_{FP}=0.2mm present two electric field peaks very close to each other while by

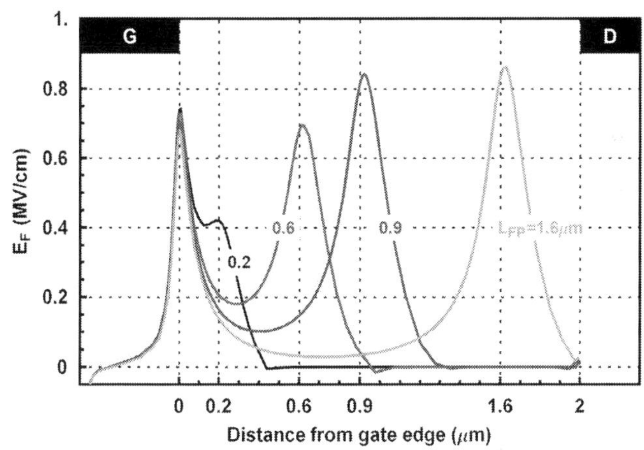

Fig. 10. Simulated off state breakdown measurements at a drain current level of 1mA/mm for a device without field plate and a device with LFP=0.2mm and tSiN=50nm. The highest drain voltage reached during the measurement (BV_{DS}) represents the maximum drain-source voltage that can be applied before reaching breakdown condition.

increasing L_{FP}=0.6 μm splits the two electric field peaks inducing an increase in the electric field profile area which results into an increase in device breakdown voltage. Increasing further the L_{FP} value shifts the peak at the field plate edge away from the gate one thus increasing the electric field area when the device reaches breakdown condition. This results into an increase of the breakdown voltage. However, at the increase of L_{FP}, the electric field in the region between the two peaks (i.e. gate edge and field plate edge) decreases. This explains the decreases of the derivative of breakdown voltage versus field plate length at the increasing of L_{FP}. In fact, the electric field area (and thus the breakdown voltage) does not increase significantly once the two peaks are far away from each other.

4. Output power and small signal parameters dependence from field plate geometry

By combining the results obtained for the device breakdown voltage by emulating the breakdown measurement technique by means of numerical simulations, and by simulating open channel I-V characteristics from which the optimum I_{MAX} and V_{KNEE} parameters can be extracted it is now possible to estimate the expected output power for the different field plate geometries that have been taken into account. Moreover, since the field plate terminal adds a parasitic capacitance between the gate and the device channel, s-parameter data have also been simulated in order to extract the device current gain cutoff frequency f_t and the power gain cutoff frequency f_{max}. The benefits of the field plate geometry will thus be evaluated in terms of absolute power levels and in terms of power-frequency product both by considering f_t and f_{max} as the frequency terms. When looking at the absolute power level that can be reached by adding the field plate structure to the simulated pHEMT device it can be seen that they follow the results previously obtained for the breakdown voltage values. In fact, since the field plate action typically takes place at high V_{DG} voltages the parameters I_{MAX} and V_{KNEE} are almost unaffected from the field plate geometry. As a consequence the only term in equation 2 that strongly depends for the field plate geometry is the device breakdown voltage. As can be seen in figure 11 numerical simulations predicts that the output power density can be improved from 0.9W/mm for the device without field plate up to a value of 4.3W/mm in the best case which corresponds to t_{SiN}=70nm and L_{FP}=1.6μm. The

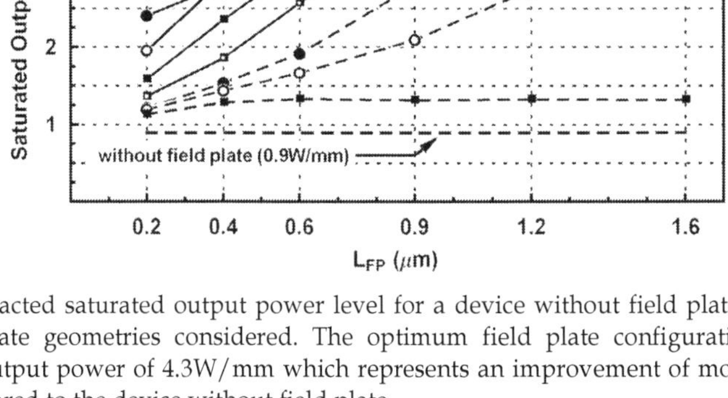

Fig. 11. Extracted saturated output power level for a device without field plate and for all the field plate geometries considered. The optimum field plate configuration yields a saturated output power of 4.3W/mm which represents an improvement of more than four times compared to the device without field plate.

value of 4.3W/mm is quite impressive compared that is obtained with a GaAs-based device but it should be noted that power densities in the 2 to 3.5 W/mm range have been reported in the literature for GaAs pHEMT (Fanning et al., 2007; Chini et al., 2008), while GaAs pHEMT without field plate typically operate in the 0.7-1W/mm range (Ross et al., 1996;

Chini et al., 2008). The obtained 4.3W/mm value is of course optimistic since it does not take into account other phenomena, such as device self-heating, that might degrade device operation, nevertheless it gives us an important information in terms of which is the "boosting" factor of a gate connected single field plate structure. In terms of absolute power an optimized field plate device can reach power densities up to 4.7 times higher than the device without field plate.

While everything seems to be very exciting in terms of output power it is now mandatory to evaluate the effects of the added field plate structure to the small-signal performances of the device. Several authors have reported a decrease in device power gain when adding field plate structures (Asano et al., 1998; Ando et al., 2003; Wu et al., 2004), and the main reason has been related to the added parasitic capacitance between the field plate and channel capacitance which gives rise to an increase in the device gate drain capacitance. The effect of increasing the gate drain capacitance is to reduce both the current gain cutoff frequency and the power gain cutoff frequency whose expression are given by (Ross et al., 1996):

$$f_t = g_m / [2\pi(C_{GS} + C_{GD})] \quad (3)$$

$$f_{max} = f_t [4\, g_o\, (R_S + R_i + R_G) + 2(C_{GD}/C_{GS})((C_{GD}/C_{GS}) + g_m(R_S + R_i))]^{-1/2} \quad (4)$$

From equations 3 and 4 it is straightforward to notice that in order to improve the current gain and power gain cutoff frequency all parameters have to be as low as possible except for the device trasconductance g_m which has to be as large as possible (Ross et al., 1996). Let us move now to the evaluation of field plate geometry on the device current gain cutoff frequency. The small signal parameters for all the field plate geometries previously considered have been extracted at a gate-source voltage of 0V and at a drain voltage of 3V. With these values the device is biased into saturation and its current is approximately 0.2A/mm which corresponds to approximately half the maximum current considered for the estimation of the maximum saturated output power. In order to take into account the effect of gate resistance, which is affecting the extraction of the power gain cutoff frequency, the simulated device has been modelled as a 10x100μm wide device. Since the field plate terminal contributes in reducing the device fingers resistance, the total gate resistance used during simulation has been scaled accord ling to $L_{FP}+L_G$. Particularly for the device without field plate, where L_G=0.5μm, a total gate resistance of 0.7 Ohm has been considered while for the devices with L_{FP}=0.2μm a total gate resistance of 0.5 Ohm has been used. The value used for R_G are reasonably comparable with those of commercially available pHEMT devices with comparable gate lengths. As can be seen in figure 12 the devices with field plate show always lower f_t values that the device without field plate. This decrease in f_t is due to the added gate capacitance that forms between the field plate terminal and the device channel. In fact by considering the simulated g_m and C_G values, see figures 13 and 14 it is straight forward to notice that field plated devices have higher gate capacitance, up to 9 times higher than the device without field plate, while the trasconductance value experiences only a

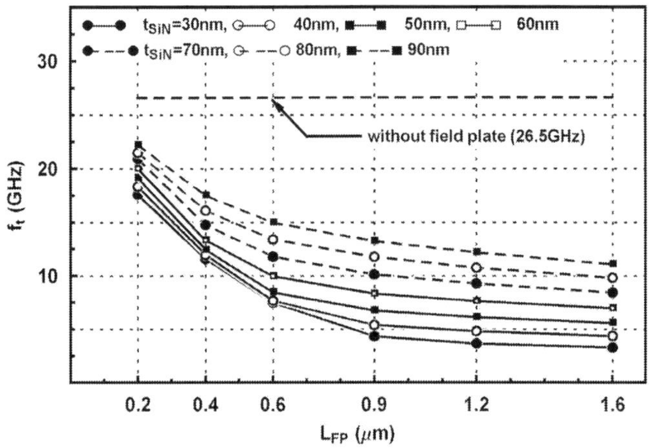

Fig. 12. Simulated current gain cutoff frequency for different field plate geometries. Using thin dielectric layer and/or large field plate extensions result in a large reduction of the device current gain cutoff frequency.

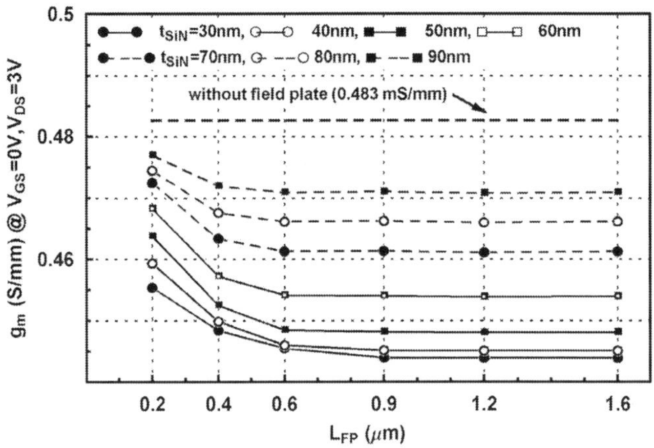

Fig. 13. Simulated intrinsic device trasconductance as obtained by small-signal parameters for different field plate geometries. Changes observed are within 10% of the trasconductance value of the device without field plate.

small decrease which is less than 10%. The decrease in f_t is thus due to the increase in the total gate capacitance which increases at the increasing of the field plate extension L_{FP} and at the decreasing of the dielectric layer thickness t_{SiN}. The reader might notice that the dependence of C_G from L_{FP} is not linear. This is due to the fact that since the device is biased

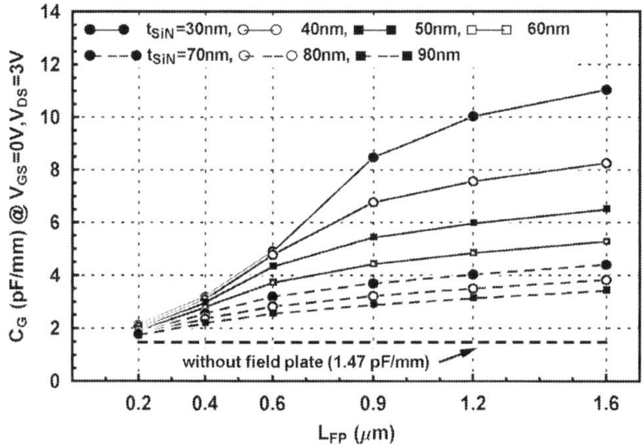

Fig. 14. Simulated total gate capacitance as obtained by small-signal parameters for different field plate geometries for V_{GS}=0V and V_{DS}=3V. Due to the field plate terminal the gate capacitance can increase up to 9 times compared to the device without field-plate.

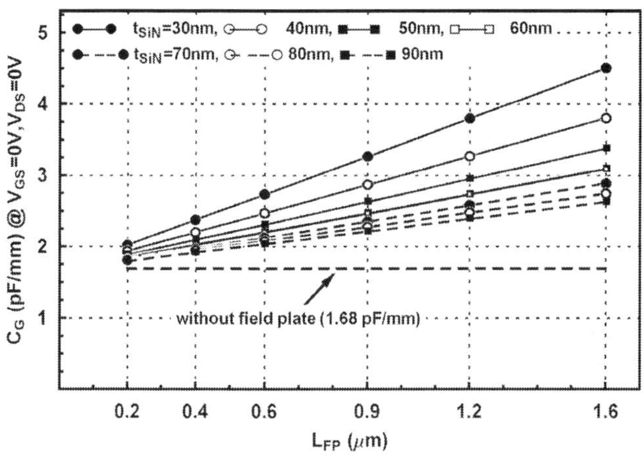

Fig. 15. Simulated total gate capacitance as obtained by small-signal parameters for different field plate geometries for V_{GS}=0V and V_{DS}=0V. The total gate capacitance increases linearly with the field plate extension parameter (L_{FP}).

into saturation a portion of the gate drain access region is depleted. For this reason C_G does not scale linearly with L_{FP}. If the C_G values are extracted by keeping gate, source and drain terminal all at 0V one can obtain the C_G value depicted in figure 15 where the linear dependence of C_G from L_{FP} is clearly visible.

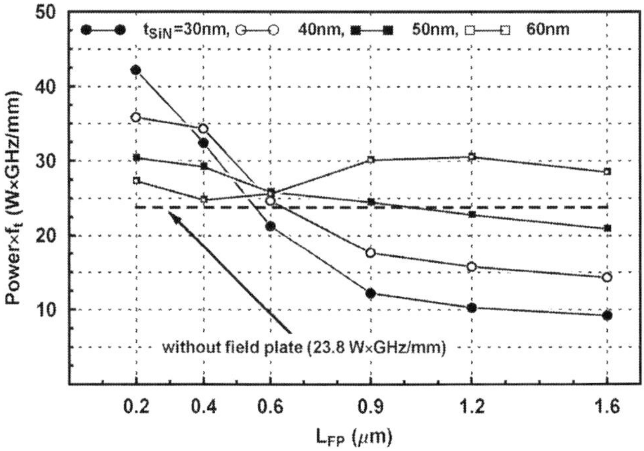

Fig. 16. Simulated power - current gain cutoff frequency product for some of the field plate geometries considered. The best result of the field plated devices is obtained with t_{SiN}=30nm and L_{FP}=0.2μm.

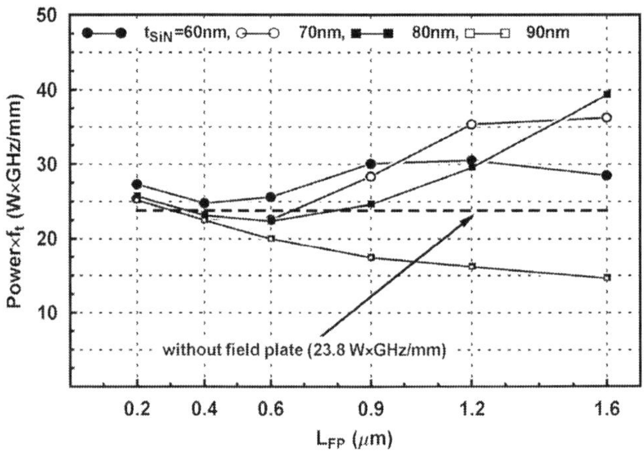

Fig. 17. Simulated power - current gain cutoff frequency product for some of the field plate geometries considered. The best result of the field plated devices is obtained with t_{SiN}=30nm and L_{FP}=0.2μm.

After having extracted the f_t values, it is possible now to calculate the power - current gain cutoff frequency product for the various geometries here considered. Has can be seen in figures 16 and 17, where the power - current gain cutoff frequency product are depicted, the best field plate geometry for improving this figure of merit is represented by the

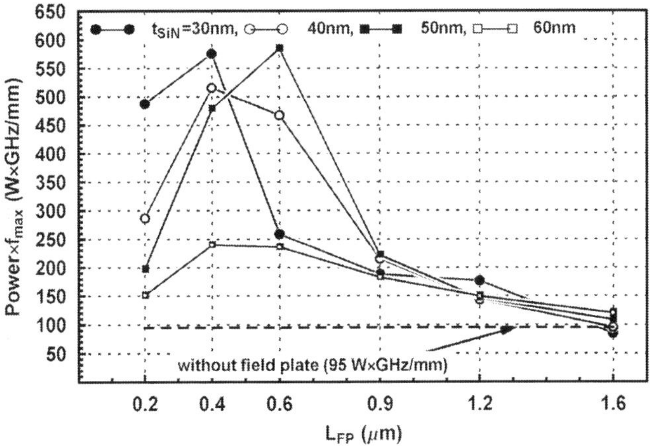

Fig. 18. Simulated power - power gain cutoff frequency product for some of the field plate geometries considered. The best results of the field plated devices are obtained with t_{SiN} ranging from 30 to 50nm and L_{FP} ranging from 0.4 and 0.6μm.

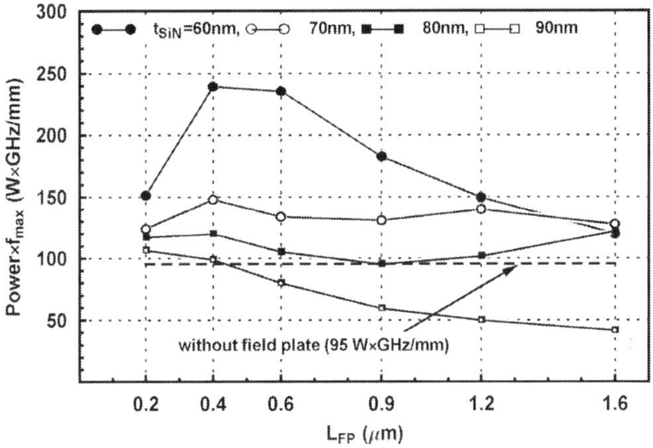

Fig. 19. Simulated power - power gain cutoff frequency product for some of the field plate geometries considered. The best results of the field plated devices are obtained with t_{SiN} ranging from 30 to 50nm and L_{FP} ranging from 0.4 and 0.6μm.

combination of L_{FP}=0.2μm and t_{SiN}=30nm. High values of this FOM are also reached for the field plate geometries with large L_{FP} (1.2 and 1.6μm) and t_{SiN} values (70 and 80nm).
Since the simulated structures represent typical power devices it is also interesting to evaluate the device performance in terms of another figure of merit. Particularly, the f_{max} values have been calculated from the small-signal simulations and another FOM defined as the power - power gain cutoff frequency product has been considered. As can be seen in

figures 18 and 19 the best FOM values are obtained with L_{FP} values in the 0.4-0.6μm range for t_{SiN} values in the 30 to 50nm range where power - power gain cutoff frequency products higher than 500 WxGHz/mm can be obtained. Thus, concerning this FOM, an optimum field plate geometry can increase by up to 5 times the power - power gain cutoff frequency product. Some comments are however mandatory on the results obtained. The numerical simulations carried out might not take into account all the real device parasitics, so the absolute value of the parameters extracted might not be too realistic. On the other hand, their ratio should instead be quite reasonable giving us an approximate value for the improvements that one can expect by adopting an optimized field plate geometry. But where does the improvement come from? As written in equation 4 the f_{max} can be improved by decreasing all the parasitics. In our case it might look like parasitics are increasing since the field plate structure increases the gate drain capacitance. However, by looking into the small signal parameter data, one can notice that the device output conductance g_o decreases when the field plate structure is adopted. As can be seen in figure 20, the device go can be reduced by up to 25 times for the case of L_{FP}=1.6μm and t_{SiN}=30nm. Beside the reduction of g_o, another term that is improving with the increase of L_{FP} is the total gate resistance which value decreases at the increasing of L_{FP}. The effects of the reduction of g_o and R_G are thus counter balancing the increase in the added capacitance at least within a certain range of L_{FP} and t_{SiN} giving rise to a 5 times improvement of the power - power gain cutoff frequency product when an optimum field plate geometry is adopted.

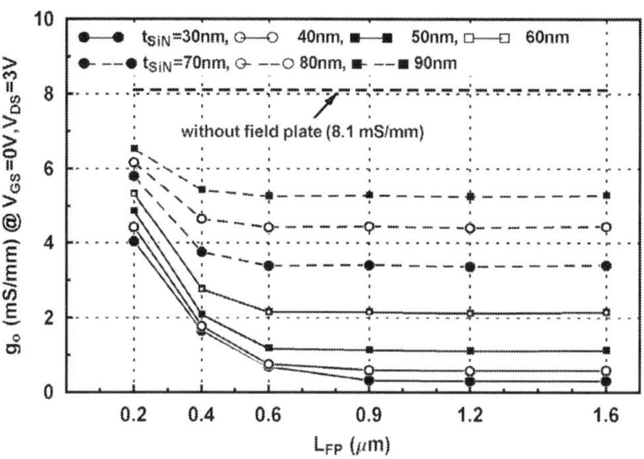

Fig. 20. Dependence of the device output conductance from the field plate geometry. Thin dielectric layers yield lower value of the device output conductance.

5. Advanced field plate structures

When analyzing the performance of the gate connected single field plate it has been shown that the field plate is able to increase the device breakdown by splitting the electric field peak at the gate edge into two peak located at the gate and the field plate edges. On the other hand, the added parasitic capacitance decreases or limit the small-signal performance. In order to further improve the device performance in terms of breakdown and/or small signal characteristics two different advanced field plate structure can be implemented.
The first one is represented by a double field plate structure (Y.-F. Wu et al., 2006) where basically two field plate terminal are placed in the gate drain access region, see figure 21.

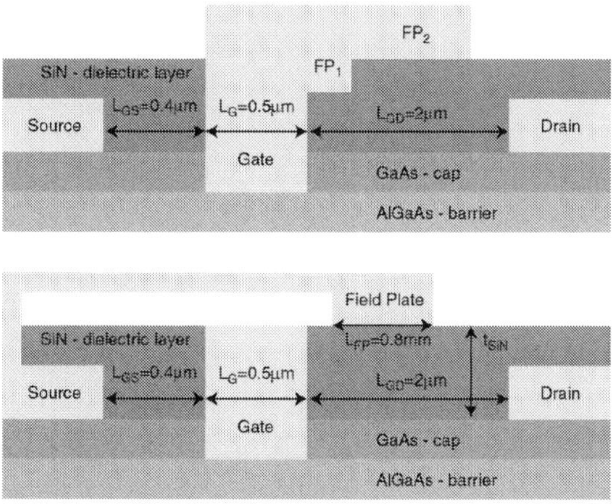

Fig. 21. Simulated advanced field plate structures. Top: double field plate structure connected to the gate terminal. Bottom: source connected single field plate structure.

The benefits arising from this structure are that the breakdown voltage can be further improved compared to the maximum achievable by a single field plate structure, and that it is possible to decrease the added capacitance while obtaining higher or comparable breakdown voltages with respect to the single field plate geometry. The first field plate increases the breakdown voltage of the device while the second field plate can be placed on a dielectric thickness larger than those needed for single field plate operation. Nevertheless the second field plate further increase the device breakdown voltage. Numerical simulations have been carried out by adding a double field plate structure to the pHEMT device. A first field-plate with t_{SiN}=30nm and L_{FP}=0.2µm and a second field plate with t_{SiN}=90nm and L_{FP}=1.4µm have been adopted. As can be seen in figure 22 the breakdown voltage reaches 59.6V, while when looking at the electric field profile at breakdown, see figure 23, three peaks are clearly visible. One is located at the gate edge, the second one at the edge of the first field plate and finally the second one is located at the edge of the second field plate. The double field plates device has reached a power - current gain cutoff frequency product of 50.7 which is larger than 42.2 that represents the best achievable power - current gain cutoff frequency product from a single field plate structure.

A second advanced structure is represented by the source-connected field plate, see figure 21. Basically, instead of connecting the field plate terminal to the gate one, it is connected to the source which remains usually grounded during normal operation.

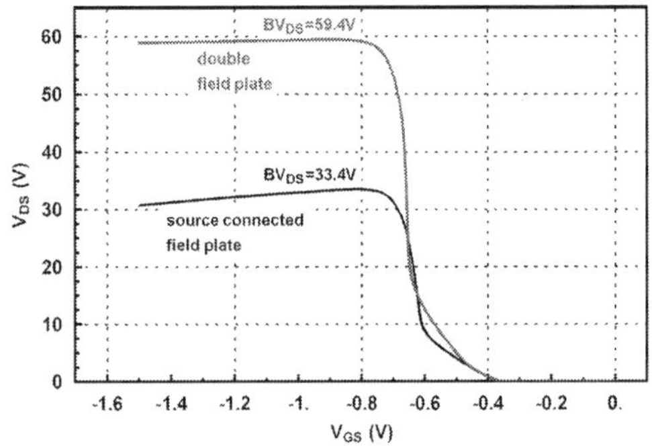

Fig. 22. Simulated off state breakdown measurements at a drain current level of 1mA/mm for a device with a source connected single field plate structure and for a double field plate structure device.

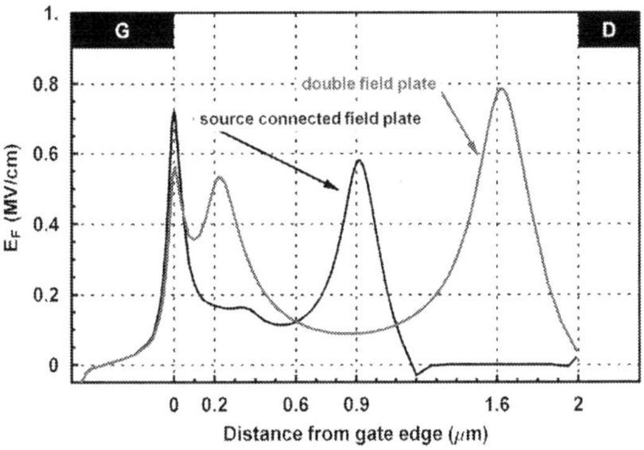

Fig. 23. Electric field profiles at breakdown in the device InGaAs channel. The device with a source connected field plate shows two electric peaks while for the double field plate device three peaks are clearly visible.

At the increasing of the drain voltage, since V_{DS} will become more positive, the field plate will start to deplete the gate-drain access region in a similar way to what happens when the field plate is connected to the gate. As long as gate voltages do not differ too much from the source voltage (Y.-F. Wu et al., 2004), field plate operation should remain almost unchanged in terms of device breakdown. The main advantage of this structure is however related to the fact that the field plate parasitic capacitance does not behave as a gate-drain one. Instead an increase of C_{DS} will happen, but as can be seen in equations 3 and 4, drain-source capacitance does not affect neither f_t or f_{max}. A pHEMT with a source connected field plate has been thus simulated in order to evaluate the benefits of this structure. A SiN thickness of 50nm and an L_{FP}=0.8μm while leaving a space of 0.1μm from the gate to the field plate contact was implemented, see figure 21. An off-state breakdown voltage of 33.4V was obtained, see figure 22, while the electric field profile at breakdown behaves in a similar way to those achievable with a gate connected field plate, see figure 23. The most interesting results are however related to the improvements in terms of small signal performance. The source connected field plate device has reached a power - current gain cutoff frequency product of 80.5 which is almost twice than 42.2 that represents the best achievable power - current gain cutoff frequency product from a single field plate structure.

6. Conclusions

Numerical simulations carried out on different field plate geometries applied to a GaAs-based pHEMT have yielded some insights on the improvements and the limitations of said structures. Field plate structure has been and will be used for the fabrication of power devices due to the extremely high benefits that can arise from this structure. It has been shown that although parasitic capacitance are added to the device, its performance can greatly improve both in terms of the power - current gain cutoff frequency and the power - power gain cutoff frequency products. The main parameter for some of the field plate geometries considered and for the two advanced structure discussed in the previous section are reported in table 1.

Device#	1	2	3	4	5	6
t_{SiN} (nm)	0	300	500	700	500 (Source)	300/900
L_{FP} (mm)	0	0.2	0.6	1.6	0.9	0.2/1.4
BV_{DS} (V)	10.8	27	34	46.6	33.4	59.4
P_{out} (W/mm)	0.9	2.4	3	4.3	3	5.5
f_t (GHz)	26.6	17.6	8.5	8.4	27.2	9.3
FOM1 (W GHz/mm)	23.8	42.2	25.8	36.2	80.5	50.7
FOM2 (W GHz/mm)	95	480	585	130	920	180

Table 1. Breakdown voltage (BV_{DS}), maximum output power (P_{out}), current gain cutoff frequency (f_t), power-current gain cutoff frequency product (FOM1) and power-power gain cutoff frequency product (FOM2) for some of the field plate structures and geometries that have been simulated.

From the value reported in table 1 it is clear that field plate parameters have of course to be tailored to the specific application in order to achieve the best results: higher operating voltages and/or output power densities can be obtained at the expenses of power gain. Advanced structures such as source connected field plate and double field plates can further improve devices performances by dramatically improving the small signal performance in the case of the source connected field plate or by further increasing the device operating voltages in the case of the double field plate structure.

7. References

Ando Y., Okamoto Y., Miyamoto H., Nakayama T., Inoue T., and Kuzuhara M. (2003) "10-W/mm AlGaN-GaN HFET With a Field Modulatine Plate", IEEE Electron Device Letters, Vol. 24, No. 5, pp. 289-291, May 2003.

Asano K., Miyoshi Y., Ishikura K., Nashimoto Y., Kuzuhara M., and Mizuta M. (1998) "Novel high power AlGaAs/GaAs HFET with a field-modulating plate operated at 35 V drain voltage", Proceedins of the International Electron Devices Meeting (IEDM '98), pp. 59-62, 6-9 Dec. 1998.

Bahl S. R., and del Alamo J. A. (1993), "A New Drain Injection Technique for the Measurement of Off-State Breakdown Voltage in FET's", IEEE Transactions on Electron Devices, Vol. 40, No. 8, pp. 1558-1560, Aug. 1993.

Bhattacharya P. K., Das U., Juang F. Y., Nashimoto Y., and S. Dhar, (1986), "Material properties and optical guiding in InGaAs-GaAs strained layer superlattices-a brief review", Solid-State Electronics, Vol. 29, No. 2, pp. 261-267, February 1986.

Chini A., Buttari D., Coffie R., Heikman S., Keller S., and Mishra U. K., (2004), "12 W/mm power density AlGaN/GaN HEMTs on sapphire substrate", IET Electronics Letters, Vol. 40, No. 1, pp. 73-74, Jan. 2004.

Chini A., S. Lavanga, M. Peroni, C. Lanzieri, A. Cetronio, V. Teppati, V. Camarchia, G. Ghione, and G. Verzellesi (2006) "Fabrication Characetrization and Numerical Simulation of High Breakdown Voltage pHEMTs", Proceedings of the 1st European Microwave Integrated Circuits Conference, pp. 50-54, Sept. 2006.

Chini A., Esposto M., verzellesi G., Lavanga S., C. Lanzieri, A. Cetronio (2008) "Characterization and Numerical Simulations of High Power Field-Plated pHEMTs", Proceedings of the 3rd European Microwave Integrated Circuits Conference, pp. 218-222, Oct. 2008.

Cripps S. C. (1999) "RF Power Amplifiers for Wireless Communications", Artech House, ISBN 0-89006-989-1.

Fanning D., Balistreri A., Beam III E., Decker K., Evans S., Eye R., Gaiewski W., Nagle T., Saunier P., and Tserng H.-Q. (2007) "High Voltage GaAs pHEMT Technology for S-band High Power Amplifiers", CS MANTECH Conference, Austin, Texas, pp. 173-176, 14-17 May 2007

Johnson E. O., (1965), "Physical limitation on frequency and power parameters of transistors ", RCA Rev., pp. 163–176, Jun. 1965.

Meneghesso G., Chini A., Maretto M., and Zanoni E. (2003), "Pulsed Measurements and Circuit Modeling of Weak and Strong Avalanche Effects in GaAs MESFETs and HEMTs", IEEE Transactions on Electron Devices, Vol. 50, No. 2, pp. 324-332, Feb. 2003.

Robbins V. M., Smith S. C., and Stillman G. E. (1988), "Impact ionization in AlxGa1-xAs for x=0.1-0.4", Applied Physics Letters, Vol. 52, No. 4, pp. 296-298, 1988.

Ross R. L., Svensson S. P., and Lugli P. (1996) "Pseudomorphic Hemt Technolgy and Applications", Springer, ISBN 0792339150.

Sung Y.-K., Kim J.-K., Ju Y.-J., and Ryoo J.-H. (1994), "The Passivation of GaAs by laser CVD", Proceedings of the 44th Electronic and Technolgy Conference, pp. 635-639, 1-4 May 1994.

Wu Y.-F., Saxler A., Moore M., Smith R.P., Sheppard S., Chavarkar P.M., Wisleder T., Mishra U.K., and Parikh P. (2004) "30-W/mm GaN HEMTs by field plate optimization", IEEE Electron Device Letters, Vol. 25, No. 3, pp. 117-119, Mar. 2004.

Wu Y.-F., Moore M., Wisleder T., Chavarkar P.M., Mishra U.K. and Parikh P. (2004) "High-gain Microwave GaN HEMTs with Source-terminated Field-plates", Proceedins of the International Electron Device Meeting (IEDM '04), pp. 1078-1079, 13-15 Dec. 2004.

Wu Y.-F., Moore M., Wisleder T., and Parikh P. (2006) "40-W/mm Double Field-plated GaN HEMTs", Proceedins of the 64th Device Research Conference, pp. 151-152, June 2006.

10

Implementation of Low Phase Noise Wide-Band VCO with Digital Switching Capacitors

Meng-Ting Hsu, Chien-Ta Chiu and Shiao-Hui Chen
Microwave Communication and Radio Frequency Integrated Circuit Lab
National Yunlin University of Science and Technology,
Department of Electronic Engineering
Yunlin, Taiwan, Republic of China

1. Introduction

In present fast-growing wireless communications, requires wide bandwidth, low-power and low-cost RF circuits [1]. In Fig. 1, it is a simple super-heterodyne transceiver [2], and in this diagram, VCO (voltage-controlled oscillator) is one of the most important building blocks in the wireless communication system. An optimum performance VCO should include low phase noise and wide bandwidth to support several communication standards of wireless transceiver, and low power design technique to enhance the battery lifetime. Recently, the standard CMOS process technology is better choice to overcome low-cost challenge. The choice is also favored by the possibility of system-on-chip integration with digital parts, which should save the total chip area and cost. The VCO with multi-band and wideband are the current trend [3] - [8]. The methods of increasing tuning range are classified as follows, switching inductors or variable inductor [3], switching capacitor modules [4], [5], varactors in parallel [6], [7] and capacitive source degeneration [8]. It is a well-known fact that the Lesson's model of the single-sideband power spectral density is given by [9]:

$$L\{\Delta w\} = 10\log\left[\frac{2FKT}{P_s}\left[1+\left(\frac{\omega_0}{2Q_L\Delta\omega}\right)^2\right]\cdot\left(1+\frac{\omega_{1/f^3}}{|\Delta\omega|}\right)\right] \tag{1}$$

Where *FKT* is the effective thermal noise with the multiplicative factor F, Boltzmann's constant *K*, the absolute temperature *T*; *PS* is the average power dissipated in the resistive part of the tank; Aft is the offset frequency; *QL* is the effective quality factor of the tank and is dominated by quality factor of spiral inductor; ω_0 is the center frequency and ω_{1/f^3} is the corner frequency of the flicker noise. The model describes well the shape of the spectrum, and realizes that many parameters affect phase noise performance. Circuit design tradeoff of the device parameters is required.

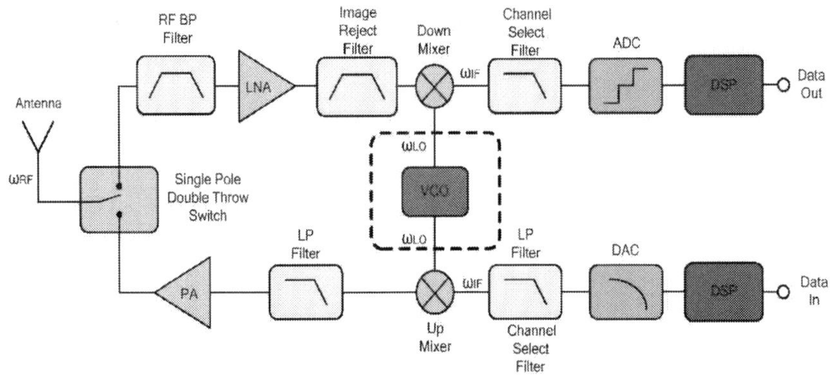

Fig. 1. Building blocks of super-heterodyne transceiver [2]

2. VCO Design

VCO must be designed carefully, its performance affects the stability of the VCO in the transceiver. This section studies how to optimize the circuit design and establish the design procedure for a voltage-controlled oscillator (VCO) in the front end of a transceiver. It promotes the better quality of communication by decreasing the power dissipation and phase noise. This VCO has good data performance between the simulation and measurement.

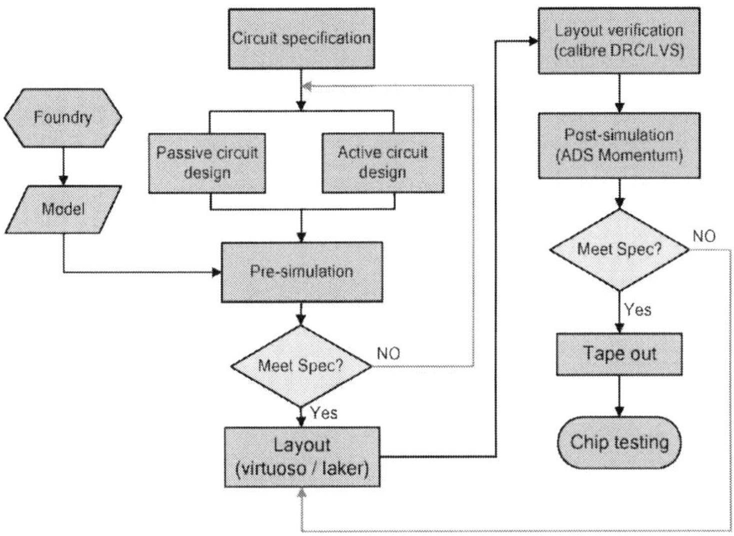

Fig. 2. Design flow of VCO

2.1 Design flow of VCO

This VCO was made by TSMC (Taiwan Semiconductor Manufacturing Company) standard 0.18 | μm 1P6M CMOS process technology. In Fig. 2, design process can be divided into the following steps: Step 1: Review of related literature, and make design specification. Step 2: Design passive and active circuit of the VCO topology.

Step 3: It is pre-simulated by Agilent Advanced Design System (ADS) with TSMC 0.18 | μm RF CMOS process model and fabricated by TSMC 0.18 | μm CMOS technology. It is need to redesign if the pre-simulation result and design goal are different.

Step 4: IC layout design using cadence virtuoso and laker.

Step 5: Layout verification using Calibre DRC (Design Rule Check) and LVS (Layout Versus Schematic).

Step 6: Using the EM simulator with ADS Momentum to perform a numerical electromagnetic analysis of the layout. It is

need to re-layout if the post-simulation and pre-simulation results are different.

Step 7: The chip is fabricated by TSMC (Taiwan Semiconductor Manufacturing Company) 0.18 | μm 1P6M standard CMOS process technology.

Step 8: The chips are measured on PCB board or on-wafer.

2.2 Simple LC Tank VCO Structure

In Fig. 3, we can analyze several important parts of this simple LC tank VCO structure:

The part A-LC tank: The tank circuit consists of a high-Q inductor and varactor components. Select the model values of inductor and varactor for control oscillatory frequency. Where, *Rp* denotes the passive element loss of LC tank. The part B-Active circuit: Active circuit is used to provide negative resistance to compensate for the loss of the LC tank. The part C-Buffer: The buffer is designed to drive the 50 ohm load of the testing instruments.

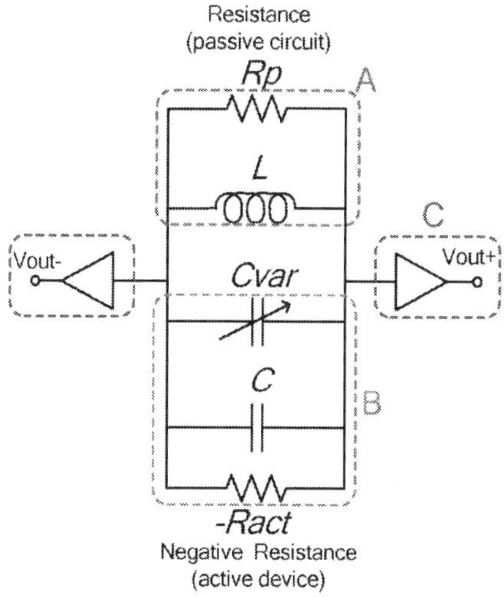

Fig. 3. Simple LC tank VCO structure

2.3 Schematic of the proposed VCO

In Fig. 4(a) shows the narrowband VCO which is composed of the complementary cross-coupled pair MOSFETs, LC tank and switching tail current transistors. In addition, we add the switching capacitor modules for wideband application in Fig. 4(b). A wide-tuning range VCO usually accompanies large Kvco (gain of VCO, Kvco=dω/dVtune). But large Kvco of VCO will amplify noise on the control node (Vtune) and hence will degrade the phase noise performance. We design the small size of PMOS varactors which are capable of providing a small gain of VCO, an array of binary switching capacitor modules were used to extend the tuning range. In this section we discuss several components such as complementary cross-couple pair, LC tank, switching tail current and switching capacitor modules.

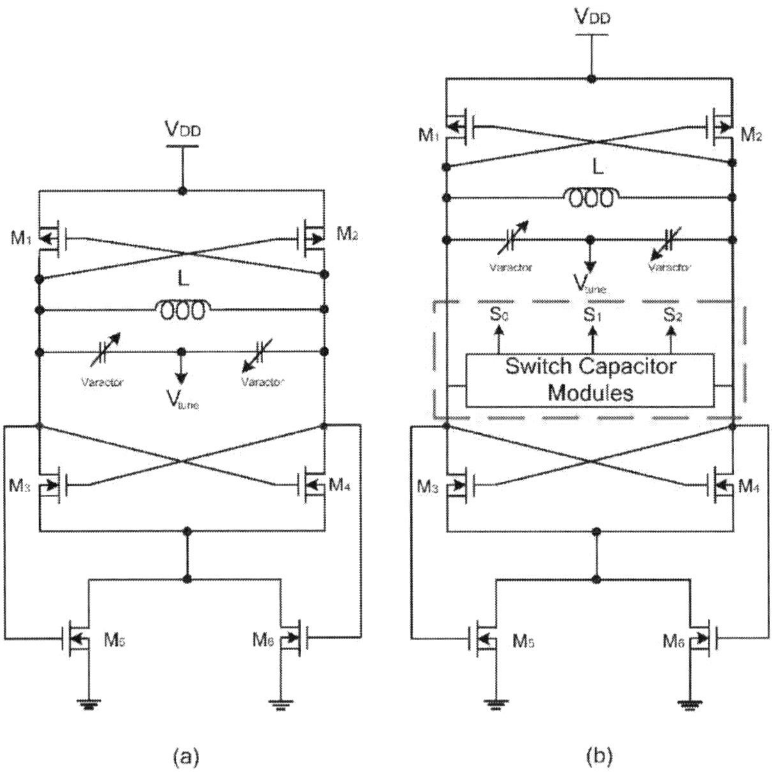

Fig. 4. Circuit schematics of (a) narrow band VCO without switching capacitor modules; (b) wide band VCO with switching capacitor modules

2.3.1 Complementary cross-couple pair

There are three merits in the complementary cross-coupled pair which described as follows [10]:

A. Same current existing, the complementary cross-coupled pair offers higher transconductance and faster switching speed on each side.

B. The output wave are more symmetrical on each other for rise-time and the fall-time, as debate the noise which comes from low frequency noise, 1/f, transferring to high frequency.
C. In all NMOS structure, the channel voltage is larger than complementary case, Therefore, it causes faster saturation speed and larger y value [11].
The simple schematic of NMOS cross-coupled pair is shown in Fig. 5. T1 and T2 indicate NMOS transistors. The high frequency equivalent circuit with capacitive parasitic is shown in Fig. 6(a). And the calculation of input impedance or admittance of the simplified equivalent circuit is shown in Fig. 6(b).

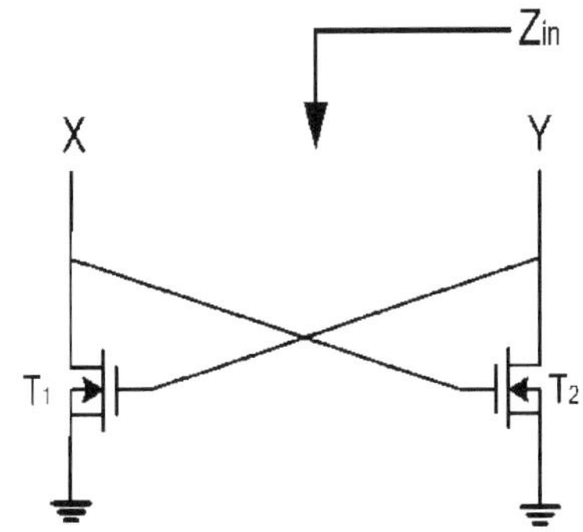

Fig. 5. Simplified schematic of NMOS cross coupled pair

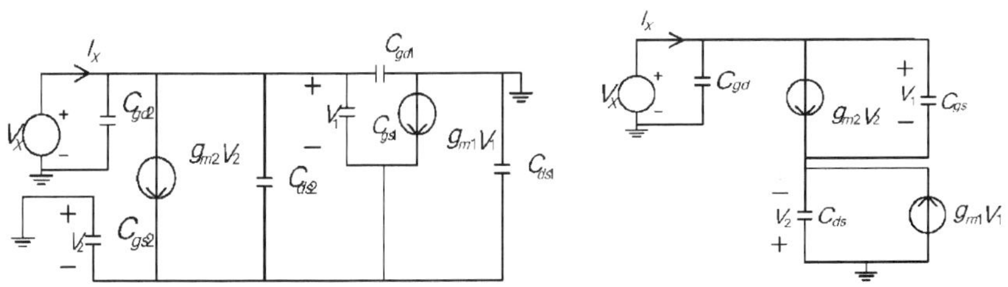

(a) (b)

Fig. 6. Small signal model of Fig. 5 for (a) high frequency equivalent circuit; (b) equivalent circuit of impedance calculations

Let $C_{gd} = C_{gd1} // C_{gd2}, C_{gs} = C_{gs1} // C_{gs2}, C_{ds} = C_{ds1} // C_{ds2}$ for simple calculation of the circuit.
We can obtain the input impedance Z_{in} as following:

$$Z_{in} = \frac{V_X}{I_X} = \frac{\dfrac{1}{g_{m1} + sC_{ds}} + \dfrac{1}{g_{m2} + sC_{gs}}}{1 + \dfrac{sC_{gd} - g_{m1}}{g_{m1} + sC_{ds}} + \dfrac{sC_{gd} - g_{m2}}{g_{m2} + sC_{gs}}} \tag{2}$$

If the transistors size are the same, we can assume that $g_{m1} = g_{m2} = g_m$ and $C_{gs} \approx C_{ds}$ for microwave range in simplified calculation with small dimension device [12]. The Eq. (2) becomes as following:

$$Z_{in} = \frac{2}{-gm + 2sC_{gd} + sC_{ds}} \tag{3}$$

If $s = j\omega$ is used, then Eq. (3) can be written as following:

$$Z_{in} = \frac{-2gm}{gm^2 + \omega^2\left(2C_{gd} + C_{ds}\right)^2} - j\frac{2\omega\left(2C_{gd} + C_{ds}\right)}{gm^2 + \omega^2\left(2C_{gd} + C_{ds}\right)^2} \tag{4}$$

If Eq. (4) $Z_{in} = R_a - jC_a$, then R_a and C_a can be expressed as :

$$R_a = \frac{-2gm}{gm^2 + \omega^2(2C_{gd} + C_{ds})^2}, \quad C_a = \frac{2\omega(2C_{gd} + C_{ds})}{gm^2 + \omega^2(2C_{gd} + C_{ds})^2},$$

where, R_a is the real part and C_a is the imaginary part, respectively. And the parameters of active device are represented in Fig. 7.

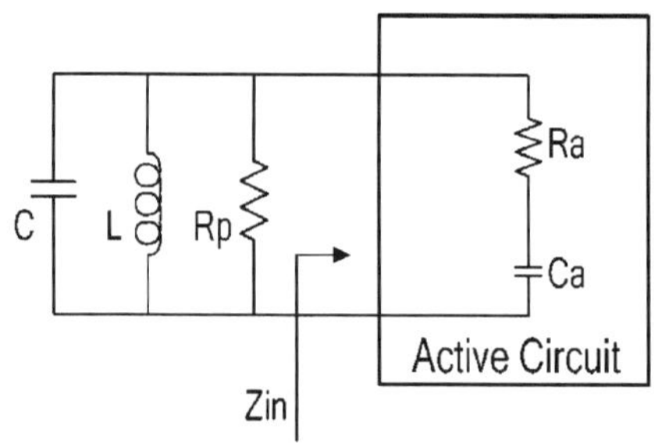

Fig. 7. Parallel LC oscillator model

When the parasitic is ignored, the traditional negative resistance of the input port is indicated by *-2/gm*. Although the complementary topology has more devices than the NMOS pair, the differential voltage swing is larger for the same current consumption resulting in reduce phase noise. The M1 ~ M4 transistors of a complementary cross-coupled pair are shown in Fig. 4, which yield $-\left(\frac{2}{g_{mn}}//\frac{2}{g_{mp}}\right)$ negative resistance to compensate the passive element loss of LC tank.

It can be achieved to start up for oscillation [13] and output signals of the circuit are differential.

2.3.2 Switching tail current

The circuit with a tail current can improve the effect of various noise sources and supply sensitivity [11], and some researchers discovered that a square wave cycling a MOS transistor from strong inversion to accumulation reduces its intrinsic 1/f noise [14]. Therefore, switched biasing can be useful in many circuits to reduce the up-conversion of noise 1/f [15]. The flicker noise from tail current source, especially in MOSFET transistors, makes a great deal of phase noise. Gradually switching tail transistors can release trapped electrons in FET channel, which results in decreasing flicker noise. Moreover, this technique can not only reduce 1/f noise up-conversion but also save power as well. The bias of tail current source was replaced by switched bias without extra DC bias [15] [16]. Utilizing the output voltage swing V1, V2 control M5, M6 which is switched turn on. The output voltage swing is 1.16~1.18V in Fig. 8. In order to determine behavior of the switching, the tail current can't too small. If it is too large, the power consumption is increased, so we need to tradeoff switching behavior, power consumption and phase noise.

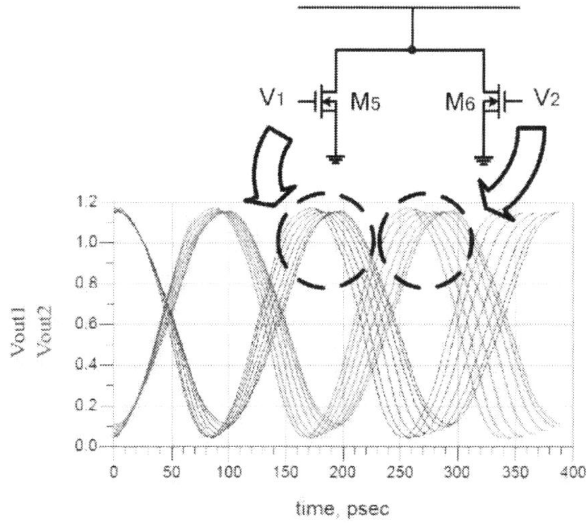

Fig. 8. The output voltage swing with switching tail transistors

The comparison of simulated phase noise performance between fixed bias and switched bias of different tail current topology is shown in Fig. 9.

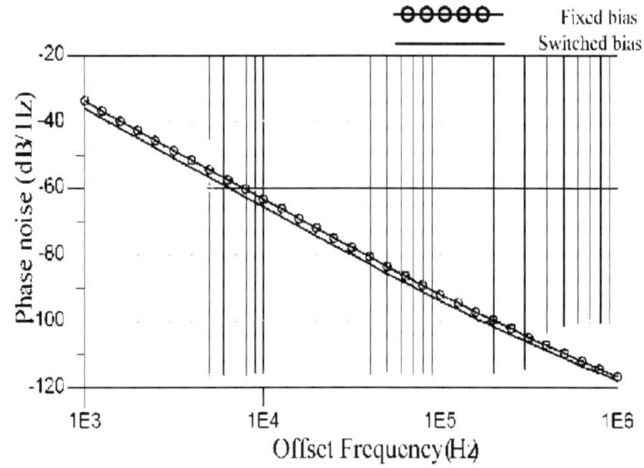

Fig. 9. Phase noise comparison between fixed bias and switched bias at 5 GHz

2.3.3 LC tank

We establish the simulation parameters of Si-substrate and the circuit models of inductors. The resonating tank causes the current in the tank to be Q times larger. Hence the metal lines connecting the LC tank need to be sufficiently large to withstand the large current [17]. In Fig. 10, the quality factor of inductor in this chip is approximately 11 over the working frequency range. The capacitance range of MOS varactor is wider than junction varactor and the equivalent series resistance of the former is smaller than that of the latter. Because using NMOS varactor that drawback is apt to be disturbed in substrate. NMOS capacitor could not implemented in the separate P-well, so NMOS capacitor has high sensitivity of noise that induced by substrate than PMOS capacitor. In view of this, we adopted PMOS varactor.

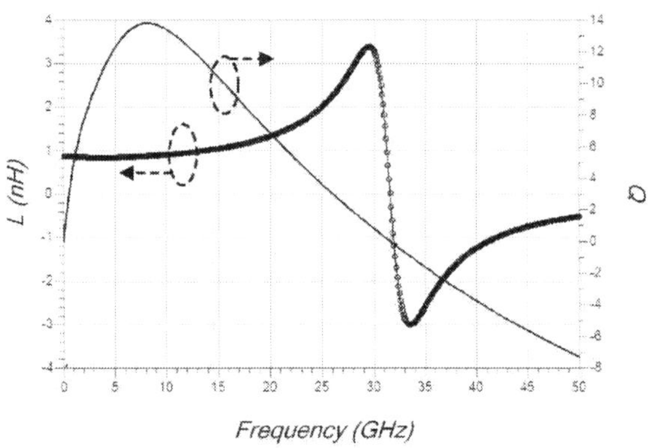

Fig. 10. Inductance and quality factor (Q)

2.3.4 Switching capacitor modules

We usually use band switching techniques to expand the tuning range. The gain of VCO (KVCO) can be reduced to improve the phase noise performance. Making use of switching capacitor modules, eight frequency channels are able to be selected. In order to enable eight channels to connect continually, we design the ratio of the capacitance C2, C1, C0 is 4.45:2.09:1. The S2, S1 and S0, digital pads of the chip, connect digital lines so as to switch different channels. The logical high is 1.8V and the logic low is 0V. The switching has less power dissipation by using NMOSFET within 0.3 mW in our practical work. The whole circuit of switching capacitor modules is shown in Fig. 11. Furthermore, the MOS varactor pair tunes the wideband operation within continuous frequency in each channel [18].

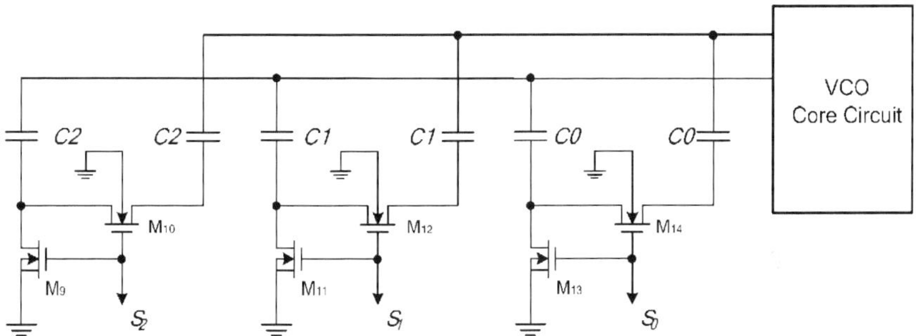

Fig. 11. A switching capacitor module

2.3.5 Output buffers

The VCO is sensitive to loading effect, and it output oscillation frequency would be changed by loading variation. If we insert the buffer between oscillator and loading, it can isolate between them, and the variation of the loading will not influence oscillator directly.

The load of the instrument for measurement is 50Q such as spectrum analyzer. Without buffers, the chip cannot directly drive instrument. The buffer is shown in Fig. 12. [16].

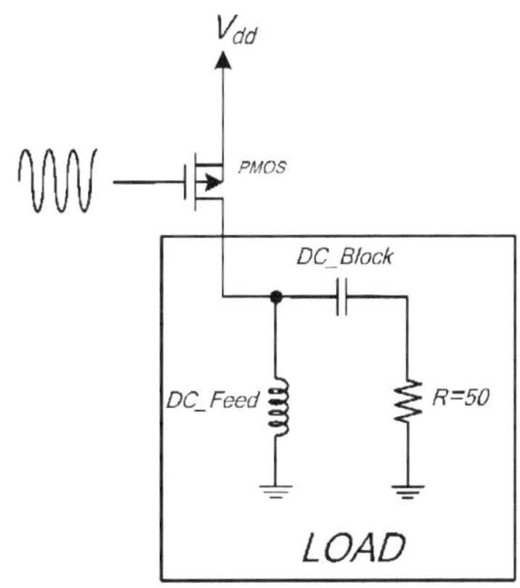

Fig. 12. A buffer schematic

2.3.6 Devices Size of the Circuit

The devices size of our proposed VCO circuit is shown in Table 1, the devices size that we take an optimization to achieve maximize quality factor and generate a negative resistance enough to oscillation, they improve the performance of this proposed VCO.

3. Experimental results

3.1 Measurement setup

A. Agilent E3631A is used as a DC source for digital switching High/Low.

B. Agilent E5052A is used as signal source analyzer and DC sources for DC supply and tuning voltage.

C. The photo of chip with pads is shown in Fig. 13(a).

D. Above a gold plated FR4 PCB is glued the chip which is bonded aluminum wires, shown in Fig. 13(b).

E. The differential outputs of PCB connect a Bias-Tee on each side and then connect two loads, Agilent E5052A and 50Q, shown in Fig. 13(c).

F. The wires which connect to instruments are shielded well and properly matched.

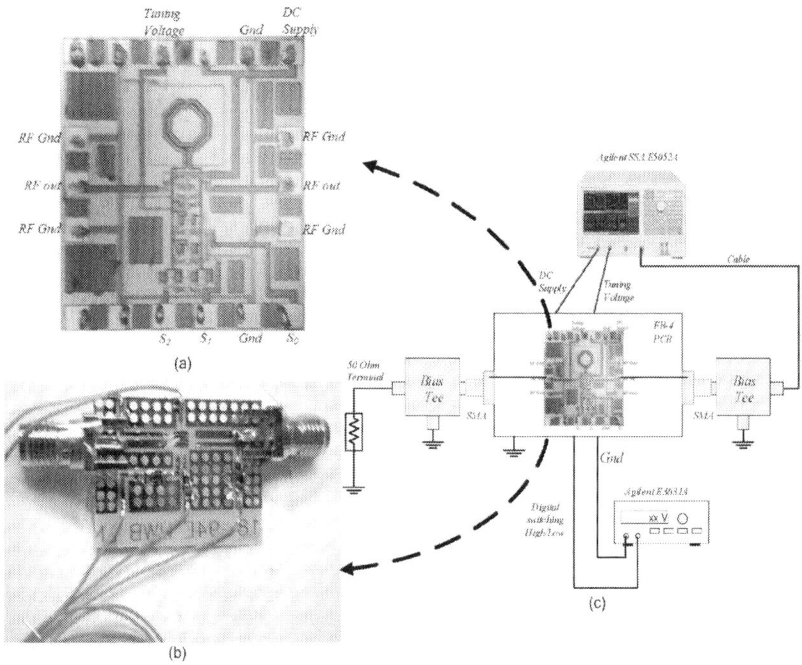

Fig. 13. Measurement setup (a) Die photo (b) Bonding on PCB (c) PCB Measurement

3.2 Measurement result

A. When switching channel is set for $S_2S_1S_0$ = "100", DC supply at 1.8V, tuning voltage from -

0.5V to 1.8V, Fig. 13 shows that the frequency range, the magnitude of carrier and the current from supply in different value of tuning voltage. From Fig. 14, we know that MOS varactor pair is able to adjust 0.24 GHz and the magnitude of carrier is -5.97 dBm at 1.15V tuning voltage.

B. Fig. 15 shows phase noise, -128 dBc/Hz with 1 MHz offset at 4.13 GHz when switching channel is set for S_2SiS_o = "100", DC supply at 1.8V, tuning voltage at 0V.

C. According to the steps above, the frequency range, phase noise, the magnitude of carrier and the current from supply in different channels are listed in Table 2. Table 2 shows that each channel works well and the current of each channel is almost the same, which means that the circuit operates in high stability within switching operation. Therefore, we may well say that the usage of switching capacitor modules is a good way to design the wide-band VCO.

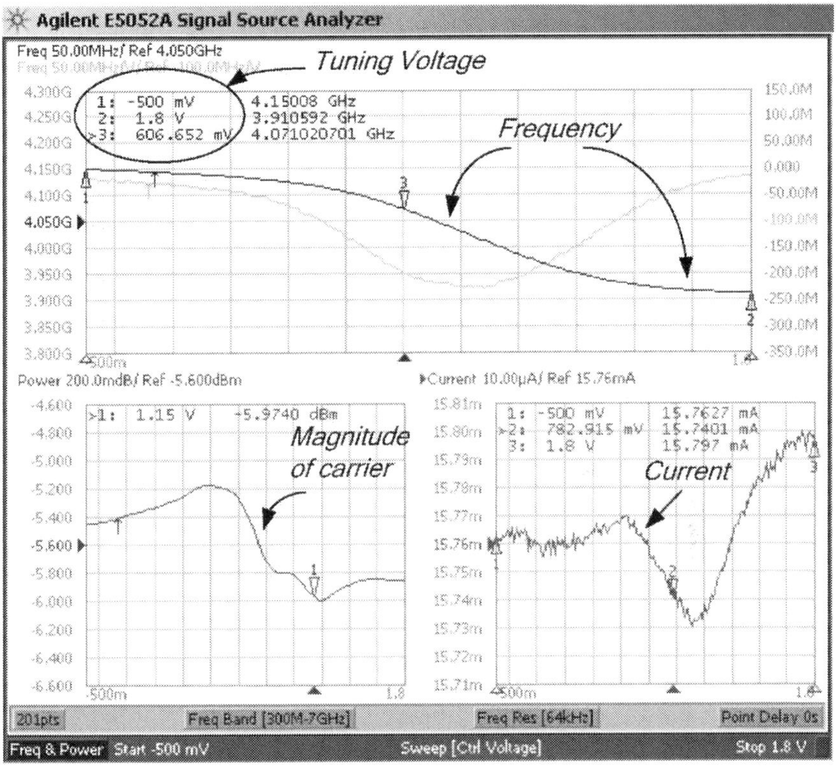

Fig. 14. S_2SiS_o = "100"; Y axes: frequency range, the magnitude of carrier and the current from supply; X axis: tuning voltage

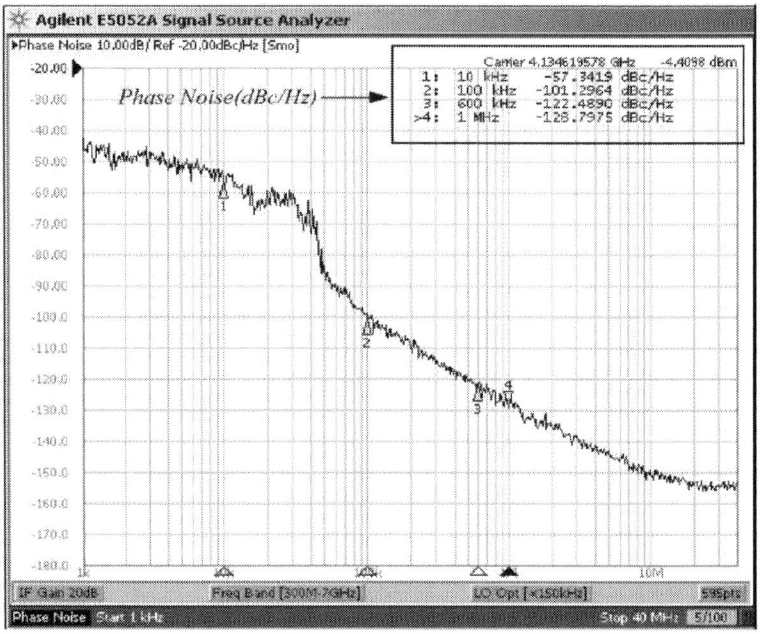

Fig. 15. Phase noise when $S_2S_1S_0$ = "100", tuning voltage = 0V

$S_2S_1S_0$	Frequency (GHz)	Phase Noise at 1MHz Offset (dBc/Hz)	Magnitude of carrier (dBm)	Current (mA)
000	5.37-4.84	-124.2 at 5.33GHz	-1.67	15.69
001	5.16-4.69	-122.1 at 5.13GHz	-1.72	15.69
010	4.80-4.43	-121.8 at 4.78GHz	-2.77	15.83
011	4.67-4.55	-124.4 at 4.64GHz	-2.68	15.90
100	4.15-3.91	-128.8 at 4.13GHz	-5.97	15.78
101	4.07-3.84	-126.4 at 4.05GHz	-6.06	15.85
110	3.89-3.69	-126.3 at 3.88GHz	-6.92	15.83
111	3.82-3.64	-122.8 at 3.81GHz	-6.78	15.84

Table 2. Performance of eight channels of the proposed VCO

The supply voltage is set at 1.8V and $S_2S_1S_0$ = "111", we attained 1.8V x 15.8mA = 28.5mW. Disconnecting two loads, we get the core power dissipation 13.7 mW at DC supply 1.8V. It is a well-known that figure of merit (FOM) is an index between different VCOs. FOM is defined as [10]

$$FOM = L\{\Delta f\} - 20\log\left(\frac{f_0}{\Delta f}\right) + 10\log\left(\frac{P}{1mW}\right) \tag{5}$$

Where $L\{\Delta f\}$ is the phase noise at Δf offset from the carrier f_o and P is the core power dissipation. Table 3 shows the comparison with recently reported papers VCOs.

	This work	[3]	[4]	[5]	[6]	[7]	[19]
Process (um)	0.18	0.18	0.18	0.18	0.18	0.13	0.09
Center Freq. (GHz)	4.50	2.02	4.40	1.80	5.15	4.75	5.63
Tuning Range (%)	38	72	41	73	29	40	45
Supply voltage (V)	1.8	1.8	1.8	1.5	0.8	1	1
Core power diss. (mW)	13.7	17.7	4.9	4.8	1.2	2.5	14
Phase noise (dBc/Hz)	-121.8 - 128.8	-135	-114	-126.5	-109.7	-121.7	-108.5
FOM (dBc/Hz)	-183 -189	-188	-181	-184	-183	-189	-171.5

Table 3. Comparison of VCOs performance

4. Conclusion

This VCO presents a technique of operating narrowband into wideband, employs switching tail current technique and maintains the good phase noise performance. The switching capacitor modules offered multi-channels can enhance oscillator frequency range and the KVCO is still small. This VCO operated from 3.64 to 5.37 GHz with 38% tuning range. The power consumption is 13.7 mW by a 1.8 V supply voltage and measured phase noise in all tuning range is less than -122 dBc/Hz at 1 MHz offset.

5. Acknowledgment

This project is support by National Science Council, (NSC 95-2221-E-224-102). We would like to thank the Taiwan Semiconductor Manufacture Company (TSMC) and Chip Implementation Center (CIC) for the wafer fabrications. We are grateful to National Nano Device Laboratories (NDL) for on-wafer measurements and National Chung Cheng University for PCB measurements by Dr. Ting-Yueh Chih.

6. References

[1] Craninckx, Michiel S. J. Steyaert, "A 1.8-GHz low-phase-noise CMOS VCO using optimized hollow spiral inductors,"Solid-State Circuits, IEEE Journal of Volume 32, Issue 5, May 1997 Page(s):736 - 744

[2] Frank Ellinger, 2008, Radio Frequency Integrated Circuits and Technologies, Springer.

[3] Ito, Y.; Yoshihara, Y.; Sugawara, H.; Okada, K.; Masu, K.;"A 1.3-2.8 GHz Wide Range CMOS LC-VCO Using Variable Inductor". Asian Solid-State Circuits Conference, 2005 Nov. 2005 Page(s):265 - 268

[4] Fard, A.; Johnson, T.; Aberg, D.;" A low power wide band CMOS VCO for multi-standard radios". Radio and Wireless Conference, 2004 IEEE 19-22 Sept. 2004 Page(s):79 - 82

[5] Berny, A.D.; Niknejad, A.M.; Meyer, R.G.;" A 1.8-GHz LC VCO with 1.3-GHz tuning range and digital amplitude calibration". Solid-State Circuits, IEEE Journal of Volume 40, Issue 4, April 2005 Page(s):909 - 917

[6] Chung-Yu Wu; Chi-Yao Yu;" A 0.8 V 5.9 GHz wide tuning range CMOS VCO using inversion-mode bandswitching varactors". Circuits and Systems, 2005. ISCAS 2005. IEEE International Symposium on 23-26 May 2005 Page(s):5079 - 5082 Vol. 5

[7] Neric H. W. Fong, Jean-Olivier Plouchart, Noah Zamdmer, Duixian Liu, Lawrence F. Wagner, Calvin Plett and N. Garry Tarr, "A 1-V 3.8-5.7-GHz Wide-Band VCO with Differentially Tuned Accumulation MOS Varactors for Common-Mode Noise Rejection in CMOS SOI Technology", IEEE Transactions on Microwave Theory And Techniques, Vol. 51, No. 8, August 2003, pp.1952-1959

[8] Byunghoo Jung; Harjani, R.;" A wide tuning range VCO using capacitive source degeneration". Circuits and Systems, 2004. ISCAS '04. Proceedings of the 2004 International Symposium on Volume 4, 23-26 May 2004 Page(s):IV - 145-8 Vol.4.

[9] Yao-Huang Kao, Meng-Ting Hsu, Min-Chieh Hsu, and Pi-An Wu, "A Systematic Approach for Low Phase Noise CMOS VCO Design", IEICE Trans. Electron., Vol. E86-C, No.8, pp.1427-1432, August 2003

[10] Donhee Ham and Ali Hajimiri, "Concepts And Methods in Optimization of Integrated LC VCOs", IEEE Journal of Solid-State Circuits, Vol.36, Issue.6, Jun 2001, pp.896-909

[11] A. Hajimiri and T. H. Lee, "Design issues in CMOS differential LC oscillators," IEEE J. Solid-State Circuits, vol. 34, no. 5, May 1999, pp. 717-724.

[12] Huang, P.-C.; Tsai, M.-D.; Vendelin, G. D.; Wang, H.; Chen, C.-H.; Chang, C.-S., "A Low-Power 114-GHz Push-Push CMOS VCO Using LC Source Degeneration", Solid-State Circuits, IEEE Journal, Vol.42, Issue 6, June 2007, pp.1230 - 1239

[13] Razavi, Behzad, "Design of Integrated Circuits for Optical Communications"-1st ed

[14] Eric A. M. Klumperink, Sander L. J. Gierkink, Amoud P. van der Wel, Bram Nauta, "Reducing MOSFET 1/f Noise and Power Consumption by Switch Biasing", IEEE Journal of Solid-State Circuits, Vol.35, Issue 7, July 2000, pp.994-1001

[15] C. C. Boon, M. A. Do, K. S. Yeo, J. G. Ma, and X. L. Zhang, "RF CMOS Low-Phase-Noise LC Oscillator through Memory Reduction Tail Transistor," IEEE Transactions on Circuits and Systems, Vol. 51, Feb. 2004, pp. 85-89

[16] Meng-Ting Hsu, Chung-Yu Chiang, and Ting-Yueh Chih, "Design of Low Power with Low Phase Noise of VCO by CMOS Process", IEEE International Asia-Pacific Microwave Conference 2005, December 4-7, 2005, pp. 880~883

[17] T. H. Lee, "The Design of CMOS Radio-FrequencyIntegrated Circuits", 2nd ed., Cambridge University Press, 2004

[18] Meng-Ting Hsu, Shiao-Hui Chen, Wei-Jhih Li, "Implementation of Low Phase Noise Wide-Band VCO with Digital Switching Capacitors", Microwave Conference, 2007. APMC 2007. Asia-Pacific 11-14 Dec. 2007 Page(s):1 - 4

[19] Soltanian, B.; Ainspan, H.; Woogeun Rhee; Friedman, D.; Kinget, P.R.;" An Ultra-Compact Differentially Tuned 6-GHz CMOS LC-VCO With Dynamic Common-Mode Feedback", IEEE Journal of Solid-State Circuits, Vol.42, Issue8, Aug. 2007, pp.l63S - 16418

11

Intercavity Stimulated Scattering in Planar FEM as a Base for Two-Stage Generation of Submillimeter Radiation

Andrey Arzhannikov
Budker Institute of Nuclear Physics SB RAS, Novosibirsk State University
Russian Federation

1. Introduction

Far infrared and submillimeter radiation with wavelength from 30 μm up to 300 μm reveals possibilities for new technologies and registration methods inaccessible earlier. One can use this terahertz radiation (THR) to investigate properties of substances and materials such as semiconductors, paper, plastics, which are opaque in the visible range. The other important moment is that the eigenfrequencies of characteristic vibrations of complex molecules belong to the terahertz region. It means that application of THR opens up possibilities of purposive influence upon organic molecules including DNA and RNA. In medicine the terahertz radiation can be used for visualization of healthy and defective tissues, as well as an instrument of therapy and surgery.

There are various methods of generation of the terahertz radiation in the pointed wavelength band and a choice of one of them strongly depends on requirements of users for parameters of the radiation. From one hand, for the case of small generated power it can be done by solid structure lasers (Kohler et al., 2002) or by back wave oscillators (Dobroiu et al., 2004). From other hand, to generate the terahertz radiation of high level power one has to create very huge installations with multi-megavolt electron accelerators (Minehara et al., 2005) and (Vinokurov et al., 2006). As one of appropriate solving the problem of generation of the high power terahertz radiation we proposed (Arzhannikov et al., 2006) to use a two-stage scheme of generation of short wavelength radiation by scattering an EM-wave on a beam of relativistic electrons for the case when at the first stage a high current sheet beam drives a free electron maser of planar geometry operated with two-dimensional distributed feedback at 4-mm wavelength (Arzhannikov et al., 1992, 1995, 2003). Theoretical analysis (Ginzburg et al., 1999) and experimental investigations (Arzhannikov et al., 2008) clear demonstrated that the free electron maser of planar geometry is truly appropriate oscillator for 4-mm radiation band. The key feature of our proposal on two-stage generation is to use two planar generators pumped by sheet beams with a few kAmps currents which plane resonators are combined as it was described for a multichannel generator of mm-wave radiation (Ginzburg et al., 2001) .

2. Proposed process and main experimental parameters

2.1 Wavelength bands of generated radiation

To start our analysis of opportunity of the proposed two-stage scheme we need to outline the wavelength bands that can be covered by the two-stage generation at the experimental conditions of the ELMI-device. At the first stage of the two-stage process a free electron maser has to be used at the parameters of recent experiments at the ELMI-device to generate the radiation with the wavelength λ_0 = 4 mm (Arzhannikov et al., 2008). If one assumes that this radiation will be scattered on the electrons with kinetic energy about of 1 MeV, one can estimate the output radiation wavelength at the second stage of generation.

For the double Doppler Effect the wavelength conversion is expressed by the following formula:

$$\lambda = \lambda_0 (1 - \beta \cos\theta_s)/(1 - \beta \cos\theta_i), \qquad (1)$$

where $\beta = v/c$, v - velocity of the beam electrons, c – velocity of light, is a β-factor of the beam electrons, θ_i, θ_s – angles of incident and scattered radiation respectively counted off from the direction of the electron velocity vector (see Fig.1). For the special cases of backscattering and 90^0-scattering the Doppler formula can be written as

$$\lambda \approx \lambda_0 /(4\gamma^2) \text{ and } \lambda \approx \lambda_0 /(2\gamma^2) \qquad (2)$$

respectively, where $\gamma = 1/\sqrt{1-\beta^2}$ is a E-beam relativistic factor. The expected wavelengths of the radiation at output of the second stage as the function of the γ-factor is presented in Fig.1.

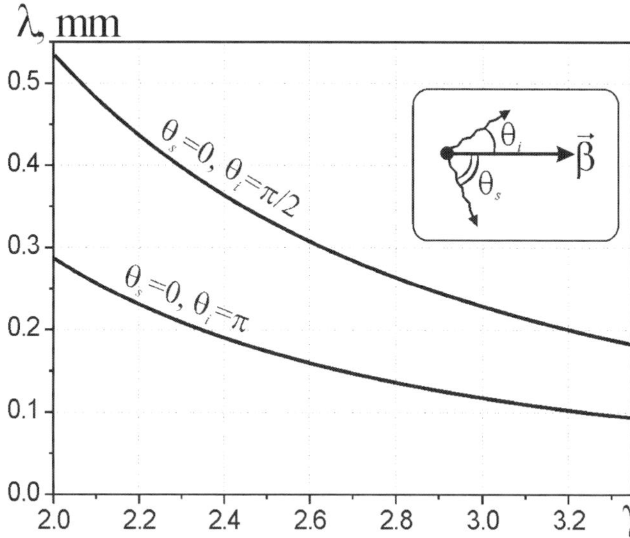

Fig. 1. Conversed wavelength due to scattering of 4-mm radiation by E-beam as the function of γ-factor of the beam electrons.

It is clear that one can obtain radiation in the band of 0.1÷0.3 mm by scattering the incident radiation in the direction opposite to the beam electron velocity at various values of the electron relativistic factor. If the incident radiation is scattered in the transverse direction to the beam electron velocity the radiation wavelength should be shifted to the band of ~0.2÷0.5 mm.

2.2 Schematic of the proposed experiments

Schematic drawings of experimental realization of submm generation for these two wavelength bands are presented in Fig.2 and Fig.3, respectively. The Fig.2 illustrates the variant of two-stage generation for the band of 0.1÷0.3 mm using backscattering of 4-mm radiation.

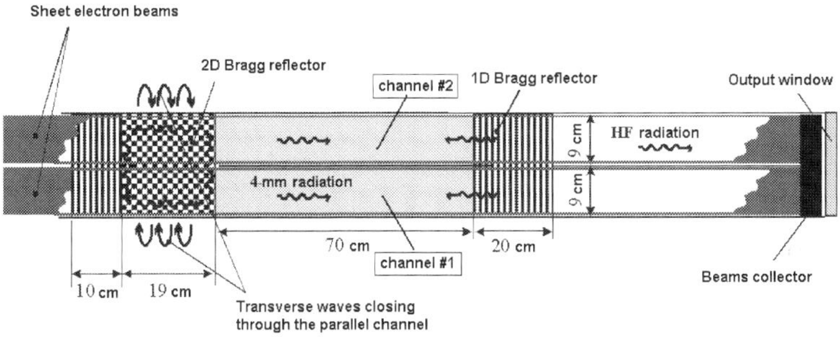

Fig. 2. Scheme of two-stage generation for the band of 0.1÷0.3 mm.

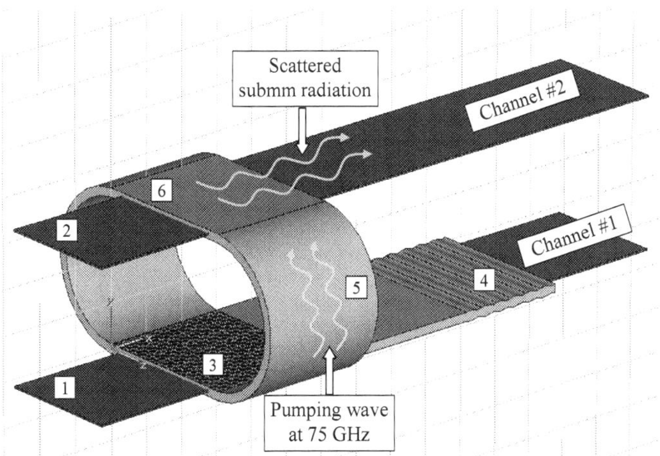

Fig. 3. Scheme of two-stage generation for the band of 0.2÷0.5 mm:

1) sheet REB for driving the planar FEM-generator; 2) sheet REB for mm-wave scattering; 3) 2-D Bragg reflector; 4) 1-D Bragg reflector; 5) feedback circuit; 6) place of scattering

Fig. 3 presents the variant of generation for the band of 0.2÷0.5 mm where the radiation is scattered at the angle 90.

For both variants we suppose to use sheet beams with 3÷4 mm thickness and 10÷20 cm width and a current density more than 1 kA/cm². The E-beams pass the slit channels at presence of longitudinal guiding magnetic field with the strength greater than 1.0T. In the channel #1 of both variants there is an undulator transverse component of the magnetic field that allows one to generate 4-mm radiation with efficiency 10÷15%. The energy density of 4-mm radiation inside the resonator of these FEM generators has a level which corresponds to the electric field strength 10^5÷10^6 V/cm (Arzhannikov et al., 2003) and the same value of the strength must be in the channels #2 of both variants.

In further analysis we shall concentrate our attention on using backscattering of 4-mm radiation that schematic is presented by Fig. 2. Main feature of the electrodynamics system for our two-stage experiments is to use Bragg reflectors in a resonator for 4-mm wave generation. Geometrical parameters of these 4-mm radiation reflectors constructed of the pair of Bragg gratings were chosen through computer simulations and their frequency selecting properties were measured on a special tested bench. Widths and lengths of the vacuum channels for passing the electron beams in were also chosen on the base of computer simulations and experimental tests.

2.3 Computer simulations and experiments on simultaneous generation and transport of two sheet beams

Before the investigations of two-stage generation by using the backscattering process, we have to design and to construct the accelerating diode suitable for simultaneous generation of two high-current sheet beams and to determine conditions for stable equilibrium transport of intense sheet electron REBs in the moderate magnetic fields inside the slit vacuum channels. Solving these two problems is described here.

2.3.1 Computer simulations

One of the key problems in generation of high power REBs suitable to produce THz-radiation in frame of the two-stage scheme is to achieve limit brightness of the beam that is proportional to the current density of the beam j and inversely proportional to the square of electron angular divergence θ^2. Simple estimations have shown that the level of the beam density $j \sim 3$ кA/cm² at the spread of longitudinal velocities of the beam electrons $\Delta V_{\|} / V_{\|} \approx \frac{\theta^2}{2} < 10^{-3}$ has to be achieved for acceptable efficiency of the wave energy transfer from the beam to the THz band radiation (Arzhannikov et al., 2006). It should be noted that to generate mm-wave radiation the value of this spread about $5 \cdot 10^{-2}$ is sufficient.

Previous analytical consideration and computer simulations (Arzhannikov & Sinitsky, 1996) showed that it was possible to reduce the angular divergence below the value $\theta \sim 2\times10^{-2}$ in case of the electron beam generated in the magnetically insulated diode with ribbon geometry at the diode voltage 1 MV and relatively low electron current density 150 A/cm² in the magnetic field 0.6 T inside the slit channel. It was achieved by proper choice of the diode geometry and configuration of the magnetic field which set conditions for subtraction of contributions to the angular electron divergence from the electric and magnetic fields inhomogeneities. In the case of four beams generated simultaneously in a single uniform accelerating diode in the results of computer simulations we have demonstrated the possibility to reach sufficiently high brightness of the beams adequate for generating mm-wave radiation. To investigate the prospects of such beams application for two-stage scheme of THz - wave generation we have performed computer modelling of simultaneous generation of two sheet beams in the magnetically insulated diode and the output of these beams in narrow slit channels. Obtained results confirmed the possibility to achieve the level of the angular divergence $\theta \sim 5{\cdot}10^{-2}$ ($\Delta V_{\|} / V_{\|} \sim 10^{-3}$) at a considerably high current density about 1 kA/cm^2 in the magnetic field 1.7 T (Arzhannikov et al., 2007). Another important problem that has to be solved is the transport of the sheet beam in the slit channel at a stable equilibrium. It was a subject of theoretical and experimental investigations described in (Arzhannikov et al., 1990, 2007) and (Sinitsky et al., 2008). For our case we simulated the beam transport by solving 2-D Poisson equation for homogeneous current and space charge densities of the beam with sharp borders inside the rectangular liner with perfectly conducting walls. When self electric and magnetic fields are small in comparison with the external guiding magnetic field directed along the channel axis, the current and charge densities remain homogeneous along the beam pass but the beam border is deformed by the drift motion of the electrons and the displacement by self magnetic field of the beam :

$$\vec{V}_{\perp} = c\frac{\vec{E}\times\vec{H}_0}{H_0^2} + v_{\|}\frac{\vec{H}}{H_0} = c\frac{\vec{E}\times\vec{H}_0}{H_0^2}\left(\frac{1}{\gamma^2} - f\right) \tag{3}$$

where $\vec{E}$ and $\vec{H}$ - self beam fields, $\vec{H}_0$ - homogeneous external field, f -neutralization degree of the beam space charge, γ - relativistic factor of the electrons. The Fig. 4 demonstrates the evolution of the cross section shape for the beam with the electron energy 0.8MeV beam current 3kA and initial cross section 0.4x6.6cm along the channel length for three distances Z from the entrance of the channel and for three values of neutralization degree f.

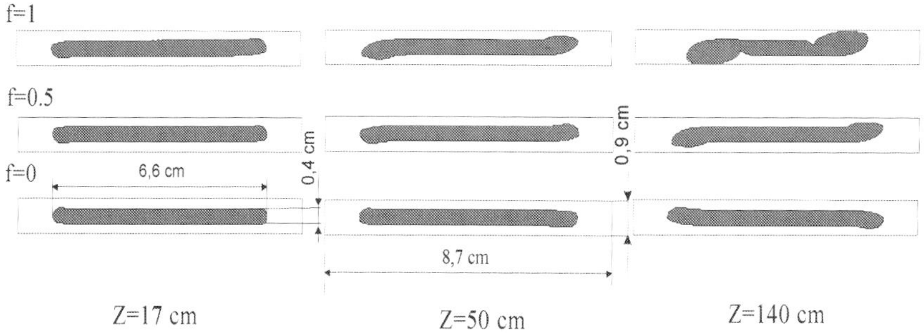

Fig. 4. Cross section shapes of the beam for three Z coordinates along the channel at three values of the space charge neutralization f.

As it is seen the substantial shape deformations for $f = 0$ and $f = 0.5$ are expected only at the end of the channel (Z=140cm) while for $f = 1$ they occur just at Z=50cm. It should be noted that to keep the beam shape unchangeable it is necessary to have beam thickness equal to ¾ of the channel gap. Unfortunately we can not satisfy this requirement because in the case of the FEM application the beam border should oscillate in the undulator field with the amplitude ~0.1cm and the electrons should have perpendicular Larmor radius ~0.1cm while the channel gap should not exceed 2-3 wavelength of the generated radiation (4mm). Thus we have advisedly chosen nonequilibrium shape of the beam assuming its deformations on the length of the FEM resonator (70cm) would be acceptable.

2.3.2 Experiments on simultaneous generation and transport of two beams

The experiments on the simultaneous generation of two sheet beams and their transport in slit vacuum channels were realized basing on the results of computer simulations. Schematic drawing of these experiments is presented in Fig. 5. (Arzhannikov et al., 2007 and Sinitsky et al., 2008). Two sheet beams are generated by two vertically elongated cathodes placed one over another (see side view). These cathodes are made of a fibrous graphite material to ensure homogeneous emission from their surfaces. The guiding magnetic field has adiabatic growth from 0.35 T in the diode up to 1.7 T in the channel that provides magnetic compression of the beam and rise of its current density up to 1-1.5 kA/cm^2. According to simulations for such magnetic field growth the pitch angle of a main part of the beam electrons should not exceed a few degrees. The outer areas of the beam cross sections are cut off in special graphite formers at the beam entrances into the slit channels. Then just central part of the beam cross sections with sizes 0.4×7 cm having minimal pitch angles of the electrons, enters the channels (see Fig. 5). The sheet beam thickness was 0.4 cm and the distance between the channel walls was 0.9 cm.

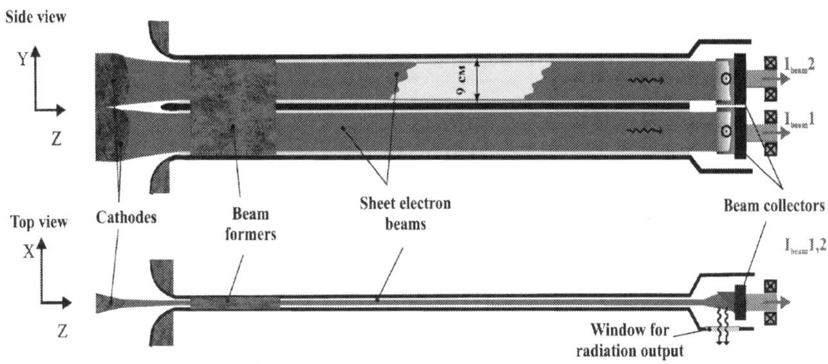

Fig. 5. Schematic drawing of the experiments on simultaneous generation of two sheet beams

Gap between the beam bounds and the channel walls should provide possibility of the beam oscillations under the transverse undulator field without contact of electrons with the channel walls. After transport through the 140 cm long channels with the magnetic field

1.7 T the beams are dumped in the graphite collectors placed in the decreased magnetic field in the described experiments.

Typical traces of the diode voltage and the beam currents measured on the collectors are presented in Fig. 6. It is clearly seen that the time dependences of the beams currents are practically the same but the values have some difference. To understand this difference and to discover possible deformation of the beams cross sections the registration of the beam cross section profile on a thin (1 mm thickness) stainless steel plates have been used. These plates were mounted on special holders inside the channels. Due to the beam exposure the material of the plate heated up to evaporation creating the trace close to the beam cross section. Really this trace was slightly large than the beam size due to the trace edges melting. The reason of the beams currents difference in the shot presented in Fig. 6, was explored by analysis of the beams traces. As a result it was discovered that this difference was caused by tilt of the guiding magnetic field lines about the direction of the channel axis at the angle ~ 0.01 rad.

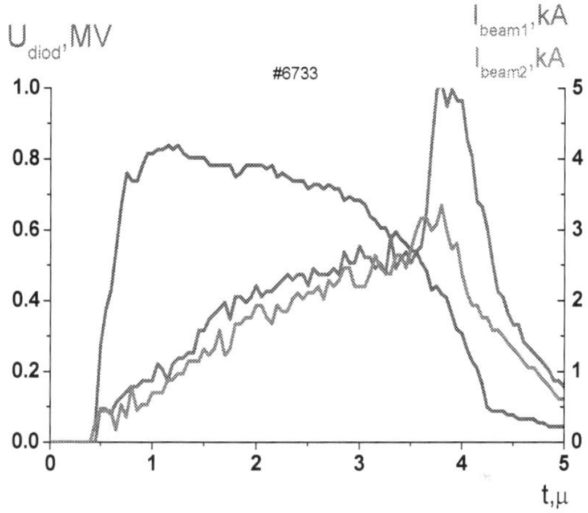

Fig. 6. Traces of the diode voltage and currents of two beams at the exit of the channels.

To eliminate this defect in the magnetic field geometry, concerned with inaccuracy in winding of magnetic coil, the special correcting coil was installed. This coil eliminated the tilt of the magnetic field without any damage in the beam cross section shape.

After that good coincidence of the beams currents has been achieved. Taking into account the results of computer simulations the analysis of drift displacements of the ends of the beams cross sections and their shape deformations (see imprints of the beam in the Fig.7) has shown that the beam space charge neutralization f is larger than $1/\gamma^2$ but far from unity. Since the initial thickness of the sheet beams was not equal to equilibrium quantity, some deformations of the beam cross sections at the transport length 140 cm have been observed in accordance with the simulation results. At the same time for the transport length 50 cm the ribbon shape of the beam cross section was good enough, and the gap between the beam border and the channel walls was still about 0.1 cm. Thus any deviations of the beam cross section shape along 70 cm channel section after the beam former, where FEM resonator will be placed, seems to be negligible.

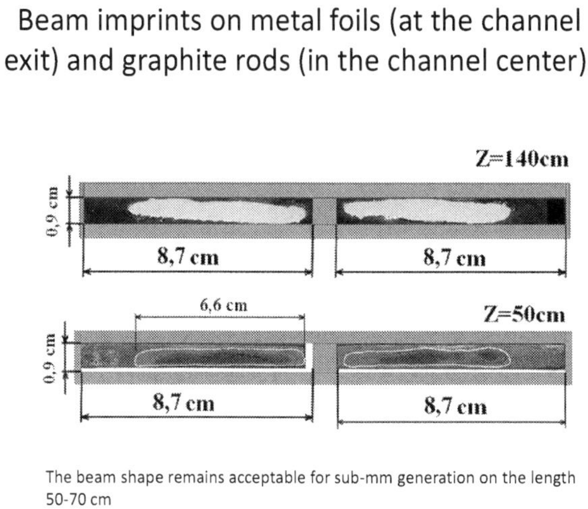

Fig. 7. Cross section shape of two beams for two positions with a different distance Z from the accelerator diode. Low pictures demonstrate the imprints of the two beams on graphite plates mounted in the channels at the distance Z=50cm. Up pictures demonstrate the imprints on titanium foils at the exit of the channels with the distance Z=140cm.

2.4 Prospects of the proposed experiments at ELMI-device

For the two-stage process of generation of submillimeter radiation we plan to use the ELMI-device where the planar free electron maser generates coherent 4-mm radiation with appropriate power (Arzhannikov et al., 2008). Typical oscillograms of voltage pulses at the accelerating diode, the electron beam current, and the signal from the detector of 4-mm radiation are shown in Fig. 8a for the experiments described in (Arzhannikov et al., 2008). The significant level of the microwave signal at the detector was observed when the beam current exceeded 1 kA, in good agreement with the calculated starting current of the generator. The use of the scatterers of transverse wave beams in the two-dimensional Bragg mirror made it possible, for a large number of pulses, to obtain a narrowband generation at a frequency close to the frequency of the mode of the hybrid Bragg resonator. For example, the heterodyne analysis in Figs. 8b and 8c shows that the radiation spectrum is localized near a frequency of 75.3 GHz during almost entire 300-ns duration of the pulse. This corresponds to the excitation of the single-resonator mode. A number of other shots showed the generation at frequencies of 74.9, 75.1, and 75.5 GHz corresponding to the excitation of other longitudinal modes of the resonator. Moreover, the simultaneous excitation of a few longitudinal modes was also observed. According to the simulation results, the possibility of exciting different modes is due to variations in the electron energies and beam current over the pulse, as well as pulse-to-pulse variations in the beam parameters. The analysis of the time behaviour of the plasma fluorescence in the channel indicates that the total duration of the microwave signal is limited due to the arrival of the collector plasma at the radiation deflector.

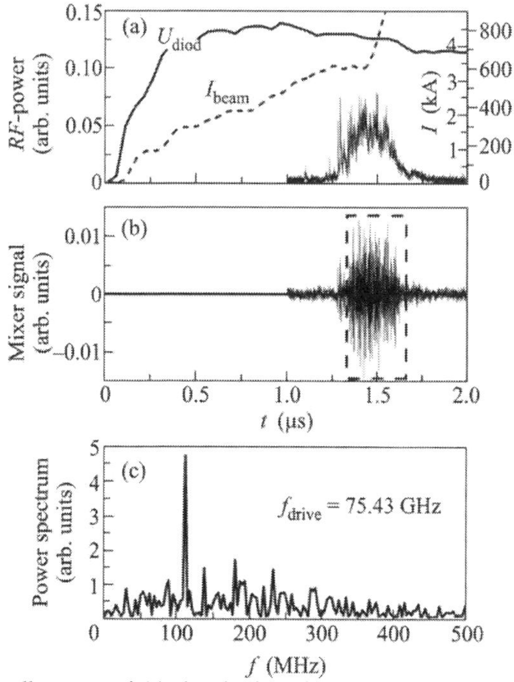

Fig. 8. Typical oscillograms of (a) the diode voltage U_{diod} and beam current I_{beam} in the microwave pulse, (b) the mixer signal, and (c) the radiation spectrum.

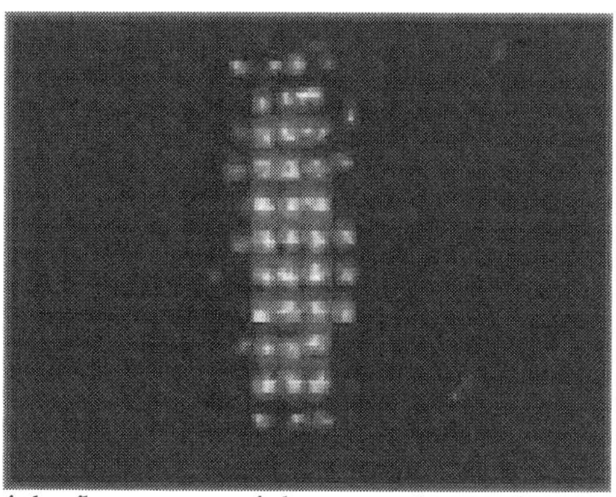

Fig. 9. Photograph of the fluorescence of the neon-tube panel exposed to the microwave pulse.

The total power of the output radiation of about a few tens of megawatts was estimated using the readings of the calorimeter and the signals from the calibrated hot-carrier detectors with allowance for the angular pattern of the output radiation determined by the fluorescence pattern of a neon-tube panel placed at various distances from the output window of the generator (see Fig. 9). The field structure similar to the *H*1, 0 wave confirms the theoretical conclusion on the uniform distribution of the comoving-wave field over the cross section at the output from the interaction space.

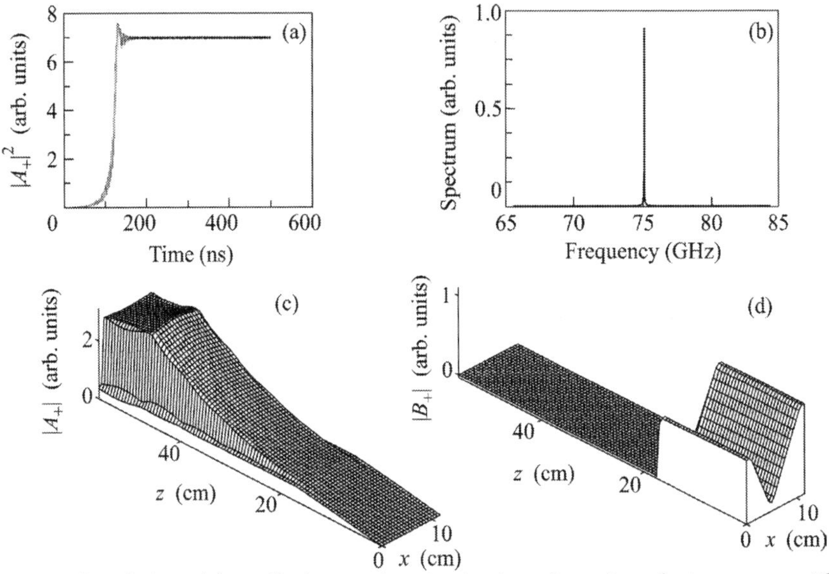

Fig. 10. Simulation of the radiation synchronization in a planar free-electron maser with the combined resonator comprising one and two-dimensional Bragg mirrors: (a) the time dependence of the normalized output power $|A+|2$ under the conditions of a singlemode, single-frequency generation, (b) output radiation spectrum, and the steady-state spatial structure of the partial-wave fields (c) $|A+|$ and (d) $|B+|$.

Fig. 10 shows the results of simulating the generation and spatial synchronization of radiation for the electrodynamic-system configuration and the electronbeam parameters close to the respective experimental conditions. The most important key result is the conclusion that a given spatial distribution of the fields, which is determined only by the system parameters and is independent of the initial conditions, is established during the development of self-excited oscillations at an arbitrary initial noise modulation of the electron beam or small initial perturbations of the electromagnetic fields. In this case, the output-radiation front associated with the wave $A+$ has a deterministic (i.e., not random) phase distribution over the transverse coordinate x. It is also important to emphasize that the partial wave $A+$ synchronous to the electron beam in the steady generation regime has an almost uniform field distribution over the x coordinate (see Fig. 10c). This ensures the same conditions for the energy extraction from all of the beam components. Variation in the electron energy, i.e., in the parameter □, results in a stepwise change in the frequency of the generated radiation, which corresponds to the excitation of modes with various numbers of field variations over the longitudinal coordinate *z*.

Now we are starting the experimental realization of the variant of the two-stage submillimeter generation presented in Fig.2 for the case that does not include any mirrors for submm radiation. In this case we can measure a level of super-radiation in the terahertz band from the channel #2 while 4-mm radiation is generated in the channel #1 by the free electron maser mechanism. It was cleared up by computer simulations and experimental studies that the conditions of the stationary state of the 4-mm generation in the ELMI-experiments can be achieved in time 100÷200ns and the frequency of generated radiation should be unchanged even when E-beam energy is varied in 50 keV. It means that we may realize a two-stage THz-generation process at the ELMI-device as soon as E-beam parameters are approximately not changed in time 200÷300ns.

The important feature of our experiments is to use high density of 4-mm waves accumulated in the resonator of the FEM-oscillator (channel #1 in Fig.2) as a pumping force in the electrodynamics undulator (channel #2 in Fig.2) of the FEL-generator. For the pulse duration about of 0.5 μs the power flow density of the 4-mm radiation in the FEM-oscillator can be achieved I_0 =0.1÷0.5 GW/cm^2 and the same level of the 4-mm radiation will be in the FEL-generator.

When the wavelength of the pumping FEM radiation is λ_0=4mm and the E-beam relativistic factor is γ =3 so that the Doppler parameter of wavelength shortening is $\lambda \approx \lambda_0 /(4\gamma^2)$ then the spatial growth rate for the second stage is estimated as (Arzhannikov et al., 2006)

$$G[cm^{-1}] \approx 7\left(\frac{j_b[kA/cm\ ]}{b[mm]\lambda_0[mm]}\right)^{1/3} \cdot \frac{a_0^{2/3}}{\gamma}, \tag{4}$$

where jb is the current density per unit transverse size of the sheet beam (linear current density), b is the gap between the vacuum channel plates, and

$$a_0 = 2.3 \cdot 10^{-8} \lambda_0[mm] \cdot E[V/cm] \tag{5}$$

is the pump wave parameter, E is the pump 4-mm wave amplitude. For our case this amplitude is estimated on the level E = 1MV/cm and the pump wave parameter a_0 =5·10^{-2} - 10^{-1}.

Thus, for the relativistic factor γ =3, the linear current density j_b = 1 kA/cm and the gap b=10mm the spatial rate of sub-mm radiation is estimated as G ≈ 0.1 cm^{-1}.

Let us look at the experimental limitations due to beam quality. A "cold" beam is defined by the following expression *S*<<1, where $S = \delta V_z /(V_{z0} - V_{wave})$, δV_z is the beam electron velocity spread, $V_{z0} - V_{wave}$ is difference between the beam velocity and the wave one. If *S*>>1 then terahertz radiation is generated incoherently. For our case the requirement *S*<<1 can be expressed in the form:

$$\frac{\delta V_z}{c} << \frac{\lambda_0 \cdot G}{4 \cdot \gamma^2} = 2 \cdot 10^{-5}. \tag{6}$$

Such a small value of the E-beam longitudinal velocity spread can be reached by a very accurate choice of geometry of magnetic field configuration in accelerator diodes and an appropriate strength of the guiding magnetic field (Arzhannikov et al., 2006).

3. Conclusion

Thus, theoretical and experimental studies demonstrate the operability of the two-dimensional distributed feedback and the possibility of use this spatial synchronization mechanism to generate the high-power mm-wave narrowband radiation.
It is important to note that the two-dimensional distributed feedback can be used for the spatial synchronization of radiation generated by several electron beams in multichannel planar FEM devices to produce microwave radiation of GW power level.
High level of the power density with its homogeneous distribution in very large volume for the case of the planar FEM with the two-dimensional distributed feedback allows one to use this electrodynamics system to generate THZ-band radiation on the base intercavity stimulated scattering of mm-wave.
Experiments carried out at the ELMI-device have shown that the sheet beams of 1-Mev electrons are produced with parameters appropriate not only for the generation of high-power 4-mm wave but also for production of submillimeter radiation through backscattering process.

4. References

Arzhannikov A.V., V.T.Astrelin, V.A.Kapitonov, M.P.Lyamzin, S.L.Sinitsky, M.V.Yushkov (1990) Sudies of microsecond ribbon REB generation and transport, *Proceedings of the 9 International Conference on High-Power Particle Beams*, Novosibirsk, USSR, 1990, Vol.1, pp.256-263

Arzhannikov A.V., S.L.Sinitsky (1996) Reduction of angular spread at nonadiabatic electron motion in magnetically insulated diode. *Proceedings of the 11-th International Conference on High Power Particle Beams*, Prague, Chech Republic, 1996. Vol.1, pp.367-370

Arzhannikov A.V., Ginzburg N.S., Nikolaev V.S., Peskov N.Yu., Sergeev A.S., Sinitsky S.L.,.Zotkin R.P (1992) FEL driven by high current ribbon REB and operated with two dimensional feedback, *Technical Digest of the 14th Intern. FEL Conference*, p.214, August 1992, Kobe, Japan.

Arzhannikov A.V., Ginzburg N.S., Peskov N.Yu., ., Sergeev A.S, Sinitsky S.L. (1995).Super-power free-electron lasers with two-dimension distributed feedback,. *Nuclear Instruments and Methods in Physics Research*, v.A358, pp.189-192, 1995.

Arzhannikov A.V.; Bobylev V.B.; Ginzburg N.S.; Ivanenko V.G.; Kalinin P.V.; Kuznetsov S.A.; Peskov N.Y.; Sergeev A.S.; Sinitsky S.L.; Stepanov V.D. (2003), Single-Channel and Multi-Channel Planar Free-Electron Masers, *Radiophysics and Quantum Electronics*, 46 (2003), 10, pp. 810-815.

Arzhannikov A.V., Ginzburg N.S., Kalinin P.V., Kuznetsov S.A., Peskov N.Yu., Rozental R.M., Sergeev A. S., Sinitsky S.L., Stepanov V.D., Thumm M., Zaslavsky V. Yu., Zotova I.V.. Intercavity Scattering Scheme for Two-stage Generation of Submillimeter Radiation on the Base of Planar 2D Bragg FEM (2006), *Proceeding of International Workshop "Strong Microwaves in Plasmas"*, Nizhny Novgorod, Russia, 25 July- 1 August 2006, v. 1, pp. 228–232.

Arzhannikov A.V., Astrelin V.T., Kalinin P.V., Sinitsky S.L., Stepanov V.D. (2007) Simultaneous generation and transport of two high-current sheet beams. *Vestnik NGU, Seria: Fizika, (Novosibirsk State University bulletin, Serial: Physics)*,Vol. 2, No 4, 2007, pp.125-131 (in Russian).

Arzhannikov A. V., Astrelin V. T., Ginzburg N. S., Kalinin P. V., Kuznetsov A. S., Kuznetsov S. A., Peskov N. Yu., Sergeev A. S., Sinitsky S. L., Stepanov V. D., Zaslavsky V. Yu., Zotova I. V. (2007) Submillimeter radiation production by intercavity stimulated scattering in planar FEM at the ELMI-device, *Conference Digest of the Joint 32nd International Conference on Infrared and Millimetre Waves, and 15th International Conference on Terahertz Electronics «IRMMW-THz 2007»*, Cardiff, UK, 3rd - 7th September, 2007, Vol. 2, pp. 835-836.

Arzhannikov A.V., Ginzburg N.S., Zaslavsky V.Yu., Ivanenko V.G., Ivanov I.A., Kalinin P.V., Kuznetsov A.S., Kuznetsov S.A., Peskov N.Yu., Sergeev A.S., Sinitsky S.L., Stepanov V.D. (2008), Generation of Spatially Coherent Radiation in Free-Electron Masers with Two-Dimensional Distributed Feedback, ISSN 0021-3640, JETP Letters, 2008, Vol. 87, No. 11, pp. 618–622. © Pleiades Publishing, Ltd., 2008.

Ginzburg N.S., Peskov N.Yu., Sergeev A.S., Phelps A.D.R., Konoplev I.V., Robb G.R.M., Cross A.W. Arzhannikov A.V., Sinitsky S.L. Theory and design of a free-electron maser with two-dimensional feedback driven by a sheet electron beam. *Physical Review E (Statistical Physics, Plasmas, Fluids, and Related Interdisciplinary Topics)*, Volume 60, Issue 1, July 1999, pp.935-945

N.S.Ginzburg, N.Yu.Peskov, A.S.Sergeev, A.V.Arzhannikov, S.L. Sinitsky (2001) Novel scheme of multi-beam FEL synchronized by two-dimensional distributed feedback, *Nuclear Instruments and Methods in Physics Research Section A: Accelerators, Spectrometers, Detectors and Associated Equipment,* v. A475 pp.173-177, December 2001.

Dobroiu A., Yamashita M., Ohshima Y.N., Morita Y., Otani C., Kawase K. (2004) The backward wave oscillator as a radiation source in terahertz imaging, *Conference Digest of the 2004 Joint 29th International Conference on Infrared and Millimeter Waves, 2004 and 12th International Conference on Terahertz Electronics,* pp. 825 - 826, Karlsruhe, Germany, University of Karlsruhe (TH), Oct. 2004. ISBN-0-7803-8490-3.

Kohler R., Tredicucci A., Beltram F., Beere H., Linfield E., Davies G., Ritchie D., Iotti R., Rossi F. (2002) Terahertz semiconducting-heterostructure laser, *Nature* **417,** 156–159 (9 May 2002), Macmillan Magazines Ltd.

Minehara E., Hajima R., Iijima H., Kikuzawa N., Nagai R., Nishimori N., Nishitani T., Sawamura M., Yamauchi T., (2005). Jaery 10KW High Power ERL-FEL and Its Applications in Nuclear Energy Industries, *Proceedings of the 27th International Free Electron Laser Conference,* , pp. 305 - 308, 21-26 August 2005, Stanford, California, USA.

Vinokurov N.A., Gavrilov N.G., Knyazev B.A., Kolobanov E.I., Kotenkov V.V., Kubarev V.V., Kulipanov G.N., Matveenko A.N., Medvedev L.E., Miginsky S.V., Mironenko L.A., Oreshkov A.D., Ovchar V.K., Popik V.M., Salikova T.V., Scheglov M.A., Serednyakov S.S., Shevchenko O.A., Skrinsky A.N., Tcheskidov V.G. (2006), Status of the Novosibirsk High Power Terahertz FEL, *Proceedings of the 28th International Free Electron Laser Conference FEL 2006,* pp. 492 - 495, August 27 - September 1, 2006, BESSY, Berlin, Germany.

Sinitsky S.L., Arzhannikov A.V., Astrelin V.T., Kalinin P.V., Stepanov V.D. (2008) Simultaneous Generation and Transport of Two Microsecond Sheet REBs in Application to Multichannel FEM. *Proceedings of the 17th International Conference on High-Power Particle Beams*, Mianyang, Sichuan, P.R. China, July 6-11, 2008, pp. 235-238.

12

Complementary high-speed SiGe and CMOS buffers

Esa Tiiliharju
University of Turku
Finland

1. Introduction

This chapter on high speed buffers in complementary SiGe and CMOS technologies studies three different buffer application areas, that is PA-driving, balun buffers, and finally LNAs. The underlaying idea of this text is to point out the benefits obtainable from the application of complementary analog signal processing techniques. More specifically, the text will study applications of the inverter-like continuously biased current-reuse stage in different buffering purposes. One implementation example is shown in the attached Fig. 1, where a continuously current-biased gain cell uses the complementary PMOS to generate extra transconductance. A challenge for microwave applications of this current-reuse cell is to find ways to deal with the added parasitic capacitance associated with the complementary device.

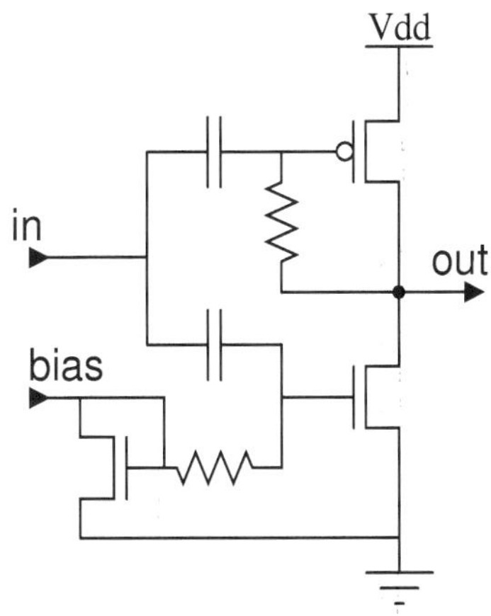

Fig. 1. A CMOS current-reuse cell.

First the reader will be introduced to the subject through a review on complementary bipolar devices in Section 2, including a discussion on a 10-GHz SiGe version of the "compound" emitter-follower. Integrated buffers with balun functionality will follow next in Section 3, where the reader focus is directed towards FET technologies. Albeit most of the baluns presented in this section have originally been implemented in GaAs, they are realizable in CMOS technologies which have lately emerged as a viable microwave technology due to the radical scaling of the minimum realizable linewidths in this planar technology. The balun section serves as an introduction to Section 4 on different CMOS low-noise amplifiers (LNA), and this includes results on a 130-nm CMOS LNA as realized by the author. Summary of the findings will conclude this chapter.

2. Complementary bipolar devices and applications

Different complementary bipolar technologies have been known for decades, e.g., Davis used a complementary compound emitter-follower stage as an inter-stage push-pull OpAmp buffer in Davis et al. (1974). Despite this early success, interest in this niche of semiconductor business really picked up after Texas Instruments introduced a modern 0.4-μm complementary SiGe process variant by El-Kareh et al. (2003). Author and his student Mr. Pellikka were able to experiment with this technology in 2006, when they realized a high dynamic range (DR) current reuse mixer and a linear push-pull buffer with an active area of 0.11mm$\times$0.1mm for base station applications in this 5-V complementary 0.4-μm SiGe process. At 2 GHz, the realized mixer achieves a nominal measured dynamic range (DR) of +154 dB, while drawing 29 mA from a 3.3-V supply, whereas the push-pull buffer has an output-refered 1-dB compression point (OP_{1dB}) of +9 dBm, while drawing 33 mA from a 5-V supply. This text will be limited to a discussion of this push-pull buffering circuit.
The initial reason for this study has been the fact that base station transmitters for 3G wireless standards require linear buffer stages to be installed after the IQ-modulator to boost the transmitted signal before the PA so as to produce high power transmissions without unacceptable spectral splatter. This leads to an increased bill of materials (BOM) and to increased power dissipation. To minimize the number of the expensive discrete buffers, it is advantageous to maximize linear output power of the IQ-modulator itself. High linearity is of increased importance due to the variable envelope modulations adopted for high bitrate performance in the emerging wireless standards. One such modulation format is OFDM, which entails an envelope variation with a peak-to-average ratio (PAR) of 8-13 dB. Transmitted noise should also be kept as low as possible, since it defines the smallest possible signal which can be transmitted. A complementary bipolar technology could help the designer to meet these requirements: a recently developed SiGe variant by El-Kareh et al. (2003) on a silicon-on-insulator (SOI) substrate has the necessary speed for most commercial applications below 3 GHz, as its f_T's are at 18 and 19 GHz for the NPN- and PNP-transistors, respectively. The quoted f_T's, in combination with a SOI-substrate make possible the definition of additional design goals: 1) broadband performance for use in multi-purpose radios; 2) use of integrated coils should be avoided so as to minimize RFIC-area, and, 3) operation at low supply voltages should be possible. The last point is somewhat contradictionary, since on the other hand it is good to dissipate less power as this makes design of the system power budget more relaxed, but on the other hand it should be noted that a high output power and robust circuit operation is fundamentally important in basestations. In this regard, the chosen 5-V complementary SiGe process has an advantage over the predominant complementary technology, the submicron CMOS, which currently typically widthstands 1.2-V supplies. Also no reports on negative bias temperature instability (NBTI) has been filed on the SiGe technology, and lack of this CMOS hazard as reported by, e.g., Peters (2004) should improve the reliability of the designs in general. In response to these demands, this paper complements the abridged description

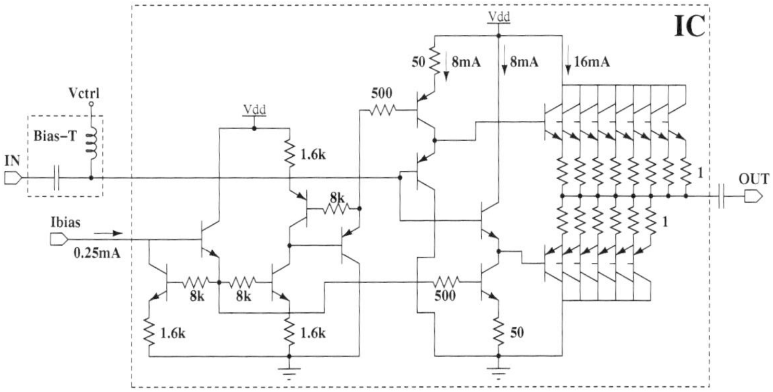

Fig. 2. Microwave push-pull buffer with ballasting resistors shown Tiiliharju & Pellikka (2007) (© 2007 IEEE).

given in Tiiliharju et al. (2006) of a broadband push-pull buffer realized in the complementary 0.4-μm SiGe process.

2.1 Push-pull Buffer

The 0.4-μm complementary SiGe technology can be used to extend the bandwidth of an emitter-follower push-pull stage from VHF to microwave frequencies with a high current drive capability similar to its predecessor which was reported as a *compound emitter-follower* in Davis et al. (1974) for inter-stage buffer use in a VHF operational amplifier. However, because of the insulating SOI substrate, there is a risk of thermal run-off in stages with high currents as noted by Monticelli (2004). This risk is a combination of two factors: 1) SOI-substrate reduces heat radiation away from the transistor, and 2) bipolar transistors have a *positive* thermal coefficient. Therefore, without feedback a warming bipolar transistor will draw more current, which makes the device warmer until this cycle leads to its destruction. To prevent this, local feedback has been introduced as each output stage emitter stripe has been ballasted with a single 1-Ω metal path resistor. This is shown in the schematic of Fig. 2 including biasing details and emitter-follower drivers for this complementary output stage. A bias-T and a dc-blocking capacitor have been used for circuit measurements, and these have been depicted in the Fig. outside the IC dashed box. A pair of these buffering circuits have been realized on the final IC to accommodate for possibly needed differential measurements with external connectorized baluns, and this is shown in the micrograph of Fig. 3. However, the results reported herewith have been extracted using a single device, unless otherwise noted.

Design simulations for this circuit indicate thermal stability, and with the 250-μA biasing current, the nominal performance includes a three decibel bandwidth (BW_{3dB}) of 11 GHz, and an OP_{1dB} of +11 dBm, while the circuit dissipates 33 mA from the 5-V supply used. Next section will give measurement results and compares those with the simulated performance to validate the design methods used.

2.2 Push-pull IC realization

Micrograph of the realized complementary SiGe push-pull buffer pair is shown in Fig. 3, and the standard 150-μm pitch gsgsg probe pad layout depicted underlines the compact size

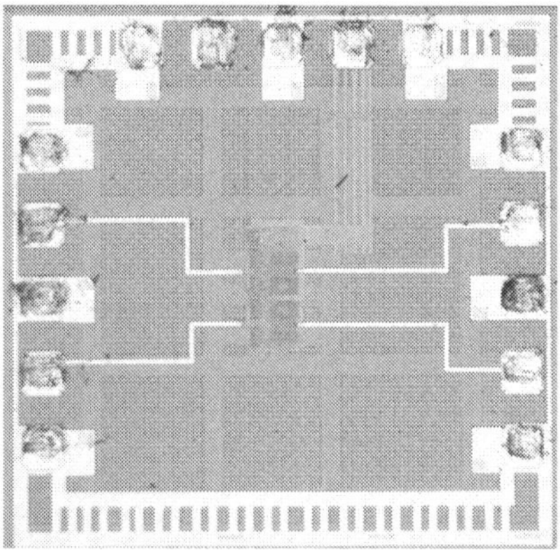

Fig. 3. Micrograph of the realized 10-GHz highly linear push-pull buffer pair.

achieved with this inductorless broadband circuit technique. In this microwave frequency measurement technique, g stands for a ground, s for a signal pad, and the deep probing marks on the pads have been inflicted during a thorough testing of the chip.

The measured insertion gain S_{21} of the push-pull stage at the chosen nominal biasing point of 33 mA from a 5-V supply is shown in Fig. 4 with a realized BW_{3dB}=9.5 GHz. Other measured data extracted using a 2 GHz signal frequency includes: OP_{1dB}=+9 dBm, and NF=5 dB, whereas the second and third harmonic products lie at -48 dBc and -54 dBc in relation to the chosen nominal 0 dBm output power. Two tones with a 10-MHz separation have been used to extract an ouput-referred third-order intercept point OIP3=+22 dBm. Combined these values reveal that good linearity and noise performance have been achieved without coils, since typically this level of performance requires the use of a distributed amplifier with a multitude of integrated coils, which increases the IC-area to 4-80 times that of the push-pull stage. In fact, the $0.11\text{mm}\times0.24\text{mm}$ active area prototype contains two push-pull stages and their shared biasing. A single push-pull measures $0.11\text{mm}\times0.1\text{mm}$. In-all, small IC area and high linearity (OIP3=+22 dBm) of the push-pull stage suggest use in broadband high output power IC's.

The measured push-pull data matches simulated values reasonably well at the same operating point with exact matches on S_{21}=6 dB, P_{OUT}=0 dBm, and 2ND-rej=-48 dBc; the remaining differences can be listed with simulated values in parentheses as: BW=9.5(11) GHz, 2ND-rej=-54(-57) dBc, OP_{1dB}=+9(+11) dBm, and NF=5(4) dB. Therefore the simulated values can be used to compare the proposed current-reuse push-pull buffer to the more commonly used emitter-follower buffer realized with NPN-transistors. This emitter-follower, or common-collector (CC), reference stage has also been realized as a cascade of two CC-stages at a similar biasing point and P_{OUT}. The data is shown in Table 1 with separate columns for the simulated values. Comparison of the simulated results predicts a +12-dB improvement in harmonics attenuation for the push-pull stage at the 2-GHz test frequency. This improvement in linearity is confirmed by the predicted increase in OP_{1dB} of +4 dB for the push-pull stage, and it is due to the increased current drive capability of the implemented complementary circuitry. Penalty for this linearity increase is a simulated 4-GHz drop in BW as the push-pull BW of 11 GHz is compared to the CC-reference BW of 15 GHz.

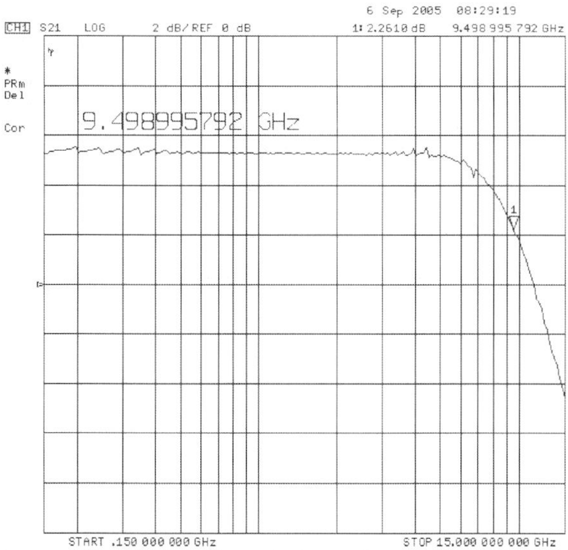

Fig. 4. Measured push pull stage insertion gain S_{21} at 33 mA/5 V_{DD} nominal operating point Tiiliharju & Pellikka (2007) (© 2007 IEEE).

To test for possible thermal run-off, a pair of close-lying push-pull stages was biased at almost twice the nominal current of 33 mA to dissipate 300 mW each from the 5-V supply: as a result a gain stability of $\pm$0.1 dB was measured during a period of 18 h with no traces of thermal run-off. Since a corresponding 12-h current dissipation measurement gave I_{DD} variation at 59.8$\pm$0.3 mA, it is safe to conclude that thermal run-off can be prevented by ballasting despite the insulating SOI substrate used.

So to conclude this section, it can be said that a successful extension of the push-pull buffer to gigahertz frequencies has been accomplished while high linearity and output drive have been maintained. However, transforming this design to other technologies such as CMOS is not a straightforward task and it is not clear whether this circuit technique could be utilized in modern nanometer CMOS designs. That said, the following section will shift focus towards CMOS technologies with a discussion on buffers with balun functionality.

3. Buffers with balun functionality

Most, if not all, communication transceivers utilize differential signaling, whereas antenna connectivity is single-ended. This means that a balun circuit needs to be employed to transform between these two signaling forms, and for economical reasons such a device should be implemented on-chip as connectrorized balun hybrids operating at micro/millimeter wave frequencies are expensive devices. Also integrated balun performance might be better, as good intra-die element matching is one of the major driving forces behing the success of the integrated circuit technologies. This is illustrated by measured amplitude (ΔA) and phase ($\Delta\phi$) errors shown in Fig. 5, where differential signaling deviations from ideal have been recorded. Three high-performance hybrids for the frequency bands of **#1**: 0.05-1 GHz, **#2**: 1-2 GHz, and

Table 1. Comparison of push-pull to a cascaded CC-stage.

		CC-CC*	Push-pull*	Push-pull
A	dB	5	6	6
BW_{3dB}	GHz	15	11	9.5
P_{OUT}	dBm	0	0	0
2ND-rej	dBc	-36	-48	-48
3RD-rej	dBc	-42	-57	-54
OP_{1dB}	dBm	+7	+11	+9
OIP_3	dBm	-	-	+22
NF	dB	4	4	5
I_{DD}	mA	32	33	33
V_{DD}	V	5	5	5

* *simulated results*

#3: 2-4 GHz have been measured for this plot. In this Fig., the phase deviation from the ideal 180° is already fairly pronounced in the band of 1-2 GHz. Therefore, this section first catalogues known differential pair based balun implementations, then moves towards CMOS baluns via a discussion on different FET-based baluns. The latter mainly discusses known work in GaAs-technologies, but these are fully realizable in current CMOS technologies. This material already includes the basic topology used in one of the recently most-reported ultra-wideband (UWB) LNA topologies, that is, the noise-canceling LNA. That said, it should be noted that prior theoretical as well as intuitive proof on the superior balun performance of the differential pair baluns has been given by Altes et al. (1986); Tiiliharju & Halonen (2005), but that its noise and distortion performance is often unsatisfactory and forces use of the other balun techniques detailed herewith.

	ΔA [dB]	$\Delta\phi$ [deg.]	BW [GHz]	S_{21} [dB]	I_{DD} [mA]	V_{DD} [V]	Process [μm]	Year
Tripodi & Brekelmans (2007)	0.9	3.5	< 1	23	33	1.2	0.09 (CMOS)	2007
Tiiliharju & Halonen (2005)	-1	1	0.4-3.7	7	23	2.5	0.8 (SiGe)	2003
Kawashima et al. (2003)	0.5	2	0.5-4	-5	NA	NA	0.3 (GaAs)	2003
Ma et al. (1998)	1	1	0.5-4	NA	3.8	3	0.5 (GaAs)	1998
Kobayashi (1996)	1	6	0.5-5	NA	NA	5	2 (GaAs)	1996
Altes et al. (1986)	0.25	1	0.2-5	NA	NA	NA	1.0 (GaAs)	1986

Table 2. Comparison of measured balun performances.

3.1 Differential pair baluns

Since any differential pair offers the possibility of implementing gain in the signal path, it should be the chosen starting-point for a lower microwave range phase-splitter implementation. Also its inherent balun operation, which is sometimes referred to as its common-mode rejection ratio (CMRR), makes it a natural reference for the balun performance that is realiz-

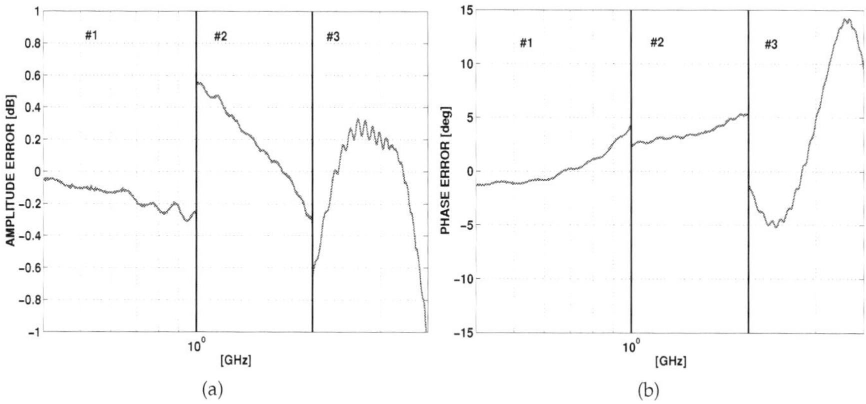

Fig. 5. Measured performance of three connectorized off-chip baluns in 0.4-4GHz.

able. A good bipolar implementation is the emitter-follower-driven differential pair shown in Fig. 6, with a reported performance of: ΔA=1 dB and ϕ=180.7°-186° in 0.5-5 GHz.

A GaAs MESFET implementation is proposed by Ma et al. (1998), where an asymmetrical feedback LCR network is used as a means for improving differential pair phase-splitting performance to within ΔA=±1 dB and ϕ=180°±1°. However, the proposed asymmetrical LCR feedback implementation shown in Fig. 7 makes achieving good broadband performance an intensive design task, as the LC feedback operates at a single resonance frequency, and location of this resonance is dependent on circuit parasitics, which depend on the supply voltages and the biasing point of the circuit.

To improve common-mode or even-order signal rejection, differential pairs can be cascaded: Altes et al. (1986) uses a single-transistor balun to drive a cascade of two differential pairs, as shown in Fig. 8. The circuit implementation achieves good phase-splitter performance at ΔA=±0.25 dB and ϕ=180°±1° in 0.2-5 GHz. For this performance, a 1 μm epitaxial GaAs MESFET technology with air bridges and MIM capacitors was used. This is a very interesting implementation as each cascaded balun theoretically improves differential signaling quality as shown by Tiiliharju & Halonen (2005), and as it deals with the known bad noise figure (NF) performance of the differential pair by driving a pair of them with a single transistor gain stage. The trick here is best revealed with the classic Friis' noise figure equation for cascaded systems, as it ties a cascade system noise figure to individual block noise figures as in:

$$NF = NF_1 + \frac{NF_2 - 1}{A_{P1}} + \frac{NF_3 - 1}{A_{P1}A_{P2}}, \tag{1}$$

where all factors are real, not in decibels, and the result is given in decibels. Numbering refers to stages from-left-right in Fig. 8, and factors $A_{P(1,2,3)}$ stand for the available power gains of three amplifying stages shown.

The Friis' equation explicitly shows that by optimizing the first stage NF and gain $A_{P(1)}$, it can be made to dominate system noise characteristics whereas high cascaded gain nulls contribution from the following stages. However, at higher frequencies each active stage usually adds notably to power dissipation, and without feedback a high number of gain stages in a cascade

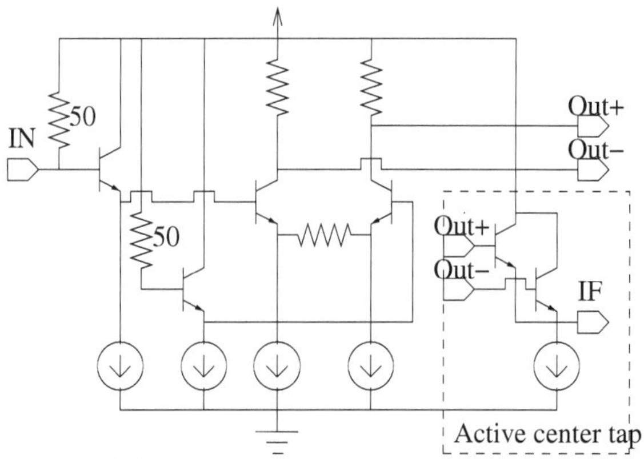

Fig. 6. Linearized emitter-follower driven differential pair reported as an integrated balun by Kobayashi (1996).

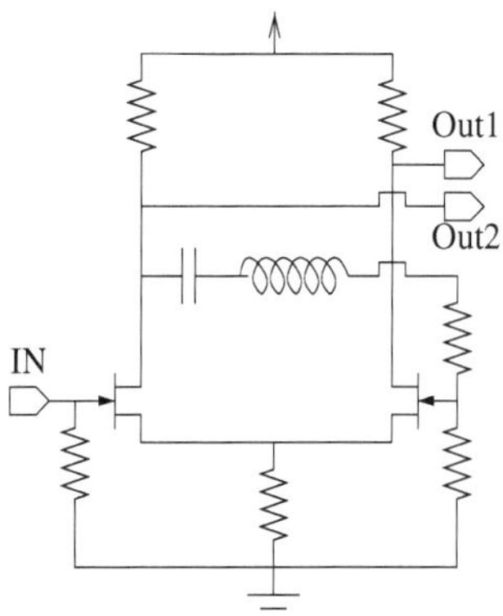

Fig. 7. A differential pair balun with a correcting feedback LCR-network by Ma et al. (1998).

will decrease circuit linearity. Furthermore, it has been shown by Fong & Meyer (1998) that a differential pair always has inferior linearity when compared to a similarly biased common-source stage, so this linearity-power tradeoff limits application of cascading in balun accuracy improvement.

3.2 Modified CGCS topology

A CGCS topology is defined as having a good broadband amplitude balance by Kawashima et al. (2003), while its phase difference is usually poor. In contrast, the modified CGCS topology shown in Fig. 9 realizes both good amplitude and phase balances with measured performance at: ΔA=0.5 dB and ϕ=178°-180.2°. A problem with this implementation is the reported -5 dB loss performance per branch, despite the 0.3-μm GaAs MESFET technology used. Nevertheless, this balun form has been known for decades, and it has also been used as a class-AB mixer input stage by Gilbert (1997), and in noise-canceling LNAs originally proposed by

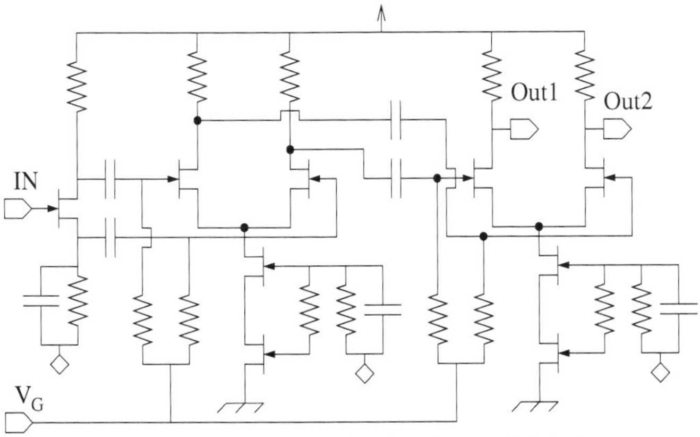

Fig. 8. Cascaded differential pairs used as a balun by Altes et al. (1986).

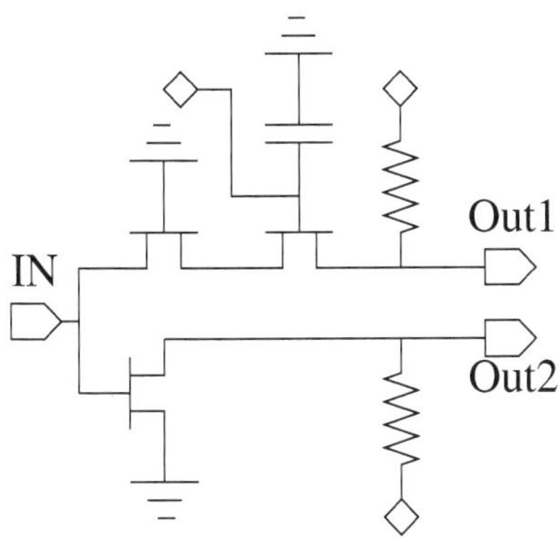

Fig. 9. The CGCS-balun principle developed by Kawashima et al. (2003).

Bruccoleri et al. (2002), a technique which has recently been favored by the UWB design community.

3.3 Single-transistor phase-splitters

Usually, single-transistor phase-splitters have too much phase error as a result of circuit parasitics. Fig. 10 depicts an example of a single transistor balun, where FET drain and source nodes are used as outputs; as is well-known, the inverting (drain) and non-inverting (source) node impedances differ substantially. The impedance seen at the drain node is formed by the parallel connection of the channel conductance g_o and several parasitic capacitances, such as the gate-to-drain C_{gd}, and drain-to-bulk C_{db} capacitances. In contrast, the device transconductance g_m seen at the source node is in most cases so high that it dominates circuit source impedance estimates. This inherent imbalance has a deleterious effect on the performance of a single FET circuit as a balun, and the reported ΔA=1 dB (gain error) and ϕ=176° (phase difference) values by Koizumi et al. (1995) are in agreement with this, as these *simulated* re-

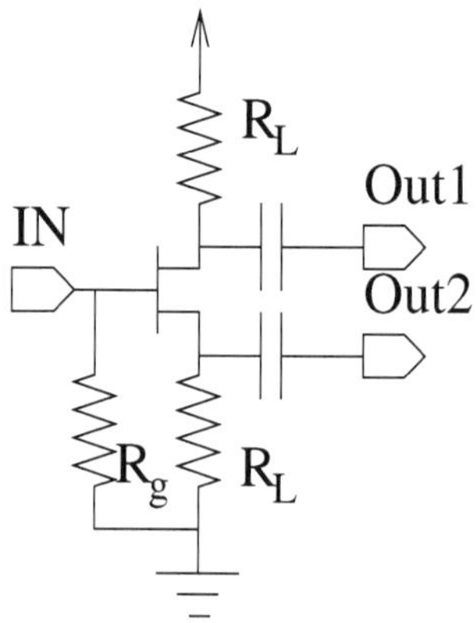

Fig. 10. One FET as a balun by Koizumi et al. (1995).

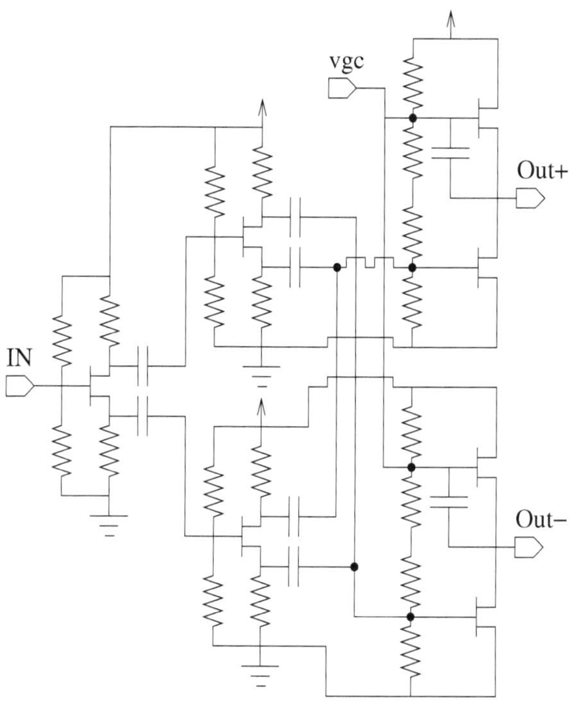

Fig. 11. Cross-connected FETs correct single FET balun response by Goldfarb et al. (1994).

sults were reported as best possible for a single FET balun optimized for use in a Personal Communication Systems (PCS) application at 950 MHz.

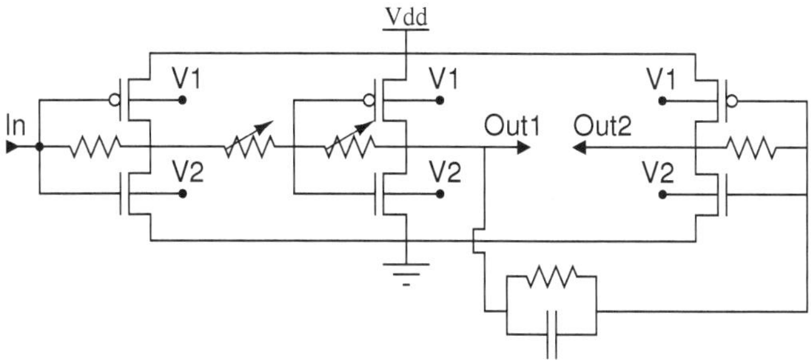

Fig. 12. One of the few balun amplifiers which exploits complementary circuit techniques made by Tripodi & Brekelmans (2007).

To correct the inherent imbalance of the single FET balun, Goldfarb et al. (1994) proposed the use of a pair of cross-connected correcting transistors, as shown in Fig. 11. However, this is not a promising candidate for a broadband balun solution, for two reasons: 1) seven integrated capacitors (excluding output buffering) are needed to make it work, so there is an increase in IC area and a great deal of parasitic capacitance involved, and 2) for better balance each transistor should have equal gain from the gate to the source/drain terminals: this limits the topology to use in low-gain applications. The combination of these two points does not imply good broadband performance for this topology, but the reported *simulated* gain and phase values from 1 GHz to 2 GHz support claims for improved accuracy: ΔA=-0.2 dB and ϕ=178°.

3.4 Inverter balun amplifier

CMOS inverters have been succesfully used to implement a balun variable gain amplifier (VGA) LNA for handheld mobile-TV applications by Tripodi & Brekelmans (2007). Schematic of this 90-nm CMOS LNA is shown in Fig. 12, and its use of inverters as gain stages has similarities with the feedback amplifier proposed by the author later in this chapter. However, important differences exist such as: a) biasing current control in Fig. 12 has been realized using the back-gate biasing voltages V1 and V2, b) use of a series-connected resistor has been avoided by the author to keep NF low, and c) the amplifier shown does not use a global feedback but uses local inverter stage feedback resistors instead. Tabulated results for this and the other baluns of this section in Table 2 miss many of the fine characteristics such as the rather low NF=2.5 dB realized for this low-noise VGA, but reference to the table is still an approapriate conclusion for this section and a fitting preliminary to the next section on LNAs. Based on the table, it is good to point out that well-used complementary stages not only give the best dynamic range performance but they also produce comparable functionality for most applications.

4. CMOS Low-noise amplifiers

It was long widely believed that short-channel CMOS application in high-speed low-noise amplifiers (LNA) would not be successful, as these devices have a higher than usual excess noise factor γ. However, that belief has been shown premeditated, first with sub-1dB noise figure LNA implementations for narrowband applications, then with emerging ultra-wideband (UWB) LNAs. Since ultra-wideband (UWB) technologies are currently gaining acceptance also in European standardization bodies, this niche of communications is under active development worldwide.

Nevertheless, before plunging forward it is approapriate to limit our broadband LNA discussion to inductorless fully integrated designs according to the general layout of this chapter. It is also necessary to mention two specific items of interest: 1) the term LNA will be limited to low-noise amplifiers which have a gain higher than 10 dB, preferably more, and 2) noise figures are only acceptable in the band where the circuit's input has been matched to 50 Ω. The first item stems from the very function of any LNA as defined by the Friis's formula: a low-noise amplifier has to have sufficient gain to isolate and to improve system noise figure, i.e., to make its own low NF the dominating factor in the system NF. The second item stems from the fact that it is trivial to achieve near GaAs-like NF-performances with large WL-area CMOS transistors which have not been matched to 50 Ω, but this is a bit unrealistic, as applications usually dictate mandatory matching to 50 Ω.

This section will first discuss existing feedback LNA solutions, then performance enhancing design techniques such as noise-canceling and current-reuse inputs will be presented, and this section will be concluded with implementation detail on an LNA by the author which uses current-reuse gain-stages in combination with a semi-active dual feedback loop to achieve low noise, high gain and good isolation in a 130-nm bulk digital CMOS technology.

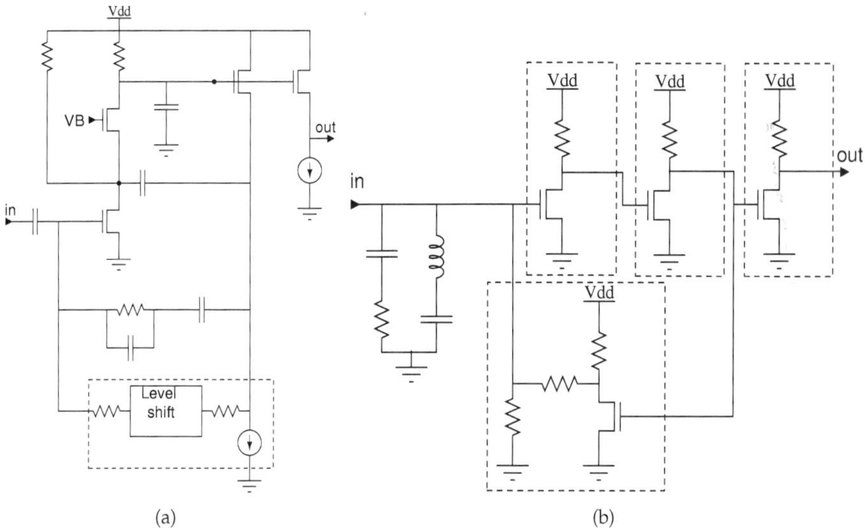

Fig. 13. Two noteworthy feedback LNAs.

4.1 Feedback LNAs

For economical reasons a bulk CMOS process mainly intended for integration of digital circuitry should be used for the purpose of implementing LNAs. Sufficient bandwidth with little gain variation could be guaranteed with three alternative techniques: 1) distributed amplification, 2) use of a complex filtering network at circuit input/output, or 3) feedback amplification. First choice is generally limited by its higher power dissipation and possibly intensive design effort, whereas the second choice includes an increased IC area, high design effort and resistive losses from parasitics. These considerations therefore suggest use of the third alternative, where a feedback network is used to swap amplifier gain for a wideband frequency response. Advantageously, this stabilizes gain and port impedances as well, and this well-known technology is compatible with low-cost integration in digital CMOS.

However, the amount of applicable feedback is limited by stability considerations, and this has traditionally been dealt with by using different compensation networks which aim at incresing the amount of available stable feedback. Conventional microwave feedback designs use complex compensating capacitor networks for the purpose, but this approach is area-consuming, sensitive to parasitics, and time-consuming to design. An example of a very complex feedback network is seen in Fig. 13(a) which is the single-stage UWB low-noise amplifier (LNA) design reported by Zhan & Taylor (2006). This high-performance low-noise amplifier (LNA) in 90-nm CMOS achieved inspiring performance with a best possible NF=2.5 dB performance over the UWB bands. However, this particular implementation uses a 2.5-V supply voltage, and is therefore really not applicable for designs in standard digital CMOS as these use 1.2-V for 130-nm and as low as 1.0-V supplies for newer process nodes, as its use of stacked transistors limits the available dynamic range (DR), and its complex feedback network requires an involved design effort. Fundamentally limiting is the low intrinsic gain of digital transistors, which decreases a single stage gain to an unacceptably low level.

A possible alternative which uses three cascaded gain stages is shown in Fig. 13(b) as reported by Janssens et al. (1997), where the main idea is to improve isolation of the circuit by driving a resistive feedback network with a gain stage. The circuit in Fig. 13(b) is in fact a variation of a well-known bipolar amplifier connection where an emitter-follower is used to drive the feedback resistors connected to the input port. However, although the depicted connection is simple on the surface, its use for e.g. UWB applications is problematic as the feedback amplifier gain roll-off introduces difficult high frequency poles to the single feedback circuit. As a testimony to this the original circuit shown in Fig. 13(b) uses two additional impedance networks at its input to compensate for parasitic effects: an inductor and its dc-block have been applied to null parasitics, and a resistor-capacitor (RC) network has also been applied to ensure stability.

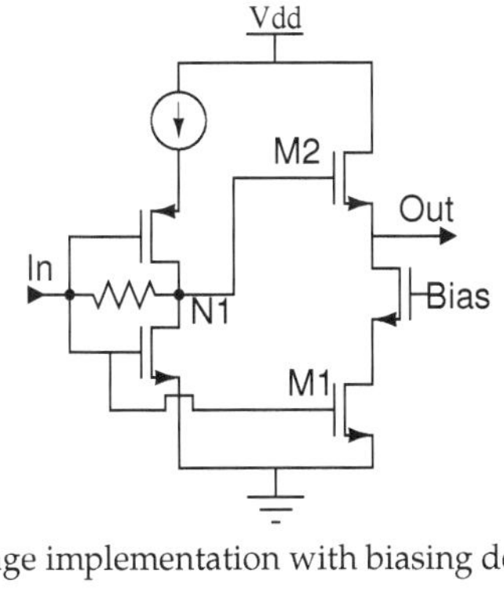

Fig. 14. A noise-canceling stage implementation with biasing details omitted for clarity.

4.2 Noise-canceling LNAs

A very popular broadband low-noise amplifier technique was proposed by Bruccoleri et al. (2002) to break the connection between input resistive matching and noise figure by exploiting two feedforward paths for the input-referred noise with matching transfer characteristics but opposite signs. A better understanding of the technique is possible with reference to Fig. 14, where the inverting noise feedforward path is via NMOS transistor M1, whereas the non-inverting noise path is via NMOS transistor M2. According to inventors, the trick here is that noise in nodes **In** and **N1** is in-phase as the same noise current flows through the feedback resistor R to the source impedance Rs (not shown). This is in contrast to signal phase, which gets inverted by the input stage, and therefore adds at circuit output.

Originally reported performance supports the proposed noise-canceling theory, as sub-2dB NF values with matched input have been reported in the band of 250-1100 MHz with good all-around performance. This performance is limited by the accuracy by which the two opposite phasing noise feedforward paths match both in magnitude and phase domain. Indeed, later implementations for higher frequencies tend to show worse NF value performance, e.g., best noise-canceling ultra-wideband LNAs reported in 2006-2007 (tabulated in the last subsection in Table 3) reach NF values of 2.7-5.5 dB. To understand this drop from expected low NF performance in many cases, it should be noted that matching of the two noise feedforward paths comes increasingly difficult at higher frequencies. Also use of nanometer CMOS devices, which have high channel conductances, makes it difficult to hold on to the assumption that M2 acts as a perfect 1/1 voltage-follower. Significance of this is better understood if the original matching condition is re-printed with the channel conductances taken into account:

$$A_{M1} = A_{M2} \Longleftrightarrow \frac{g_{m1}}{g_{m2} + g_{d2}} = \left(1 + \frac{R}{Rs}\right) \frac{g_{m2}}{g_{m2} + g_{d2}}, \tag{2}$$

where $A_{M(1,2)}$ are FET M1 and M2 associated signal path gains, $g_{m(1,2)}$ are FET transconductances, g_{d2} represents all impedances at the output node, and the feedback and source impedance have been labeled as R and Rs, respectively.

A simple practical interpretation for this matching condition is as follows: since gain is needed to make the LNA noise performance the dominant one, both paths need to have a medium-to-high gain, a condition which dictates matching of a source-follower M2 transfer function with that of a common-source stage M1, including its Miller capacitance. This is clearly a very demanding task for broadband amplifiers.

Therefore, rest of the chapter will discuss possibilities to overcome feedback stability problems so as to fully utilize cascaded current-reuse amplifiers' gain in an ultra-wideband LNA application. This approach is somewhat prone to dissipate higher currents, but its application band should increase in direct relation to decreasing parasitics, i.e., this approach should scale well for nanometer CMOS use.

4.3 Current-reuse LNA with semi-active feedback

This section proposes a current-reuse LNA implementation with a semi-active dual feedback loop as reported by the author in (Tiiliharju & Koivisto (2008)) for the lower UWB band. The proposed LNA topology scalability to nanometer CMOS processes is good, and as a proof-of-concept it has been integrated in a 130-nm digital CMOS process. The proposed LNA can be mass-produced at a negligible cost with extremely small die area, as it utilizes an area-saving inductorless topology. Furthermore, its novel feedback stage improves isolation, increases stability, and slightly improves circuit noise performance with no discernible extra cost.

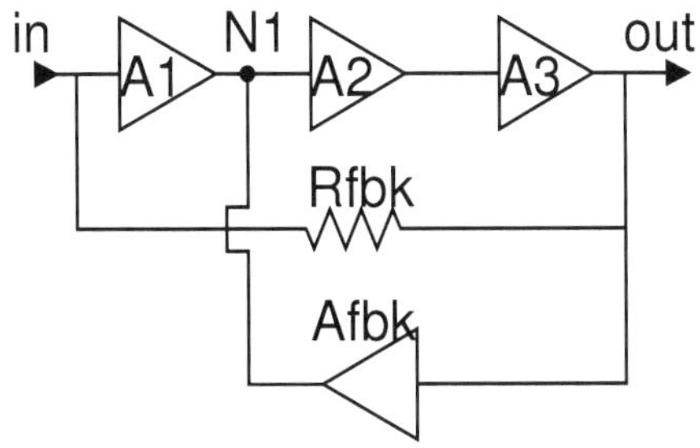

Fig. 15. Proposed feedback network application in a cascade amplifier.

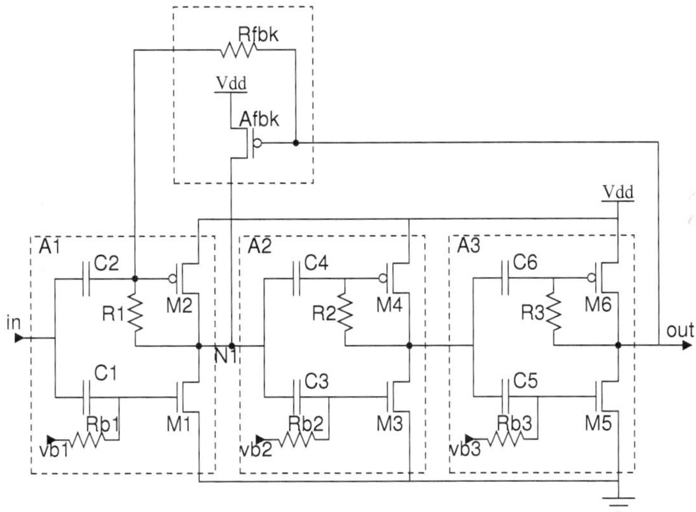

Fig. 16. Transistor level realization of the proposed feedback network application in an UWB LNA.

4.3.1 Design and Architecture

Generally the amount of applicable feedback is limited by stability considerations, but the amount of available stable feedback can be increased by using an active stage Afbk to feed output signaling back to a first internal node N1 at the output of the first amplifier stage A1 of the cascade A1-A3, and also to its input port via a resistor connection as shown in Fig. 15. A copy of the last amplifier stage, or part thereof, could be used as the proposed active feedback stage as this allows accurate setting of the amount of feedback used by simple scaling of said dc-connected feedback stage. The proposed use of a copy of the last amplifier stage is the key behind increased amount of stable feedback available, as this inherently realizes frequency compensation by duplicating single amplifier pole and zero locations. Thus the well-known stability condition reported by Sedra & Smith (2003), which denies exceeding a 20-dB difference between the slopes of the amplifier and feedback frequency response curves at the point of their Bode-plot intersection is naturally easier to meet. This preferred embodiment also avoids prior art (Janssens et al. (1997)) problem of loading the amplifier input port with feedback amplifer poles and zeros, and the designer can opt for the added flexibility of two feedback paths by realizing part of the desired feedback with a feedback resistor Rfbk, which is connected between the cascade amplifier input and output ports. Isolation is also increased and noise slightly decreased, since feedback resistor Rfbk values can be made larger or practically infinite for the same amount of feedback. This is a direct consequence of the smaller amount of feedback which has to be realized resistively for a given desired amount of feedback.

Fig. 16 shows proposed transistor-level realization of the wideband cascade amplifier implementation wherein feedback network (Afbk, Rfbk) has been arranged to trade signal gain arising from the three amplifying stages A1-A3 to a wideband frequency response. Technology used for this implementation is a bulk 130-nm digital CMOS process with optional MIM capacitors used for dc-blocking, and a nominal supply of 1.2 volts. High-speed transistors with low threshold voltages at V_{TN0}=380 mV for NMOS, and V_{TP0}=-390 mV for PMOS variants have been used to build the three near identical core amplifier blocks A1, A2, and A3. All capacitors are 1.25-pF integrated MIMs except input capacitor C2 which has been realized as an off-chip capacitor. Local feedback and biasing resistors R1 and R3 at the input and output buffering amplifiers A1 and A3 have been set at a low value of 400 Ω to improve input match and to linearize the device at its output, whereas the second stage local feedback resistor R2 has been set to 1200 Ω to increase gain. Transistor M1-M6 areas have been set quite high to keep the noise figure floor of each stage at a low value; thus $16 \times 8\mu m/0.13\mu m$ has been given to each device, notwithstanding whether the device in question is a N- or a PMOS transistor. Traditionally PMOS-transistors with similar channel lenghts L were allocated as much as three times the channel width W of their NMOS counterparts, but to cut down circuit parasitics this approach has now been avoided.

Based on previous knowledge and simulations each 8-μm wide unit transistor has been realized in 4 fingers, as this configuration should help to minimize noise by keeping channel resistances at bay. The biasing resistors Rb1, Rb2, and Rb3 have no effect on broadband noise figure, as they have been given a high value at 9.2 kΩ to exclude biasing chain from signal path and maximize gain. The feedback network devices have been set at Afbk=$8\mu m/0.13\mu m$/PMOS, and Rfbk=1.2 kΩ.

4.3.2 Simulated performance

The advantages of the proposed feedback network show more clearly with increasing amounts of feedback. To demonstrate this Fig. 17 depicts simulation results for two feedback amplifiers which trade gain from identical similarly biased core amplifiers for extended bandwidths at ca. 9 GHz with equal remaining 15-dB midband/dc-gains. Thus both amplifiers use a similar amount of feedback with the results simulated for the proposed dual-loop feedback ticked with ◇. Results simulated for the prior-art resistive-only feedback amplifier have been ticked with ✚, respectively.

Upper sub-picture of Fig. 17 depicts voltage gains for the amplifiers. Small-signal simulation allows extraction of gain as circuit output voltages (VDB(out)), as a (1-V_p~0 dB) input signal can be used without distortion effects. The plotted data is used to compare peaking near amplifier 3-dB points, where application of the present invention is shown to reduce peaking noticeably for this 15-dB amplifier example. To put this result in perspective two things will be disclosed next: 1) with different element values of the feedback network the improvement

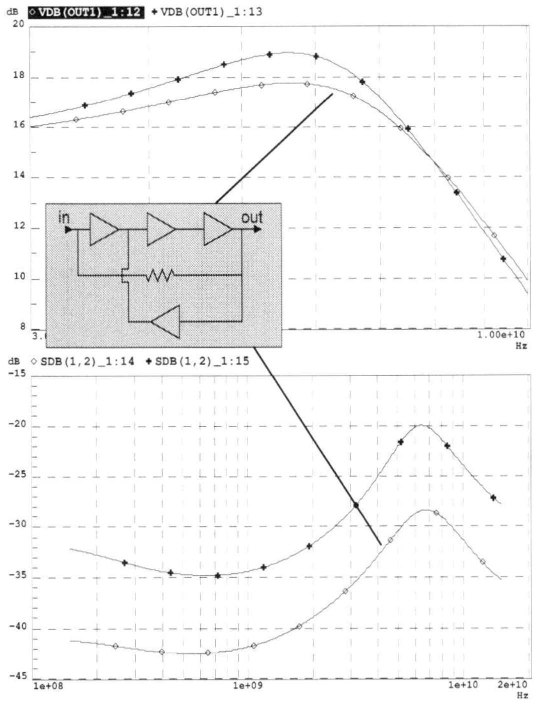

Fig. 17. Simulated comparison of feedback techniques (proposed active feedback=◇, prior art resistive-only=✚) show a) voltage gain peaking near amplifier 3-dB points, and b) amplifier isolation performances.

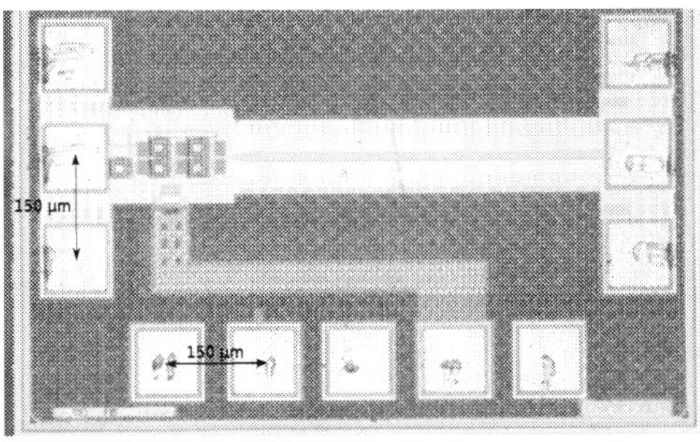

Fig. 18. Microphotograph of the realized UWB LNA shows an active area of $193\mu m \times 124\mu m$.

obtainable can be increased to ca. 3 dB for this 15-dB amplifier example; and 2) when feedback is increased to produce over 10-GHz bandwidths at 13-dB midband voltage gains, simulation results for the resistor-only feedback amplifier indicate *instability* whereas the proposed circuit maintains stable behavior. Lower sub-picture of Fig. 17 compares simulated two-port isola-

tion parameters S12 for the implemented 15-dB amplifiers with a clear 7-dB improvement indicated for the proposed feedback network technology.
Simulated characteristics for the implemented LNA in Fig. 16 at the nominal biasing point of 14.5 mA from a 1.2-V supply predicts good performance: midband gain is 23.7 dB, bandwidth (BW) reaches 7.2 GHz with good input matching of S_{11}=-20.8 dB at 4 GHz. Simulated noise figures remain below 2.3 dB, and LNA figure-of-merit (FOM) characteristics peaks at 23. The FOM has been used as defined by Borremans et al. (2007):

$$FOM = 20\log_{10}\left(\frac{Gain(real)\,BW(GHz)}{Power(mW)\,(NF(real)-1)}\right), \tag{3}$$

where Gain stands for insertion gain S_{21}, BW for amplifier 3-dB bandwidth (in GHz), Power stands for DC power dissipated by the circuit (in milliwatts), and NF is the noise figure given as a real number, i.e., the noise factor of the circuit.

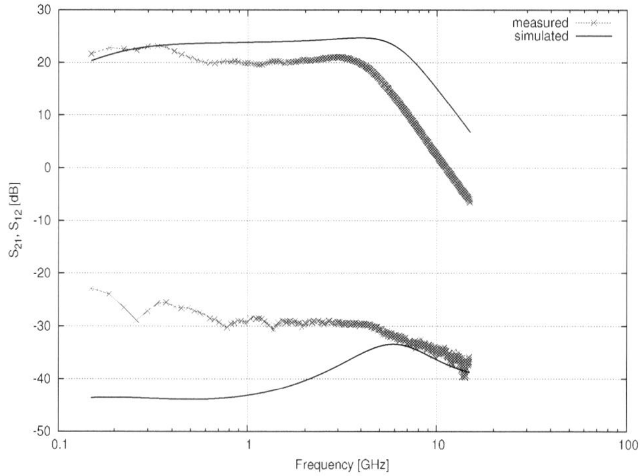

Fig. 19. Comparison of measured and simulated insertion gain (S_{21}) and isolation (S_{12}) values at the 1.2-V biasing point Tiiliharju & Koivisto (2009) (© 2009 IEEE).

4.3.3 Experimental results

The circuit has been tested in nominal conditions using a supply voltage of 1.2 volts, and a biasing current of 14.5 mA. Testing of the IC shown in Fig. 18 has been done using co-planar wafer probes with a pitch of 150 μm. Measured frequency response performance has been compared to simulated values in Figs. 19-20. Latter of the figures also shows that matching performance is acceptable up to ca. 3 GHz as input return loss values stay below -10 dB. However, the depicted measured values differ from the simulated ones, and this is also seen from tabulated characteristics in Table 3 where noise figures topping 4 dB have been recorded together with $|S_{11}|$=7 dB as measured at 4 GHz. The 2-dB NF-value increase from the simulated ones has been verified up to 5 GHz at the three different tabulated operating points, and the measured results have been depicted in Fig. 21. An extra low-noise instrumentation amplifier has been used to drive the spectrum analyzer during the noise measurements as

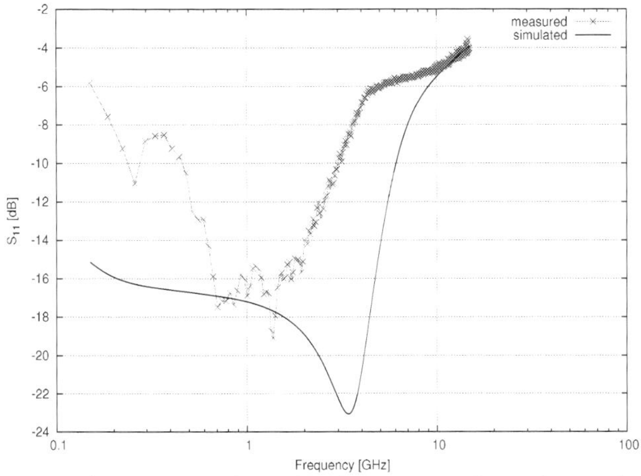

Fig. 20. Comparison of measured and simulated input return loss values at the 1.2-V biasing point Tiiliharju & Koivisto (2009) (© 2009 IEEE).

Tech. CMOS	Gain dB	BW GHz	S11 dB	NF dB	IIP3 dBm	freq. GHz	VDD V	Power mW	Area mm2	FOM	Type	Ref.expl.
130-nm	20	4.9	-7	4.2	-13	4	1.2	17.4	0.0239	5	feedback	This work
	19.4	4.5	-7	4.1	-13		1	12.3		6.4		
	17.8	4	-6	5.9	-15		0.8	7.9		2.7		
90-nm	25	0.5-8.2	-7	2	-11	4	2.5	39.0	0.025	15.6	feedback	Zhan & Taylor (2006)
130-nm	17	1-7	-10	2.7	-4	3	1.4	25.1	0.019	5.9	noise cancel	Ramzan et al. (2007)
90-nm	15.3	0-6	-10	3.7	NA	4	1	3.4	0.0017	17.7	feedback	Borremans et al. (2007)
90-nm	24	0.5-6.2	-15	2.7	-5	4	2.7	42.0	0.016	7.9	feedback	Perumana et al. (2007)
65-nm	15.6	0.2-5.2	-13	3.2	3	4	1.2	21.0	0.01	2.4	noise cancel	Blaakmeer et al. (2007)
90-nm	12	2-11	-10	5.5	-4	4	1.2	17.0	0.7	-2	noise cancel	Wang & Wang (2006)

Table 3. Comparison of LNA performances.

this increases reliability of the Y-parameter noise measurements. The measurement setup has also been verified by measuring another amplifier with known noise performance. All other measurements have been done unbuffered, i.e., the proposed LNA has been used to directly drive the equipment.

The plotted NF data together with the recorded gains hints at a layout error at amplifier input, as any noisy resistive parasitics at the LNA output should be masked by its high gain. Nevertheless, the proposed amplifier FOM-performance compares well to state-of-the-art, as it peaks at the 1.0-V biasing point at 6.4. Only one design uses such a low supply voltage, but this has been realized with a more advanced process node. Measured frequency responses at all biasing points shown in Fig. 22 also confirms the claims on stability and good isolation. Only a uniform gain decrease has been recorded with lowering supply voltages, with no discernible degradation in isolation or peaking at passband edge.

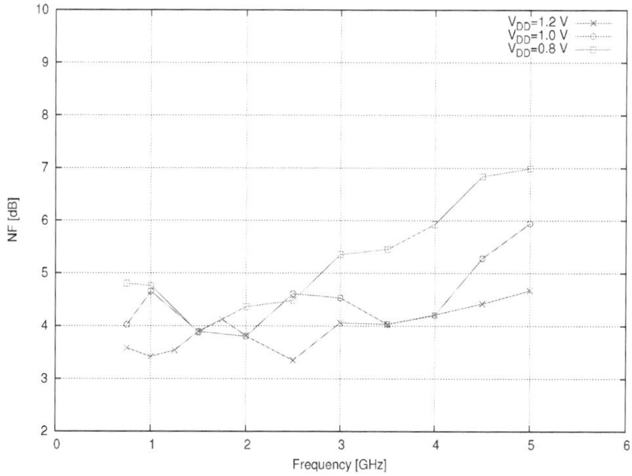

Fig. 21. Comparison of measured NF performance at the 1.2-V, 1.0-V and 0.8-V biasing points.

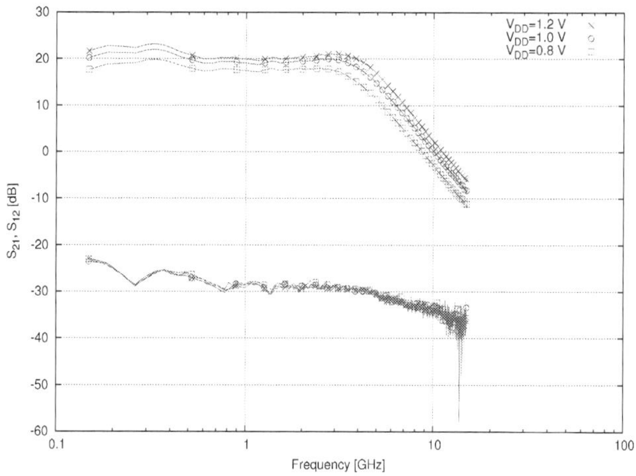

Fig. 22. Comparison of measured insertion gain (S_{21}) and isolation (S_{12}) performances at the 1.2-V, 1.0-V and 0.8-V biasing points.

5. Summary and future work

Successful applications of complementary signal processing to microwave buffers have been studied in this chapter with special emphasis on CMOS. This approach is justified by CMOS scaling to the nanometer domain, which makes it possible to use this very economical technology in the microwave domain. However, first section has elaborated on a complementary bipolar process and its possible application for basestation buffering purposes, an application which is perhaps better served with this high-voltage process. Second section has discussed integrated baluns, which naturally has taken this text to the third section on LNAs where different topologies compatible with modern nanoscale CMOS technologies have been studied. To summarize, it seems that there is a substantial benefit in using complementary analog signal processing techniques, however, parasitics compensation is a demanding design task in the higher operating bands.

6. References

Altes, S. K., Chen, T.-H. & Ragonese, L. J. (1986). Monolithic RC all-pass networks with constant-phase-difference outputs, *IEEE Trans. Microw. Theory Tech.* **34**(12): 1533–1537.

Blaakmeer, S. C., Klumperink, E. A. M., Nauta, B. & Leenaerts, D. M. W. (2007). An inductor-less wideband balun-LNA in 65nm CMOS with balanced output, *33rd European Solid State Circuits Conference, 2007. ESSCIRC*, pp. 364–367.

Borremans, J., Wambacq, P. & Linten, D. (2007). An ESD-protected DC-to-6GHz 9.7mW LNA in 90nm digital CMOS, *Solid-State Circuits, 2007 IEEE International Conference Digest of Technical Papers*, pp. 422–423, 613.

Bruccoleri, F., Klumperink, E. A. M. & Nauta, B. (2002). Noise cancelling in wideband CMOS LANs, *Solid-State Circuits Conference, 2002. Digest of Technical Papers. ISSCC. 2002 IEEE International*, Vol. 2, pp. 330–533.

Davis, P. C., Moyer, S. F. & Saari, V. R. (1974). High slew rate monolithic operational amplifier using compatible complementary P-N-P's, *IEEE J. Solid-State Circuits* **9**(6): 340–347.

El-Kareh, B., Balster, S., Leitz, W. andSteinrnannl, P., Yasudal, H., Corsi, M., Dawoodi, K., Dirnyke, C., Foglietti, P., Haeusle, A., Menz, P., Ramin, M., Schamagl, T., Schiekofe, M., Schober, M., Schulz, U., Swanson, L., Tatman, D., Waitschul, M., Weijtmans, J. & Willis, C. (2003). A 5 V complementary-SiGe BiCMOS technology for high-speed precision analog circuits, *Bipolar/BiCMOS Circuits and Technology Proceedings of the 2003 Meeting*, IEEE, pp. 211–214.

Fong, K. L. & Meyer, R. G. (1998). High-frequency nonlinearity analysis of common-emitter and differential-pair transconductance stages, *IEEE J. Solid-State Circuits* **33**(4): 548–555.

Gilbert, B. (1997). The MICROMIXER: A highly linear variant of the gilbert mixer using a bisymmetric Class-AB input stage, *IEEE Journal of Solid-State Circuits* **32**(9): 1412–1423.

Goldfarb, M., Cole, J. & Platzker, A. (1994). A novel MMIC biphase modulator with variable gain using enhancement-mode FETS suitable for 3 V wireless applications, *Microwave and Millimeter-Wave Monolithic Circuits Symposium, 1994. Digest of Papers.*, Vol. I, IEEE, pp. 99–102.

Janssens, J., Steyaert, M. & Miyakawa, H. (1997). A 2.7 Volt CMOS broadband low noise amplifier, *VLSI Circuits, 1997. Digest of Technical Papers., 1997 Symposium on*, pp. 87–88.

Kawashima, M., Nakagawa, T. & Araki, K. (2003). A novel broadband active balun, *33rd European Microwave Conference*, München, Germany, pp. 495–498.

Kobayashi, K. W. (1996). A novel HBT active transformer balanced Schottky diode mixer, *IEEE MTT-S International Microwave Symposium Digest*, Vol. 2, IEEE, pp. 947–950.

Koizumi, H., Nagata, S., Tateoka, K., Kanazawa, K. & Ueda, D. (1995). A GaAs single balanced mixer MMIC with built-in active balun for personal communication systems,

Microwave and Millimeter-Wave Monolithic Circuits Symposium, 1995. Digest of Papers., IEEE, pp. 77–80.

Ma, H., Fang, S. J., Fujiang, L. & Nakamura, H. (1998). Novel active differential phase splitters in RFIC for wireless applications, *IEEE Trans. Microw. Theory Tech.* **46**(12): 2597–2603.

Monticelli, D. (2004). The future of complementary bipolar, *Bipolar/BiCMOS Circuits and Technology Proceedings of the 2004 Meeting*, IEEE, pp. 21–25.

Perumana, B. G., Zhan, J.-H. C., Taylor, S. S. & Laskar, J. (2007). A 5 GHz, 21 dBm output-IP3 resistive feedback LNA in 90-nm CMOS, *33rd European Solid State Circuits Conference, 2007. ESSCIRC*, pp. 372–375.

Peters, L. (2004). NBTI: A growing threat to device reliability, *Semiconductor international*.

Ramzan, R., Andersson, S., Dabrowski, J. & Svensson, C. (2007). A 1.4V 25mW inductorless wideband LNA in 0.13μm CMOS, *Solid-State Circuits, 2007 IEEE International Conference Digest of Technical Papers*, pp. 424–425, 613.

Sedra, A. S. & Smith, K. C. (2003). *Microelectronic Circuits*, fifth edn, Oxford University Press.

Tiiliharju, E. & Halonen, K. (2005). Active differential broadband phase-splitter for quadrature-modulator applications, *Microwave Theory and Techniques, IEEE Transactions on* **53**(2): 679–686.

Tiiliharju, E. & Koivisto, T. (2008). Feedback network for cascaded ultra-wideband amplifiers, *Ultra-Wideband, 2008. ICUWB 2008. IEEE International Conference on*, Vol. 1, pp. 21–24.

Tiiliharju, E. & Koivisto, T. (2009). A dual feedback loop low-noise amplifier, *Proceedings of the 27th Norchip Conference*, Trondheim, Norway, pp. xxx–xxx. submitted to be accepted.

Tiiliharju, E. & Pellikka, H. (2007). Complementary bipolar devices for base station applications, *Proceedings of the 2007 IEEE International Microwave Symposium (IMS-07)*, IEEE, Honolulu, Hawaii, USA, pp. 263–266.

Tiiliharju, E., Pellikka, H. & Halonen, K. (2006). A current re-use mixer and a push-pull buffer for base station applications, *IEEE Microw. Wireless Compon. Lett.* **16**(9): 514–516.

Tripodi, L. & Brekelmans, H. (2007). Low-noise variable-gain amplifier in 90-nm CMOS for TV on mobile, *33rd European Solid State Circuits Conference, 2007. ESSCIRC*, pp. 368–371.

Wang, C.-S. & Wang, C.-K. (2006). A 90nm CMOS low noise amplifier using noise neutralizing for 3.1-10.6GHz UWB system, *Solid-State Circuits Conference, 2006. ESSCIRC 2006. Proceedings of the 32nd European*, pp. 251–254.

Zhan, J.-H. C. & Taylor, S. S. (2006). A 5GHz resistive-feedback CMOS LNA for low-cost multi-standard applications, *Solid-State Circuits, 2006 IEEE International Conference Digest of Technical Papers*, pp. 721–722, 730.

13

Integrated Passives for High-Frequency Applications

Xiaoyu Mi and Satoshi Ueda
Fujitsu Laboratories Ltd.
Japan

1. Introduction

1.1 Definitions of Integrated Passives

Passive elements are indispensable in RF systems and are used for matching networks, LC tank circuits, attenuators, filtering, decoupling purposes and so on (Tilmans H. A C et al., 2003). Passive elements can be simply classified into distributed elements including transmission lines and waveguides, and lumped elements including inductors, capacitors and resistors. The distributed circuits take into account the phase shift occurring when the signal wave propagates along the circuits. As the operating frequency moves into the microwave spectrum, the distributed circuits have a higher Q factor, and thus they are usually used for high-frequency applications. Lumped elements are zero-dimensional by definition. In other words, the lumped elements have no physical dimensions which are significant with respect to the wavelength at the operating frequency, so that the phase shift that arises can be ignored. Discrete lumped elements are conventionally used in electronic circuits that work at a lower frequency. This is because the sizes of the discrete lumped elements become comparable to the wavelength at microwave frequencies.

With the advent of new photolithography and passive integration technologies, the three basic building blocks for circuit design-inductors, capacitors, and resistors can be made small enough to be available in lumped form (Tummala R. R. et al., 2000). Lumped passive components may be discrete, integrated or embedded. The discrete is a singular device in a leaded or surface mount technology (SMT) case. This includes screen-printed resistors, capacitors, and inductors. Passive integration technologies allow several passive components to be integrated, either into a substrate (embedded) or onto a substrate (integrated). Integrated passive devices usually come in a compact SMT package or chip-scale package (CSP) as a stand-alone component with input, output and ground terminations, which is much smaller than the operating wavelength providing some complete circuit functions, such as impedance-matching, filtering and so on, for high-frequency applications up to several tens of GHz (Tilmans H. A C et al., 2003). The lumped element circuits have the advantage of a smaller size, lower cost, and wide-band characteristics, though the Q factor is generally lower than distributed circuits. Integrated lumped passive circuits with a small form factor are especially suitable for some RF and microwave applications where real estate or wide-band requirements are of prime importance, for example mobile phones or other handheld wireless products. The choice between lumped and distributed element depends on the circuit functions to be fulfilled, operating frequency, size and cost requirements, and performance targets. Sometime these factors must be considered generally before making a trade off between performance and cost or size. Lumped elements can be integrated together with distributed circuits to construct so-called half-lumped circuits enabling more flexible and complex circuit designs. The lumped-element circuits can also be integrated or combined (attached) with microwave integrated circuits (MICs) to construct RF modules (Bahl I. & Bhartia P., 2003). Embedded passives are buried into the substrate itself as an integral part of the substrate along with multiple layers of conductors and do not need to be mounted or connected to the substrate. The multiple inner layers of conductors are separated by a dielectric material with local metal vias to provide interconnects among these embedded passives. Ceramic substrates or printed circuit boards (PCBs) are used as the embedded substrates, since it is easy to build multiple interconnects inside these substrate. This chapter will focus on integration technologies for passive elements. The integration technologies can be classified into three categories according to the construction method used:

- Laminate-based passive integration technology
- LTCC (low-temperature-co-fired-ceramics) based passive integration technology
- Thin-film-based passive integration technology

Laminate-based technology and LTCC-based technology are technologies that allow the passive elements to be embedded or built in the LTCC or polymer substrate. In contrast, the thin-film-based technology is used to integrate passive circuits on the surface of a substrate by performing thin-film deposition of multiple layers of metals and dielectrics. Section 2 will give a comparison of the configuration and performance of these three technological approaches, from the viewpoints of microwave and milli-metre wave applications, and system miniaturization.

1.2 Reasons and Applications for Integrating Passive Devices

Recently, there has been an explosion of growth in the wireless telecom industry. There is a strong market-driven demand to increase the functionality of internal electronics while drastically reducing the total size and cost, particularly in mobile radio frequency applications. This demand has been satisfied to date by major advances in integrated circuits and continuing reductions in the size of discrete surface-mounted passive components. The continuing reduction in the size of surface-mounted passive components is reaching its limit and producing diminishing returns because of the incompatibility of printed circuit board (PCB) technology as well as the high cost of assembly for those tiny discrete components. Nowadays, the 0603 or 0402 size surface-mounted devices are commonly used for printed circuit boards. The assembly cost usually includes the price of the discrete components and the conversion cost consisting of the cost of placement, soldering, and inspection. The typical conversion cost for installing one piece of 0603 or 0402 size SMD (surface-mounted devices) component is $0.02 which is typically more than the price of the SMD itself. SMD components smaller than 0402 will have a significant higher installation cost compared to the 0603 or 0420 size. Therefore further reductions in size and cost will come from integrating the passive components to reduce the component count. A typical mobile phone has hundreds of passive components and only 20 to 40 ICs. The discrete passive components account for 90% of the component count, 80% of the size and 70% of the cost in a handset. As mobile phones come with an increasing array of functions, their active component count will likely remain stable. Therefore, the designers of compact electronics systems, especially handheld and wireless devices, who are being faced with more and more stringent board space constraints, are looking for alternative technologies to integrate these passive components into devices or to remove these passives from the PCB surface (Dougherty J. P. et al., 2003; Doyle L., 2005).

Functional integration (system integration) is another key to miniaturizing handsets. RF modules are moving to higher levels of integration. The modules are required to offer more functionality and higher performance, incorporate the passive circuit inside, and occupy a smaller footprint (Norlyng S., 2003; Pulsford N., 2002). Despite many years of research, the IC industry is facing a technological barrier preventing the integration of bulky, expensive, off-chip passive RF components, such as high-Q inductors, capacitors, varactor diodes and ceramic filters. These components are limiting reduction in size. On-chip passive components, fabricated along with the active elements, as part of the semiconductor wafer in various RFIC technologies have failed to provide adequately high-quality factors compared to the off-chip passives. The typical Q factor of an integrated inductor using (Bi) CMOS or bipolar technologies is usually around 10 (Tilmans H. A C et al., 2003). It can be increased up to between 20 and 30 by introducing some special processing steps that are usually complex and costly, such as etching away the Si under the inductors (Jiang H. et al., 2000) or placing a very thick insulation layer between the inductor and the Si wafer (Kim D. et al., 2003). However, these processes are still not enough for the many important circuit functions in wireless communications systems. For RF front end and radio transceiver applications, it is preferable for the inductor to have a Q factor of at least 30. System-in-package solutions (SIP) are promising as a means of combining these passives and actives together in a single package. The SIP solutions require the passives be small and easy to combine with other devices. The evolution of module technology strongly depends on improvements in passive integration and 3D assembly technologies.

Moreover, the continuing scaling of IC technology affects the required interconnection and packaging technologies significantly. Improvements in the density of standard interconnection and packaging technologies have not kept pace with IC scaling trends, resulting in a so-called "interconnect technology gap" (Wojnowski M. et al., 2008). The peripheral pads pitch of IC will trend to less than 30 μm in the near future. In contrast, standard PCB technology commonly provides a coarse contact pitch of 400-1000 μm. An interposer enabling high-density interconnection has to be used between the high-density IC technology and the coarse standard PCB technology. The future MCM (Multi-Chip-Module) substrates and packages are required to function as so-called interposers. To incorporate the passive circuit into the interposer is attractive and powerful for constructing next-generation SIP modules (Carchon G. et al, 2008).

Over the past 10 years, passive integration technology has gone through a significant evolution to meet the requirements for lower cost solutions, system miniaturization, and high levels of functionality integration, improved reliability, and high-volume applications. In addition, the passive integration technologies have been leading to the benefits show below:

- Smaller size, weight, and volume
- Improved electrical performance due to the proximity of the passives to the active devices reducing parasitic and increasing switching speeds
- Improved reliability through a reduction in the number of solder connections

- Lower total cost due to reduced costs for procurement, logistics and installation

Passive integration technologies can be used in both digital and analog/RF applications. Some of these applications include mobile phones, personal digital assistants (PDAs), wireless computer networks, radar systems, and phased array antennas. Integrated passive circuits with high-performance characteristics function in these systems as:

- RF front end modules
- RF power amplifier couplers
- Filters (low pass, high pass and band pass)
- Functional interposers between ICs and the primary interconnect substrate
- Multi-band transceivers

1.3 General Design Considerations for Integrated Passives

Inductor

One of the most critical elements in RF and microwave circuits for high-frequency wireless applications is the inductor. If the Q value is too low, the lumped circuit will not reach the desired performance targets. Spiral inductors providing a high Q factor and inductance value are commonly used for high-density circuits. The important characteristics of an inductor are its inductance value and its parasitic capacitance and resistance, which determine its Q factor and self-resonant frequency. The Q factor values can be obtained from one-port or two-port scattering parameter data. A simplified one-port lumped-element equivalent circuit model used to characterize inductors is shown in Fig. 1.1. Accurate inductor models using measured two-port scattering parameters will be discussed in section 3. In this one-port model, L, R, and C represent the total inductance, series resistance, and parasitic capacitance of the inductor, respectively. The admittance of an inductor is expressed as

$$\begin{aligned} Y &= j\omega C + \frac{1}{R + j\omega L} \\ &= \frac{R}{R^2 + \omega^2 L^2} + j(\omega C - \frac{\omega L}{R^2 + \omega^2 L^2}). \end{aligned} \tag{1.1}$$

The series resistance R used to model the dissipative loss is given by

$$R = R_{dc} + R_{ac}\sqrt{f} + R_d f \tag{1.2}$$

where R_{dc} represents DC resistance of the inductor, R_{ac} models resistance due to skin effect in the conductive trace, and R_d represents resistance due to eddy current excitation and dielectric loss in the substrate.

The Y parameters can be obtained from one-port S parameter. The Q factor is then calculated from

$$Q = \left|\mathrm{Im}(Y)/\mathrm{Re}(Y)\right|. \tag{1.3}$$

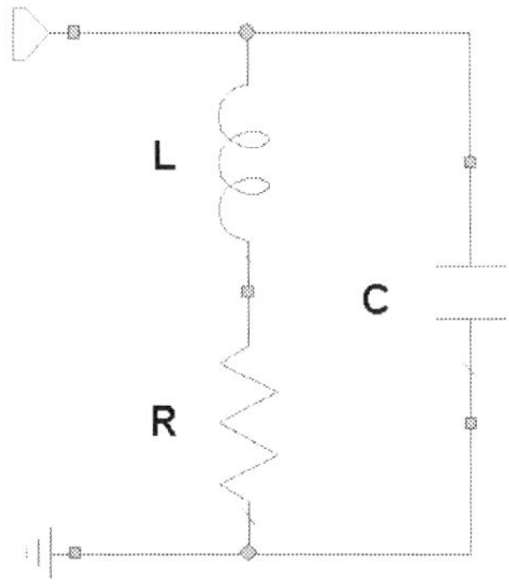

Fig.1.1 Simplified one-port lumped-element equivalent circuit model of an inductor

When the parasitic capacitance C is very small and ignorable, Q factor can be given by

$$Q = \frac{\omega L}{R}. \tag{1.4}$$

As can be seen in the above-mentioned equation (1.4), achieving a predetermined inductance L at a small resistance R contributes to an increase in the Q-factor. The self-resonant frequency (SRF) of an inductor is determined by the Y parameter when $\mathrm{Im}(Y) = 0$, that is to say,

$$\omega_0 C - \frac{\omega_0 L}{R^2 + \omega_0^2 L^2} = 0. \tag{1.5}$$

Using the self-resonance condition, equation (1.5), the self-resonant frequency f_o is then given by the following equation.

$$f_o = \frac{1}{2\pi}\sqrt{\frac{1}{LC} - \frac{R^2}{L^2}}. \tag{1.6}$$

Usually the R is small, so self-resonant frequency f_o can be estimated by

$$f_o = \frac{1}{2\pi}\sqrt{\frac{1}{LC}}. \tag{1.7}$$

When self-resonance occurs, the inductive reactance and the parasitic capacitive reactance become equal. Beyond the self-resonant frequency, the inductor becomes capacitive. The self-resonant frequency decides the frequency from which the inductor cannot work well as an inductor any more. The self-resonant frequency of an inductor is supposed to be much higher than its operating frequency. To increase the self-resonant frequency, the parasitic capacitance C in an inductor has to be suppressed.

The maximum diameter of an inductor should be less than $\lambda/30$ in order to avoid distributed effects. High-frequency applications require a smaller size and higher self-resonant frequency, so the inductance density also becomes more and more important. Therefore a major design goal for inductor components is to increase the Q factor, density of inductors and self-resonant frequency.

Capacitor

There are two types of passive capacitors generally used in RF and microwave circuits: interdigital, and metal-insulator-metal (MIM). The choice between the interdigital and MIM capacitors mainly depends on the capacitance value to be made. Usually interdigital capacitors are only used to realize capacitance values less than 1 pF. For capacitance values greater than 1 pF, MIM structures are generally used to minimize the overall size and to avoid the distributed effects. For a capacitance value greater than 200 pF, usually surface-mounted devices are necessary. The capacitor performance is strongly associated with the Q-factor and parasitic inductance of the capacitor. The parasitic inductance L caused by the connection to the capacitor electrodes must be accounted for. The effective capacitance C_e is given by

$$C_e = C\left(1 + \frac{f^2}{f_0^2}\right) \tag{1.8}$$

where $f_o = \dfrac{1}{2\pi}\sqrt{\dfrac{1}{LC}}$ is the self-resonant frequency of the capacitor and f is the operating frequency. When the capacitor operates at the self-resonant frequency, the capacitance will become zero. To have effective capacitance reach the designed capacitance C, the parasitic inductance in the capacitor has to be suppressed.

The quality factor of MIM capacitors is given by

$$Q = \frac{1}{1/Q_c + 1/Q_d} = \frac{Q_c Q_d}{Q_c Q_d}. \tag{1.9}$$

Where $Q_c = 1/\omega CR$ accounting for conducting loss resulted from the wiring and electrode resistor R, and $Q_d = 1/\tan\delta$ accounting for dielectric loss in the capacitor. $\tan\delta$ is the loss tangent of the insulator material of the capacitor. To achieve a high Q-factor, it is essential to reduce the conducting loss in the wiring and electrode and to use dielectric material with a small loss tangent.

The dimension of capacitors should be less than 0.1 λ in dielectric film high-frequency applications. To increase high-frequency performance and the passive circuit density and reduce the cost, a large capacitance density is highly desirable. Silicon oxide and nitride are commonly used in conventional MIM capacitors. They can provide good voltage linearity and low-temperature coefficients. Their capacitance density will be limited by their low dielectric permittivity. The capacitance density can be given by $\varepsilon_0 k / t_d$. Attempts to increase the capacitance density by reducing the dielectric thickness (t_d) usually cause an undesired high leakage current and poor loss tangent. Therefore, high-k dielectric materials are necessary to provide good electrical performances and increase the circuit density.

Resistor

Integrated resistors can be produced either by depositing a thin film of lossy metal on a dielectric substrate or by screen-printing a resistive paste to form a thick-film resistor on or in a ceramics or PCB substrate. Nichrome and tantalum nitride are the most popular film materials for thin-film resistors. SiCr and poly-silicon thin films also used for thin-film resistors. TaN is preferred to NiCr for RF applications, due to the presence of undesirable magnetic material, i.e., nickel, in NiCr, which is believed to introduce unwanted inter-modulation products in multi-carrier wireless systems. Ruthenium dioxide paste and carbon-filled polymer paste are widely used for thick-film resistors. A common problem with planar film resistors is the parasitic capacitance arising form the underlying dielectric region and the distributed inductance. These parasitics make the resistors have a frequency dependence at high frequencies. To shorten the resistor length by introducing films having a larger sheet resistivity is helpful for suppressing the parasitics.

Desirable characteristics of resistors for high-frequency applications are summarized below.

- Stable resistance value without changing with time
- Low temperature coefficient of resistance (TCR)
- Large sheet resistivity (kΩ/square to MΩ/square) to minimize the parasitics and to guarantee the resistor length less than 0.1 λ so that distribution effects can be ignored
- Adequate power dissipation capability

The required tolerances for passive components are roughly summarized in Table 1-1. Analog and RF applications typically necessitate small tolerances of less than±5% and high-performance characteristics such as high Q factors and high self-resonance frequency.

Application	Element Type	Required Tolerance
Damping	Resistor (10-33 Ω)	±30%
Bypass	Capacitor (50 pF-1 μ F)	±30%
Pull-up, Pull-down	Resistor (500-1 M Ω)	±10%
Integral calculus circuit	Capacitor (100 pF-1 μ F)	±15%
Differential circuit	Capacitor (10 pF-10 μ F)	±5%
Oscillation circuit	Capacitor (10 pF-10 μ F)	±5%
Bias circuit	Resistor(1 k-10 M Ω)	±1%
IC controlling	Resistor(≧10 k Ω)	±1%
Filter	Capacitor(≦1 μ F); Inductor(≦100 nH)	±5%
Impedance matching	Resistor(50-100 Ω); Capacitor(≦10 nF)& Inductor(≦100 nH)	±5%

Table 1.1 Required tolerances for passive components

2. Current Research Status and Trend of Passive Integration

2.1 Laminate-Based Passive Integration Technology

Laminate-based passive integration technology is extended from printed circuit board (PCB) technology which has been extensively used for all electronic applications. The primary function of conventional PCBs is to provide mechanical carrier and multilevel electrical interconnections for packaged solid state devices and passive components. Embedded passive technologies on organic substrates were introduced by Packaging Research Centre, Georgia Institute of Technology in 1993. Since then several embedded passive technologies have been developed (Dougherty J. P. et al., 2003; Jung E. et al., 2009; Chason M. et al., 2006). System integration based on embedded passive in PCBs is illustrated in Fig. 1.2 (a) and (b). The embedded substrate is constructed by laminating together polymer-based dielectric or resistive films on which metal conductor are defined on one or both sides. The discrete components can also be buried into the organic substrate for process simplicity. Recently, some advanced PCB technologies have offered embedded passives for low-GHz RF applications, such as a Bluetooth transceiver module, power amplifier module or sensors (Li L. et al., 2003; Li L. et al., 2004; Vatanparast R. et al., 2007).

Embedded Discrete Passives Technology

The simplest embedded passive solution is to embed discrete SMT components into a multilayer organic substrate. The SMT passives are buried in the individual layers that usually are the core-layer (Sugaya Y. et al., 2001; Wu S. M. et al., 2007). On one or both sides of the core-layer, the build-up layers are laminated and vias are formed to provide interconnections for the embedded SMT passives (Shibasaki S. et al., 2004). This solution uses SMD components and offers a relatively simple fabrication process and high reliability and accuracy. For example, SMD film resistors can be made with inter-metallic semi-amorphous alloys with a low thermal coefficient of resistance. The resistance can be laser-trimmed to within 0.1 % accuracy. SMD inductors are optimized according to value and

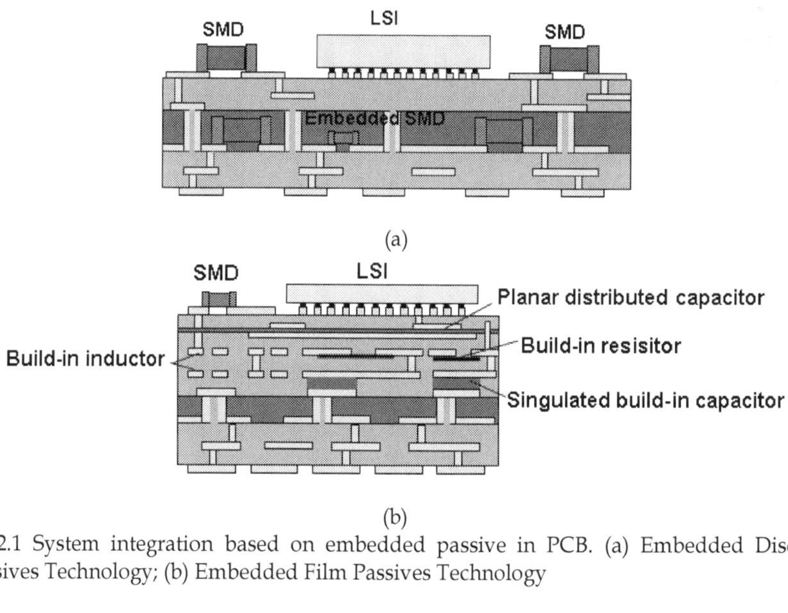

(b)

Fig.2.1 System integration based on embedded passive in PCB. (a) Embedded Discrete Passives Technology; (b) Embedded Film Passives Technology performance with thin-film, thick-film or wire wound technology. With high-tolerance SMD capacitors, components can be selected in terms of which capacitors to embed, and this can lead to improved final capacitance accuracy. An embedded substrate using a discrete SMD is usually thick and big compared to film elements due to the large size of the buried SMD components. To embed SMD components contradicts the minimization purpose to some extent. Several companies have checked the reliabilities of this solution and have used it for mass production (Shibasaki S. et al., 2004; Kamiya H. et al., 2005; Kondou K. & Kamimura R., 2002).

Embedded Film Resistor Technology

Three material technologies have been developed for embedded resistors: thin-film metal (such as NiP), thick-film ceramics, and polymer thick-film (PTF) materials. Their performance comparison is listed in Table 2.1. Copper foils supplied with a resistive thin-film coating are available (Ohmega Technologies, Inc.; Ticer Technologies; Norlyng S., 2003). The thin film metals typically comprise Ni, NiP, NiCr, or NiCrAlSi. These kinds of metal or metal alloy are deposited onto a copper foil by sputtering, evaporation, CVD or plating, depending on the required composition. The deposited thickness ranges from 50 nm to 400 nm. The resistive foils can be laminated onto a core layer. Then photolithography and etching are conducted to define the copper electrodes and to remove the unwanted resistive material. Highly accurate resistors necessitate a laser trimming process (Fjeldsted K., 2004). After resistors have been defined, the following multilayer processing steps like stacking and lamination can be conducted. Thin-film metals have a good TCR performance and are usually used to form small and accurate embedded resistors.
Selective plating of NiP can also be used after the copper pattern is defined and etched (M-Pass, MacDermid Inc.). The plating time defines the thickness and resistivity. Due to the relative low sheet resistivity, this technology is not suited for high resistor values.

	Material Classes		
	Thin-Film Metal	Thick-Film Ceramics	Polymer Thick-Film
Materials	Ni-P; Doped Pt; Ni-Cr; Ni-Cr-Al-Si	Metal oxide powder + Glass powder	Carbon powder + Polymer
Sheet resistance (Ω/square)	10-1 k	10-1 M	10-1 M
Tolerance	±5-10%	±10-20%	±15-20%
TCR (ppm/℃)	±50-100	±150-250	±350
Formation	Etching or Plating	Screen Printing Firing (500-900℃)	Screen Printing Heat hardening (≦200℃)

Table 2.1 Performance comparison of resistive film materials

Polymer thick-film (PTF) materials can be screen-printed and cured at temperatures from 100 to 200℃ (Jillek, W. et al., 2005). The typical cured film thicknesses range form 15-20 μ m. Aerosol-based deposition and Ink-jet printing for resistor-film formation were also reported (Hong T. K. & Kheng L. T., 2005; Shah V. G. & Hayes D. J., 2003). PTF materials are an attractive option and popularly used by the PCB industry due to their flexibility in sheet resistivity (10 Ω～1 MΩ/Square), the ability to print multiple sheet resistivity inks on one layer, and their low cost compared with thin-film metals and thick-film ceramics materials. The challenges of using PTF resistors have been accurate printing and finished resistor stability. Screen-printing is less precise in controlling dimensions than print-and-etch processes used for thin-film metals. Usually screen printing has a tolerance in the order of ±30 μ m, whereas etching a PCB feature allows dimensional control in the order of about 10 μ m.

Additionally, the PTF resistors must be cured, and the resistivity of the finished materials exhibits dependence on the cure profile. Therefore, tolerances on PTF resistors are considered to be higher than those of the thin-film metals. Resistor stability is another concern when using PTF. PTF termination directly on copper results in a relatively poor resistor stability under environmental stress (exposure to 85% RH, 85℃) due to corrosion at the copper/carbon ink interface (Chason M. et al., 2006). Using immersion silver on etched copper pads as the corrosion barrier mitigates the resistor drift under environmental stress while preserving the precise dimensions of the photo-lithographically patterned copper (Savic J. et al., 2002; Dunn G. et al., 2004). PTF resistors are relatively poor in TCR compared to thin-film metal and thick-film ceramics materials.

Thick-film ceramic resistive paste is usually used with copper foils. The pastes are screen printed on the copper foil and fired in an N_2 atmosphere at a high temperature(500～900℃). Then the copper foil is laminated to a core layer or pre-preg. The conductor pattern is defined by photolithography and copper etching processes (Bauer W., 2003; Borland W. et al., 2002). A wide range of sheet resistance values are available with thick film ceramics pastes. Typical fired thicknesses range from 10 to 15 μ m.

Embedded Film Capacitor Technology

Embedded film capacitors have restricted ranges of capacitance values based on the choice of dielectric material, capacitor structure and area allowed. Some popular dielectric materials used in embedded capacitors and their properties are listed in Table 2.2. The challenges for embedded film capacitors have been developing stable high-k materials and forming thin dielectric films. The modern PCB industry uses unloaded epoxy resin as An insulation layer between copper conductors. Typical unloaded epoxy resin has a relatively low dielectric constant (~4) and the resin thickness is approximately 50 μ m, resulting in a capacitance density of 0.7 pF/mm^2, too small to embed typical capacitor values in a reasonable area. Many ceramic-filled resins, such as BaTiO3 in epoxy or Polyimide, have been developed for high-capacitance-density dielectrics (Ulrich R. & Schaper L. W., 2003). Non-photosensitive-type ceramic-filled resins usually come in laminate sheet form compatible with typical PCB fabrication techniques (Norlyng S., 2003; Oak-Mitsui Technologies Technical Data; 3M C-ply Technical Data; Kumashiro Y. et al., 2004). The non-photosensitive-type ceramic-filled resin is coated on a copper foil and after curing the ceramic-filled resin film, another piece of copper foil is stuck to the other side of the ceramic-filled resin film. The electrodes of film capacitors are formed by a photolithography process and the copper foils are etched on ether one side or both sides of an insulator film. Non-photosensitive type ceramic-filled resins are also available in the paste form (Norlyng S., 2003; DuPon Microcircuit Materials, USA). The insulator paste and the upper electrode patterns are screen-printed onto the lower electrode patterns prepared in advance on a substrate or resin film. The ceramic-filled resins can also be made photosensitive, to produce a so called ceramic-filled photo-dielectric (CFP) (Chason M. et al., 2006; Croswell R. et al., 2002). The CFP material is coated onto a planar copper surface, and a second sheet of copper is laminated on top of the dielectric. After electrode patterning, this copper layer becomes a self-aligned mask for dielectric exposure and development. The dielectric constant of the ceramic-filled resin is limited by the density of the ceramics filler, thus the resulting capacitance density is less than 50 pF/mm^2, promising for small to medium capacitor values in a reasonable area. The materials in film form usually used for planar-distributed film capacitors are difficult to use for discrete capacitor formation because of unwanted dielectric removal and layer registration problems. Photosensitive and paste materials are necessary for singulated film capacitors.

<table>
<tr><th rowspan="3"></th><th colspan="5">Material Classes</th></tr>
<tr><th colspan="3">Ceramic-filled Resin</th><th colspan="2">Ceramic</th></tr>
<tr><th>Photosensitive</th><th colspan="2">Non-photosensitive</th><th>Paste</th><th>Nano-particle</th></tr>
<tr><td>Dielectric materials</td><td>$BaTiO_3$ in Photosensitive epoxy</td><td colspan="2">$BaTiO_3$ in Polyimide/Polymer/ Epoxy/Other resin</td><td colspan="2">$BaTiO_3$, $BaSrTiO_3$</td></tr>
<tr><td>Materials form</td><td>Paste</td><td>Film on copper foil</td><td>Paste</td><td>Film on copper foil</td><td>Nano-particle</td></tr>
<tr><td>Capacitor formation</td><td>Exposure and development</td><td>Photolithography and etching</td><td>Screen printing and cure</td><td>Photolithography and etching</td><td>Aerosol deposition</td></tr>
<tr><td>Dielectric constant</td><td>21 (1 MHz)</td><td colspan="2">Up to 60 (1 kHz)</td><td>Up to 1000</td><td>Up to 210</td></tr>
<tr><td>Thickness (μm)</td><td>Up to 11</td><td>4-50</td><td>15-20</td><td>0.6-40</td><td>0.5-10</td></tr>
<tr><td>C-density (pF/mm^2)</td><td>17</td><td>0.7-48</td><td>24-32</td><td>16-15000</td><td>Up to 3000</td></tr>
<tr><td>Dielectric loss</td><td>1.4% (1 kHz)</td><td colspan="2">Up to 5% (1 kHz)</td><td colspan="2">1-1.5%</td></tr>
</table>

Table 2.2 Dielectric materials used for embedded capacitors

Some companies have developed solutions to use thin $BaTiO_3$ Ceramics film as the insulation layer of embedded film capacitors to offer an extremely high capacitance density. The ceramics film is sandwiched with copper (2-20 μm) and nickel (20-50 μm) electrode layer and sintered at high temperature (600-900℃). The ceramics film can be made as thin as 0.6 μm. The sandwiched ceramics sheet is easy to be incorporated into a standard PCB structure by patterning and etching the conductor layers on the both sides, and laminating the sheet into the board. A capacitance density of up to 15 nF/mm^2 is available for a high-k ceramics film (Tanaka H. et al., 2008). But it is difficult to form a singulated capacitor which restricts the range of applications of this kind of high-k film material. Aerosol deposition technology has been developed to form a high-k ceramics layer by spraying ceramic nano-particles directly onto the metal electrode surface (Imanaka Y. et al., 2005; Imanaka Y. et al., 2007). The available film thickness ranges form 0.6 to 10 μm. Aerosol deposition technology

enables both use of high-k ceramic materials and the singulated capacitor structure. Moreover, a multilayered capacitor structure is also possible with this technology.
The capacitance tolerances of embedded capacitors are affected by the precision of dielectric thickness and electrode dimensions. Laminate based technologies have typical layer thickness tolerances of up to 10%. Thick-film patterning technologies are less precise in controlling dimensions. Usually screen printing has a tolerance in the order of ±30um and etching thick metal foils allows dimensional control in the order of about 10um. Additionally, the embedded capacitors commonly undergo a heat treatment processes for hardening or sintering dielectric materials or metal electrodes. So the heat treatment profile also will affect the resulting capacitance deviation. The above-stated factors in general will result in a relatively high capacitance tolerance. Motorola reported that a capacitance tolerance of ±21% (7% standard deviation in the worst cases) can be achieved for CFP capacitors with a mezzanine micro via contact (Croswell R. et al., 2002). When ferroelectric materials such as $BaTiO_3$ are used in the capacitor dielectric layer, the capacitor commonly exhibits a large temperature drift in the order of several hundreds of ppm in capacitance (Kawasaki M. et al., 2004; Popielarz R. et al., 2001; Kuo D. H. et al., 2001; Lee S. et al., 2006) and a loss tangent as large as 1-5%. It is difficult to obtain a resulting Q factor of the capacitors above 30 at the GHz frequency band.

Embedded Inductor Technology

Single and multilayer spiral inductors with inductance up to 30 nH can be printed with standard PCB printing and etching techniques (Chason M. et al., 2006). Transmission line structures can also be used to produce small inductances under about 3 nH. The tolerance of the embedded spiral inductors is between 15-20% and is affected by the registration, print-and-etch and HDI (High-Density-Integration) dielectric thickness control capability of the individual PCB fabricator (Savic J. et al., 2002).
The quality factor of the built-in inductor will be low and the inductor size will be large due to a relatively large dielectric constant of the substrate materials. The size of the embedded coils in PCB is usually larger than 1 mm. To obtain a large Q-factor and a high self-resonance frequency (SRF) for RF applications, the traces of the coils have to be separated a lot and more layers have to be used. The distance between the coil traces and ground plate has also to be large. Moreover, low-k materials with a low loss are preferable for use in the insulation materials between the coil traces. A liquid crystal polymer (LCP) possessing low and stable dielectric constant and loss tangent up to a high frequency is attractive for use as a PCB insulation material, especially for embedded inductor applications at high frequency (Lu H. et al., 2007; Stratigos J., 2007; Govind V. et al., 2006).

2.2 LTCC-Based Passive Integration Technology

For embedded passives, LTCC (low temperature co-fired ceramic) is the preferred technology (Sutono A. et al., 2001). Most LTCC's are based on glass (Ca-B-Si-O glass) with alumina as a filler (Brown R. L. et al., 1994). LTCC allows the use of highly conductive metals such as copper or silver due to its low firing temperature so that good coil inductors with a high-quality factor and low loss interconnects can be built in the substrate. The LTCC process flow is simply illustrated in Fig. 2.2. Thin unfired ceramic tapes (green sheets) are punched, vias are filled and resistors or conductors are screen-printed. The individual sheets are aligned, laminated and co-fired around 850-900℃. LTCC-based integrated passive devices are illustrated in Fig 2.3. The conductors can be the electrodes of parallel plate capacitors or windings of an inductor. For small capacitance values, the normal LTCC tapes are usually used for the capacitor's dielectric layer. The tape thickness can be made as thin as 12.5 μ m. For large capacitance values, high-k LTCC tapes need to be introduced. To form embedded singulated capacitors, high-k ceramic paste materials are needed. The high-k paste is screen-printed on unfired tapes and co-fired with LTCC tapes after laminating all the tapes together. The resistors are usually printed on the surface of the outer layers so that laser trimming can be conducted to achieve a high resistance tolerance of up to $\pm 1\%$. Ruthenium (RuO_2)-doped glass is commonly used as the resistor materials. Film resistors can be formed by either co-firing or post-firing technology. In the co-firing process the resistor paste is fired along with the LTCC. And in the post-firing process, the resistor paste is printed onto the fired LTCC surface and then sintered again at a temperature lower than that of LTCC's firing.

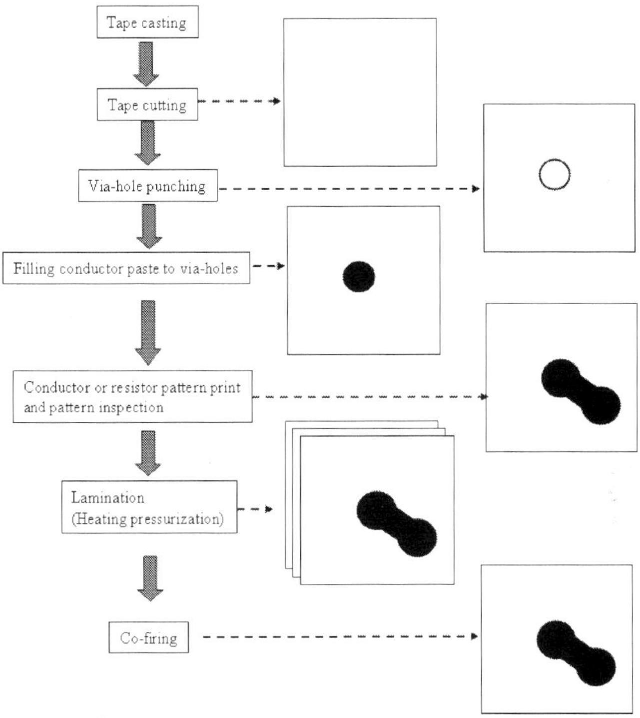

Fig. 2.2 LTCC process flow

HTCC uses standard ceramic materials such as alumina (Al_2O_3) or aluminum nitride (AlN) that are fired at high temperature (1600℃). The firing temperature of HTCC precludes the use of highly conductive metals as inner electrical interconnects, which are necessary for high-frequency applications to realize low insertion loss. Refractory metals such as tungsten or molybdenum with a low conductivity must be used in HTCC, resulting in additional

resistive losses. No suitable co-friable resistive and capacitive materials for embedding passive components in HTCC exist. The integration of passives for HTCC is limited to the surface by using post-firing technology.
Material properties of LTCC, HTCC and FR4/glass are listed in Table 2.3. LTCC succeeded in producing some good characteristics for HTCC including high thermal conductivity, low dielectric loss, and high resistance against humidity and heat. Recently, some new LTCC materials with high mechanical strength have been developed (muRata LTCC). One newly developed material has a high flexural strength of up to 400MPa, almost the same as HTCC, whereas the conventional LTCC has a flexural strength of only about 200 MPa. High mechanical strength helps protect the substrate from cracking when it receives strong mechanical or thermal shocks. High mechanical strength also allows for the use of a thinner substrate.
One more advantage that LTCC technology has over HTCC technology is that a high level of dimension precision, less than ±0.1% can be obtained by using no shrinkage firing techniques, which are not available with its rival, HTCC technology. This high dimension precision allows the module substrate to achieve high density integration and assembly.

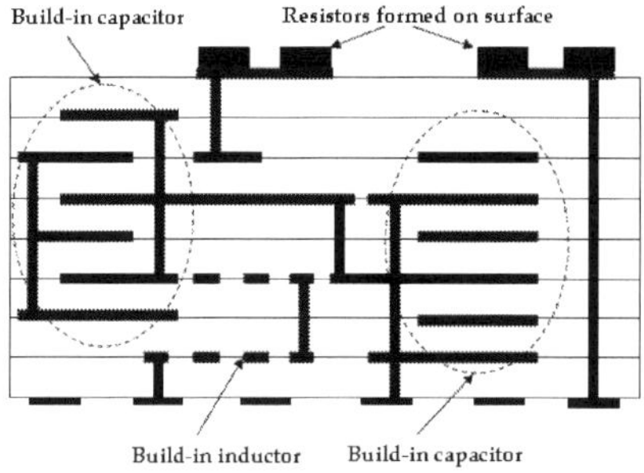

Fig. 2.3 LTCC-based integrated passive devices

Similar to laminate-based passive integration technology, the built-in components are defined by a screen-print technique having less precision in the pattern dimension and thickness. This results in a relatively high production deviation in property values of embedded passive components. Usually embedded inductors can have inductance values of up to 10 nH at a reasonable area and thickness. Capacitors of up to 10 pF can be built in at a reasonable area without introducing additional high-k dielectric layers. A capacitance tolerance less than ±20% can be expected. If ferroelectric ceramic materials are used to obtain large capacitance density, the capacitor will show a great temperature drift in capacitance and the loss tangent will also increase in the order of several percentage points.
Low electrical and dielectric losses, a good thermal conductivity, and high levels of dimension precision and passives integration make LTCC technology attractive for RF module applications. LTCC is now popular for constructing compact passive circuits ranging from traditional baluns (Lew D. W. et al., 2001), couplers (Fujiki Y. et al., 1999), and filters (Yeung L. K. & Wu K. L., 2003; Piatnitsa V. et al., 2004) to more sophisticated diplexers (Sheen J. W., 1999) and balanced filters (Yeung L. K. & Wu K. L., 2006), for different wireless communication systems such as mobile phones, Bluetooth, and wireless LAN equipped terminals. LTCC technology is also attracting a great deal of interest for produce highly

integrated RF Front-End-Modules (FEM) where embedded passive circuits are combined with active devices to make complete functional modules, like those encompassing a Tx/Rx switch, SAW filters, and/or power amplifier (Marksteiner S. et al., 2006).

	Electrical			Thermal		Mechanical	
	ε	tan δ (10^{-4})	Resistivity (Ω cm)	CTE (ppm/K)	Thermal Conductivity (W/mK)	Flexural Strength (MPa)	Young's Modulus (GPa)
LTCC	5 - 80	2.5 - 40	$>10^{14}$	3-12	1.2-5	170-400	74-188
HTCC	8.5 - 10	5-25	$>10^{14}$	6.9-7.2	10-25	400-460	260-310
ALN	8.7	170	$>10^{14}$	4.7	150-230	400	320
FR4/ glass	4.5 - 5.5	200 - 300	$>10^{14}$	xy:16-20 z:50-70	0.2	430	–

Table 2.3 Dielectric materials used for embedded capacitors

2.3 Thin-Film-Based Passive Integration Technology

Thin-film-based passive integration is used to integrate passive elemnets onto a substrate's surface using photolithography and thin-film technology, enabling a fine structure and high integration density. Such thin-film-based integrated passive devices are usually called as IPDs (integrated passiv devices). The passive elements and high-density interconnects are fabricated by depositing thin metal and dielectric materials onto a high-resistivity or dielectric substrate by using standard IC fabrication technologies such as evaporation, sputtering, electroplating, chemical vapor deposition (CVD) and spin-coating. The layers are defined by standard photolithographic etching or selective deposition such as lift-off process which are used in the semiconductor IC industry. A typical IPD structure is illustrated in Fig.2.5. Usually resistors, capacitors and inductors are formed directly on the substrate surface. A interconnection layer is formed above the passive elements to connect these elements. A dielectric interlayer is placed between the interconnection layer and passive elements as an insulation layer. A passivation layer is usually formed on the surface to protect the passive circuts from the atmosphere. Pads are formed on the top of the passivation layer to provide acess between the IPD and the outside.

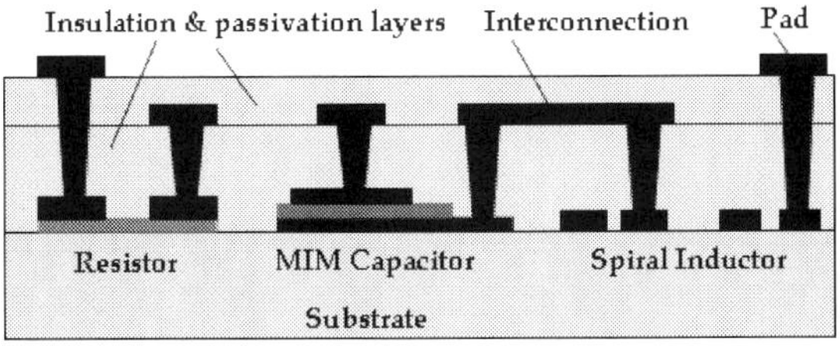

Fig. 2.4 A tipical IPD structure

When semiconductor materials are used as the IPD's substrate, an additional dielectric layer is needed to insulate the passive elements and the substrate, and the inductors are favarourablly formed at the interconnection layer so as to separate the inductor from the substrate to reduce the eddy current loss in the semiconductor substrate.

	Electrical			Thermal		Mechanical	
	ε	tan δ (10^{-4})	Resistivity (Ωcm)	CTE (ppm/K)	Thermal Conductivity (W/mK)	Flexural Strength (MPa)	Young's Modulus (GPa)
Quartz	3.8	0.2	>10^{18}	0.56	1.5	Up to100	74
Glass	5.3	10	>10^{17}	3.8	1-2.5	<94	Up to 100
Al_2O_3 99.9%	9.9	1	>10^{15}	6.8	38	660	390
HR-Si	12	10-50	1000-5000	2.6	150	2800	129-190
HR-GaAs	12.9	6	10^7-10^9	5	46	-	-

Table 2.4 Commonly used materials for IPD substrate

IPD Substrates

The substrate materials usually used in IPD are list in Table 2.4. For high-frequency applications, the substrates have to be selected to have low dielctric losses and low permitivity to reduce the RF power dissipation in the substrates and to increase the self-resonant-frequency. Glass, fused quartz, high-resistivity silicon and ceramics are usually used as IPD's substrate for RF applications. Glass and fused quartz have both a low dielectric constant and low dielectric tangent which makes them preferrable for achieving a high self-resonant frequency and reducing the eddy loss in the substarate which is asociated with the qualty factor of passive circuits. From the viewpoint of high-frewuency applications, glass and fused quartz are the most suitable substarte materials. The semiconductor substrates are attractive for allowing active devices to be integrate with the passive circuit. For example, combining ESD Protection and a low pass filter to attenuate the RF noise, which may otherwise interere with the internal base band circuitry of a mobile phone (Doyle L., 2005). Normal semiconductor substrates used for IC fabrication have such a low resistivity (≦50 Ωcm) as to be very lossy for RF signals, as magnetic fields penetrate deeply into the substrate causing losses and reducing both the inductanse and Q-factor. Inductors formed on such substrates merely provide a Q-factors of around 10 (Tilmans H. A C et al., 2003; Chong K. et al., 2005). High resistivity in the order of 1000-10000 Ωcm is in general required for the semiconductor substrates used for IPD to suppress these parasitic phenomena (Tilmans H. A C et al., 2003). Use of high-resistivity Si is not cost effective for most CMOS applications. Moreover, when high-resistivity silicon is used for the substrate, the surface of the silicon has to be oxidized. In addition, fixed charges occurring in the oxide layer will cause DC dependency and performance spread. Ar implantation on the silicon surface has been proposed to migarate these negative factors (Carchon G. et al., 2008). If a normal Si substarte with low resistivity is used for IPD, some special processes have to be taken to separate the IPD from the substrate or to reduce the substrate loss. A number of new techniques have been proposed to reduce the substrate loss. Placing a 25um porous silicon dioxide layer between the IPD and silicon substrate has been proposed by Telephus

(Kim D. et al., 2003). MEMS (Microelectromechanical System) technology has been used to remove the substrate under the inductors (Jiang H. et al., 2000; Chang J. Y-C et al., 1993). And there have been reports on suspending large-sized inductors that were integrated with the RF mixer using post-CMOS-based techniques (Wu J. C. & Zaghloul M. E., 2008). Using a ground shielding metal layer between the inductor and Si substrate to prevent electrical coupling with the substrate can improve the Q factor by up to 50% (Yue C. P. & Wong S. S., 1998). Another approach is to introduce an air gap into the Si substarte using Si deep-RIE and a Si thermal oxidization technique (Erzgraber H. B. et al., 1998). These alternative solutions can increase the Q-factor up to 20-30 at a low GHz frequency, but usually they are complex, costly or not compatible with standard CMOS lines. For many important circuit functions in wireless communications systems, such as RF front end and radio transceiver applications, it is preferable for the inductor to have a Q factor of at least 30. Furthermore, since the wireless systems are moving to a much higher frequency, a high Q, high self-resonant-frequency and small size are required for integrated passive devices.

IPD Resistors

IPD resistors are made by sputtering or evaporating resistive material onto the substrate, like NiCr, Mo, Ti, Cr, or TaN and CrSi (Tilmans H. A C et al., 2003; Bahl I. & Bhartia P., 2003). Some popularly used thin-film resistive materials are listed in Table 2.5. For large sheet resistivity, Cr, Ti and CrSi are favorable. NiCr, Ta and TaN provide good stability. Cr is poor in terms of stability. Resistor values ranging from 0.1 Ω to several tens of M Ω can be acheived with 10% tolerance (Doyle L., 2005). It is also easy to conduct laser trimming to further tighter the tolerance because the resistor films are formed on the surface. To shorten the resistor length by introducing films having a larger sheet resistivity is helpful for suppressing the parasitic L and C, and to guarantee a resistor length less than 0.1λ so that distribution effects can be ignored, which is important for high- frequency applications.

Resistive Material	Resistivity (Ω/square)
NiCr	40-400
Cr	10-1000
Ti	5-2000
Ta	5-100
TaN	4-200
CrSi	Up to 600

Table 2.5 Thin-film resistive materials for IPDs

IPD Capacitors

IPD's capacitors are typically MIM or interdigitated capacitors with dielectric materials between the electrodes. The size of an integrated capacitor depends the dielectric constant and thickness of the dielectric material used in the capacitor. Since the capacitors are formed on the substrate's surface, an ultra-thin insulator film can be used for capacitors so that a relatively high capacitance density can be achieved. Since the capacitor area is defined by a standard photolithographic etching or lift-off process, very high accuracy can be obtained. A capacitance density of 200 pF/mm^2 has been realized with a tolerance less than ±3% (Mi X. et al, 2008). The dielectric materials usually used in IPD capacitors are listed in Table 2.5. A good dielectric material should have a high dielectric constant, a high band gap to limit leakage currents, and a high dielctric strength to meet reliability reqirements.

	Dielectric Constant	Dielctric Loss (10^{-4})	Breakdown Field (MV/cm)	Demonstrated C-dencity (nF/mm^2)
SiO_2	4.2	10	10	1
Si_3N_4	7.6	11	7	2
Al_2O_3	7.9	30	8	3.5
HfO_2	17-21	500	6	5, 13
Ta_2O_5	22-25	100	5	5
ZrO_2	45	-	4	-
$SrTiO_3$	150	200	1	10
BaTiO3	800	<60		80
PZT	900			

Table 2.6. Dielectric materials for used in IPD capacitors

Although SiO_2, Si_3N_4, and Al_2O_3 have small dielctric constant, thay are typically used in IPD capacitors due to their low dielectric loss, high dielctric strength (breakdown field) and good film qualty (Huylenbroeck S. V. et al., 2002; Zurcher P. et al., 2000; Jeannot S. et al., 2007; Allers K. -H. et al., 2003). A thin insulation film is favorable for achieving a large capacitance density, but presents a risk in terms of breakdown voltage. Achieving a high-quality thin-film formation and high dielctric strength are the key points to realizing a high-capacitance density. When the dielectric film thickness is reduced to below a few tens of nm, Si_3N_4 is more suitable to use. The reason is that unacceptable leakage currents due to tunneling conduction and a low breakdown voltage will arise in such a thin SiO_2 film. Highly C(V) nonlinear properties observed in thin Al_2O_3 need to be taken into account for lineaity-critical applications (Jeannot S. et al., 2007). Ta_2O_5 has excellent C(V) linearity compared with Al_2O_3 and a high breakdown field, but it also has an extremely high leakage current (Giraudin J.-C. et al., 2007; Thomas M. et al., 2007). HfO_2 has a relatively low breakdown voltage and worse C(V) linearity compared to Al_2O_3, but presents low leakage currents due to its high band gap (Yu X. et al., 2003). To further improve electrical performances, a combination of different dielectric materials such as a $HfO_2/Ta_2O_5/HfO_2$(HTH) multi-layer has led to good results (Jeannot S. et al., 2007). ZrO_2 has been demonstrated to be promising, exhibting a lower dielectric leakage than Al_2O_3 and HfO_2 and simmilar braekdown field with Al_2O_3 (Berthelot A. et al., 2007). These high-k materials have been facing the limitation that a capacitance above 5 nF/mm^2 can hardly be reached with a planar MIM architecture. Besides the evolution of high-k dielectrics, new developemnts in capacitor architectures have also been put forward to further increase capacitance density. High density trench capacitor (HiDTC) architecture has been demonstrated to be feasible for extremely high capacitance density (Giraudin JC. et al., 2007; Giraudin JC. et al., 2006). Based on such architecture, a 35 nF/mm^2 MIM capacitor has been developed with an Al_2O_3 dielectric of 20 nm, whereas the capacitance density is only 3.5 nF/mm^2 in planar MIM architecture.

MIM capacitors using ferroelctric materials such as STO, BTO and PZT have also been studied intensively, and they have a very high dielectric constant favourable for achieving a very high capacitance dencity (Ouajji H. et al., 2005; Defaÿ E. et al., 2006; Wang S. et al., 2006, Banieki J. D. et al., 1998). These materials usually need high-temperature processing and noble metals for the electrodes. Recently, STMicroelectronics has reported the dry etching of high-k dielectric PZT stacks for integrated passive devices. (Beique G. et al., 2006)

IPD Inductors and Interconnects

IPD inductors are usually formed in conductive interconnection layers on the substrate or insulation layer as shown in Fig. 2.4. A fine trace width and space less than 10 μm can be realized by lithography and electroplating technologies with extremely high accuracy and low manufacturing costs. In addition, an inductance tolerance of less than 2% can be expected(Mi X. et al., 2007; Mi X., 2008).

The conductive materials for IPD inductors and interconnects should have high conductivity, a low temperature coefficient of resistance, low RF resistance, good adhesion to the substrate or insulation materials, and be easy to deposit or electroplate. The RF resistance is determined by the surface resistivity and shin depth. The conductor thickness should be at least three to four times the skin depth, to increase the section area of the conductor where the RF current will flow.

Table 2.6 shows the properties of some normally used conductor materials for IPDs. In general, these conductors, such as Au, Cu, Al, and Ag, have good electrical conductivity but also have poor substrate adhesion. Conversely, some conductors having poor electrical conductivity such as Cr, Pt, Ti and Ta possess good substrate adhesion. To obtain a good adhesion to the substrate and high conductivity at same time, a very thin adhesion layer of a poor conductor has to be deposited between the substrate and the good conductor. This thin adhesion layer does not contribute to any RF loss due to its extremely thin thickness. Since the electroplating is widely used to form a thick conductive layer, the compatibility with the plating process should be taken into account when choosing the conductor materials.

Considerable research is focused on developing high-Q on-chip inductors. Various MEMS technologies have been used to construct a 3D-inductor. Out-of-plane coil structures have been realized by surface micromachining and sacrificial layer techniques (Dahlmann G. W. et al., 2001; Zou J. et al., 2001). These out-of-plane coils vertical to the substrate help reduce the substrate loss and parasitics, but the reported Q-factors do not exceed 20. Moreover the vertical coil is too high (several hundred μm) to use in practical applications, though they do not occupy footprints. Palo Alto Research Center has reported a 3D solenoid inductor in the air constructed using stressed metal technology (Chua C. L. et al., 2002; Chua C. L. et al., 2003). A release metal layer was placed under the stress-engineered metal layer and a release photo-resist layer above the stress-engineered metal layer. When the release metal layer and photo-resist layer were removed, the traces curled up and interlocked with each other to form a coil. Similar structures and fabrication techniques have also been reported by Purdue University (Kim J. et al., 2005). These solenoid inductors in the air show a high Q-factor and self-resonant frequency and are attractive for high-frequency applications. The large size appears to be a drawback for this solenoid type inductor from the viewpoint of miniaturization. Some other reported solenoid type inductors did not show a sufficiently high Q-factor due to the small inductor core area (Yoon Y. K. et al., 2001; Yoon Y. K. & Allen M. G., 2005). Integration of magnetic materials into inductors can significantly increase inductance while keeping similar Q-factor at the frequencies of up to several hundred of MHz (Gardner D. S. et al., 2007). It is difficult nowadays to enable magnetic materials to have both a high permeability and resistivity at high frequencies of above 1GHz. Spiral coil architecture is widely used for IPD inductors due to its high inductance density, compact size (Wu J. C. & Zaghloul M. E., 2008; Tilmans H. A C et al., 2003; Yoon J. B. et al., 2002). Optimized 2-layered spiral coils in the air have been demonstrated for IPDs to offer high quality factor and self-resonant-frequency (Mi X. et al., 2007; Mi X. et al., 2008), which will be explained in section 3 and 4 in detail.

	Surface Resistivity $(\Omega / squre \times 10^{-7}\sqrt{f})$	Skin depth @2 GHz (um)	CTE (ppm/K)	Adherence to dielectrics	Deposition technique
Ag	2.5	1.4	21	Poor	Evaporation, sputtering or plating
Cu	2.6	1.5	18	Poor	
Au	3	1.7	15	Poor	
Al	3.3	1.9	26	Poor	Evaporation, sputtering, EB-evaporation, EB-sputtering
Cr	4.7	2.7	9	Good	
Ta	7.2	4	6.6	Good	
Ti	13.9	7.9	8.4	Good	
Mo	4.7	2.7	6	Fair	
W	4.7	2.6	4.6	Fair	

Table 2.7 Properties of some conductor materials used in IPDs.

Dielectric Materials for Insulation and Passivation layers

Photosensitive polymer dielectric materials are usually used to form insulative interlayers and passivation layers in IPDs. These dielectric materials insulate or protect the integrated passive elements and conductive interconnects so that they are critial for IPD performance, especially for high-frequency performance and reliability (Pieters P. et al., 2000; Li H. Y. et al., 2006). Since these materials cover all the elements, magnetic fields occuring in the passive circuits will penetrate the polymer dielectric material causing losses and reducing the Q-factor and self-resonant-frequency of the passive circuit. The dielctric materials have to be selected with low dielctric constant, low dielectric loss and good electrical performance. Some good polymer dielctric materials suitable for IPD and their propertis are listed in Table 2.8. BCB has good dielectric properties and mechanical chericteristics and verified realiability, and so it has been the most widely used insulation material for IPD or other passivation applications (Tilmans H. A C et al., 2003; Carchon G. J. et al., 2005).

Based on the above-stated device achitecture, materials and fabrication technologies, many IPD devices have been developed and practically used due to their good RF performances, compact size and high manufacture tolerance at serval companies like IMEC (Carchon G. J. et al., 2005; Carchon G. et al., 2001; Tilmans H. A C et al., 2003), Sychip (Davis P. et al., 1998), Telephus (Jeong I-H. et al., 2002; Kim D. et al., 2003), Philips (Graauw A. et al., 2000; Pulsford N., 2002), TDK (CHEN R. et al., 2005), Fujitsu (Mi X. et al., 2007; Mi X. et al., 2008). IMEC IPD used borosilicate glass, TaN resistors, Ta_2O_5 capacitors and electropalated Cu for coils and interconnects. Telephus IPD based on a 25 μm thick oxide grown on low-cost silicon wafers, NiCr resistors, SiNx capasitors, BCB dielectric and electroplated Cu conductors. IPDs have been used for constructing compact passive circuits such as couplers (Carchon G. J. et al., 2001), and filters (Paulsen R. & Spencer M., 2008; Frye R. C. et al., 2008), diplexers (CHEN R. et al., 2005) and impedance matching circuits (Nishihara T. et al., 2008; Tilmans H. A C et al., 2003), for different wireless communication systems such as mobile phones, Bluetooth, and wireless LAN equipped terminals. The IPD technologies are also powerful for realizing highly integrated RF Front-End-Modules (FEM) where IPD can be combined with active devices to make complete functional modules, like those encompassing a Tx/Rx switch, SAW filters, and/or power amplifier (CHEN R. et al., 2005; Jones R. E. et al., 2005).

	BCB	AL-Polymer	Polymide	Epoxy
Dielectric constant	2.7	2.6-2.7	3.2-3.5	4.2
Dielectric loss	0.0008	0.003	0.002	0.02
Water absorption (%)	0.24	0.1-0.3	Up to 3	>0.3
Elongation at break (%)	8	20	10-40	6
Young's modulus (Gpa)	2.9	1.3	2-4	4
Tensil strength (Mpa)	87	90	100-150	100
CTE (ppm/k)	52	60	40	69
Photodefinition	Negative/ Solvent Positive/ Aqueous	Negative/ Solvent	Various	Negative/ Aqueous
Cure temperature (℃)	225-250	180-250	350-400	190-200

Table 2.8 Some good polymer dielctric materials suitable for IPDs and their properties

2.4 Comparison Among Laminate, LTCC and Thin Film Based Approaches

Capacitance values and the corresponding capacitor areas are compared in Fig. 2.5 between various dielectric materials typically used for integrated film capacitors in laminate, LTCC and thin-film based passive integration approaches respectively. Unloaded epoxy resin normally used in PCB having a relatively low dielectric constant (about 4) and minimum thickness of approximate 50um, offers a small capacitance density of 0.7 pF/mm^2. Normal LTCC tape with a dielectric constant around 7~9 and a minimum thickness of 12.5 μm can offers a capacitance density of about 6pF/mm^2. If high-k LTCC tape with a dielectric constant of 80 and thickness of 12.5 μm is used, a capacitance density of 57 pF/mm^2 can be expected, which is still low from the viewpoint of module miniaturization. By introducing ferroelectric ceramic-filled-polymer materials such as CFP (ceramic-filled photo-dielectric), laminate-based embedded film capacitors can increase the capacitance density to several tens of pF/mm^2. Furthermore, an embedded film capacitor with a capacitance density of 15 nF/mm^2 using ferroelectric $BaSrTiO_3$ foil has been reported as a bypass capacitor (Tanaka H. et al., 2008). Although laminate-based film capacitors can provide similar or bigger capacitance densities compared to LTCC by introducing ferroelectric materials, they exhibit a large temperature coefficient of capacitance (TCC). The reason is that the ferroelectric ceramics used in film capacitors, such as $BaTiO_3$, $SrTiO_3$ show large changes in their dielectric constant around the phase transition temperatures (Kawasaki M. et al., 2004; Popielarz R. et al., 2001; Kuo D. H. et al., 2001; Lee S. et al., 2006). Moreover, polymer materials such as epoxy usually have a large change in dielectric constant around the glass transition temperature (Tg). LTCC-based film capacitors have little temperature dependence and superior reliability because the LTCC materials having dielectric constants up to 100 are usually paraelectric material.

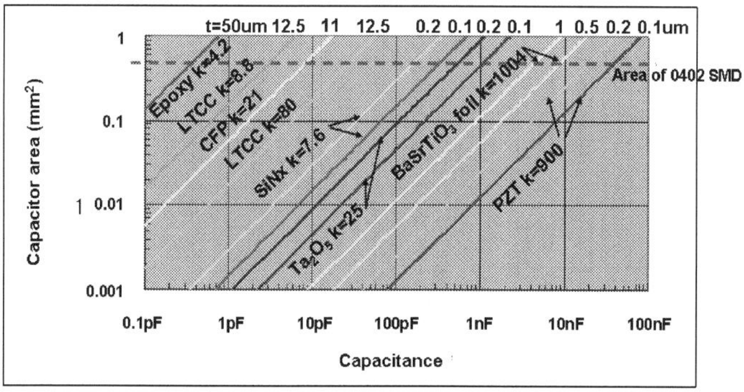

Fig. 2.5 Comparison of capacitance values and corresponding capacitor areas between various dielectric materials used for integrated film capacitors

Since with thin-film-based passive integration technology the capacitors are formed on the substrate surface, ultra-thin high-k dielectric film with good film quality can also be deposited for capacitors so that a relatively high capacitance density can be achieved. A 0.1 μ m SiNx film can offer a capacitance density over 600 pF/mm^2. Higher capacitance densities are also possible at the expense of the breakdown voltage reduction by using an ultrathin film (less than 50 nm). Based on HiDTC (high-density-trench capacitor) architecture, a 35 nF/mm^2 MIM capacitor has been developed with an Al_2O_3 dielectric of 20 nm.

Embedded inductors in an organic subatrate offering an inductance of up to 30 nH have been reported (Govind V. et al., 2006). LTCC are usefule for integating inductors less than 10nH. For IPD (thin-film-based solution), inductance up to 30nH at a size of less than φ0.6 mm has been reported (Mi X. et al., 2008). Mike Gaynor provided a good case study, in which performances of integrated inductors constructed by laminate, LTCC and Si-IPD respectively are compared in detail (Gaynor M., 2007). For the same inductance, laminate inductors need an area of 5 times that required by IPD. LTCC inductors, which have increasing thickness because more layers are used, have a steep area vs. inductance slope and a slightly increased area compared to IPD inductors.

Since the laminate or LTCC-based methods build the passives into the substrate, the quality factor of the built-in inductor will be low and the inductor size will be large due to a relatively large dielectric constant and loss tangent of the substrate materials. The typical inductor diameters are in the order of 1mm. To obtain a large Q-factor and a high self-resonance frequency (SRF) for RF applications, the traces of the coils have to be separated a lot and more layers have to be used. That is to say, low-k materials are preferable for inductor performance. Elsewhere, the typical capacitance density of built-in capacitors is a few pF/mm^2 at present, which is still low from the viewpoint of module miniaturization. It is usually difficult to introduce a very thin insulation layer into these two technologies, so high-k insulators are favorable for obtaining a large capacitance. However introducing low-k and high-k insulators into the same substrate will drastically increase the fabrication complexity and cost.

IPDs are fabricated using photolithography and thin-film technology, enabling a fine structure and high integration density. The inductors and capacitors are formed on the substrate's surface. Some low-k materials can be used easily for inductors so that a small inductor can provide a large inductance and high SRF compared to laminate- or LTCC-based technology (Mi X. et al., 2008). In general, IPD inductors can also provide the best Q-factor for a given size, if low-loss substrates are used.

The most important advantage of IPD is the high production precision. The inductor's tolerance is less than ±2% and the capacitor's tolerance is less than ±3% (Mi X. et al., 2008). This degree of production precision is not available in laminate- or LTCC-based technologies. The embedded inductors in organic or LTCC substrate usually have manufacture tolerance around ± 10%. The embedded film capacitors usually shows manufacture tolerance around ±20%. Since resistors can be formed on the substrate surface and thus laser trimming can be introduced, IPD and LTCC solutions provide excellent resistor precision of ±1%. Embedded resistors in organic substrates usually show a high manufacture tolerance between ±5% and ±20%.

The LTCC has a good thermal dissipation and is preferable for power modules when compared to laminate-based technology. IPD can also have good thermal dissipation capabilities, if their substrates are made of Si or ceramics providing high thermal conductivity.

Element / Method	Inductor	Capacitor	Inner Wiring	Cost/mm^2
Laminate (PWB) LTCC	• Low Q-factor • Low SRF (Due to high ε) • Big and thick	• Low density (<50pF/mm^2)	O	Low-moderate (Low density)
IPD on Si/Glass	• High Q (>40) • High SRF (>8GHz) • Small and thin	• High density ≥ 600pF/mm^2	× Through-wafer via is possible	Depending on wafer size (High density)

Table 2.9 Performance comparison among laminate, LTCC and thin-film based technologies

Laminate-based capacitors are now at the early development stage with many materials and processes in development. The yields and reliability of laminate-based capacitors also need to be evaluated. The high tolerance of embedded passive elements will limit them to coarse applications or digital applications. Thin-film-based passive integration technology provides the highest integration density with the best dimensional accuracy and smallest feature size, which makes it the best alternative for passive circuits in SIP solution at high frequencies. The drawback remains the higher cost compared to the laminate- and LTCC-based technologies. It is now generally accepted that laminate-based passive integration shows the lowest cost per unit area, but occupies the largest area; LTCC-based passive integration has a medium cost per unit area and can integrate more functionality in a smaller size than laminate-based; and thin-film based passive integration has the highest cost per unit area but the smallest size, thus offsetting the cost for the same functionality. Moreover, since thin-film-based passive integration is based on a wafer process, the cost per unit area strongly depends on the wafer size. If 8-inch wafers are used, the thin-film-based solution will cheaper than the LTCC-based one. A performance comparison of these three technologies is summarized in Table 2.9. The disadvantages of conventional thin-film-based technology such as glass/Si IPD compared to laminate- or LTCC-based technologies are that the inner wiring is not available and, while a through-wafer via is possible for a Si or glass substrate, it is expensive (Bhatt D. et al., 2007; Beyne E., 2008). Fujitsu demonstrated IPD-on-LTCC technology. IPD-on-LTCC technology combines the advantages of IPD and LTCC and provides a technical platform for future RF-modules, having all the technical elements necessary for module construction, including integrated passives, dense interconnection, and package substrate. Section 3 and 4 will explain the details of this technology.

3. Design Consideration for High Q, High SRF (Self-Resonant Frequency)

Inductors are one of the most important passive components in RF circuits. The quality factor and self-resonance frequency (SRF) of an integrated inductor are the most important characteristics for high-frequency applications, which decide the working frequency band and the insertion loss of the integrated passive circuits. For a conventional spiral coil inductor structure, the Q-factor is usually below 30, which is not enough for the RF front end and radio transceiver sections. How to construct an inductor having a high Q and high SRF and small size is the key point for integrated passive circuits.

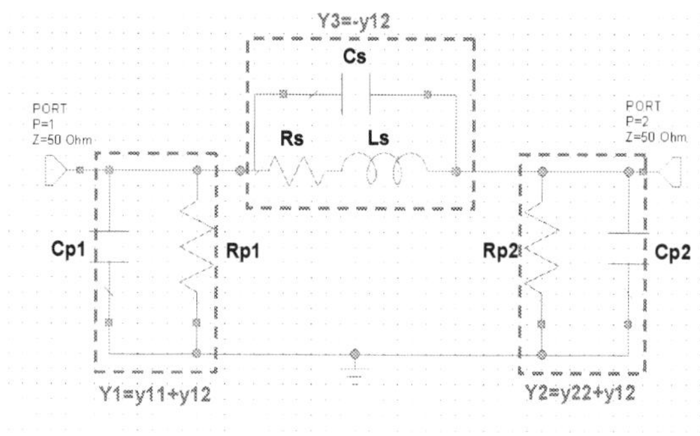

Fig. 3.1 Two-port lumped physical model for an on-chip inductor

To understand the inductor well, an equivalent circuit model and a method of extracting the Q-factor, inductance and parasitic components have to be clarified first. A two-port lumped physical model for an on-chip inductor is shown in Fig. 3.1 (Niknejad A. M. &, Meyer R. G., 1998). The physical model is a two-port π network of series and shunt components of inductors and capacitors, where *Ls, Rs, Cs,* represent the inductance, series resistance, and parasitic capacitance of the inductor, and *Rp, Cp* represent the substrate resistivity and parasitic capacitance in the substrate, respectively.
The Y parameters can be obtained from the measured two-port S parameter. The Q-factor of the inductor can be extracted from two-port Y parameter as shown in equation (3.1).

$$Q = \left|\text{Im}(Y_{11})/\text{Re}(Y_{11})\right| \quad (3.1)$$

According to the two-port lumped physical model, Ls, Rs, Cp, Rp can be extracted from the two-port Y parameter as shown in the following equations (3.2 to 3.5) respectively.

$$L_s = \frac{1}{2\pi f}\text{Im}\left(\frac{1}{Y_{12}}\right) \tag{3.2}$$

$$R_s = \text{Re}\left(-\frac{1}{Y_{12}}\right) \tag{3.3}$$

$$C_p = \frac{1}{2\pi f}\text{Im}(Y_{11} + Y_{12}) \tag{3.4}$$

$$R_p = \text{Re}\left(\frac{1}{Y_{11} + Y_{12}}\right). \tag{3.5}$$

The quality of an inductor is evaluated by its Q factor, which is generally defined as

$$Q = 2\pi \cdot \frac{energy \quad stored}{energy \quad loss \quad in \quad one \quad oscillation \quad cycle}. \tag{3.6}$$

It is inevitable that some parasitic capacitances will occur in a real inductor. For an inductor, only the energy stored in the magnetic field is of interest. Any energy stored in the electric field due to the parasitic capacitances is counterproductive. Thus the Q-factor of an inductor should be given by equation (3.7) (Yue C. P. & Wong S. S., 1998)

$$Q = 2\pi \cdot \frac{peak \quad magnetic \quad energy - peak \quad electric \quad energy}{energy \quad loss \quad in \quad one \quad oscillation \quad cycle} = \frac{R}{2\pi f L}\cdot\left[1-\left(\frac{f}{f_0}\right)^2\right]. \tag{3.7}$$

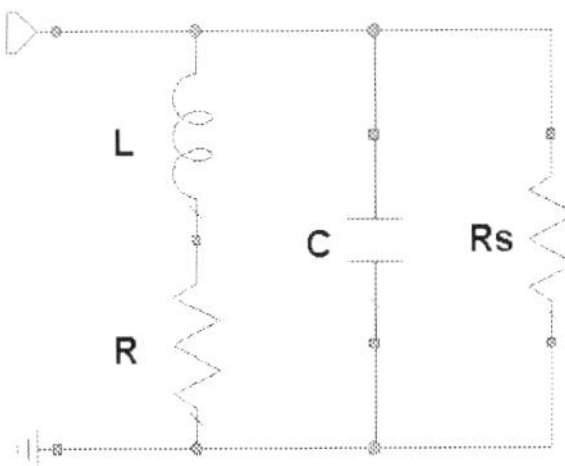

Fig 3.2 One-port physical model of an inductor including parasitic effects

For simplicity, an one-port physical model including parasitic effect as shown in Fig. 3.2, is used to derive the expression of the Q value. According to one-port physical model including parasitic effects, the energies stored in the magnetic and electric fields and lost can be expresses in eqation (3.8 to 3.10) respectively.

$$E_{peak \quad magnetic} = \frac{V_0^2 L}{2 \cdot \left[(\omega L)^2 + R^2 \right]} \tag{3.8}$$

$$E_{peak \quad electric} = \frac{V_0^2 C}{2} \tag{3.9}$$

$$E_{loss \quad in \quad one \quad oscillation \quad cycle} = \frac{2\pi}{\omega} \cdot \frac{V_0^2}{2} \cdot \left[\frac{1}{R_s} + \frac{R}{(\omega L)^2 + R^2} \right]. \tag{3.10}$$

Where R represents the series resistance of the inductor; C represents the total parasitic capacitance including that of the inductor and substrate; Rs represents the substrate resistivity and V_0 denotes the peak voltage at the inductor port. The inductor Q-factor can be obtained as shown in equation 3.11 by substituting equations 3.8 to 3.10 into 3.7.

$$Q = \frac{\omega L}{R} \cdot \frac{R_s}{R_s + \left[(\omega L / R)^2 + 1 \right] R} \cdot \left[1 - \frac{R^2 C}{L} - \omega^2 LC \right]. \tag{3.11}$$

The Q-factor in the equation 3.11 is expressed as a product of three factors, where the first factor is called the ideal Q factor, the second factor is called the substrate loss factor, and the third factor is called the self-resonance factor.

The inductance L is defined as $L=\varphi/I$, here φ is the magnetic flux crossing the inductor coil and I is the current flowing through the coil. The multi-layered coil inductor produces a large inductance, L, as an entire inductor because the multi-layered coil shows mutual inductance due to mutual electromagnetic induction between the multiple coils connected in series and thus increase the magnetic flux crossing the inductor coil. For this reason, according to the multi-layered coil inductor, the total length of conductive wire necessary for achieving a given inductance L tends to be short. The shorter the total length of the conductive wire for constituting the multi-layered coil inductor, the smaller the resistance R in the multi-layered coil inductor tends to be. As can be seen in the above-mentioned first factor, achieving a predetermined inductance L at a small resistance R contributes to an increase in the Q-factor. The inductor coils have to be constructed with good conductivity. The width and height of the coil traces have to be designed carefully to ensure low RF resistance at operating frequencies.

The second factor in equation 3.11 suggests that a substrate having high resistivity Rs should be used to lower the loss from the substrate and to increase the second factor so that it is 1. For this reason, ceramic, glass, or fused quartz are suitable for the substrate.

The third factor of equation 3.11 suggests that the parasitic capacitance C should be lowered. As can be seen from the above-mentioned third factor, making the parasitic capacitance C zero brings this factor close to 1 and contributes to an increase in the Q-factor. Further, lowering of the parasitic capacitance is also favourable for achieving a good high-frequency performance. The self-resonant frequency (SRF) f_0 of an inductor can be determined when the third factor of equation 3.11 becomes zero. Using this self-resonance condition, the same result as equation 1.6 is then obtained.

$$f_o = \frac{1}{2\pi}\sqrt{\frac{1}{LC} - \frac{R^2}{L^2}}. \tag{3.12}$$

Generally, the smaller the parasitic capacitance of the inductor, the greater is the inductor's SRF shift toward the high frequency side, making it easier to achieve a good high-frequency characteristic. For these reasons, we recommend using the two-layered coil in the air, as no material has a dielectric constant that is lower than air.

As discussed above, the optimized structure of the IPD is illustrated in Fig. 3.3 (Mi X. et al., 2007). The lower spiral coil is directly formed on the substrate and the upper coil is freestanding in the air. Air is used as the insulation layer between the two coils to minimize the parasitic capacitance in the coil inductor. The two spiral coils are connected in series by a metal via and the direction of the electric current flowing through the two coils is the same. The direction of the magnetic flux occurring in the two coils agrees, and the total magnetic flux crossing the two-layered coil increases. The two coils are arranged to overlap with each other to further maximize the mutual induction. There are no support poles under the upper coil, nor are there intersections between the wiring and the coils in the two-layered coil structure. It also helps to prevent an extra eddy current loss from occurring in these sections and maximizing the Q-factor of the coil inductor. The capacitor is of a metal-insulator-metal structure, as shown in Fig. 3.3 (c). A thick metal is used for the lower and upper electrodes of the capacitor to suppress the loss and parasitic inductance arising in the electrodes and thus to enlarge the Q-factor of the capacitor and its self-resonant-frequency. A 3D interconnection in the air is introduced for the upper electrode to help eliminate the parasitic capacitance that results from the wiring to the MIM capacitor.

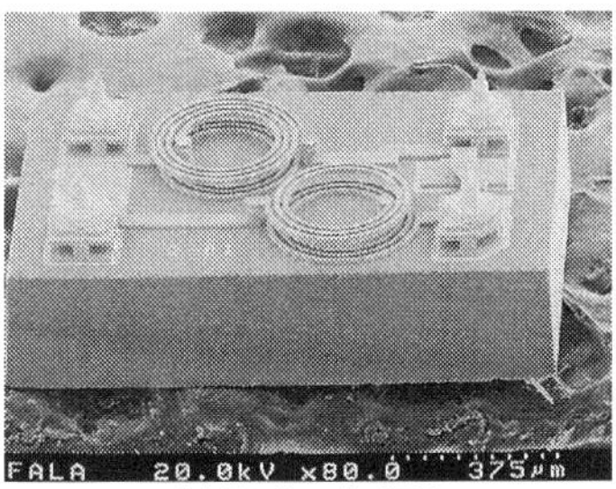

(a) Fujitsu's IPD

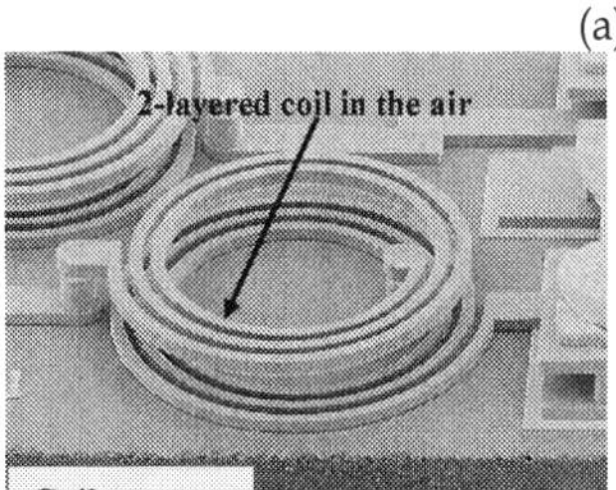

(b) Coil part

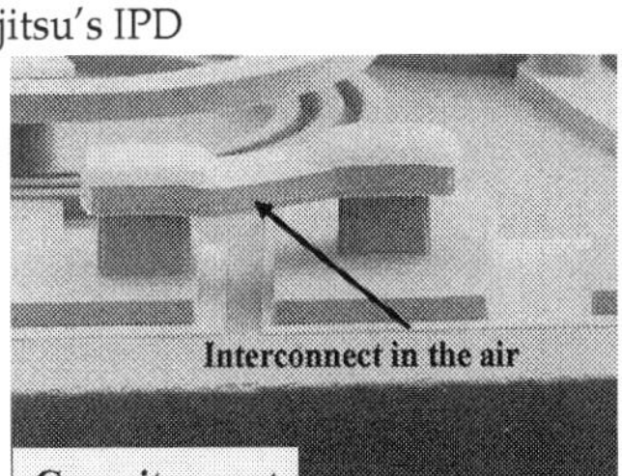

(c) Capacitor part

Fig. 3.3 High-Q IPD configuration

4. IPD on LTCC technology

4.1 Concept

We propose a new technology to combine the advantages of LTCC and IPD technology. High-Q passive circuits using a two-layered aerial spiral coil structure and 3D interconnection in the air are constructed directly on an LTCC wiring wafer. This technology is a promising means of miniaturizing the next generation of RF-modules

A conceptual schematic diagram of the proposed IPD on the LTCC for the RF module applications is illustrated in Fig. 4.1. The above-mentioned high-Q IPD is directly formed on the LTCC wiring wafer. Functional devices such as the ICs are mounted above the IPD, while the LTCC wiring wafer has metal vias on the surface for electrical interconnection between the wiring wafer and the integrated passive circuit or the mounted function device chips. Pads are formed on the reverse side of the LTCC wafer to provide input and output paths to the motherboard. The inner wiring of the LTCC wafer can provide very dense interconnects between the passive circuit and the functional devices. Because the function device chips are installed above the integrated passives, the chip-mounting efficiency can approach 100%, which means a chip-sized module can be realized.

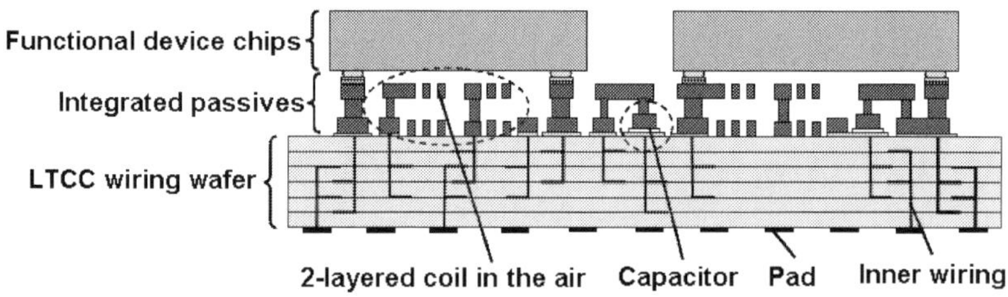

Fig.4.1. Conceptual schematic diagram of the proposed IPD on LTCC for RF module applications

4.2 Development

The fabrication technology of the high-Q IPD on LTCC is shown in Fig. 4.2. The basic concept is to form a large-size LTCC wiring wafer, and then to form the IPD directly on the wafer surface.

First, an LTCC wiring wafer is fabricated, and the surface of the wafer is subject to a smoothing process. The surface roughness needs to be reduced to ensure that the wafers can go through the following photolithography and thin-film formation processes. The capacitors, lower coils and interconnects are then formed by thin-film technology and electrical plating technology. Next, a sacrificial layer is formed, which has the same height as that of the lower metal structure. At the via positions, windows are opened in the sacrificial layer to facilitate an electrical connection between the upper and lower metal structure. On the sacrifice layer, a metal seed layer is formed for the following electrical plating process. After that, the upper coils and interconnects as well as the bumps for interconnection between the function device chips and IPD wafer are formed by electrical plating technology. The metal seed layer and the sacrificial layer are then removed to release the integrated passives. The upper and lower metal structures are made of copper and the bumps are gold-plated. Function device chips such as the IC can be mounted onto the bumps by flip-chip bonding technology. If necessary, a sealing or under-filling process can be conducted. Finally, the module units are created by cutting the wafer. All of the fabrication processes are carried out at wafer level, which leads to high productivity. The fabricated high-Q IPD on LTCC wiring substrate is shown in Fig. 4.3.

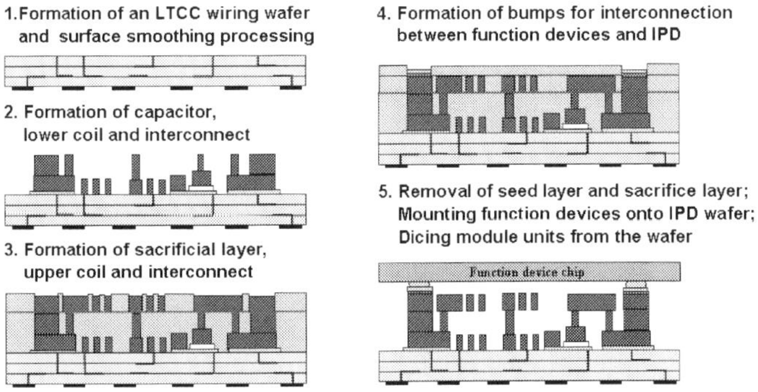

Fig. 4.2. Fabrication technology of the high-Q IPD on LTCC

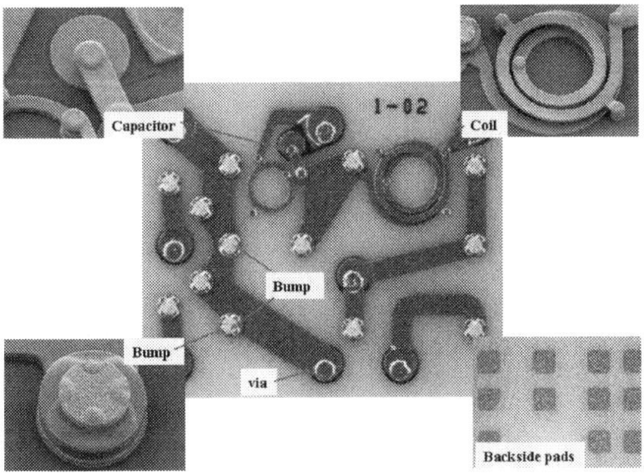

Fig. 4.3. Fabricated high-Q IPD on LTCC wiring substrate

We inspected the performance of the fabricated high-Q IPD on LTCC wiring wafer. Figure 4.4. shows a performance comparison between a two-layer coil in the air and a one-layer coil in resin. The two coils have the same inductance of 12 nH, but differ in coil size. The two-layer coil is 350um in diameter, and the one-layer coil in resin is 400um in diameter. The two-layer coil can represent a 30% saving in area while providing the same inductance. The developed two-layered coils can achieve an inductance of up to 30nH at a size of less than φ0.6 mm. For a given size, the two-layer coil in the air improves the Q-factor by 1.7 to 2 times that of the conventional one-layer coil in the resin. Moreover, the SRF also increases from 7.5 GHz to 8.5 GHz. It indicates that the two-layered coil in the air is more suitable for high-frequency applications exceeding 3GHz where hardly any surface-mounting devices (SMD) usually work well due to the low SRF.

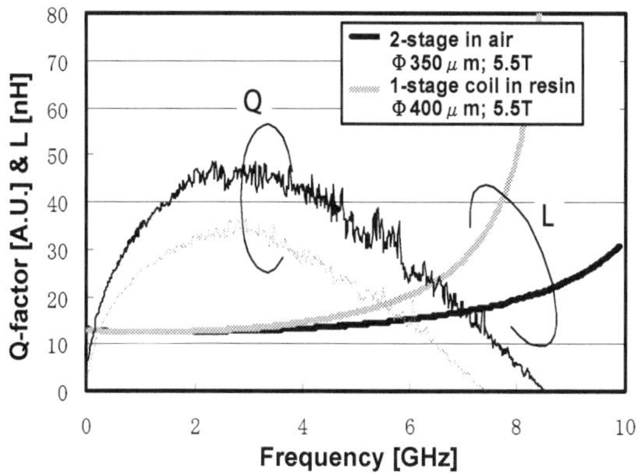

Fig. 4.4. Performance comparison between a two-layer coil in the air and a one-layer coil embedded in resin

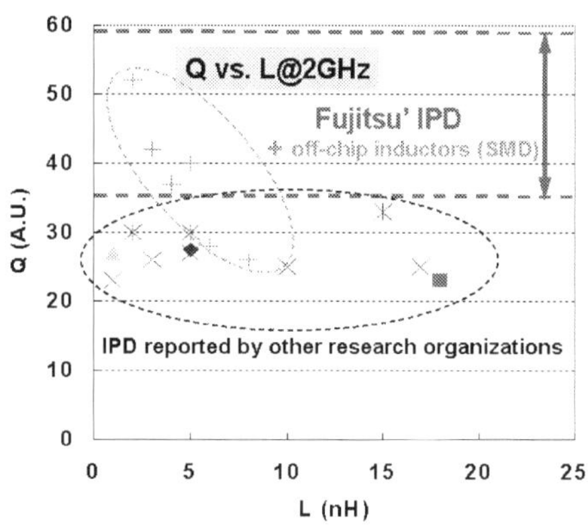

Fig. 4.5. Q-factor comparison between Fujitsu's 2-layered IPD and those made by other companies and SMD inductors on a similar size basis

Q-factor comparison between Fujitsu's 2-layered aerial spiral coil and those made by other companies and SMD inductors are compared at 2 GHz in Fig. 4.5. In general, the Q-factor of an inductor strongly depends on inductor size. Fujitsu IPD inductors have an outer diameter smaller than 0.6 mm. Those points for the integrated inductors reported by other research organizations have a similar size to Fujitsu's IPD inductor. The SMD inductors compared in Fig. 4.5. have the a size of 0.6 mm×0.3 mm. The conventional 1-layer integrated spiral coils in resin with a size less than 0.6 mm square can only offer a Q-factor of less than 30. These off-chip inductors (SMD) with the similar a size of 0.6 mm×0.3 mm can offer a Q-factor higher than 40 only when the inductance is less than 5 nH. When the inductance increases, the Q-factor rapidly declines to less than 30. As a result, Fujitsu's 2-layered aerial spiral coil can provide a performance that is superior to its rivals of a similar size.

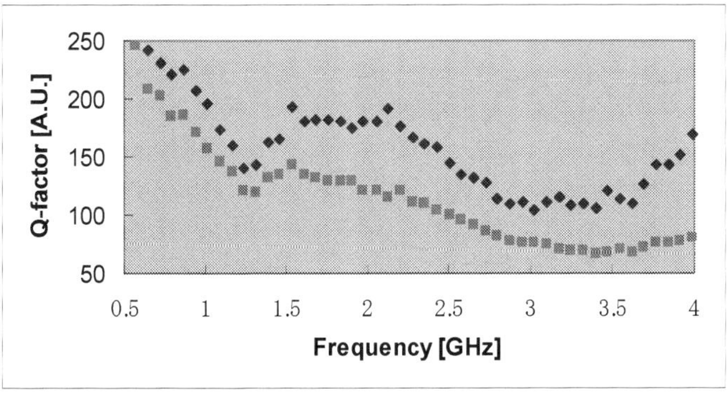

Fig. 4.6. Performance comparison between a capacitor using a 3D interconnect in the air and a capacitor embedded in resin

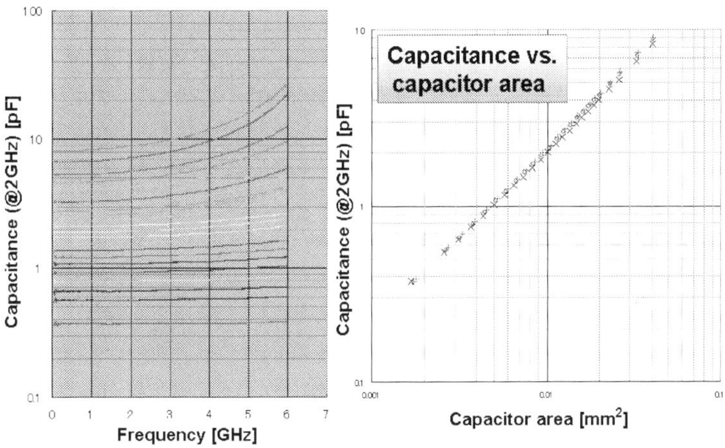

Fig. 4.7. Capacitor performances of Fujitsu's IPD (a) Dependence of capacitance on frequency (b) Relationship between capacitance and capacitor area

The Q-factor comparison between the newly developed capacitor using 3D interconnection in the air and the conventional capacitor embedded in resin is shown in Fig. 4.6. The Q-factor is improved from 110 to 180 at 2 GHz. The other performances of Fujitsu's IPD capacitors are shown in Fig. 4.7. and Fig. 4.8. Figure 4.7. (a) shows the dependence of the capacitance on frequency and Fig. 4.7. (b) shows the relationship between the capacitance and the capacitor area. As shown in Fig.4.7. (a), the capacitance stays flat up to several GHz, indicating that the developed capacitor has small parasitic inductance and high self-resonance frequency. The capacitance density of the developed integrated capacitors reaches 200 pF/mm2, which makes it possible to reduce the size while covering almost all RF applications. Ultra-thin insulation film is favorable for achieving a large capacitance density, but has a risk in terms of breakdown voltage. The breakdown voltage characteristic depends strongly on the substrate roughness and quality of the dielectric film used for the capacitor. High-quality thin-film formation technology is the key to realizing high capacitance density. We also checked the breakdown voltage of the integrated capacitors and the result is shown in Fig. 4.7. The average breakdown voltage exceeds 200 V, which is enough for RF module applications.

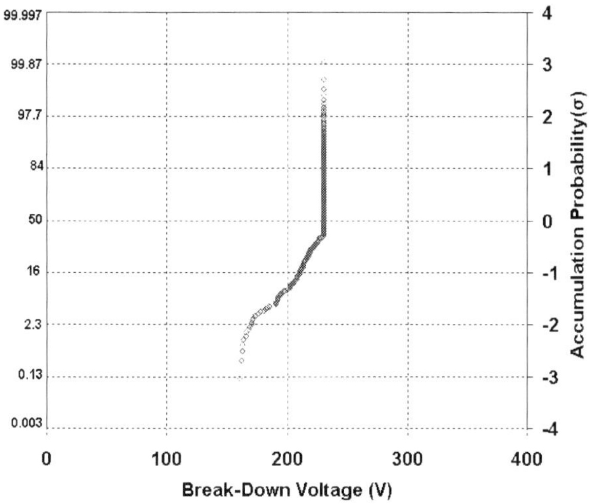

Fig. 4.8. Break-down voltage of Fujitsu's IPD capacitors

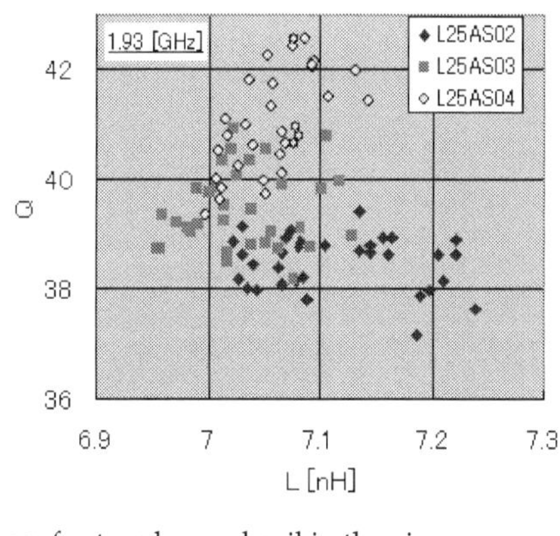

Fig. 4.9. Production tolerance for two-layered coil in the air

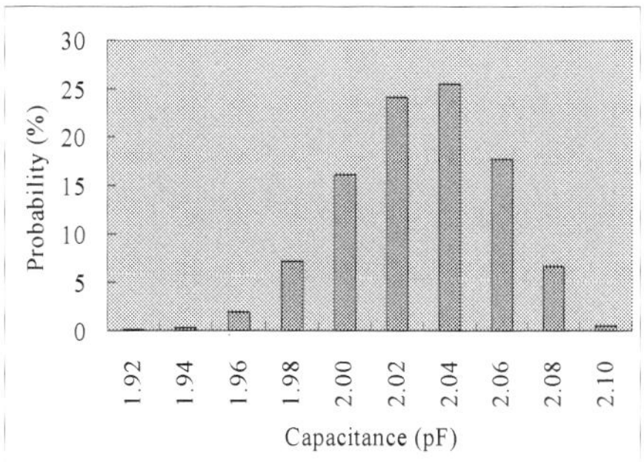

Fig. 4.10. Production tolerance for capacitor using 3D interconnect in the air

We inspected the production tolerance of the developed integrated passives. The inductance and Q-factor of 7.1 nH coils fabricated in different production batches were measured. The results are shown in Fig. 4.9. The deviation in inductance is less than ±2%. The deviation in the Q-factor is about ±5%. The capacitance deviation of 2 pF capacitors fabricated in different production batches was evaluated and the result is shown in Fig. 4.10. The deviation in capacitance is less than ±3%. The above-mentioned production tolerance includes the wafer deviation and batch deviation, which is not available in the case for its rivals, namely laminate-based and LTCC-based technologies.

High-Q IPD on LTCC technology has been demonstrated for RF-module applications using the newly developed multistage plating technology based on a sacrifice layer. A two-layered aerial spiral coil structure and 3D interconnection in the air are used to increase the quality factor and to reduce the parasitic capacitance. This configuration enables us to achieve a Q-factor of 40 to 6o at 2 GHz for the integrated spiral inductors of a size smaller than φ0.6 mm, while providing a high self-resonance frequency of over 8 GHz. The Q-factor of the capacitors has been improved from 110 to 180 at 2 GHz. Very high production precision has been achieved: less than ±2% for inductors and less than ±3% for capacitors. This technology combines the advantages of LTCC and IPD. The function device chips can be mounted above the IPD. The inner wiring built in the LTCC wafer provides dense interconnection. And the pads on the reverse side allow easy access to the motherboard. This technology combines the advantages of IPD and LTCC and provides a technical platform for future RF-modules, which has all the technical elements necessary for module construction, including integrated passives, dense interconnection, package substrate. These advantages are promising for the miniaturization of RF-modules and the realization of a chip-sized-module.

5. Summary and Discussions

In this chapter, we have concisely reviewed the recent developments in passive integration technologies and design considerations for system miniaturization and high-frequency applications. Over the past 10 years, passive integration technologies, laminate-, LTCC- and thin-film based technologies have gone through a significant evolution to meet the requirements of lower cost solutions, system miniaturization, and high levels of functionality integration, improved reliability, and high-volume applications. Some of them have enabled miniaturized or modularized wireless telecommunication products to be manufactured.

Developments in new materials and technologies for laminate-based technology have been significantly advanced. This makes possible the lowest cost integration of embedded resistors, capacitors, and inductors. Embedded discrete passives technology has been used for mass production. The materials and processes of laminate-based film capacitors are now immature and the yields and reliability also need to be evaluated. The large production tolerance due to instabilities in the materials and the fabrication processes remains the drawback. LTCC-based passive integration has high material reliability, good thermal dissipation and relatively high integration density compared to laminate-based technologies, but has the common drawback of a large production tolerance due to the screen-printed conductors and the shrinkage during the firing process. The high tolerance of embedded passive elements in organic or LTCC substrate limits their use to coarse applications or digital applications. The thin film based passive integration, usually is called as integrated passive device (IPD) provides the highest integration density with the best dimensional accuracy and smallest feature size, which makes it the most powerful technology for passives integration in SIP solution at high frequencies. When a large wafer size is used for IPD, the cost per unit area will be drastically reduced and can compete with laminate- and LTCC-based technologies at the same functionality.

A small size, high Q-factor, high SRF, and large inductance are required for integrated inductors to meet the demands for high-frequency performances and low cost. Conventional spiral coils cannot meet these requirements at the same time. We have established process technology to produce IPD using 2-layered coil in the air and confirmed its good performance.

- 2-layered coil in air : $Q \geqq 40$@2 GHz; $Q \geqq 30$@0.85 GHz with a coil size less than 0.6 mm
- Capacitor: 200 pF/mm2 density and break-down voltage over 200 v.

For integrated capacitors, the capacitance density should be increased by introducing a high-k thin film with good film quality. This will help increase the capabilities of integrating large capacitance or scaling-down the capacitor size.

Current IPD technologies such as IPD on glass/Si, have disadvantages compared to laminate- or LTCC-based technologies, namely the inner wiring is not available and, while a through-wafer via is possible for a Si or glass substrate, it is expensive. This will result in limitations for future system level integration including size, complexity and cost. We demonstrated IPD-on-LTCC technology, which combines the advantages of IPD and LTCC and provides a technical platform for future RF-modules, and which has all the technical elements necessary for module construction, including integrated passives, dense interconnection, and package substrate. These advantages are promising for the miniaturization of RF-modules and the realization of a chip-sized-module to meet the future market demand for higher levels of integration and miniaturization.

In the future, system integration will become more complicated and involve more and more functions of the package, such as sensors, actuators, MEMS, or power supply components. For example, decoupling, filtering and switching are all electrical functions which cannot be effectively integrated on active silicon nowadays, but which are required for the generic circuit blocks of high-frequency radio front ends. Moreover, tunable capabilities are strongly expected to offer more flexible radio front-ends for future software-defined-radio or cognitive radio systems. MEMS devices have shown promise for realizing tuning functions.

Incorporating RF-MEMS components such as switches, variable capacitors and tunable filters, in RF-module platforms will drastically increase the functionality and will be the next challenging development. When constructing such complicated 3D built-up systems, system electro-magnetic field modeling will become more difficult and challenging. In addition, thermal and current as well as mechanical stress management will have to be taken into account from the beginning of the system concept. Setting up a well-established design methodology with capabilities to design and optimize extensive components including active, passive and MEMS devices is also important and is a future task.

6. References

Allers K. -H.; Brenner P. & Schrenk M. (2003). Dielectric Reliability and Material Properties of Al203 in Metal Insulator Metal capacitors (MIMCAP) for RF Bipolar technologies in comparison to SiO2, SiN and Ta205, Proceedings of BCTM, Toulouse, France, 2003, pp. 35-38

Arnold R. G.; Faulkner C. C. & Pedder D. J. (1997). Silicon MCM-D Technology for RF Integration, *Proceedings of Ineternational Conference on Multichip Modules IEEE*, pp. 340-344, 2-4 April 1997

Bahl I. & Bhartia P. (2003). *Microwave Solid State Circuit Design*, John Wiley & Sons, ISBN 0-471-20755-1, Hoboken, New Jersey

Banieki J. D. et al. (1998). Dielectric relaxation of Ba0.7Sr0.3TiO3 thin film from 1 mHz to 20 GHz, Appl. Phys. Lett., 72(4), 26 January 1998, pp. 498-500

Bauer W.; Purger S. & Schrittwieser W. (2003). Embedded Passive Technology form a PCB Maker's Perspective, *Circuitree*, October 1, 2003

Beique G. et al. (2006). Dry Etching of High-K Dielectric PZT Stacks for Integrated Passive Devices, Integrated Ferroelectrics, Vol. 86, 2006, pp. 49–56, Taylor & Francis Group, LLC, ISSN 1058-4587

Berthelot A. et al. (2007), Highly Reliable TiN/ZrO2/TiN 3D Stacked Capacitors for 45 nm Embedded DRAM Technologies, Proceedings of IEEE ESSDERC, Montreux, Switzerland, 2006, pp. 343-346

Beyne E. (2008). Solving Technical and Economical Barriers to the Adoption of Through-Si-Via 3D Integration Technologies, Proceedings of 10th Electronics Packaging Technology Conference, 2008, pp. 29-34

Bhatt D. et al. (2007). Process Optimisation and Characterization of Excimer Laser Drilling of Microvias in Glass, Proceedings of 9th Electronics Packaging Technology Conference, 2007, pp. 196-201

Borland W. et al. (2002). Embedded Singulated Ceramic Passives in Printed Wring Boards, *Proceeding of the IMAPS-Advanced Technology Workshop on Passive Integration*, Ogunquit, Maine, June 2002

Brown R. L.; Polinski P. W. & Shaikh A. S. (1994). MANUFACTURING OF MICROWAVE MODULES USING LOW-TEMPERATURE COFIRED CERAMICS, Proceedings of IEEE MTT-S International Microwave Symposium (IMS), pp. 1727-1730, 1994.

Carchon G. et al. (2001). Multi-layer thin-film MCM-D for the integration of high performance wireless front-end systems, Microw. J., Vol. 44, 2001, pp. 96–110

Carchon G.; Brebels S. & Vasylchenko A. (2008). Thin film Technologies for Millimeter-Wave Passives and Antenna Integration, *Proceedings of Workshop: System in Package Technologies for Microwave and Millimeter Wave Integration, European Microwave Week 2008*, WFR-14-1, Amsterdam, The Netherlands, 27-31 Oct. 2008

Carchon G. J. et al. (2001), A direct Ku-band linear subharmonically pumped BPSK and I/Q vector modulator in multilayer thin-film MCM-D, *IEEE Trans. Microw. Theory Tech.*, Vol. 49, 2001, pp. 1374–1382

Carchon G. J.; Raedt W. R. & Beyne E. (2005). Wafer-Level Packaging Technology for High-Q On-Chip Inductors and Transmission Lines, *IEEE TRANSACTIONS ON MICROWAVE THEORY AND TECHNIQUES*, Vol. 52, No. 4, APRIL 2004, pp. 1244-1251

Chang J. Y-C; Abidi A. A. & Gaitan M. (1993). Large Suspended Inductors on Silicon and their Use in a 2-µm CMOS RF Amplifier, *IEEE Electron Device Lett.*, Vol. 14, 1993, pp. 246-248

Chason M. et al. (2006). Toward Manufacturing Low-Cost, Large-Area Electronics, MRS BULLETIN, Vol. 31, June 2006, pp471-475

CHEN R. et al. (2005). A COMPACT THIN-FILM WLAN ANTENNA SWITCHING MODULE, *MICROWAVE J.*, January 2005 issue

Chong K. et al. (2005). High-Performance Inductors Integrated on Porous Silicon, IEEE Electron Device Letters, Vol. 26, No. 2, Feb. 2005, pp. 93-95

Chua C. L. et al. (2002). SELF-ASSEMBLED OUT-OF-PLANE HIGH Q INDUCTORS, Proceedings of Solid-State Sensor, Actuator and Microsystems Workshop, Hilton Head Island, South Carolina, June 2-6, 2002, pp. 372-373

Chua C. L. et al. (2003). Out-of-Plane High-Q Inductors on Low-Resistance Silicon, J. Microeletromech. Syst., Vol. 12, No. 6, Dec. 2003, pp. 989-995

C-ply/3M :

http://solutions.3m.com/wps/portal/3M/en_US/

Croswell R. et al.(2002). Embedded Mezzanine Capacitor Technology for Printed Wiring Boards, *CircuiTree*, August, 2002

Dahlmann G. W. et al. (2001). High Q Achieved in Microwave Inductors Fabricated by Parallel Self-Assembly, Proceedings of The 11th International Conference on Solid-State Sensors and Actuators, Munich, Germany, June 10–14, 2001, Vol. 2, pp. 1098-1101

Davis P. et al. (1998). Silicon-on-silicon integration of a GSM transceiver with VCO resonator, Proceedings of Int. Solid-State Circuits Conf. 1998, pp. 248–249

Defaÿ E. et al. (2006). Above IC integrated SrTIO3 high K MIM capacitors, Proceedings of ESSDERC, Montreux, Swithzerland, 2006, pp. 186-189

Doyle L. (2005) Integrated Passive and Active Devices Using CSP, DFN and QFN Packaging for Wireless Applications, *Microwave Engineering Europe*, June 2005, pp 36-40

Dougherty J. P. et al. (2003). The NEMI Roadmap: Integrated Passives Technology and Economics, *Proceedings of the Capacitor and Resistor Technology Symposium (CARTS)* ,Scottsdale AZ, April 2003

Dunn G. et al. (2004). New Developments in Polymer Thick Film Resistor Technology, *IPC Printed Circuit Expo*, Anaheim, CA, 21-25 February, 2004

DuPon Microcircuit Materials, USA (http://www2.dupont.com/MCM/en_US/products/resistors/)

Erzgraber H. B. et al. (1998). A Novel Buried Oxide Isolation for Monolithic RF Inductors on Silicon, Proceedings of IEDM, San Francisco, USA, 1998, pp. 535-539

Fjeldsted K. (2004). Trimming Embedded Resistors: Third Party Processing, *CircuiTree*, February, 2004

Frye R. C.; Liu K. & Lin Y. (2008). Three-Stage Bandpass Filters Implemented in Silicon IPD Technology Using Magnetic Coupling between Resonators, *Proceedings of 2008 IEEE MTT-S International Microwave Symposium*, ATLANTA, USA, 15–20 June 2008, pp. 783–786

Gardner D. S. et al. (2007). Integrated On-Chip Inductors With Magnetic Films, *IEEE TRANSACTIONS ON MAGNETICS*, Vol. 43, No. 6, June 2007, pp. 2615-2617

Gaynor M. (2007). RF SiP Technology and Capability Overview, Proceeding of Workshop: System-in-Package Technologies for Cost, Size, and Performance, at IEEE MTT-S International Microwave Symposium 2007

Giraudin JC. et al. (2006). Demonstration of three-dimensional 35nF/mm2 MIM Capacitor integrated in BiCMOS Circuits, Proceedings of BCTM, 2006, 11.3

Giraudin JC. et al. (2007). Development of Embedded Three-Dimensional 35-nF/mm2 MIM Capacitor and BiCMOS Circuits Characterization, IEEE JOURNAL OF SOLID-STATE CIRCUITS, VOL. 42, NO. 9, SEPTEMBER 2007, pp. 1842-1850

Govind V. et al. (2006). Design of Novel Highly Integrated Passive Devices for Digital Broadcasting Satellite / 802.11 Home Networking Solution in Liquid Crystal Polymer (LCP) Based Organic Substrates, *Proceedings of IEEE MTT-S International Microwave Symposium (IMS)*, pp. 1157-1160, San Francisco, CA, Jun. 2006

Graauw A.; Copetti C. & Weekamp W . (2000). A new thin film passive integration technology for miniaturisation of mobile phone front end modules, *Proceedings of 2000 IEEE MTT-S International Microwave Symposium*, Boston, MA, USA, 11–16 June 2000, pp. 1925–1928

Fujiki Y.; Mandai H. & Morikawa T. (1999). Chip Type Spiral Broadside-Coupled Directional Couplers and Baluns Using Low Temperature Co-fired Ceramic, *Proceedings of IEEE Electronic Comp. Tech. Conf.*, pp. 105-110, San Diego, CA, June 1999

Hong T. K. & Kheng L. T.(2005). A Novel Approach of Depositing Embedded Resistors, *Proceedings of 2005 Electronics Packaging Technology Conference*, pp. 144-147

Huylenbroeck S. V. et al. (2002). Investigation of PECVD Dielectrics for Nondispersive Metal–Insulator–Metal Capacitors, IEEE ELECTRON DEVICE LETTERS, Vol. 23, No. 4, APRIL 2002, pp. 191-193

Imanaka Y.; Takenouchi M. & Akedo J. (2005). Ceramic dielectric film for microwave filter deposited at room temperature, *J. Crystal Growth*, Vol. 275, Issues 1-2, 15 Feb. 2005, pp. e1313-e1319

Imanaka Y. et al. (2007). Aerosol deposition for post-LTCC, *Journal of the European Ceramic Society*, 27 (2007), pp. 2789-2795

Jeannot S. et al. (2007). Toward next high performances MIM generation: up to 30fF/Wm2 with 3D architecture and high-k materials, Proceedings of IEDM, Washington, DC, USA, 2007, pp. 997-1000

Jeong I-H. et al. (2002). High quality RF passive integration using 35 μm thick oxide manufacturing technology, *Proceedings of ECTC*, San Diego, USA, 2002, pp. 1007–1111

Jiang H. et al. (2000). On-Chip Spiral Inductors Suspended over Deep Copper-Lined Cavities, *IEEE Trans. Microwave Theory Tech.*, vol. 48, NO. 12, 2000, pp2415-2423

Jillek, W. et al.(2005). Embedded components in printed circuit boards: a processing technology review, *Int J. Adv Manuf. Technol.*, 2005, pp. 350-360.

Jones R. E. et al. (2005). System-in-a-Package Integration of SAW RF Rx Filter Stacked on a Transceiver Chip, *IEEE TRANSACTIONS ON ADVANCED PACKAGING*, VOL. 28, NO. 2, MAY 2005, pp. 310-319

Jung E.; Potter H. & John L. G. (2009). Packaging, Interconnection, Assembly Packaging Innovations for Novel Products-Leveraging PCB Technology for System Level Integration, International Magazine on Smart System Technology-mst news, No. 3/09, June 2009, pp4-8

Kamiya H. et al. (2005). Development of the Embedded LSI Technology in PALAPTM, *Proceedings of 2005 International Symposium on Electronics Materials and Packaging (EMAP2005), IEEE*, pp183-186

Kawasaki M. et al. (2004). Development of High-k Inorganic/Organic Composite Material for Embedded Capacitors, Proceeding of 54th Electronic Components and Technology Conference, pp. 525-530, Las Vegas, NV, May, 2004

Kim D. et al. (2003). High Performance RF Passive Integration on a Si Smart Substrate for Wireless Applications, *ETRI Journal*, vol. 25, Number 2, 2003, pp. 65-72

Kim J. et al. (2005). Design of Toroidal Inductors Using Stressed Metal Technology, *Proceedings of IEEE MTT-S International Microwave Symposium*, 12-17 June 2005, pp. 705-708

Kondou K. & Kamimura R. (2002). Thermoplastic Film Based Multilayer Printed Circuit Board and Corresponding Material Recycle System, DENSO Technocal Review, 7, 91 (2002), pp. 91-95

Kumashiro Y. et al. (2004). High Dielectric Constant Insulation Sheet for PCBs with Embedded Capacitors, Hitachi Chemical Technical Report, Vol. 43, 15 (2004), pp. 15-18

Kuo D. H. et al. (2001). Dielectric behaviours of multi-doped BaTiO3/epoxy composites, *Journal of the European Ceramic Society*, 21 (2001), pp. 1171-1177

Lee S. et al. (2006). Multi-Functional Epoxy/SrTiO3 Ceramic Powder Embedded Capacitor Films(ECFs) for Organic Substrates, Proceedings of 2006 Electronics Packaging Technology Conference, pp. 549-552

Lew D. W. et al. (2001). A Design of Ceramic Chip Balun Using Multilayer Configuration, IEEE Trans. Microwave Theory Tech., Vol. 49, No. 1, Jan. 2001, pp. 220-224

Li H. Y. et al. (2006). BCB (Benzocylcobutene) Process Integration for the RF Passive Device, Proceedings of 2006 Electronics Packaging Technology Conference, 2006, pp.40-45

Li L. et al. (2003). Embedded Passives in Organic Substrate for Bluetooth Transceiver Module, Proceedings of 53rd Electronic Components and Technology Conference, pp. 464-469, New Orleans, LA, May 2003

Li L.; Kapur A. & Heames K. B. (2004). Characterization of Transfer Molding Effects on RF Performance of Power Amplifier Module, Proceedings of 54th Electronic Components and Technology Conference, pp. 1671-1677, Las Vegas, NV, June 2004

Lu H. et al. (2007). Fabrication and Characterization of Multi-layer Liquid Crystal Polymer (LCP) Substrate, Proceedings of 29th Electronics Packaging Technology Conference, pp. 514-517, 2007

Marksteiner S. et al. (2006). Solidly Mounted BAW Resonator Technology for Use in Mobile Communication Systems, *Proceedings of BAW workshop at APMC 2006*, WS12-5, Yokohama, Japan, Dec. 2006

Mi X.; Takahashi T. & Ueda S. (2008). Integrated Passives on LTCC for Achieving Chip-Sized-Modules, *Proceedings of the 38th European Microwave Conference*, pp. 607-610, October 2008, Amsterdam, The Netherlands

Mi X. et al. (2007). Integrated Passive Devices for RF-Module, Proceedings of International Workshop on Piezo-Devices Based on Latest MEMS Technologies, Tokyo, Japan, 26-27 Nov. 2007, pp. 87-93

M-Pass, MacDermid Inc. (http://www.macdermid.com/electronics/adv_embedded.html)

muRata LTCC: http://www.murata.co.jp/

Niknejad A. M. &, Meyer R. G. (1998). Analysis, Design, and Optimization of Spiral Inductors and Transformers for Si RF IC's, *IEEE JOURNAL OF SOLID-STATE CIRCUITS*, VOL. 33, NO. 10, OCTOBER 1998, pp. 1470-1481

Nishihara T. et al. (2008). BAW/SAW/IPD hybrid type duplexer with Rx balanced output for WCDMA Band I, *Proceedings of 2008 IEEE MTT-S International Microwave Symposium*, ATLANTA, USA, 15–20 June 2008, pp. 831–834

Norlyng S. (2003). An overview of Integrated Component Technologies, *Advancing Microelectronics*, May/June 2003, pp. 9-13

Oak-Mitsui Technologies Technical Data: http://www.faradflex.com/materialproperties/materialproperties.aspx

Ohmega Technologies, Inc. (www.ohmega.com/pdfs/)

Ouajji H. et al. (2005). Dielctric Properties in Thin Film SrTIO3 Capacitor, Proceedings of 2005 Electronics Packaging Technology Conference, pp. 740-743

Paulsen R. & Spencer M. (2008). MINIATURIZATION OF FIXED AND TUNABLE FILTERS – WHERE AND WHEN TO USE INTEGRATED PASSIVE DEVICES, *Proceeding of the SDR 07 Technical Conference and Product Exposition*

Piatnitsa V.; Jakku E. & Leppaevuori S. (2004). Design of a 2-pole LTCC Filter for Wireless Communications, *IEEE Trans. Wireless Comm.*, Vol. 3, Mar. 2004, pp. 379-381

Pieters P. et al. (2000). Integration of passive components in thin multilayer MCM-D Technology for wireless frontend application, Proceedings of Asia-Pacific Microwave Conference, 2000, pp. 221 - 224

Popielarz R. et al. (2001). Dielectric Properties of Polymer/Ferroelectric Ceramic Composites from 100 Hz to 10 GHz, *Macromolecules*, 34 (2001), pp. 5910-5915

Pulsford N. (2002). Passive integration technology: Targeting small, accurate RF parts, *RF Design Magazine*, Nov. 2002, pp. 40-48, Online Available: http://www.rfdesign.com

Savic J. et al. (2002). Embedded Passives Technology Implementation in RF Applications, *CircuiTree*, June, 2002

Shah V. G. & Hayes D. J. (2003). Trimming and Printing of Embedded resistors Using Demand-Mode Ink-Jet Technology and Conductive Polymer, *CircuiTree*, March, 2003

Sheen J. W. (1999). LTCC-MLC Duplexer for DCS-1800, IEEE Trans. Microwave Theory Tech., Vol. 47, Sept. 1999, pp. 1883-1890

Shibasaki S. et al. (2004), Development of Buried Bump Interconnection Technology with Embedded Passive Devices, *Proceedings of International Conference on Electronic Packaging, 2004*, pp.113-118, Tokyo, Japan, April 2004

Stratigos J. (2007). Capabilities of Multi-Layer Organic Packaging, Microwave Journal, September 14, 2007

Sugaya Y. et al. (2001). A new 3-D module using embedded actives and passives, Proceedings of IMAPS-US, Baltimore,2001

Sutono A. et al. (2001). High-Q LTCC-Based Passive Library for Wireless System-on-Package (SOP) Module Development, IEEE Trans. Microwave Theory Tech., Vol. 49, No. 10, Oct. 2001, pp. 1715-1724

Tanaka H. et al. (2008). Embedded High-k Thin Film Capacitor in Organic Package, *proceeding of 10th Electronics Packaging Technology Conference, IEEE*, 2008, pp.988-993

Thomas M. et al. (2007). Reliable 3D Damascene MIM architecture embedded into Cu interconnect for a Ta2O5 capacitor record density of 17 fF/µm2, Proceeding of 2007 Symposium on VLSI Technology, Kyoto, Japan, 2007, pp. 58-59

Ticer Technologies (www.TICERTECHNOLOGIES.com)

Tilmans H. A C; Raedt W. D. & Beyne E. (2003). MEMS for wireless communications: 'from RF-MEMS components to RF-MEMS-SIP', *J. Micromech. Microeng.*, vol. 13(2003), pp. S139-S163

Tummala R. R. et al. (2000). Sop: The microelectronics for 21st century with integral passive integration. Adv. Microelectron., vol. 27 (2000), pp.13-19

Ulrich R. & Schaper L. W. (2003). Integrated Passive Component Technology, Wiley-IEEE Press, ISBN: 0-471-244-317, New York, 2003

Vatanparast R. et al. (2007). Flexible Wireless Wearable Sensor with Embedded Passive Components, Proceedings of 9th Electronics Packaging Technology Conference, pp.8-13, 2007

Wang S.; Kawase A. & Ogawa H. (2006). Preparation and Characterization of Multilayer Capacitor with SrTIO3 Thin Films by Aerosol Chemical Vapor Deposition, Japanese J. Appl. Phys., Vol. 45, No. 9B, 2006, pp. 7252-7257

Wojnowski M. et al. (2008). Package Trends for Today's and Future mm-Wave Applications, *Proceedings of Workshop: System in Package Technologies for Microwave and Millimeter Wave Integration, European Microwave Week 2008*, WFR-14-1, Amsterdam, The Netherlands, 27-31 Oct. 2008

Wu J. C. & Zaghloul M. E. (2008). CMOS Micromachined Inductors With Structure Supports for RF Mixer Matching Networks, IEEE Electron Device Lett., Vol. 29, No. 11, Nov. 2008, pp. 1209-1211

Wu S. M. et al. (2007). Study of Discrete Capacitor Embedded Process and Characterization Analysis in Organic-Base Substrate, *proceedings of 9th Electronics Packaging Technology Conference, IEEE*, pp.125-129, Singapore, 10-12 Dec. 2007

Yeung L. K. & Wu K. L. (2003). A Compact Second-order LTCC Bandpass Filter with Two Finite Transmission Zeros, IEEE Trans. Microwave Theory Tech., Vol. 51, No. 2, Feb. 2003, pp. 337-341

Yeung L. K. & Wu K. L. (2006). An LTCC Balanced-to-unbalanced Extracted-pole Bandpass Filter with Complex Load, *IEEE Trans. Microwave Theory Tech.*, Vol. 54, No. 4, Apr. 2006, pp. 1512-1518

Yoon J. B. et al. (2002). CMOS-Compatible Surface-Micromachined Suspended-Spiral Inductors for Multi-GHz Silicon RF ICs, IEEE Electron Device Lett., Vol. 23, No. 10, Oct. 2002, pp. 591-593

Yoon Y. K. et al. (2001). Embedded Solenoid Inductors for RF CMOS Power Amplifiers, *Proceedings of The 11th International Conference on Solid-State Sensors and Actuators*, Munich, Germany, June 10–14, 2001, Vol. 2, pp. 1114-1117

Yoon Y. K. & Allen M. G. (2005). Embedded Conductor Technology for Micromachined RF Elements, *J. Micromech. Microeng.*, Vol. 15, 2005, pp. 1317-1326

Yue C. P. & Wong S. S. (1998). On-Chip Spiral Inductors with Patterned Ground Shields for Si-Based RF IC's, IEEE JOURNAL OF SOLID-STATE CIRCUITS, Vol. 33, No. 5, MAY 1998, pp. 743-752

Yu X. et al. (2003). A High-Density MIM Capacitor (13 fF/ m2) Using ALD HfO2 Dielectrics, IEEE ELECTRON DEVICE LETTERS, Vol. 24, No. 2, FEBRUARY 2003, pp. 63-65

Zou J.; Chen J. & Liu C. (2001). Plastic Deformation Magnetic Assembly (PDMA) Of 3D Microstructures: Technology Development and Application, Proceedings of The 11th International Conference on Solid-State Sensors and Actuators, Munich, Germany, June 10–14, 2001, Vol. 2, pp. 1582-1585

Zurcher P. et al. (2000). Integration of Thin Film MIM Capacitors and Resistors into Copper Metallization based RF-CMOS and Bi-CMOS Technologies, *Proceedings of IEDM*, San Francisco, USA, 2000, pp. 153-156

14

Modeling of Spiral Inductors

Kenichi Okada and Kazuya Masu
Tokyo Institute of Technology
Japan

1. On-Chip Inductors

This section explains overview of the on-chip inductors. Demand for on-chip inductors has been increasing especially for CMOS RF circuits. An on-chip inductor is a key component because its performance directly affects the circuit performance and the chip cost. The on-chip inductors have been used, for example, as a matching network component, as baluns, and as an inductive load. The requirements for on-chip inductors are to have an appropriate inductance value, high self-resonance frequency, and high-quality factor Q, and to occupy a small layout area. In this section, several types of on-chip inductors are introduced.

1.1 Classification of on-chip inductors

A spiral inductor is one of the most common structures among on-chip inductors, and there are some variants of spiral inductor. Figure 1 shows spiral inductors commonly used as an on-chip inductor. Figure 1(a) is a single-ended type of spiral inductor, which has a spiral trace with an underpass. The spiral trace is made on top metal layer, and the underpass is made on lower metal layer. Spiral inductors are usually characterized by the diameter, the line width, the number of turns, and the line-to-line space.

Symmetric inductors are often used for differential circuits, *e.g.*, voltage controlled oscillator, low noise amplifier, mixer, *etc*. Figures 1(b) and (c) show symmetric spiral inductors with/without a center tap. The center tap is used to obtain a connection to the center node of symmetric inductor. The symmetric inductor achieves chip area reduction rather than using two single-ended spiral inductors to realize a symmetric structure. The single-ended inductor usually has only one underpass while the symmetric one has some underpasses depending on the number of turns. The underpass has larger resistance caused by inter-layer vias, which degrades quality factor.

Patterned ground shield (PSG) is sometimes employed to reduce substrate loss as shown in Fig. 2. As a drawback, capacitance between the spiral metal and the shield is increased, and self-resonance frequency is also decreased.

Figure 3 shows a meander inductor and a solenoid inductor, which are sometimes employed. The meander inductor does not have underpasses while area efficiency is not good due to negative line-to-line mutual inductance. The solenoid inductor can be implemented vertically by using multiple metal layers. Area efficiency is high even though quality factor is not good due to multiple via connections.

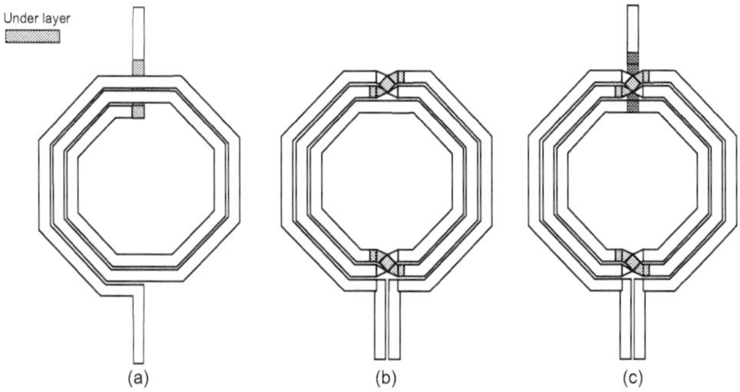

Fig. 1. On-chip spiral inductors. (a) single-ended (b) symmetric (c) symmetric with a center tap

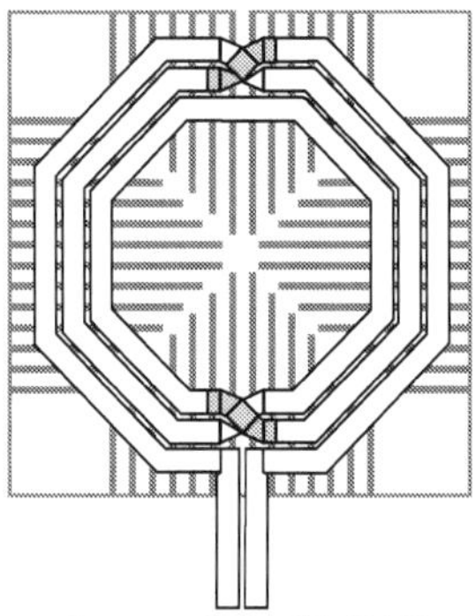

Fig. 2. Spiral inductor with patterned ground shield (PSG).

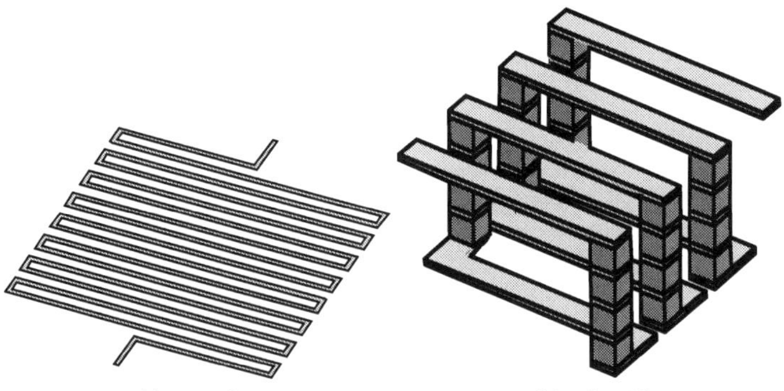

(a) meander (b) solenoid

Fig. 3. On-chip inductors.

2. Modeling of 2-Port and 3-Port Inductors

This section explains how to derive inductance and quality factor from measured S-parameters or Y-parameters for various excitation modes.

Definition of inductance and quality factor of on-chip inductors is not unique, and there are actually several definitions. The reason is that inductance and quality factor depend on excitation mode of input ports. In this section, an equivalent circuit model is shown, and derivation methods of the parameters in various excitation modes are explained. First, derivation for 2-port inductors is explained. Next, generic 3-port characterization is explained, and then, experimental results using measurements are shown.

2.1 Modeling of 2-port inductors

Here, a traditional π-type equivalent circuit is introduced for simple two-port inductors, and the derivation of inductance and quality factor is explained.

Figure 4(a) shows a common equivalent circuit model for 2-port inductors, utilized for CMOS LSIs, and Fig. 4(b) shows a physical structure of the inductor. Each parameter in Fig. 4(a) is related to the structure[1]. Actually, the π-type lumped equivalent circuit is usually utilized even if these RLC components are distributed. L_s is inductance of the spiral wire, and R_s means resistance of it. C_s is line-to-line capacitance of the spiral wire. C_{oxn} means capacitance between the wire and substrate. C_{Sin} and R_{Sin} mean capacitance and resistance in the Si substrate, respectively.

The equivalent circuit in Fig. 4(a) can express characteristics of on-chip inductors with frequency dependence, and each part of the equivalent circuit can be derived from Y-parameters according to the definition of Y-parameter as shown in Fig. 5(a). In this case, Y_{12} and Y_{21} are supposed to be equal to each other.

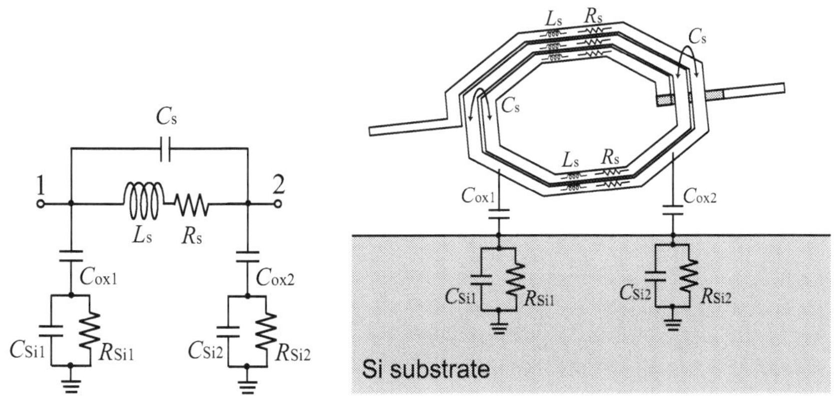

(a) Equivalent circuit (b) Physical structure

Fig. 4. An equivalent circuit model for a 2-port inductor.

[1] Unfortunately, the parameters are not exactly agreed with values calculated from the physical structure. R_s sometimes becomes almost twice because of eddy current in Si substrate, which is not characterized by the equivalent circuit model shown in Fig. 4(a).

L_s can be derived from Y_{12}. However, L_s is usually not utilized to evaluate an on-chip inductor because it is not an effective value used in a circuit. Actually, the inductance of on-chip inductor becomes zero at high frequency due to parasitic capacitances. To express this frequency dependence, inductance defined by Y_{11} is commonly employed. The reason is explained as followed. As explained, inductance and quality factor depend on each port impedance, and inductors are often used at a shunt part as shown in Fig. 6(a). In this case, input impedance of the inductor can be calculated by $1/Y_{11}$ as shown in Fig. 5(b).

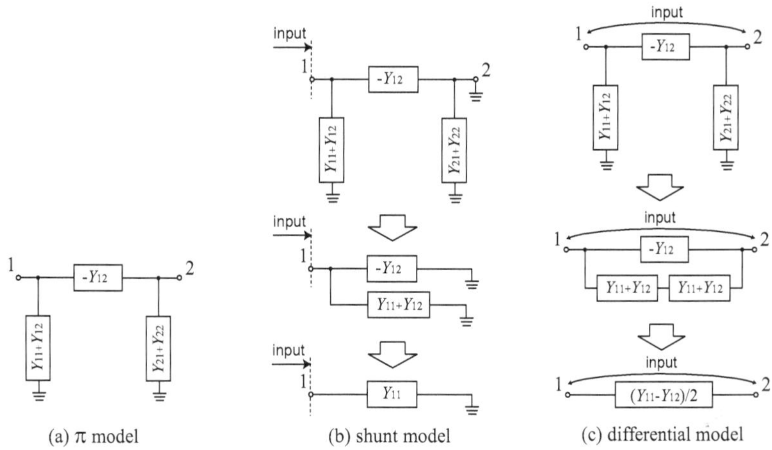

Fig. 5. Y-parameter calculation.

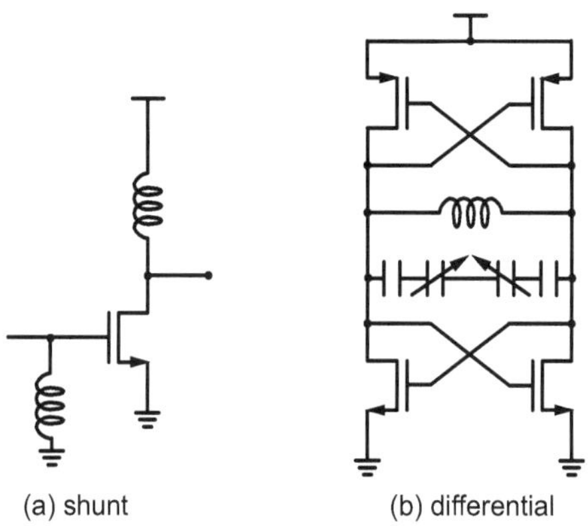

Fig. 6. Inductor usage.

L_{shunt} and Q_{shunt} are defined by the following equations.

$$L_{\text{shunt}} = \frac{\text{Im}\left(\frac{1}{Y_{11}}\right)}{\omega} \tag{1}$$

$$Q_{\text{shunt}} = \frac{\text{Im}\left(\frac{1}{Y_{11}}\right)}{\text{Re}\left(\frac{1}{Y_{11}}\right)} \tag{2}$$

$$= -\frac{\text{Im}\left(Y_{11}\right)}{\text{Re}\left(Y_{11}\right)} \tag{3}$$

This definition (1)(3) is widely used because the definition does not depend on equivalent circuits and only Y_{11} is required to calculate them.
In case using the equivalent circuit in Fig. 4(a), Y_{11} can be derived by the following equation.

$$Y_{11} = \frac{1}{R_s + j\omega L_s} + j\omega C_s + j\omega C_{\text{ox1}} // \left(\frac{1}{R_{\text{Si1}}} + j\omega C_{\text{Si1}}\right), \tag{4}$$

and it can be approximated at lower frequency as follows.

$$1/Y_{11} \approx R_s + j\omega L_s \tag{5}$$

This means that L_{shunt} and Q_{shunt} are close to L_s and $\omega L_s / R_s$ at lower frequency, respectively, and they are decreased by the parasitic capacitances at higher frequency.
On the other hand, on-chip inductors are often used in differential circuits as shown in Fig. 6(b). In this case, the inductor has different characteristics from the shunt case shown in Fig. 6(a), and the input impedance in differential mode becomes $\frac{4}{Y_{11}+Y_{22}-Y_{12}-Y_{21}}$ while the input impedance in single-ended mode is $1/Y_{11}$. The detailed calculation is explained in Sect. 2.2. Thus, effective inductance L_{diff} and effective quality factor Q_{diff} in differential mode can be calculated by using the differential input impedance $\frac{4}{Y_{11}+Y_{22}-Y_{12}-Y_{21}}$ as follows.

$$L_{\text{diff}} = \frac{\text{Im}\left(\frac{4}{Y_{11}+Y_{22}-Y_{12}-Y_{21}}\right)}{\omega} \tag{6}$$

$$Q_{\text{diff}} = \frac{\text{Im}\left(\frac{4}{Y_{11}+Y_{22}-Y_{12}-Y_{21}}\right)}{\text{Re}\left(\frac{4}{Y_{11}+Y_{22}-Y_{12}-Y_{21}}\right)} \tag{7}$$

$$= -\frac{\text{Im}\left(Y_{11}+Y_{22}-Y_{12}-Y_{21}\right)}{\text{Re}\left(Y_{11}+Y_{22}-Y_{12}-Y_{21}\right)} \tag{8}$$

For symmetric inductors, Y_{22} and Y_{21} are approximately equal to Y_{11} and Y_{12}, respectively, so the following approximated equations can also be utilized as shown in Fig. 5(c).

$$L_{\text{diff}} \approx \frac{\text{Im}\left(\frac{2}{Y_{11}-Y_{12}}\right)}{\omega} \tag{9}$$

$$Q_{\text{diff}} \approx -\frac{\text{Im}\left(Y_{11}-Y_{12}\right)}{\text{Re}\left(Y_{11}-Y_{12}\right)} \tag{10}$$

In a similar way to Eq.(4), the following equation can also be derived from Fig. 4. Examples of calculation of the above parameters will be explained in Sect. 2.5.

$$\frac{Y_{11}+Y_{22}-Y_{12}-Y_{21}}{4} \tag{11}$$

$$= \frac{1}{R_s+j\omega L_s} + j\omega C_s + \left\{ j\omega C_{ox1} // \left(\frac{1}{R_{Si1}} + j\omega C_{Si1} \right) \right\} // \left\{ j\omega C_{ox2} // \left(\frac{1}{R_{Si2}} + j\omega C_{Si2} \right) \right\} \tag{12}$$

2.2 Equivalent circuit model for 3-port inductors

A symmetric inductor with a center tap has 3 input ports as shown in Fig. 1(c). The characteristics of symmetric inductor depend on excitation modes and load impedance of center-tap, *i.e.*, single-ended mode, differential mode, common mode, center-tapped and non-center-tapped. Unfortunately, 2-port measurement of the 3-port inductors is insufficient to characterize the 3-port ones in all operation modes. Common-mode impedance of center-tapped inductor has influence on circuit performance, especially about CMRR of differential amplifiers, pushing of differential oscillators, *etc*, so 3-port characterization is indispensable to simulate common-mode response in consideration of the center-tap impedance. The characteristics of symmetric inductor can be expressed in all operation modes by using the measured S parameters of the 3-port inductor. In this section, derivation method using 3-port S-parameters is explained to characterize it with the center-tap impedance.

2.3 Derivation using Y-parameters

Inductance L and quality factor Q of 3-port and 2-port inductors can be calculated by using measured Y parameters. The detailed procedure is explained as follows. First, input impedance is calculated for each excitation mode, *i.e.*, single-ended, differential, common. In case of common mode, the impedance depends on the center-tap impedance Y_3, so the input impedance is a function of the center-tap impedance Y_3. Next, inductance L and quality factor Q are calculated from the input impedance as explained in Sect. 2.1.

In case using 3-port measurements in differential mode, differential-mode impedance Z_{diff} can be derived as follows.

$$\begin{pmatrix} I_{\text{diff}} \\ -I_{\text{diff}} \\ I_3 \end{pmatrix} = \begin{pmatrix} Y_{11} & Y_{12} & Y_{13} \\ Y_{21} & Y_{22} & Y_{23} \\ Y_{31} & Y_{32} & Y_{33} \end{pmatrix} \cdot \begin{pmatrix} V_{\text{diff}}/2 \\ -V_{\text{diff}}/2 \\ V_3 \end{pmatrix} \tag{13}$$

$$Z_{\text{diff}} = \frac{V_{\text{diff}}}{I_{\text{diff}}} = \frac{2(Y_{23}+Y_{13})}{Y_{23}(Y_{11}-Y_{12})-Y_{13}(Y_{21}-Y_{22})} \tag{14}$$

Note that this differential impedance Z_{diff} does not depend on the center-tap impedance Y_3. Inductance L_{diff} and quality factor Q_{diff} are calculated with Z_{diff} by the following equations.

$$L_{\text{diff}} = \frac{\text{Im}\left[Z_{\text{diff}}\right]}{\omega} \tag{15}$$

$$Q_{\text{diff}} = \frac{\text{Im}\left[Z_{\text{diff}}\right]}{\text{Re}\left[Z_{\text{diff}}\right]} \tag{16}$$

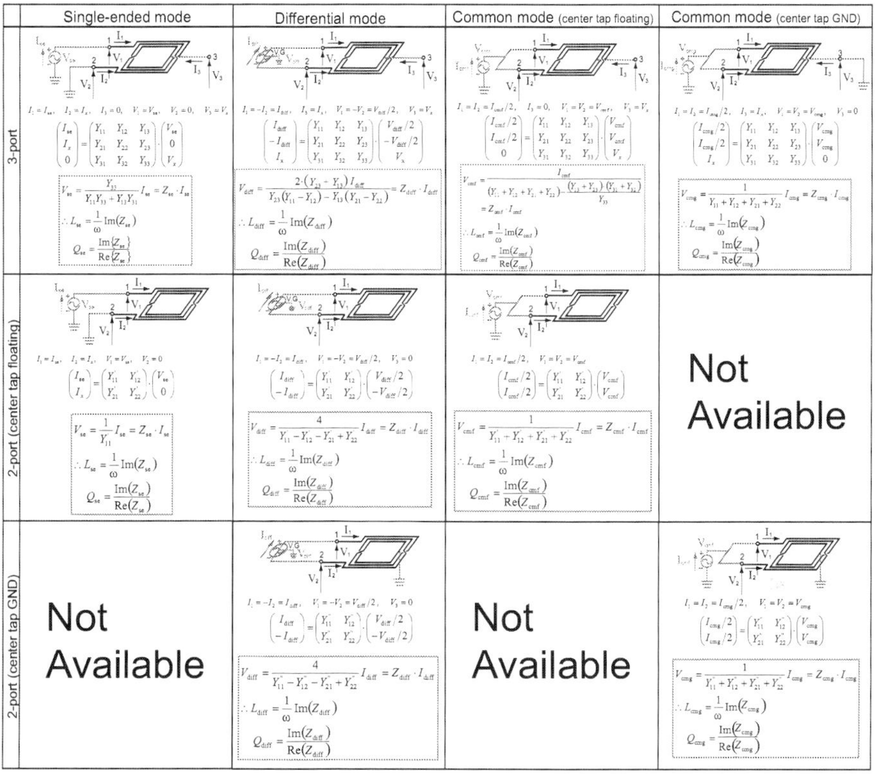

Fig. 7. Equations derived from Y parameter to evaluate L and Q of 2-port and 3-port inductors.

In case using 3-port measurements in common mode, common-mode impedance Z_{cm} can be derived as follows.

$$\begin{pmatrix} I_{cm}/2 \\ I_{cm}/2 \\ I_3 \end{pmatrix} = \begin{pmatrix} Y_{11} & Y_{12} & Y_{13} \\ Y_{21} & Y_{22} & Y_{23} \\ Y_{31} & Y_{32} & Y_{33} \end{pmatrix} \cdot \begin{pmatrix} V_{cm} \\ V_{cm} \\ V_3 \end{pmatrix} \tag{17}$$

$$Z_{cm} = \frac{V_{cm}}{I_{cm}} = \frac{1}{(Y_{11}+Y_{12}+Y_{21}+Y_{22}) + \dfrac{(Y_{13}+Y_{23})(Y_{31}+Y_{32})}{Y_3 - Y_{33}}} \tag{18}$$

where the center-tap impedance Y_3 is given by I_3/V_3. Note that the common-mode impedance Z_{cm} depends on the center-tap impedance Y_3. Inductance L_{cm} and quality factor Q_{cm} in common mode are calculated with Z_{cm} by the following equations.

$$L_{cm} = \frac{\mathrm{Im}\left[Z_{cm}\right]}{\omega} \tag{19}$$

$$Q_{cm} = \frac{\mathrm{Im}\left[Z_{cm}\right]}{\mathrm{Re}\left[Z_{cm}\right]} \tag{20}$$

Figure 7 summarizes calculation of L and Q from 2-port and 3-port Y-parameters. The 2-port symmetric inductor has two types of structures, center-tapped and non-center-tapped ones. It is impossible to characterize the center-tapped inductor only from measurement of non-center-tapped one. On the other hand, all characteristics can be extracted from the Y parameters of 3-port inductor due to its flexibility of center-tap impedance. Therefore, we need 3-port inductor to characterize all operation modes of symmetric inductors.
The definition of quality factor in Eqs. (16) and (20) uses ratio of imaginary and real parts. The definition is very useful to evaluate inductors. On the other hand, it is not convenient to evaluate LC-resonators using inductors because the imaginary part in Eqs. (16) and (20) is decreased by parasitic capacitances, *e.g.*, C_s, C_{oxn}, C_{Sin}. Quality factor of LC-resonator is higher than that defined by Eqs. (16) and (20). Thus, the following definition is utilized to evaluate quality factor of inductors used in LC-resonators.

$$Q = \frac{\omega}{2} \left| \frac{1}{Z} \frac{\partial Z}{\partial \omega} \right| \tag{21}$$

where Z is input impedance.

2.4 Derivation using S-parameters

By the same way, inductance L and quality factor Q of 3-port and 2-port inductors can also be derived from S-parameters. As explained in Fig. 8, the input impedances for each excitation mode, *e.g.*, Z_{diff}, Z_{cm}, can be derived from S-parameters as well as Y-parameters, and L and Q can also be calculated from the input impedance in a similar way.

2.5 Measurement and parameter extraction

In this subsection, measurement and parameter extraction are demonstrated. Figure 9 shows photomicrograph of the measured symmetric inductors. The symmetrical spiral inductors are fabricated by using a 0.18 μm CMOS process (5 aluminum layers). The configuration of the spiral inductor is 2.85 turns, line width of 20 μm, line space of 1.2 μm, and outer diameter of 400 μm. The center tap of 3-port inductor is connected to port-3 pad. Two types of 2-port inductors are fabricated; non-center-tapped (center tap floating) and center-tapped (center tap GND) structures.
The characteristics of inductors are measured by 4-port network analyzer (Agilent E8364B & N4421B) with on-wafer probes. An open dummy structure is used for de-embedding of probe pads.
Several equivalent circuit models for symmetric inductor have been proposed Fujumoto et al. (2003); Kamgaing et al. (2002); Tatinian et al. (2001); Watson et al. (2004). This demonstration uses 3-port equivalent circuit model of symmetric inductor as shown in Fig. 10. This model uses compact model of the skin effect (R_m, L_f and R_f) Kamgaing et al. (2002; 2004). Center tap is expressed by the series and shunt elements.
Figure 11 shows frequency dependences of the inductance L and the quality factor Q of measured 2-port and 3-port inductors and the equivalent circuit model for various excitation modes. L and Q of measured inductors can be calculated using Y parameters as shown in Fig. 7. Table 1 shows extracted model parameters of the 3-port equivalent circuit shown in Fig. 10. The parameters are extracted with numerical optimization.
In Figs. 11 (a) and (b), self-resonance frequency and Q excited in differential mode improve rather than those excited in single-ended mode due to reduction of parasitic effects in substrate Danesh & Long (2002), which is considerable especially for CMOS LSIs. In common

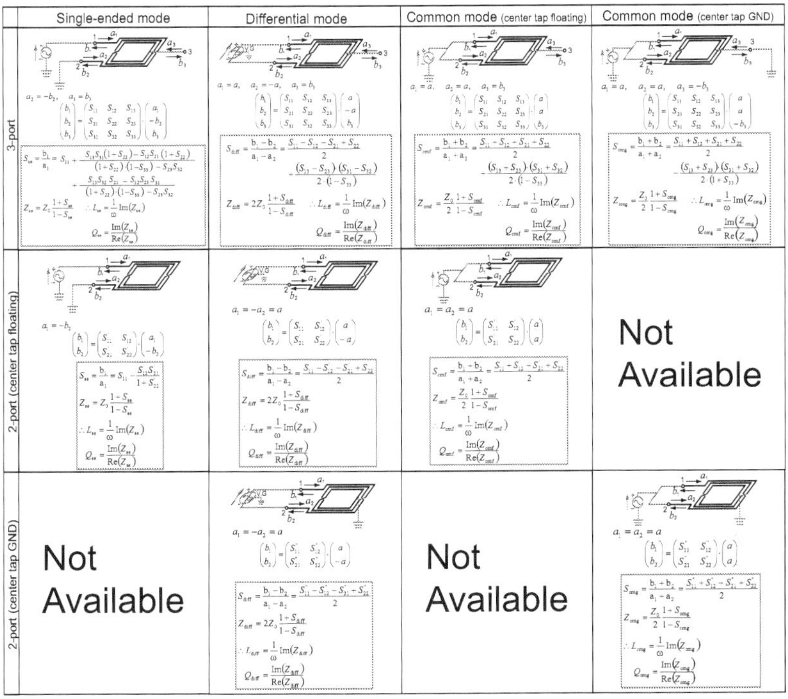

Fig. 8. Equations derived from S parameter to evaluate L and Q of 2-port and 3-port inductors.

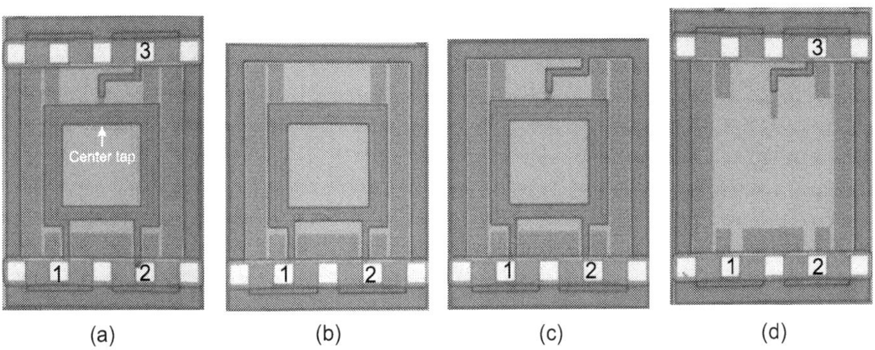

Fig. 9. Photomicrograph of the measured symmetric inductors. (a) 3-port inductor. (b) 2-port inductor (center tap floating). (c) 2-port inductor (center tap GND). (d) Open pad. The center tap of 3-port inductor is connected to port-3 pad.

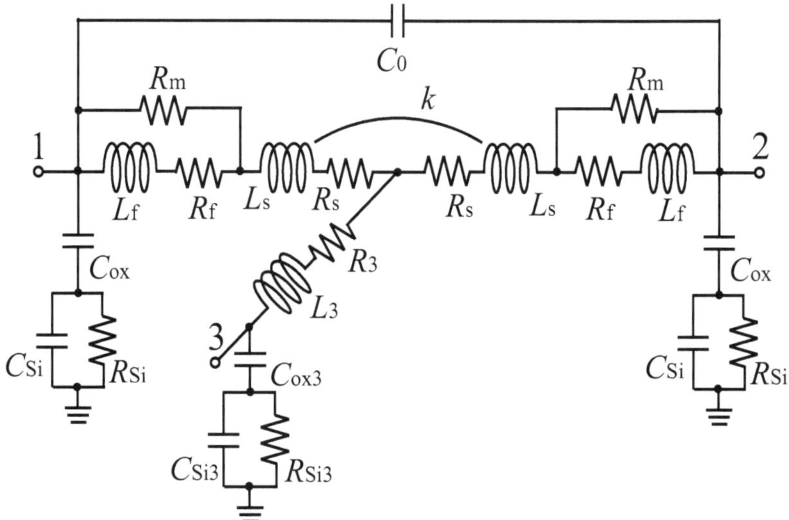

Fig. 10. An equivalent circuit model for a 3-port symmetric inductor.

L_s [nH]	R_s [Ω]	L_f [nH]	R_f [Ω]	R_m [Ω]
1.34	1.87	0.91	2.66	14.6
C_0 [pF]	L_3 [nH]	R_3 [Ω]	C_{ox} [pF]	C_{ox3} [pF]
0.08	0.00	0.25	0.18	0.19
R_{Si} [Ω]	R_{Si3} [Ω]	C_{Si} [pF]	C_{Si3} [pF]	k
24.7	3.37	0.01	0.07	0.44

Table 1. Extracted Model Parameters of 3-port Symmetric Inductor

mode (center tap floating), L is negative value because inductor behaves as open Fujumoto et al. (2003) as shown in Fig. 11 (c). These characteristics extracted from 2-port and 3-port inductors agree with each other. In Fig. 11 (d), L and Q excited in common mode (center tap GND) are smaller because interconnections between input pads and center-tap are parallel electrically. The characteristics of the equivalent circuit model are well agreed with that of measured 3-port inductor in all operation modes. These results show measured parameter of 3-port inductor and its equivalent circuit model can express characteristics of symmetric inductor in all operation modes and connection of center tap.

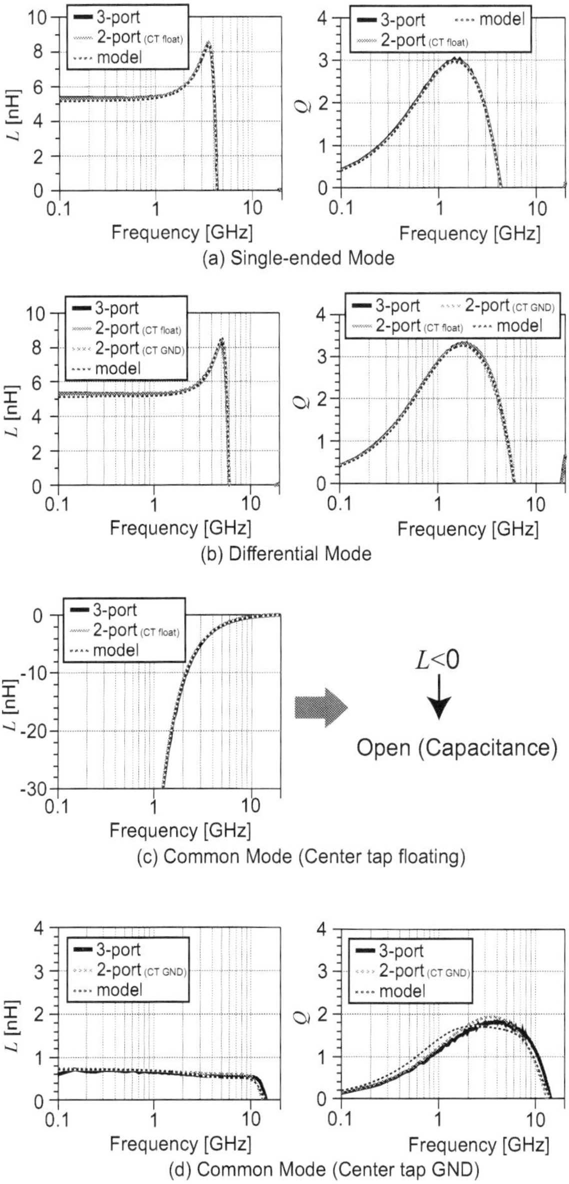

Fig. 11. Frequency dependences of the inductance L and the quality factor Q in various excitation modes. (a) Single-ended mode. (b) Differential mode. (c) Common mode (center tap floating). (d) Common mode (center tap GND).

3. Modeling of Multi-Port Inductors

Multi-port inductors, like 3-, 4-, 5-port, 2-port symmetric one with a center-tap, *etc.*, are very useful to reduce circuit area?, ?. In this section, a generic method to characterize the multi-port inductors is presented??.

As a conventional method, one of the methods to characterize the multi-port inductor is to extract each parameter of an equivalent circuit by the numerical optimization. However, the equivalent circuit has to be consisted of many circuit components characterizing self and mutual effects, and it is not easy to extract these parameters considering all the mutual effects. In this section, a method utilizing a matrix-decomposition technique is presented, and the method can extract each self and mutual parameters mathematically, which contributes to improve the extraction accuracy. The method can also be used for charactering a differential inductor with a center-tap, which is a kind of 3-port inductor.

3.1 Derivation of matrix Y_c

This section describes a method to decompose self and mutual inductances of multi-port inductors. A 5-port inductor shown in Fig. 12 is utilized as an example while the method can be also applied to generic multi-port inductors. Figure 13 shows an equivalent circuit of the 5-port inductor, which consists of core, shunt, and lead parts Long & Copeland (1997); Niknejad & Meyer (1998). The core part expresses self and mutual inductances with parasitic resistance and capacitance, which are characterized by Z_n in Fig. 13. The core part is also expressed by a matrix $\mathbf{Y_c}$. The shunt part characterizes parasitics among Inter Layer Dielectric (ILD) and Si substrate. It is expressed by a matrix $\mathbf{Y_{sub}}$. Ports 2, 3, and 4 have lead parts as shown in Fig. 12, which are modeled by Z_{short} and Y_{open} as shown in Fig. 13. On the other hand, a lead part of ports 1 and 5 is assumed to be a part of inductor as shown in Fig. 12. $\mathbf{Y_{sub}}$ consists of admittances $Y_{\text{sub}n}$ in Fig. 13. The lead part characterizes lead lines, and it is also expressed by matrix $\mathbf{Y_{open}}$ and $\mathbf{Z_{short}}$. These matrices can be combined by the following equations.

$$
\begin{aligned}
\mathbf{Z_{meas'}} &= (\mathbf{Y_{meas}} - \mathbf{Y_{open}})^{-1} \\
&\quad - (\mathbf{Z_{short}}^{-1} - \mathbf{Y_{open}})^{-1}, \qquad (22)\\
\mathbf{Y_c} &= \mathbf{Y_{meas'}} - \mathbf{Y_{sub}}, \qquad (23)
\end{aligned}
$$

where admittance matrix $\mathbf{Y_{meas}}$ is converted from measured S-parameter. To decompose each part of multi-port inductor in Fig. 13, first the matrix $\mathbf{Y_{sub}}$ is calculated. The matrix $\mathbf{Y_{sub}}$ can be expressed by admittances $Y_{\text{sub}n}$ as follows.

$$
\mathbf{Y_{sub}} = \begin{pmatrix} Y_{\text{sub1}} & 0 & 0 & 0 & 0 \\ 0 & Y_{\text{sub2}} & 0 & 0 & 0 \\ 0 & 0 & Y_{\text{sub3}} & 0 & 0 \\ 0 & 0 & 0 & Y_{\text{sub4}} & 0 \\ 0 & 0 & 0 & 0 & Y_{\text{sub5}} \end{pmatrix} \qquad (24)
$$

For the sum of the matrices $\mathbf{Y_c}$ and $\mathbf{Y_{sub}}$, the following equation can be defined by vectors v and i as defined in Fig. 14.

$$
\begin{pmatrix} i_1 \\ i_2 \\ i_3 \\ i_4 \\ i_5 \end{pmatrix} = \mathbf{Y_{meas'}} \begin{pmatrix} v_1 \\ v_2 \\ v_3 \\ v_4 \\ v_5 \end{pmatrix} = (\mathbf{Y_c} + \mathbf{Y_{sub}}) \begin{pmatrix} v_1 \\ v_2 \\ v_3 \\ v_4 \\ v_5 \end{pmatrix} \qquad (25)
$$

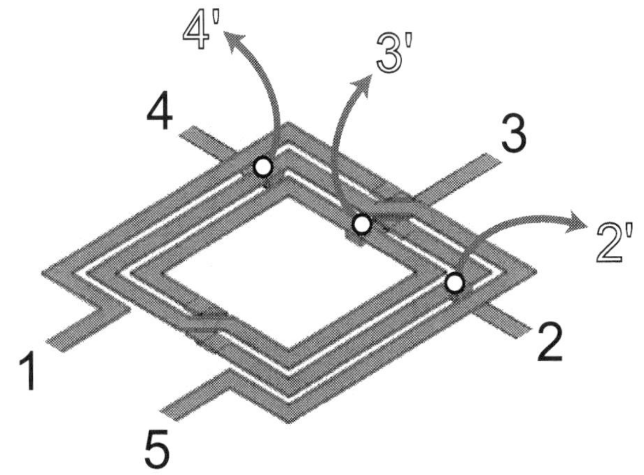

Fig. 12. Structure of the 5-port inductor.

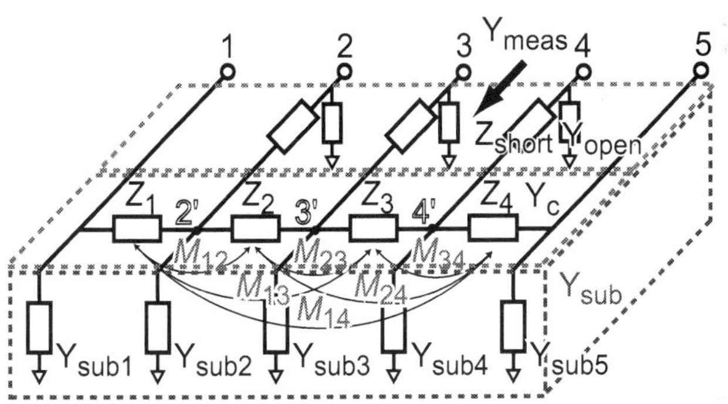

Fig. 13. Equivalent circuit of the 5-port inductor.

$Y_{\text{meas}'}$ consists of Y_c and $Y_{\text{sub}n}$. The circuit part expressed by Y_c does not have a current path to ground as shown in Fig. 13. When $v_1 = v_2 = v_3 = v_4 = v_5 = v'$, no current flows into Z_n shown in Fig. 14, because $\boldsymbol{Y}_c$ connects to only ports and not ground. This is described by the following equation.

$$\boldsymbol{Y}_c \begin{pmatrix} v' \\ v' \\ v' \\ v' \\ v' \end{pmatrix} = 0 \tag{26}$$

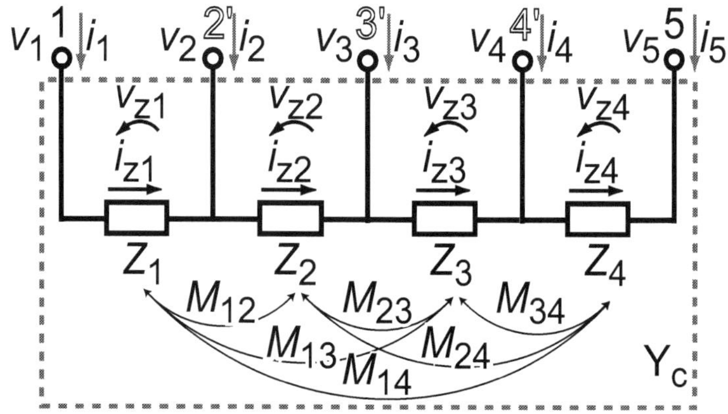

Fig. 14. Equivalent circuit of core.

The following equation is derived from Eqs.(25)(26)

$$\begin{pmatrix} i_1 \\ i_2 \\ i_3 \\ i_4 \\ i_5 \end{pmatrix} = \boldsymbol{Y}_{\mathbf{meas}'} \begin{pmatrix} v' \\ v' \\ v' \\ v' \\ v' \end{pmatrix} = \boldsymbol{Y}_{\mathbf{sub}} \begin{pmatrix} v' \\ v' \\ v' \\ v' \\ v' \end{pmatrix} = \begin{pmatrix} Y_{\text{sub}1} \\ Y_{\text{sub}2} \\ Y_{\text{sub}3} \\ Y_{\text{sub}4} \\ Y_{\text{sub}5} \end{pmatrix} v' \tag{27}$$

Therefore, each $Y_{\text{sub}n}$ in $\boldsymbol{Y}_{\mathbf{sub}}$ can be calculated by the following equation.

$$Y_{\text{sub}n} = \sum_{i=1}^{5} Y_{\text{meas}'ni} \tag{28}$$

3.2 Conversion of matrix Y_c to Z_{core}

Figure 14 shows the core part of the entire equivalent circuit in Fig. 13, which is expressed by the matrix $\boldsymbol{Y}_\mathbf{c}$. In this case, we need each parameter of Z_n and M_{nm}, so the matrix $\boldsymbol{Y}_\mathbf{c}$ is converted into a matrix $\mathbf{Z}_{\mathbf{core}}$. When $\boldsymbol{Y}_\mathbf{c}$ is a $n \times n$ matrix, $\mathbf{Z}_{\mathbf{core}}$ is a $(n-1) \times (n-1)$ matrix. The matrix $\mathbf{Z}_{\mathbf{core}}$ is defined by the following equations.

$$\boldsymbol{i} = \begin{pmatrix} i_1 \\ i_2 \\ i_3 \\ i_4 \\ i_5 \end{pmatrix} = \boldsymbol{Y}_\mathbf{c} \begin{pmatrix} v_1 \\ v_2 \\ v_3 \\ v_4 \\ v_5 \end{pmatrix} = \boldsymbol{Y}_\mathbf{c}\boldsymbol{v} \tag{29}$$

$$\boldsymbol{v}_\mathbf{z} = \begin{pmatrix} v_{z1} \\ v_{z2} \\ v_{z3} \\ v_{z4} \end{pmatrix} = \mathbf{Z}_{\mathbf{core}} \begin{pmatrix} i_{z1} \\ i_{z2} \\ i_{z3} \\ i_{z4} \end{pmatrix} = \mathbf{Z}_{\mathbf{core}}\boldsymbol{i}_\mathbf{z} \tag{30}$$

$$\mathbf{Z_{core}} = \begin{pmatrix} Z_1 & j\omega M_{12} & j\omega M_{13} & j\omega M_{14} \\ j\omega M_{21} & Z_2 & j\omega M_{23} & j\omega M_{24} \\ j\omega M_{31} & j\omega M_{32} & Z_3 & j\omega M_{34} \\ j\omega M_{41} & j\omega M_{42} & j\omega M_{43} & Z_4 \end{pmatrix} \quad (31)$$

where vectors $\boldsymbol{v}$, $\boldsymbol{i}$, $\boldsymbol{v_z}$, and $\boldsymbol{i_z}$ are defined as shown in Fig. 14. Each element of the matrix $\mathbf{Z_{core}}$ expresses self and mutual components directly. The matrix $\boldsymbol{Y_c}$ has the same numerical information as $\mathbf{Z_{core}}$, which is explained later. The self and mutual inductances in $\mathbf{Z_{core}}$ can be derived from $\boldsymbol{Y_c}$.

In this case, rank of the matrix $\mathbf{Z_{core}}$ is 4 for the 5-port inductor, and the matrix $\boldsymbol{Y_c}$ consists of the same components as shown in Fig. 14. Thus, rank of $\boldsymbol{Y_c}$ is also 4 although $\boldsymbol{Y_c}$ is a 5×5 matrix. The matrix $\boldsymbol{Y_c}$ is not a regular matrix, and it does not have an inverse matrix. Here, converting matrices $\boldsymbol{A}$ and $\boldsymbol{B}$ are utilized, which are also not a regular matrix.

The converting matrices $\boldsymbol{A}$ and $\boldsymbol{B}$ are derived by the following procedure. Vectors $iandvareconvertedintovectors\boldsymbol{i_z}and\boldsymbol{v_z}bythefollowingequations.\boldsymbol{v_z} = \boldsymbol{Av}(32)$

$$\boldsymbol{i_z} = \boldsymbol{Bi} \quad (33)$$

$$\begin{pmatrix} v_{z1} \\ v_{z2} \\ v_{z3} \\ v_{z4} \end{pmatrix} = \begin{pmatrix} v_1 - v_2 \\ v_2 - v_3 \\ v_3 - v_4 \\ v_4 - v_5 \end{pmatrix} = \boldsymbol{Av} \quad (34)$$

$$\boldsymbol{A} = \begin{pmatrix} 1 & -1 & 0 & 0 & 0 \\ 0 & 1 & -1 & 0 & 0 \\ 0 & 0 & 1 & -1 & 0 \\ 0 & 0 & 0 & 1 & -1 \end{pmatrix} \quad (35)$$

$$\begin{pmatrix} i_{z1} \\ i_{z2} \\ i_{z3} \\ i_{z4} \end{pmatrix} = \begin{pmatrix} i_1 \\ i_1 + i_2 \\ i_1 + i_2 + i_3 \\ i_1 + i_2 + i_3 + i_4 \end{pmatrix} = \boldsymbol{Bi} \quad (36)$$

Here, the follwing equation can be derived from Fig. 14.

$$\sum_{n=1}^{5} i_n = i_1 + i_2 + i_3 + i_4 + i_5 = 0 \quad (37)$$

According to Eq.(37), $i_1 + i_2 + i_3 + i_4$ is also expressed by $-i_5$ as an example.
Thus, matrix $\boldsymbol{B}$ has several solutions as follows.

$$\begin{aligned} \boldsymbol{B} \;=\; & \begin{pmatrix} 1 & 0 & 0 & 0 & 0 \\ 1 & 1 & 0 & 0 & 0 \\ 1 & 1 & 1 & 0 & 0 \\ 1 & 1 & 1 & 1 & 0 \end{pmatrix}, \begin{pmatrix} 1 & 0 & 0 & 0 & 0 \\ 1 & 1 & 0 & 0 & 0 \\ 1 & 1 & 1 & 0 & 0 \\ 0 & 0 & 0 & 0 & -1 \end{pmatrix}, \\ & \begin{pmatrix} 1 & 0 & 0 & 0 & 0 \\ 1 & 1 & 0 & 0 & 0 \\ 0 & 0 & 0 & -1 & -1 \\ 0 & 0 & 0 & 0 & -1 \end{pmatrix}, \begin{pmatrix} 1 & 0 & 0 & 0 & 0 \\ 0 & 0 & -1 & -1 & -1 \\ 0 & 0 & 0 & -1 & -1 \\ 0 & 0 & 0 & 0 & -1 \end{pmatrix}, \\ & \begin{pmatrix} 0 & -1 & -1 & -1 & -1 \\ 0 & 0 & -1 & -1 & -1 \\ 0 & 0 & 0 & -1 & -1 \\ 0 & 0 & 0 & 0 & -1 \end{pmatrix}, \cdots \end{aligned} \quad (38)$$

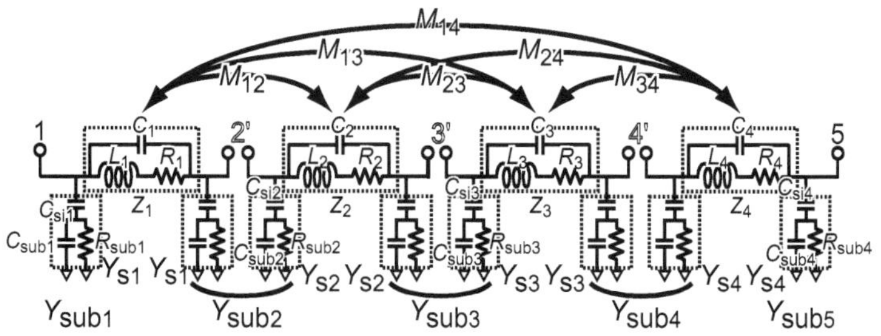

Fig. 15. π-type equivalent circuit.

Eqs. (29)(32)(33) are substituted into Eq. (30), and the following equations are obtained.

$$A \begin{pmatrix} v_1 \\ v_2 \\ v_3 \\ v_4 \\ v_5 \end{pmatrix} = \begin{pmatrix} v_{z1} \\ v_{z2} \\ v_{z3} \\ v_{z4} \end{pmatrix} = \boldsymbol{Z}_{\mathbf{core}} \begin{pmatrix} i_{z1} \\ i_{z2} \\ i_{z3} \\ i_{z4} \end{pmatrix} = \boldsymbol{Z}_{\mathbf{core}} \boldsymbol{B} \begin{pmatrix} i_1 \\ i_2 \\ i_3 \\ i_4 \\ i_5 \end{pmatrix} \tag{39}$$

$$\boldsymbol{A} = \boldsymbol{Z}_{\mathbf{core}} \boldsymbol{B} \boldsymbol{Y}_{\mathbf{c}} \tag{40}$$

Matrices $\boldsymbol{A}$ and $\boldsymbol{B}$ are not regular matrix. Matrix $\boldsymbol{Z}_{\mathbf{core}}$ is 4×4 matrix. $\boldsymbol{Y}_{\mathbf{c}}$ is shrunk to $\boldsymbol{Z}_{\mathbf{core}}$ by pseudo-inverse matrix $\boldsymbol{A}^{+}$. For example, $\boldsymbol{A}^{+}$ can be defined as follows.

$$\boldsymbol{A}^{+} = \boldsymbol{B}^{T} \tag{41}$$

$$\boldsymbol{A}\boldsymbol{A}^{+} = \begin{pmatrix} 1 & -1 & 0 & 0 & 0 \\ 0 & 1 & -1 & 0 & 0 \\ 0 & 0 & 1 & -1 & 0 \\ 0 & 0 & 0 & 1 & -1 \end{pmatrix} \begin{pmatrix} 1 & 1 & 0 & 0 \\ 0 & 1 & 0 & 0 \\ 0 & 0 & 0 & 0 \\ 0 & 0 & -1 & 0 \\ 0 & 0 & -1 & -1 \end{pmatrix} = \boldsymbol{I} \tag{42}$$

Finally, the following equations are derived. $\boldsymbol{Z}_{\mathbf{core}}$ is expressed by Eq. (44). The self and mutual inductances are calculated from S-parameter.

$$\boldsymbol{I} = \boldsymbol{Z}_{\mathbf{core}} \boldsymbol{B} Y_c \boldsymbol{A}^{+} \tag{43}$$

$$\boldsymbol{Z}_{\mathbf{core}} = (\boldsymbol{B}\boldsymbol{Y}_{\mathbf{c}}\boldsymbol{A}^{+})^{-1} \tag{44}$$

Next, each parameter of equivalent circuit is extracted. Figure 15 shows a π-ladder equivalent circuit, which is transformed from the circuit model shown in Fig. 13. Z_n and M_{nm} can directly be obtained from $\boldsymbol{Z}_{\mathbf{core}}$ shown in Eq. (44). $Y_{\text{sub}m}$ is divided to $Y_{\text{s}n}$ as shown in Fig. 15

$$Y_{\text{sub}m} = \begin{cases} Y_{\text{s}1} & (m = 1) \\ Y_{\text{s}m-1} + Y_{\text{s}m} & (2 \le m \le n-1) \\ Y_{\text{s}n-1} & (m = n) \end{cases} \tag{45}$$

L_n, C_n, and R_n in Fig. 15 are fitted to Z_n by a numerical optimization. $C_{\text{si}n}$, $C_{\text{sub}n}$, and $R_{\text{sub}n}$ in Fig. 15 are also fitted to $Y_{\text{s}n}$.

3.3 Parameter extraction of multi-port inductor

In this subsection, parameter extraction using measurement results is presented. Figure 16 shows microphotograph of the 5-port inductors, which are fabricated by using a 180 nm Si CMOS process with 6 aluminum layers. The configuration of the 5-port inductor is symmetric, 3 turns, width of 15 μm, line space of 1.2 μm, and outer diameter of 250 μm. 5-port S-parameter is obtained from two TEGs (Test Element Group) shown in Fig. 16 because common vector network analyzers have only four ports at most. Port 3 of inductor (a) is terminated by 50Ω resistor as indicated in Fig. 16. Port 4 of inductor (b) is also terminated by 50Ω resistor. Y_{meas} is obtained from measured S-parameters by the following equation.

$$\mathbf{Y_{meas}} = \begin{pmatrix} a_{11} & a_{12} & b_{31} & a_{14} & a_{15} \\ a_{21} & a_{22} & b_{32} & a_{24} & a_{25} \\ b_{31} & b_{32} & b_{33} & b_{32} & b_{35} \\ a_{41} & a_{42} & b_{23} & a_{44} & a_{45} \\ a_{51} & a_{52} & b_{53} & a_{54} & a_{55} \end{pmatrix}, \tag{46}$$

where a_{ij} and b_{ij} are measured S-parameter elements of inductors (a) and (b), respectively. In this case, $Y_{\text{meas}_{34}}$ and $Y_{\text{meas}_{43}}$ cannot be obtained, so these components are substituted by $Y_{\text{meas}_{32}}$ and $Y_{\text{meas}_{23}}$, respectively. The matrix $\mathbf{Z_{core}}$ is calculated from $\mathbf{Y_{meas}}$, $\mathbf{Y_{open}}$, and $\mathbf{Z_{short}}$ as explained in Sect. 3.1 and 3.2.

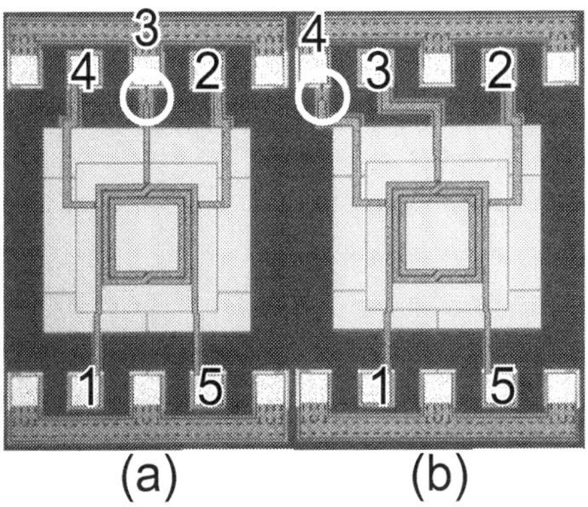

Fig. 16. Microphotograph of the 5-port inductor.

Measured results and equivalent circuit model are compared as follows. First, $L_n{}'$ are extracted from measured results by the following equations. To evaluate inductance and quality factor between port n and $(n+1)'$, Y_{nn} is utilzed. For example, Y_{11} is derived from Z_1 and Y_{s_1}.

$$Y_{11} = \frac{1}{Z_1} + Y_{s_1} \tag{47}$$

$$L_1{}' = \frac{\mathrm{Im}\left(\frac{1}{Y_{11}}\right)}{\omega} \tag{48}$$

$$Q_{L_1'} = \frac{\mathrm{Im}\left(\frac{1}{Y_{11}}\right)}{\mathrm{Re}\left(\frac{1}{Y_{11}}\right)}, \tag{49}$$

where Z_1 is derived from $\boldsymbol{Z_{\mathrm{core}}}$ in Eq. (44), and Y_{s_1} is derived from $\boldsymbol{Y_{\mathrm{sub}}}$ in Eq. (28).
On the other hand, $L_n{'}\_{\mathrm{model}}$ and $Q_{L_n{'}\_{\mathrm{model}}}$ are obtained from fitted parameters by a numerical optimization in Sect. 3.2, which are calculated as follows.

$$\begin{aligned} Y_{11\mathrm{model}} &= \frac{1}{Z_{1\mathrm{model}}} + Y_{s_1\mathrm{model}} \\ &= \frac{1}{(R_1 + j\omega L_1) // j\omega C_1} \\ &+ j\omega C_{\mathrm{si1}} // \left(\frac{1}{R_{\mathrm{sub1}}} + j\omega C_{\mathrm{sub1}}\right) \end{aligned} \tag{50}$$

$$L_1{'}\_\mathrm{model} = \frac{\mathrm{Im}\left(\frac{1}{Y_{11\mathrm{model}}}\right)}{\omega} \tag{51}$$

$$Q_{L_1{'}\_\mathrm{model}} = \frac{\mathrm{Im}\left(\frac{1}{Y_{11\mathrm{model}}}\right)}{\mathrm{Re}\left(\frac{1}{Y_{11\mathrm{model}}}\right)} \tag{52}$$

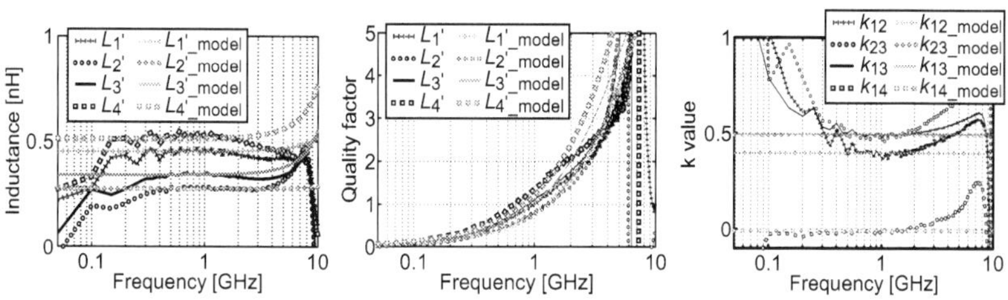

(a) Inductance (b) Quality factor (c) Coupling coefficients

Fig. 17. Parameters of the 5-port inductor

Figures 17(a)(b)(c) show measured and modeled inductances, quality factors, and coupling coefficients of the 5-port inductor, respectively. The coupling coefficients of the 5-port inductor are calculated by Eq. (53).

$$k_{nm} = \frac{M_{nm}}{\sqrt{L_n L_m}} \tag{53}$$

k_{nm} is coupling coefficient between L_n and L_m. Coupling coefficients k_{nm} have various values from -0.00 to 0.50 because line to line coupling intensity is different depending on topology of each segment. In this experiment, coupling coefficient k_{23} is larger than the others because L_2 and L_3 are arranged parallelly. Coupling coefficient k_{14} is almost zero because L_1 and L_4 are arranged orthogonally. k_{nm_model} is obtained from average coupling coefficient, because ideal coupling coefficient is independent of frequency.

3.4 Parameter extraction of 3-port symmetric inductor

The 5-port modeling has been presented in Sect. 3.1-3.3, and in this subsection calculation and parameter extraction of a 3-port inductor, *i.e.*, a 2-port inducotor with a center tap, shown in Fig. 18 are presented as a simple example. Note that the center tap is chosen as port 3 in Fig. 18.

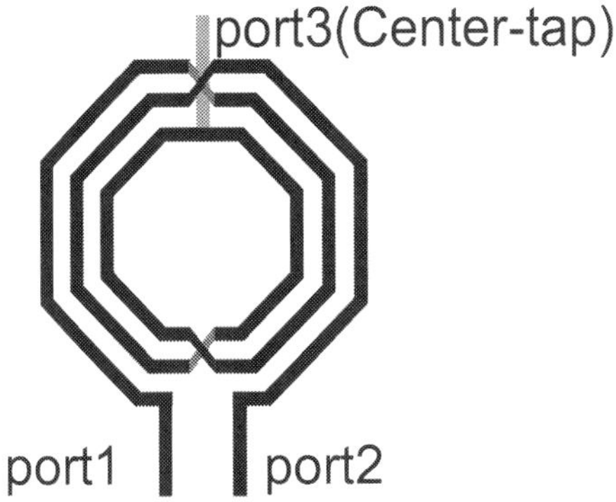

Fig. 18. A symmetric inductor with a center-tap.

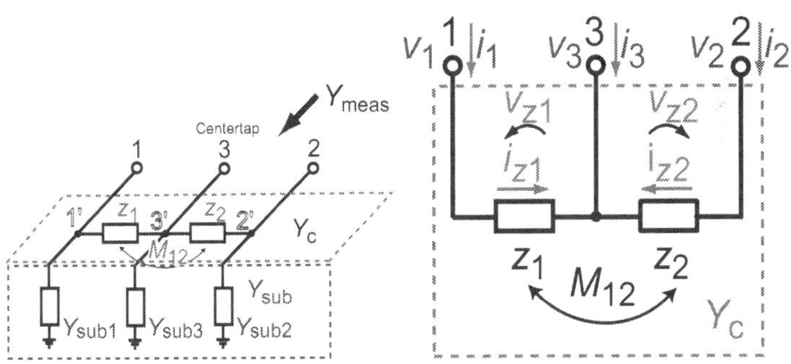

Fig. 19. An equivalent circuit of 3-port inductors (a) whole (b) core part.

Fig. 19(a) shows an equivalent circuit of 3-port inductors, and Fig. 19(b) shows core part of the equivalent circuit. In this case, $\mathbf{Z_{core}}$ can be defined by the following equation.

$$\mathbf{Z_{core}} = \begin{pmatrix} Z_1 & -j\omega M_{12} \\ -j\omega M_{21} & Z_2 \end{pmatrix} \tag{54}$$

Each element of the matrix $\mathbf{Z_{core}}$ expresses self and mutual components directly.

$$\mathbf{Y_{meas}} = \begin{pmatrix} Y_{11} & Y_{12} & Y_{13} \\ Y_{21} & Y_{22} & Y_{23} \\ Y_{31} & Y_{32} & Y_{33} \end{pmatrix} \tag{55}$$

According to Eq.(28), $Y_{\text{sub}n}$ can be calculated by the following equations.

$$Y_{\text{sub1}} = Y_{11} + Y_{12} + Y_{13} \tag{56}$$

$$Y_{\text{sub2}} = Y_{21} + Y_{22} + Y_{23} \tag{57}$$

$$Y_{\text{sub3}} = Y_{31} + Y_{32} + Y_{33} \tag{58}$$

The self and mutual inductances in $\mathbf{Z_{core}}$ can be derived from $\mathbf{Y_c}$ as follows.

$$\mathbf{Y_c} = \mathbf{Y_{meas}} - \mathbf{Y_{sub}}, \tag{59}$$

$$\mathbf{Z_{core}} = \begin{pmatrix} Z_1 & -j\omega M_{12} \\ -j\omega M_{21} & Z_2 \end{pmatrix} \tag{60}$$

$$= (\mathbf{B}\mathbf{Y_c}\mathbf{A}^{+})^{-1} = \begin{pmatrix} -Y_{12} - Y_{13} & Y_{12} \\ Y_{21} & -Y_{21} - Y_{23} \end{pmatrix}^{-1} \tag{61}$$

$$= \frac{-1}{Y_{12}Y_{23} + Y_{13}Y_{21} + Y_{13}Y_{23}} \begin{pmatrix} Y_{21} + Y_{23} & Y_{12} \\ Y_{21} & Y_{12} + Y_{13} \end{pmatrix}, \tag{62}$$

where

$$\mathbf{A} = \begin{pmatrix} 1 & 0 & -1 \\ 0 & 1 & -1 \end{pmatrix} \tag{63}$$

$$\mathbf{B} = \begin{pmatrix} 1 & 0 & 0 \\ 0 & 1 & 0 \end{pmatrix} \tag{64}$$

$$\mathbf{A}^{+} = \mathbf{B}^{T} = \begin{pmatrix} 1 & 0 \\ 0 & 1 \\ 0 & 0 \end{pmatrix}. \tag{65}$$

Here, a π-type equivalent circuit shown in Fig. 20 is utilized for the parameter extraction. Each parameter in Fig. 20, *i.e.*, Z_1, Z_2, M_{12}, Y_{sub1}, Y_{sub2}, Y_{sub3}, can be calculated by Eqs.(56)(57)(58)(60)(62).

To demomstrate this method, left-right asymmetry is evaluated for symmetric and asymmetric inductors as shown in Fig. 21. As I described, symmetric inductors are often used for differential topology of RF circuits, *e.g.*, voltage controlled oscillator, low noise amplifier, mixer. Asymmetry of inductors often cause serious degradation in performances, *e.g.*, IP_2 of LNA. The symmetric inductor shown in Fig. 21(a) is ideally symmetric. The asymmetric inductor shown in Fig. 21(b) has the same spiral structure as Fig. 21(a), but it has an asymmetric shape of ground loop.

Left- and right-half inductances are compared, which are calculated by the following equation.

$$L_n = \frac{\mathrm{Im}\,[Z_n]}{\omega} \qquad (n = 1, 2) \tag{66}$$

Fig. 22 shows the results. The ideally symmetric inductor has only 1.5% of mismatch in inductance. On the other hand, the asymmetric inductor has 4.0% of mismatch as shown in Fig. 22(b). This can be utilized to characterize symmetric inductors in consideration of asymmetry.

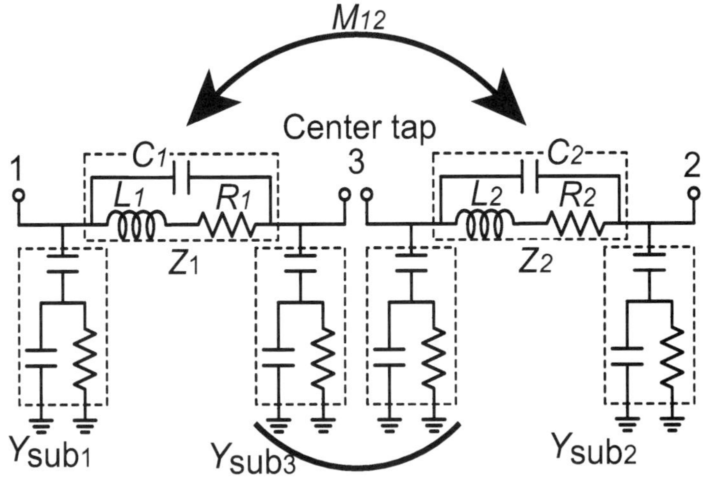

Fig. 20. π-type equivalent circuit of the 3-port inductor.

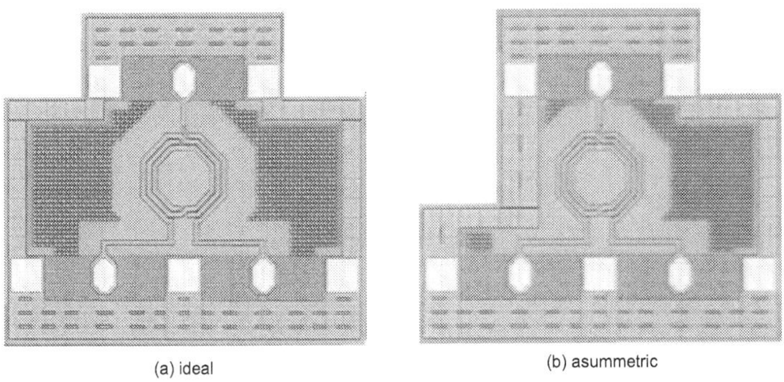

Fig. 21. Microphotograph of the center-tapped inductors.

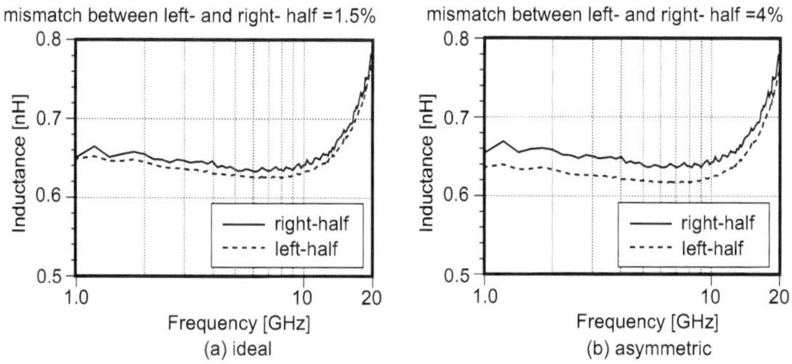

Fig. 22. Inductances of the center-tapped inductors.

4. References

Danesh, M. & Long, J. R. (2002). Differentially driven symmetric microstrip inductors, *IEEE Trans. Microwave Theory Tech.* **50**(1): 332–341.

Fujumoto, R., Yoshino, C. & Itakura, T. (2003). A simple modeling technique for symmetric inductors, *IEICE Trans. on Fundamentals of Electronics, Communications and Computer Sciences* **E86-C**(6): 1093–1097.

Kamgaing, T., Myers, T., Petras, M. & Miller, M. (2002). Modeling of frequency dependent losses in two-port and three-port inductors on silicon, *IEEE MTT-S Int. Microwave Symp. Digest*, Seattle, Washington, pp. 153–156.

Kamgaing, T., Petras, M. & Miller, M. (2004). Broadband compact models for transformers integrated on conductive silicon substrates, *Proc. IEEE Radio Frequency Integrated Circuits (RFIC) Symp.*, pp. 457–460.

Long, J. R. & Copeland, M. A. (1997). The modeling, characterization, and design of monolithic inductors for silicon RF IC's, *IEEE Journal of Solid-State Circuits* **32**(3): 357–369.

Niknejad, A. M. & Meyer, R. G. (1998). Analysis, design and optimization of spiral inductors and transformers for Si RF IC's, *IEEE Journal of Solid-State Circuits* **33**(10): 1470–1481.

Tatinian, W., Pannier, P. & Gillon, R. (2001). A new 'T' circuit topology for the broadband modeling of symmetric inductors fabricated in CMOS technology, *Proc. IEEE Radio Frequency Integrated Circuits (RFIC) Symp.*, Phoenix, Arizona, pp. 279–282.

Watson, A. C., Melendy, D., Francis, P., Hwang, K. & Weisshaar, A. (2004). A comprehensive compact-modeling methodology for spiral inductors in silicon-based RFICs, *IEEE Trans. Microwave Theory Tech.* **52**(3): 849–857.

15

Mixed-Domain Fast Simulation of RF and Microwave MEMS-based Complex Networks within Standard IC Development Frameworks

Jacopo Iannacci
Fondazione Bruno Kessler – FBK, MemSRaD Research Unit
Italy

1. Introduction

MEMS technology (MicroElectroMechanical-System) has been successfully employed since a few decades in the sensors/actuators field. Several products available on the market nowadays include MEMS-based accelerometers and gyroscopes, pressure sensors and micro-mirrors matrices. Beside such well-established exploitation of MEMS technology, its use within RF (Radio Frequency) blocks and systems/sub-systems has been attracting, in recent years, the interest of the Scientific Community for the significant RF performances boosting that MEMS devices can enable. Several significant demonstrators of entirely MEMS-based lumped components, like variable capacitors (Hyung et al., 2008), inductors (Zine-El-Abidine et al., 2003) and micro-switches (Goldsmith et al., 1998), are reported in literature, exhibiting remarkable performance in terms of large tuning-range, very high Q-Factor and low-loss, if compared with the currently used components implemented in standard semiconductor technology (Etxeberria & Gracia, 2007, Rebeiz & Muldavin, 1999). Starting from the just mentioned basic lumped components, it is possible to synthesize entire functional sub-blocks for RF applications in MEMS technology. Also in this case, highly significant demonstrators are reported and discussed in literature concerning, for example, tuneable phase shifters (Topalli et al., 2008), switching matrices (Daneshmand & Mansour, 2007), reconfigurable impedance matching networks (Larcher et al., 2009) and power attenuators (Iannacci et al., 2009, a). In all the just listed cases, the good characteristics of RF-MEMS devices lead, on one side, to very high-performance networks and, on the other hand, to enabling a large reconfigurability of the entire RF/Microwave systems employing MEMS sub-blocks. In particular, the latter feature addresses two important points, namely, the reduction of hardware redundancy, being for instance the same Power Amplifier within a mobile phone suitable both in transmission (Tx) and reception (Rx) (De Los Santos, 2002), and the usability of the same RF apparatus in compliance with different communication standards (like GSM, UMTS, WLAN and so on) (Varadan, 2003). Beside the exploitation of MEMS technology within RF transceivers, other potentially successful uses of Microsystems are in the Microwave field, concerning, e.g., very compact switching units, especially appealing to satellite applications for the very reduced weight (Chung et al., 2007), and phase shifters in order to electronically steer short

and mid-range radar systems for the homeland security and monitoring applications (Maciel et al., 2007).
Given all the examples reported above, it is straightforward that the employment of a proper strategy in aiming at the RF-MEMS devices/networks optimum design is a key-issue in order to gain the best benefits, in terms of performance, that such technology enables to address. This is not an easy task as the behaviour of RF-MEMS transversally crosses different physical domains, namely, electrical, mechanical and electromagnetic, leading to a large number of trade-offs between mechanical and electrical/electromagnetic parameters, that typically cannot be managed within a unique commercial simulation tool.
In this chapter, a complete approach for the fast simulation of single RF-MEMS devices as well as of complex networks is presented and discussed in details. The proposed method is based on a MEMS compact model library, previously developed by the author, within a commercial simulation environment for ICs (integrated circuits). Such software tool describes the electromechanical mixed-domain behaviour typical of MEMS devices. Moreover, through the chapter, the electromagnetic characteristics of RF-MEMS will be also addressed by means of extracted lumped element networks, enabling the whole electromechanical and electromagnetic design optimization of the RF-MEMS device or network of interest. In particular, significant examples about how to account for the possible non-idealities due to the employed technology as well as for post-processing steps, like the encapsulation of the MEMS within a package, will be reported. The optimization methodology, along with practical hints reported in this chapter, will help the RF-MEMS designer in the fast and proficient reaching of the optimum implementation that maximizes the performance of the device/network he wants to realize within a certain technology.

2. MEMS Compact Model Software Library

The MEMS compact model library adopted in the next pages, for the simulation of RF-MEMS devices and networks, has been previously developed by the author within the Cadence™ IC framework by using the VerilogA© HDL-based (hardware description language) syntax (Jing et al., 2002). The library features basic components, that are described by suitable mathematical models, and that connect with the surrounding elements by means of a reduced number of nodes. This enables the composition of complex MEMS devices geometries at schematic-level, as it is usually done when dealing with standard electronic circuits. The most important components available in the library are the rigid plate electrostatic transducers (realizing suspended air-gaps) and the flexible straight beam defining the elastic suspensions. Beside such main elements, the library also includes anchoring points and mechanical stimuli (like forces and displacements) in order to apply the proper boundary conditions to the analyzed MEMS structure schematic. The air-gap and flexible beam models are described more in details in the following two subsections.

2.1 Suspended Rigid Plate Electromechanical Transducer

Being this element a rigid body, the mechanical model is rather simple as it is based on the forces/torques balancing between the four plate vertexes, where the nodes are placed and where the plate is connected to other elements, and the centre of mass (Fedder, 2003). The model includes 6 DOFs (degrees of freedom) at each vertex, namely, 3 linear displacements and 3 rotation angles around the axes. Fig. 1 shows the schematic of the rigid plate in a generic position in space where all the DOFs are highlighted for each of the 4 vertexes labelled as *NW, NE, SE* and *SW* (*North-West, North-East, South-East* and *South-West*, respectively). The forces/torques applied to each node are transferred and summed into the centre of mass (*CM* in Fig. 1) according to the well-known equation of dynamics:

$$F = mA \tag{1}$$

where F is the applied force, m the mass of the plate and A its acceleration in a certain direction. The force/torque contributions are summed separately depending on the DOF/DOFs involved.

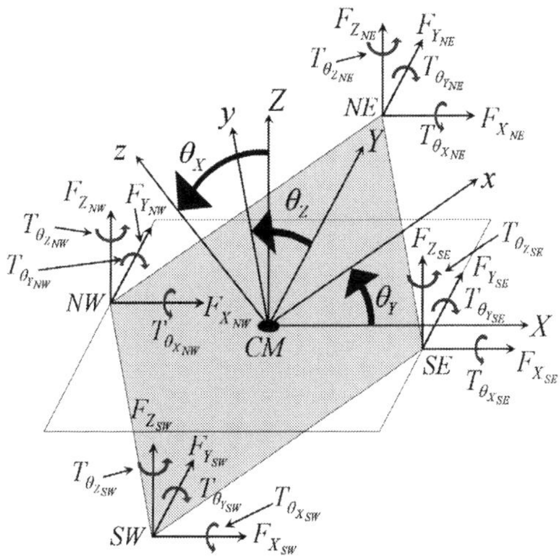

Fig. 1. Schematic of the rigid suspended plate in a generic position with all the 24 DOFs highlighted (6 DOFs per each vertex, namely, 3 linear DOFs and 3 rotational DOFs).

The rigid plate element also includes a contact model that manages the collapse onto the underneath electrode (pull-in) and the transduction between the electrical and mechanical domain, accounting for the capacitance and the electrostatic attractive force, between the suspended plate and the underneath electrode, when a biasing voltage is applied to them. Such magnitudes are calculated starting from well-known basic formulae, used in electrostatics, that have been extended to a double integral closed form, accounting for the most generic cases, when the plate assumes non-parallel positions with respect to the substrate. Given this consideration, the capacitance and electrostatic force are expressed as follows:

$$C = \frac{\varepsilon}{\sigma} \int_{-\frac{W}{2}}^{\frac{W}{2}} \int_{-\frac{L}{2}}^{\frac{L}{2}} \frac{dxdy}{Z(x, y, \theta_X, \theta_Y, \theta_Z)} \tag{2}$$

$$F = \frac{1}{2}\frac{\varepsilon V^2}{\sigma^2}\int_{-\frac{W}{2}}^{\frac{W}{2}}\int_{-\frac{L}{2}}^{\frac{L}{2}}\frac{dxdy}{Z(x,y,\theta_X,\theta_Y,\theta_Z)^2} \tag{3}$$

where ε is the permittivity of air, W and L are the plate dimensions, V is the voltage applied between the two plates and σ is a coefficient that accounts for the curvature of the electric field lines, occurring when the plate is tilted (i.e. non-parallel to the substrate). Note that the punctual distance Z between the suspended plate and the underlying electrode depends on the coordinates of each point integrated over the plate area and on the three rotation angles θ_X, θ_Y and θ_Z. The electrostatic transduction model also accounts for the effects due to the presence of holes on the plate surface, needed in order to ease the sacrificial layer removal, and to the fringing effects due to the distortion of the electric field lines in the vicinity of plate and holes edges. Finally, the description of the plate dynamics is completed by a model accounting for the viscous damping effect due to the air friction. Such model is based on the squeeze-film damping theory, and takes into account the presence of holes on the plate area. All the just listed rigid plate model features are not described here but are available in details in (Iannacci, 2007), together with their validation against FEM (Finite Element Method) simulated results and experimental data.

2.2 Flexible Straight Suspending Beam

The flexible straight beam model is based on the theory of elasticity (Przemieniecki, 1968) and the deformable suspension is characterized by two nodes, one per each end, including 6 DOFs, 3 linear and 3 angular deformations (or torques). Consequently, the beam has totally 12 DOFs as the schematic in Fig. 2 shows, and the whole static and dynamic behaviour is expressed by the following constitutive equation:

$$\mathbf{F} = \mathbf{K}\mathbf{X} + \mathbf{C}\dot{\mathbf{X}} + \mathbf{M}\ddot{\mathbf{X}} \tag{4}$$

where **F** is the 12x1 vector of forces/torques corresponding to the 12 DOFs reported in Fig. 2, **K** is the Stiffness Matrix, describing the elastic behaviour of each DOF, **M** is the Mass Matrix, accounting for the inertial behaviour of each DOF and **C** is the Damping Matrix, modelling the viscous damping effect. Moreover, it must be noticed that **K**, **C**, and **M** are multiplied by the 12x1 vector of linear/angular displacements **X**, and by its first and second time derivatives, respectively, being the latter two the vectors of velocity and acceleration. It is straightforward that (4) is a generalization of (1) accounting for the whole behaviour of the flexible beam. The **C** matrix is obtained by applying the same squeeze-film damping model adopted in the rigid plate. Finally, the beam model is completed by the electromechanical transduction model that accounts for the capacitance and electrostatic attractive force between the suspended deformable beam and the substrate. It is similar to the one reported in Subsection 2.1, even though it has been modified in order to account for the deformability of the beam. More details about the beam model and its validation are available in (Iannacci, 2007).

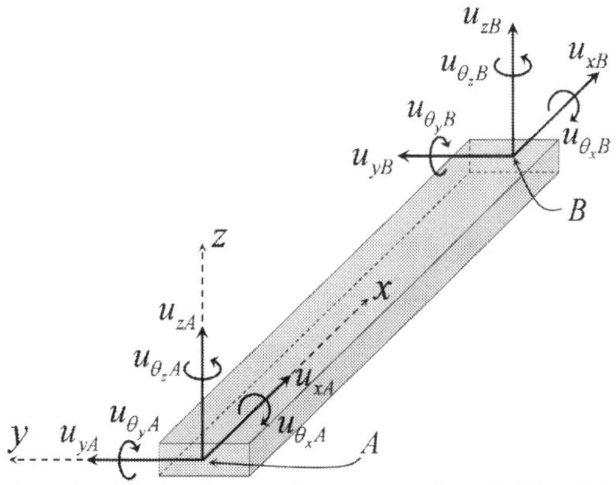

Fig. 2. Schematic of the 12 DOFs flexible straight beam. The 6 DOFs (3 linear and 3 angular) at each of the ends *A* and *B* are visible.

3. RF Modelling of a MEMS-based Variable Capacitor

In this section the complete modelling approach involving the RF and electromechanical behaviour of MEMS devices is introduced and discussed. A lumped element network describing the intrinsic RF-MEMS device and all the surrounding parasitic effects will be extracted from S-parameter measured datasets. Moreover, the MEMS device mechanical properties and electromechanical experimental characteristics will be exploited in order to prove the correctness of the RF modelling previously performed.

The specific analyzed RF-MEMS device is a variable capacitor (varactor) manufactured in the FBK RF-MEMS surface micromachining technology (Iannacci et al., 2009, a). An experimental 3D view obtained by means of an optical profiling system is reported in Fig. 3.

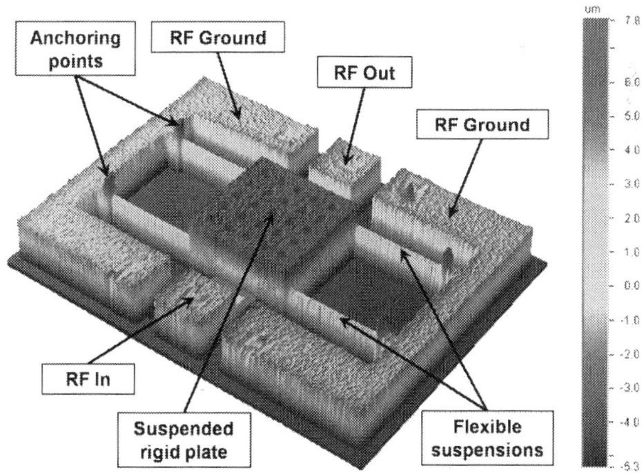

Fig. 3. 3D view of the studied RF-MEMS varactor obtained by means of an optical profiling system. The colour scale represents the vertical height of the sample.

The variable capacitance that loads the RF line (shunt-to-ground) is realized by a gold plate kept suspended over the underneath fixed electrode by four flexible straight beams. Depending on the DC bias applied between the two plates, the gold one gets closer to the substrate because of the electrostatic attraction, eventually collapsing onto it when the pull-in is reached, thus leading to the maximum capacitance value.

3.1 Equivalent Lumped Element Network Extraction

The lumped element network extraction, that is going to be discussed, starts from measured S-parameter datasets (2 ports) collected, on the same sample of Fig. 3, onto a probe station with GSG (ground-signal-ground) probes and an HP 8719C VNA (vector network analyzer) in the frequency range 200 MHz - 13.5 GHz. The controlling DC voltage that biases the suspended MEMS plate is applied directly to the RF probes by means of two bias-Tees. The DUT (device under test) is biased at different (constant) voltage levels. The performed VNA calibration is a SOLT (short, open, load, thru) (Pozar, 2004) on a commercial impedance standard substrate (ISS), i.e. the reference planes are brought to the GSG tips of the two probes. Consequently, the collected S-parameters include the behaviour of the intrinsic variable capacitor (i.e. the MEMS suspended plate) as well as the contribution due to the input/output access CPWs (see Fig. 3) plus parasitic effects, i.e. no de-embedding has been performed. Given these assumptions, we have exploited a well-known technique, usually adopted in microwave transistor modelling (Dambrine et al., 1988) based on the extraction of lumped parasitic elements that are wrapped around the intrinsic device. Fig. 4 shows the schematic of the intrinsic MEMS variable capacitor impedance and of the wrapping lumped element network accounting for the surrounding parasitic effects.

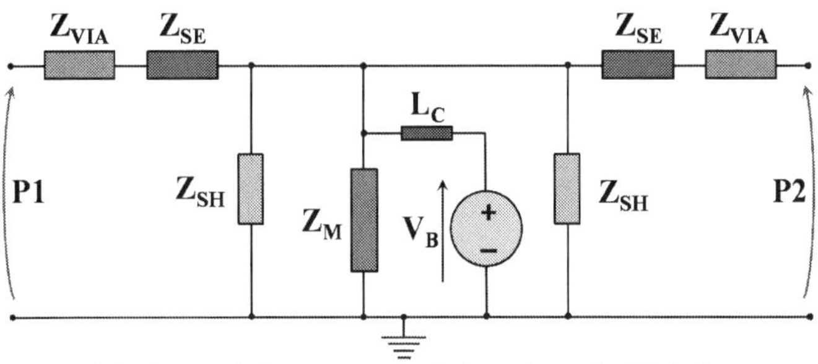

Fig. 4. Schematic of the lumped element network describing the RF behaviour of the device reported in Fig. 3. The network includes the intrinsic MEMS device and the parasitic effects.

The intrinsic MEMS impedance is indicated with $\mathbf{Z_M}$, while $\mathbf{Z_{SE}}$ and $\mathbf{Z_{SH}}$ model the impedance of the access CPWs at the ports **P1** and **P2**. Furthermore, $\mathbf{Z_{VIA}}$ models the impedance due to the parasitic effects introduced by the gold to multi-metal through vias (explained in details later) while L_C is a choke inductor (1 mH) necessary in the Spectre simulations to decouple the DC bias from the RF signal. The lumped elements composing $\mathbf{Z_M}$, $\mathbf{Z_{SE}}$, $\mathbf{Z_{SH}}$ and $\mathbf{Z_{VIA}}$ are shown in Fig. 5.

The intrinsic MEMS variable capacitor is modelled as a shunt to ground capacitance (C_{MEMS}), in parallel with a resistor accounting for small dielectric losses ($\mathbf{R_{MEMS}}$) and in series with an inductance (L_{MEMS}) accounting for the contribution of the four flexible beam suspensions (reported in Fig. 5-a). The accessing CPWs are modelled according to a well-known lumped network scheme (Pozar, 2004) shown in Fig. 5-b. It relies on a series RL section, accounting for the resistive losses within the metal and the line inductance respectively, and a parallel RC shunt section to ground, modelling the losses within the substrate and the capacitive coupling between the signal and ground planes through the air and the substrate.

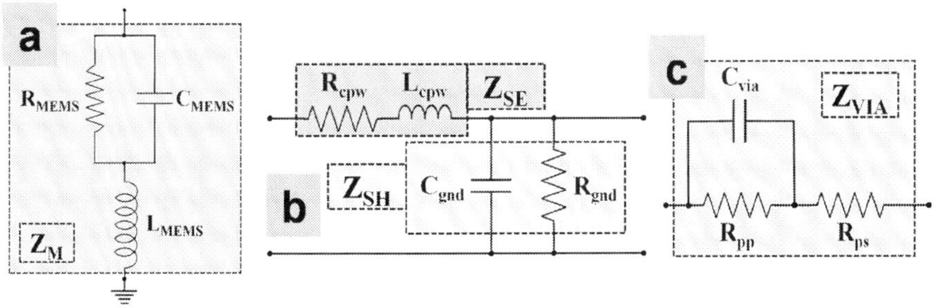

Fig. 5. Sub-networks composing the lumped element network of Fig. 4. a) Intrinsic MEMS variable capacitor; b) Input/output CPW-like line (see Fig. 3); c) Through-oxide via model accounting for the technology non-idealities.

Finally, the network of Fig. 5-c accounts for the parasitic effects due to a technology issue linked to the opening of vias through the oxide. Because of an inappropriate time end-point of the dry etching recipe performed on the batch, a very thin titanium oxide layer lays on the vias deteriorating the quality of the metal-to-metal transition (Iannacci et al., 2009, a). Such unwanted layer introduces additional losses and a series parasitic large capacitance that mainly affects the RF behaviour of the variable capacitor in the low-frequency range (as it will be discussed later in this section). Looking at Fig. 5-c, this non-ideality is modelled with a capacitance ($\mathbf{C_{via}}$) in parallel with a resistor ($\mathbf{R_{pp}}$), that models the losses in the low frequency range, plus a series resistance ($\mathbf{R_{ps}}$) accounting for the losses through the whole frequency span. Once the whole topology of the lumped element network is fixed, the specific values of all its components are tuned by using the optimization software tool available within the Agilent ADS^{TM} framework (Iannacci et al., 2007). Suitable targets aiming at the reduction of the difference between measured and modelled S-parameters are defined. The first optimization run is performed with the S-parameters measured at 0 V bias. The optimized value of the intrinsic MEMS variable capacitor is compared with the analytical one to verify the consistency of the optimizer output. Other optimization runs are performed replacing the target of the first run with the S-parameters measured at applied voltage of 1.25 V, 2.5 V, 3.75 V and so on up to 25 V, i.e. beyond the pull-in voltage of the DUT of Fig. 3 that is around 15 V (see next subsection). The consistency of the extracted lumped element values is monitored step by step. To do this, the extracted intrinsic MEMS capacitance is cross-checked with the analytical value, computed for each voltage, from the vertical displacement known after the experimental measurements (see next subsection). All the element values of the network in Fig. 4, excluded $\mathbf{C_{MEMS}}$ and $\mathbf{R_{MEMS}}$, do not show any significant change with the applied voltage. Once all the lumped element values are determined, they are kept fixed and only $\mathbf{C_{MEMS}}$ and $\mathbf{R_{MEMS}}$ are allowed to change. The extracted values for all the fixed elements composing the network of Fig. 4-5 are reported in Table 1.

Input/output CPW				Via parasitic effects			MEMS
R_{cpw}	L_{cpw}	C_{gnd}	R_{gnd}	C_{via}	R_{pp}	R_{ps}	L_{MEMS}
100 mΩ	122 pH	30 fF	840 GΩ	134 pF	2.82 Ω	700 mΩ	15 pH

Table 1. Extracted value of the fixed lumped elements composing the sub-networks of Fig. 5.

The four elements composing the CPW short lines show typical values for such a structure realized in a highly conductive metal onto a high-resistivity silicon substrate, as in the case of the DUT. Differently, the parasitic effects introduced by the non-ideal through-oxide vias are rather significant, being the resistive loss quite large (700 mΩ and 2.82 Ω) as well as the C_{via} (134 pF). Finally, the series inductance L_{MEMS} included in the intrinsic RF-MEMS device sub-network (see Fig. 5-a) is 15 pH. The two missing lumped elements in Table 1 are C_{MEMS} and R_{MEMS} as they change depending on the controlling DC voltage applied to the MEMS device. Table 2 reports their extracted values for a few applied voltages in the RF-MEMS varactor not actuated state (minimum capacitance), while Table 3 reports six cases in which the varactor is actuated (maximum capacitance).

V_B = 0 V		V_B = 3.75 V		V_B = 10 V		V_B = 11.25 V	
R_{MEMS}	C_{MEMS}	R_{MEMS}	C_{MEMS}	R_{MEMS}	C_{MEMS}	R_{MEMS}	C_{MEMS}
645 GΩ	160 fF	956 GΩ	185 fF	44.5 GΩ	190 fF	111 GΩ	192 fF

Table 2. Extracted R_{MEMS} and C_{MEMS} values (see Fig. 5-a) for different applied bias levels in the varactor not actuated state (low capacitance).

V_B = 25 V		V_B = 22.5 V		V_B = 20 V	
R_{MEMS}	C_{MEMS}	R_{MEMS}	C_{MEMS}	R_{MEMS}	C_{MEMS}
269 GΩ	862 fF	228 GΩ	804 fF	91 GΩ	755 fF
V_B = 17.5 V		V_B = 15 V		V_B = 12.5 V	
R_{MEMS}	C_{MEMS}	R_{MEMS}	C_{MEMS}	R_{MEMS}	C_{MEMS}
276 GΩ	683 fF	268 GΩ	636 fF	277 GΩ	551 fF

Table 3. Extracted R_{MEMS} and C_{MEMS} values (see Fig. 5-a) for different applied bias levels in the varactor actuated state (high capacitance).

Concerning the C_{MEMS} extracted values in the MEMS not actuated state there is a good agreement with the analytical ones. Indeed, by applying the well-known formula for a parallel plate capacitor, where the area of the DUT is 220x220 μm^2 and the distance between the electrodes is about 2.7 μm (see next subsection), the capacitance value is ~160 fF as extracted with the method here discussed. Focusing now on the C_{MEMS} extracted in the actuated state (Table 3) it has to be highlighted that the maximum capacitance is always rather low compared to the nominal one. Indeed, when the MEMS suspended plate collapses onto the substrate there is a ~400 nm thick oxide layer between it and the underlying electrode, leading to a maximum capacitance of about 4 pF. However, the extracted values show a C_{MAX} about 5 times smaller (862 fF) than the ideal one. Such reduction is mainly caused by two factors, namely, the surfaces roughness and the residual stress within the suspended gold (Iannacci et al., 2009, b). The roughness of the surfaces coming into contact (in this case the gold plate and the underneath oxide) lead to the presence of air between the two faces also when the switch is actuated. This causes the C_{MAX} to reduce as it is not anymore determined by the oxide layer only, but is given by the contribution of two series capacitors, one due to the oxide layer and the second one due to the residual air layer. Moreover, the mechanical stress that accumulates within the suspended gold during the release step (performed in plasma oxygen), usually is not uniform along the vertical dimension (stress gradient). This causes the central plate of Fig. 3 to be not perfectly planar but rather arched, thus leading to a further reduction of the contact surface and, consequently, of the C_{MAX}. The two just mentioned non-idealities are accounted for by including in the simulations and analytical calculations a constant equivalent air gap as Fig. 6 shows schematically.

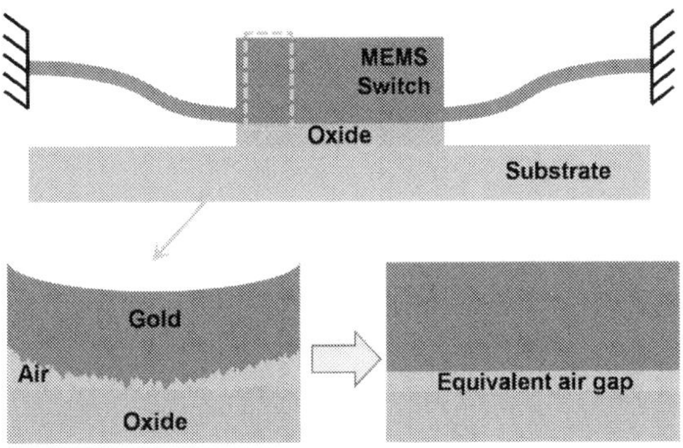

Fig. 6. *Top image*: Schematic cross-section of the actuated RF-MEMS varactor (see Fig. 3). *Bottom-left image*: Close up of one part of the actuated switch highlighting the surface roughness and the gold bowing induced by the stress gradient (both the effects are exaggerated). *Bottom-right image*: Equivalent air gap included in the simulations accounting for the just mentioned non-idealities.

After inverting the formula for the oxide and air series capacitances and using the C_{MAX} extracted value with a biasing level of 15 V (see Table 3), close to the plate release (pull-out), an equivalent air gap of 590 nm is extracted. The correctness of such value will be proven in the next subsection by means of electromechanical simulations. A final consideration has to be considered concerning the $\mathbf{R_{MEMS}}$, reported in Tables 2 and 3, that in all the studied cases shows a very large value, that indicates negligible resistive losses of the intrinsic RF-MEMS varactor. Given this assumption, the $\mathbf{R_{MEMS}}$ can be fixed to a certain value (e.g. 100 GΩ) in all the cases reported in Table 2 and 3 without any accuracy loss of the proposed network.
The network of Fig. 4 has been simulated within ADS with the extracted values reported in Tables 2 and 3, and the results are compared to the S-parameter measurements. The simulated and measured S11 and S21 parameters (reflection and transmission, respectively) are reported for an applied controlling voltage of 3.75 V (varactor not actuated) in Fig. 7 and for an applied bias of 25 V (varactor actuated) in Fig. 8, where the good superposition of the

curves is clearly visible. Concerning the not actuated state (Fig. 7), the influence of the parasitic effects introduced by the through-oxide vias affects both the S11 and S21 parameters up to about 2 GHz, where the reflection presents a minimum around 1 GHz, while it should be monotone, and the transmission increases with the frequency. This behaviour confirms the presence of a large unwanted series capacitance on the RF signal path acting as a spurious DC signals block. On the other hand, in the actuated state (Fig. 8) the isolation (S21) is never better than about 7 dB due to the small value of C_{MAX} compared to the nominal one and caused by the technology non-idealities already discussed. All the assumptions made up to now in the RF modelling are going to be verified by means of the electromechanical modelling.

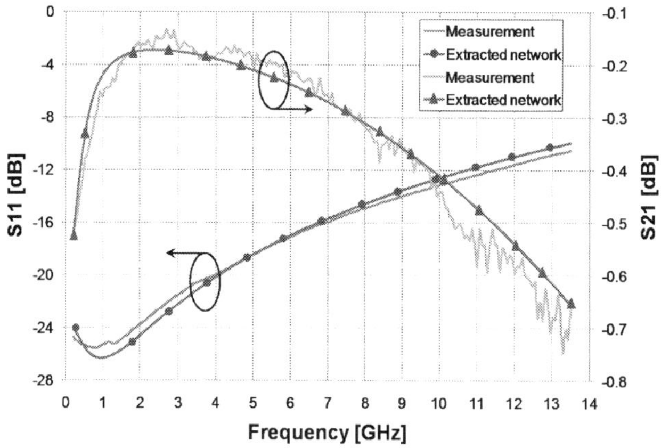

Fig. 7. Comparison of the measured and extracted (see Fig. 4) S11 and S21 parameter in the MEMS varactor not actuated state.

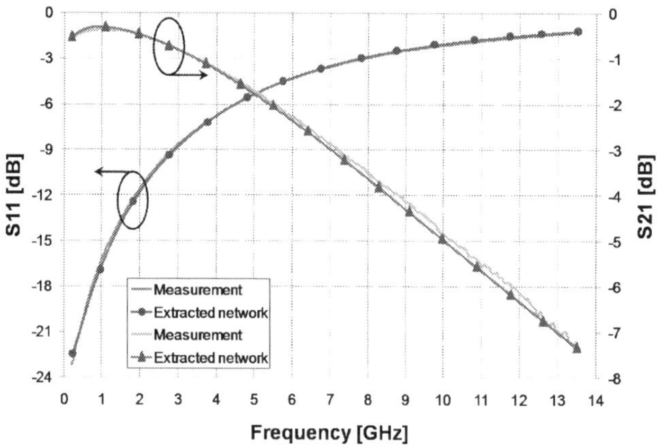

Fig. 8. Comparison of the measured and extracted (see Fig. 4) S11 and S21 parameter in the MEMS varactor actuated state.

3.2 Electromechanical Modelling and Verification

The electromechanical properties of the RF-MEMS varactor discussed up to now are observed once again, starting from experimental data, on the basis of which simulations are tuned and effective values accounting for the non-idealities are extracted. Verification and validation of the method discussed in previous subsection, concerning the RF domain, are reached, as the effective values extracted from electromechanical simulations coincide with the same values adopted in the RF simulations.

Fig. 9-top shows the Spectre schematic of the RF-MEMS varactor composed with the elementary MEMS models previously discussed in Section 2 for the simulation within Cadence. The central plate symbol is wired to four straight beams anchored at the opposite ends. The suspended plate is biased by means of a voltage source available within a Cadence library of standard components. Moreover, looking at Fig. 9-bottom, it is easy to identify the correspondence between the real MEMS device topology and the Spectre schematic.

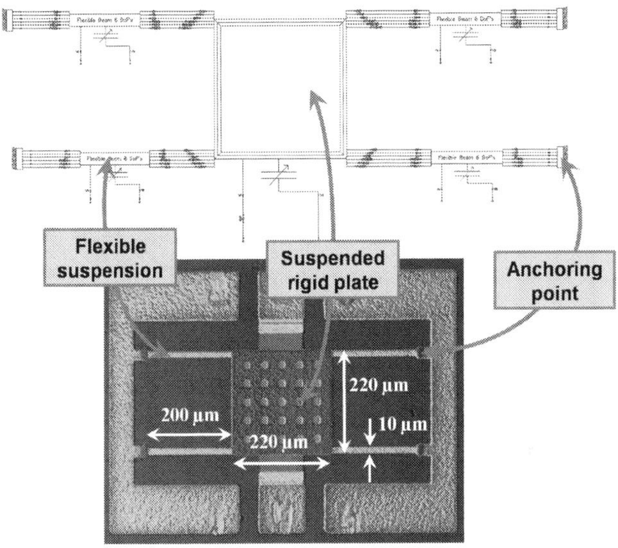

Fig. 9. Spectre schematic (top image) of the RF-MEMS varactor discussed here and assembled with the elementary components available in the software library discussed in Section 2. The correspondence between the schematic and the real device, reported in the top view measured with an optical profilometer (bottom image), is straightforward.

The RF-MEMS varactor sample of Fig. 3 and Fig. 9-bottom is measured in static regime by means of the afore-mentioned optical profilometer. A triangular symmetric voltage ranging from -20 V up to 20 V (zero mean value) with a frequency of 20 Hz is applied to the DUT. By changing the phase of the stroboscopic illuminator with respect to the biasing signal, it is possible to observe the vertical displacement of the DUT for different bias levels (Novak et al., 2003). This enables the acquisition of the whole experimental pull-in/pull-out characteristic. Subsequently, the schematic of Fig. 9-top is simulated within Spectre (DC simulation) in order to obtain the same pull-in/pull-out characteristic. A residual air gap of 590 nm is set in the simulation when the plate collapses onto the substrate. Such value comes from the extracted C_{MAX} discussed in previous subsection. Fig. 10 reports the measured and simulated pull-in/pull-out characteristic of the RF-MEMS varactor, showing a very good agreement of the two curves. In particular, the measured pull-in voltage (~15 V) and pull-out voltage (~9 V) are predicted very accurately by the compact models in Spectre. The characteristics of Fig. 10 show the typical hysteresis of MEMS devices.

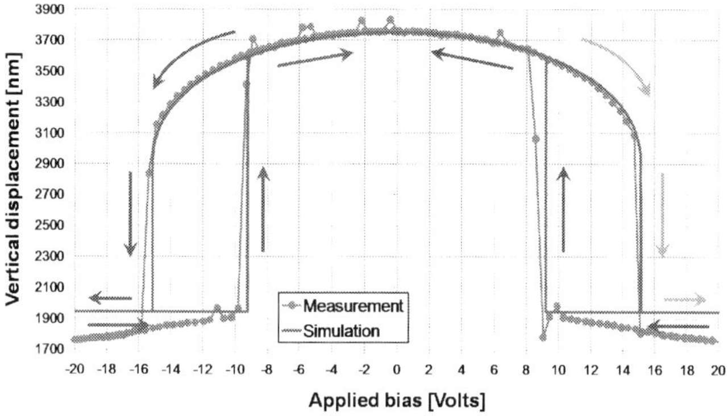

Fig. 10. Measured static pull-in/pull-out characteristic compared to the one simulated with the schematic of Fig. 9-top within Cadence (DC simulation in Spectre). Arrows help in identifying the pull-in/pull-out hysteresis.

More in details, the good agreement of the measured and simulated pull-in voltage confirms both that the elastic constant k is modelled correctly in the Spectre simulation, and that the initial air gap g is properly set (Iannacci, 2007). After this consideration, the good superposition of the measured and simulated pull-out voltage (V_{PO}) finally confirms that the residual air gap t_{air}, previously extracted from RF measurement, is correct since the V_{PO} depends on it as follows (Iannacci et al., 2009, b):

$$V_{PO} = \sqrt{\frac{2kg(t_{ox} + t_{air})(t_{air}\varepsilon_{ox} + t_{ox}\varepsilon_{air})}{A\varepsilon_{ox}\varepsilon_{air}}} \tag{5}$$

where t_{air} is the oxide layer thickness, A the electrodes area, ε_{ox} and ε_{air} the dielectric constant of the oxide and air, respectively. A further confirmation of the DUT non-idealities comes from the observation of Fig. 10. Starting from the pull-in voltage (~15 V) and rising up to 20 V, the vertical quote of the switch is not constant as it would be expected, but tends to decrease of about 200 nm. Interpretation of such an awkward behaviour is straightforward, by knowing that the profiling system determines each point of the pull-in/pull-out characteristic as the mean value of all the vertical quotes measured onto the plate surface. Because of the plate non-planarity schematically shown in Fig. 6, after the plate pulls-in, it tends to get more flat onto the underneath oxide as a result of the attractive force increase

due to the applied voltage rise. This also explains why the extracted C_{MAX} values reported in Table 3 are larger for higher applied bias levels.
In conclusion, a few more considerations are necessary to extend the applicability of the method discussed in previous pages. In the particular case discussed in this section, the electromechanical and electromagnetic simulation of the DUT was based upon an on-purpose software tool developed by the author (Iannacci et al., 2005). However, the same method that accounts for the RF-MEMS devices non-idealities here discussed, can be effectively exploited by relying on the use of commercial simulation tools (e.g. FEM-based electromechanical and electromagnetic tools like Ansys™, Coventor™, Ansoft HFSS™ and so on) as well as by simply performing analytical calculations, based on the constitutive equations describing the multi-physical behaviour of RF-MEMS. The benefits of the modelling method here discussed, when dealing with the RF-MEMS design optimization, are straightforward. First of all, in the early design stage, the designer has to deal with a large number of DOFs influencing the electromechanical and electromagnetic performances, hence leading to the identifications of several trade-offs. Availability of a fast analysis method, like the just presented one, enables the designer to quickly identify the main trends linked to the variation of the available DOFs, as well as the parameters that exhibit the most significant influence on the overall RF-MEMS device/network performances. Moreover, starting from the availability of a few experimental datasets, the discussed analysis can be tailored to the effective parameters accounting for the non-idealities of the chosen technology, rather than the nominal ones. This means that the use of FEM tools, typically very accurate but time consuming, can be reserved to the final design stage, when the fine optima are sought, while the rough optimum design can be easily and quickly addressed by following the method discussed in this chapter. Since the presented procedure can be implemented and parameterized with small effort within any software tool for mathematical calculation (e.g. MATLAB™), it is going to be synthetically reviewed and schematized as subsequent steps in the next subsection.

3.3 Summary of the Whole RF-MEMS Modelling Method

Starting from a lumped element description of the DUT (in this case an RF-MEMS varactor), like the one proposed in Fig. 4-5, the capacitance of the intrinsic MEMS device is known. In the case here discussed the experimental data are S-parameter measurements. However, the MEMS capacitance can also be determined by means of C-V (Capacitance vs. Voltage) measurements in AC regime, by exploiting an LCR-meter. In this case the wrapping network described in Fig. 4 is not necessary, and can be drastically simplified, as at low-frequency most of the lumped components there included are negligible. First of all, starting from the measured/extracted minimum capacitance C_{MIN} corresponding to a 0 V applied bias, the effective air gap g_1 can be extracted by inverting the well-known parallel plate capacitor formula, and the oxide capacitance can be considered negligible:

$$g_1 = \frac{\varepsilon_{air} A}{C_{MIN}} \qquad (6)$$

Differently, given the maximum measured/extracted capacitance in the pulled-in state (C_{MAX}), the effective air gap (t_{air1}) due to the surface roughness and gold bowing can be determined by inverting the formula of the oxide plus air series capacitance:

$$t_{air1} = \frac{\varepsilon_{air} A}{C_{MAX}} - \frac{\varepsilon_{air}}{\varepsilon_{ox}} t_{ox} \tag{7}$$

Let us now consider the cross-check of the extracted values by means of electromechanical measurements. Starting from the measured pull-in voltage V_{PI} and the maximum vertical displacement ΔZ, that in the case of Fig. 10 is the quote difference between 0 V and ±16 V applied bias (when the plate collapses onto the lower oxide layer), the effective elastic constant (k_{eff}) accounting for the influence of residual stress on the flexible suspensions is:

$$k_{eff} = \frac{27}{8} \frac{V_{PI}^2 \varepsilon_{air} A}{(\Delta Z + t_{ox})^3} \tag{8}$$

Also in this case the capacitance contribution of the oxide is neglected. Starting from the measured pull-out voltage V_{PO} and inverting its formula including the (8), the residual air gap t_{air2} is extracted as follows:

$$t_{air2} = -\frac{t_{ox}}{2}\left(1 + \frac{\varepsilon_{air}}{\varepsilon_{ox}}\right) + \sqrt{\frac{t_{ox}^2}{4}\left(1 - \frac{\varepsilon_{air}}{\varepsilon_{ox}}\right)^2 + \frac{V_{PO}^2 \varepsilon_{air} A}{2k_{eff}\Delta Z}} \tag{9}$$

Final verification of the derived effective parameters is performed by comparing their value extracted from electromagnetic/AC measurements and electromechanical experimental data. In particular, it has to be verified that:

$$g_1 \cong t_{ox} + t_{air2} + \Delta Z \tag{10}$$

$$t_{air1} \cong t_{air2} \tag{11}$$

Now that the complete method has been described, next sections will be focused on the report of a few significant examples of its application to modelling problems referred to RF-MEMS devices and network.

4. Mixed-Domain Simulation of a hybrid RF-MEMS/CMOS Voltage Controlled Oscillator (VCO)

One important feature of the discussed MEMS simulation tool is that it enables the analysis of blocks composed by different technologies, namely, RF-MEMS and standard CMOS, within the same Cadence schematic. To this purpose, the example reported in this section concerns the simulation of a hybrid Voltage Controlled Oscillator (VCO) (Tiebout, 2005).

The oscillator is designed in standard CMOS technology and implemented with the design-kit released by AMS© (0.35 µm HBT BiCMOS S35 technology, website: www.austriamicrosystems.com). Whereas, the varactors of the LC-tank are implemented in MEMS technology with the compact models previously shown. The Cadence schematic of the VCO is shown in Fig. 11.

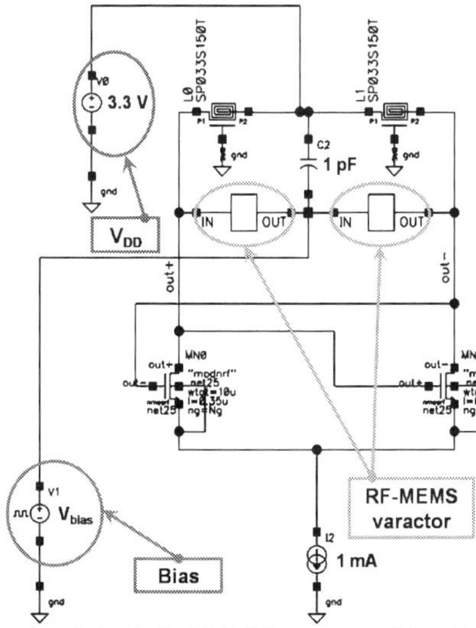

Fig. 11. Cadence schematic of the hybrid VCO composed by the CMOS oscillator in AMS technology and the RF-MEMS LC-tank.

The two symbols representing the tuneable capacitors are realized with a suspended rigid plate and four straight beams connected to its corners. Each of them corresponds to the Cadence schematic of Fig. 9-top. The two inductors within the LC-tank in Fig. 11 are also included in the design-kit provided by AMS and mentioned above. Two RF-MEMS varactors are included in the symmetric LC-tank scheme, decoupling the controlling voltage from the oscillator RF output. For the same reason, a capacitor (1 pF) is placed between the controlling voltage generator and the voltage supply (V_{DD} = 3.3 V). Depending on the bias applied to the common node between the RF-MEMS varactors, their capacitance changes and consequently the oscillation frequency of the overall VCO. Transient analysis is performed in Spectre for different bias levels lower than the pull-in of the structure in Fig. 9 top (i.e. 15.6 V). The VCO tuning characteristic (frequency vs. biasing voltage) is shown in Fig. 12. The capacitance of each RF-MEMS varactor and the corresponding VCO oscillation frequency are reported in Table 4. The just shown RF-MEMS/CMOS VCO implementation represents a meaningful example about the utilization of the mixed-domain simulation environment here proposed and discussed (Iannacci, 2007).

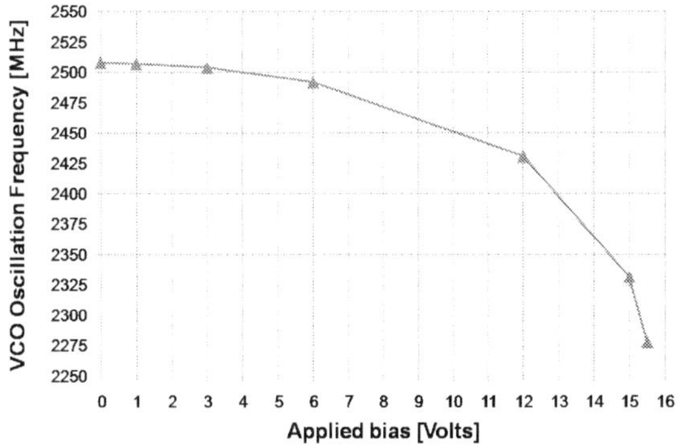

Fig. 12. VCO oscillation frequency vs. bias applied to the RF-MEMS varactors (tuning characteristic).

Bias level (V)	Capacitance (fF)	VCO Frequency (MHz)
0	597	2508
1	598	2507
3	601	2504
6	611	2492
12	671	2431
15	775	2332
15.5	838	2278

Table 4. VCO oscillation frequency depending of the bias level applied to the RF-MEMS varactors of the LC-tank.

5. Fast Simulation of a Reconfigurable RF-MEMS Power Attenuator

In this section the discussed MEMS compact model library is exploited in order to simulate the RF/electromechanical behaviour of a complex RF-MEMS network, namely, a multi-state reconfigurable RF/Microwave broad-band power attenuator. The network topology and performance have been already presented by the author (Iannacci et Al., 2009, a). The network is based on two resistive branches composed of 6 different resistances each, connected in series. Depending on the state (actuated/not-actuated) of 6 electrostatically controlled suspended gold membranes, it is possible to short selectively one or more resistances, thus modifying the power attenuation of the whole RF-MEMS network. Moreover, the two above mentioned branches can be selected/deselected by two SPDT (single pole double throw) stages in order to include one single resistive load or both in parallel, doubling, in turn, the number of achievable attenuation levels. A microphotograph of the whole fabricated network is reported in the top-left of Fig. 13, where the two resistive branches together with the SPDT sections are highlighted. Moreover, the top-right of Fig. 13 shows a 3D close-up of one branch composed of 6 resistances and 6 suspended membranes, and a further close-up of one single electrostatically controlled MEMS shorting switch. Both these images are obtained with an optical profiling system based on interferometry. The bottom part of Fig. 13 reports the schematic of the whole RF-MEMS network, composed with the compact models previously discussed, within Cadence for the Spectre simulations.

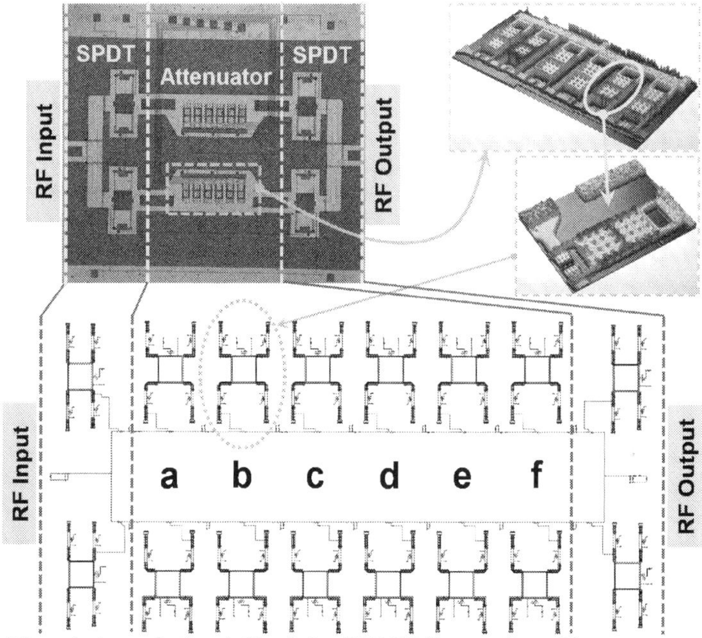

Fig. 13. Microphotograph (top-left) of the RF-MEMS reconfigurable attenuator and 3D measured profile of one of the 6 resistive loads branch and of one MEMS suspended membrane (top-right). Spectre schematic (bottom-image) of the whole network composed with the compact models discussed above. The 6 resistive loads are labelled with the letters "a,b,c,d,e,f". The correspondence between the real network and the schematic is highlighted.

The resistance value for each of the 6 loads, as it results from measurements, is reported in Table 5 (Iannacci et Al., 2009, a). The Spectre schematic is completed with extracted lumped element sections similar to the ones of Fig. 4 and 5 (too small to be distinguished in figure), accounting for the short CPW portions included in the network layout (see Fig. 13 top-left).

a	b	c	d	e	f
9.3 Ω	18.6 Ω	18.6 Ω	93 Ω	206 Ω	206 Ω

Table 5. Value of the 6 resistive loads included in each branch of the reconfigurable RF-MEMS attenuator of Fig. 13.

Mixed S-parameter/electromechanical simulations are performed in Spectre on the schematic of Fig. 13. In particular, Fig. 14 refers to the RF behaviour of the network when only one of the two branches is selected. Starting from the configuration introducing the maximum attenuation (i.e. none of the 6 membranes is actuated), the MEMS suspended membranes are actuated (pull-in) in sequence (1, 2,...6 actuated), showing that when a resistance is shorted the corresponding attenuation level decreases from DC up to 40 GHz.

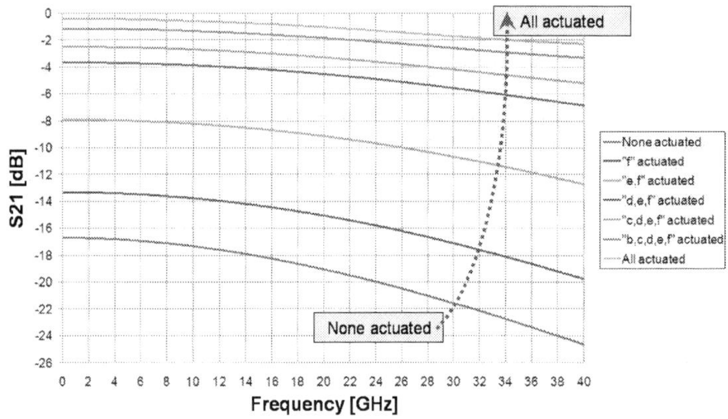

Fig. 14. S21 parameter behaviour of the RF-MEMS multi-state attenuator simulated in Spectre. When a MEMS membrane pulls-in, thus shorting the corresponding resistive load, the attenuation level decreases and the shift of the transmission parameter is proportional to the resistance value (see Table 5).

The same schematic has been simulated with both the resistive branches inserted (resistances in parallel). In this case the S-parameter simulation is performed at a single frequency (20 GHz) and the bias DC voltage, controlling each of the 6 shorting suspended membranes, is alternatively swept between 0 and 20 V. Fig. 15 shows the results highlighting the pull-in voltage of the membranes that is around 13 V.

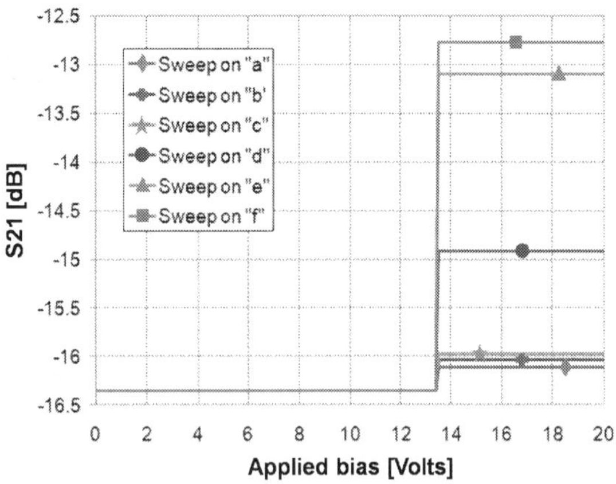

Fig. 15. S21 parameter behaviour simulated in Spectre at 20 GHz vs. the DC bias applied to the selecting suspended membranes. The attenuation shift depends on the resistance value.

The S21 parameter change depends on the value of the shorted resistive load. Moreover, it should be noted that the maximum attenuation level (i.e. none of the membranes actuated) is about 16.5 dB (as visible in Fig. 15 for applied voltage lower than the pull-in) while in Fig. 14 it is about 19 dB at 20 GHz. The reason for this difference is that the simulations reported in Fig. 15 refer to both the branches connected in parallel and, consequently, to a lower load resistance.

6. Lumped-Element Network of In-Package Coplanar Wave-Guide Structures

This last section is devoted to the description of the RF behaviour due to the package. Indeed, RF-MEMS devices (as well as MEMS in general) are very fragile against environmental factors (like moisture, dust particles, shocks and so on) due to their characteristics (Gilleo, 2005). Because of these motivations, RF-MEMS devices need to be encapsulated within a package that can just isolate them from the external environment, or even enhance their performance by ensuring specific working conditions. In the latter case, the vacuum condition within the packaged housing for a MEMS resonator increases dramatically its Q-Factor (Nguyen, 2004). In turn, application of a package to RF-MEMS devices introduces additional losses and impedance mismatch, due to the increased signal path and discontinuities, indeed affecting their performances. Given these considerations, the package design and fabrication has to be thought carefully in order to minimize its impact on the RF-MEMS devices/networks performance. The author already presented an approach to the electromagnetic (EM) optimization of the package layout for RF-MEMS within a given technology, based on the implementation of a parameterized 3D model within a commercial FEM-based EM tool, and validated against experimental data (Iannacci et Al., 2008). In this section, the focus is going to be concentrated on the RF simulation of the package based on lumped element networks, thus pushing forward the methodology discussed in previous pages, aiming at a complete description of RF-MEMS devices/networks. The structure to be analyzed is a standard CPW (Coplanar Wave-Guide) instead of complete RF-MEMS devices, as they are based on the CPW topology. To this purpose, a CPW has been simulated within the Ansoft HFSS™ EM tool in air at first, and then with the package model described in (Iannacci et Al., 2008). Both the CPW and package characteristics, as well as the wafer-to-wafer bonding technique, are based on the technology process available at the DIMES Research Centre (Technical University of Delft, the Netherlands) (Iannacci et Al., 2006). In particular, the package is based on vertical through wafer vias for the signal redistribution from the MEMS device wafer to the external world.
Fig. 16 shows the HFSS 3D schematic of an uncapped CPW (left-image) and of the same CPW with the package (right-image), where vertical vias and top CPW are visible (the package substrate was hidden to allow the vias view). The CPW reported in Fig. 16 has been first simulated within HFSS without any package. The silicon substrate thickness is 500 μm and its resistivity is 2 KΩ.cm. The CPW is 2 mm long, the signal line width, ground lines width and gap are 100 μm, 700 μm and 50 μm, respectively. Finally, the CPW is realized in a 2 μm thick electrodeposited copper layer. Subsequently, the CPW with package (Fig. 16-right) is simulated and, being the model parameterized, a few features, like vertical vias diameter and lateral distance between the signal and ground vias, were changed. The package is also realized with a 500 μm thick and 2 KΩ.cm silicon substrate and vertical through-wafer vias are opened with the deep reactive ion etching (DRIE) and filled

(electrodeposition) with copper. The top CPWs (see Fig. 16-right) are also made of copper. Their dimensions are the same of the uncapped CPW, apart from the length that is 500 μm, and have been also simulated in HFSS as standalone structures. A lumped element network describing the packaged transmission line is built and its components values are extracted with the ADS optimization tool as previously described in Subsection 3.1.

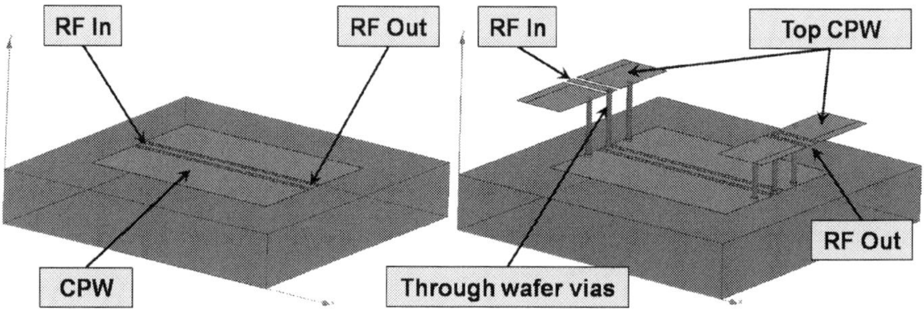

Fig. 16. HFSS schematic of an uncapped CPW (left-image) and of the same CPW with the package (right-image). The package substrate is removed to allow the view of vertical vias.

The extracted network schematic is shown in Fig. 17 where the blocks labelled as "**CPW**" and "**Top CPW**" are items available within ADS in order to link the data, simulated in HFSS and provided in Touchstone format, of the CPW of Fig. 16-left and of the top CPW (see Fig. 16-right), respectively. All the other lumped elements are placed in the schematic according to the expected behaviour of each part of the package, i.e. vertical vias, solder bumps, discontinuity between the top CPWs and vertical vias and interaction of the package with the EM field above the capped CPW.

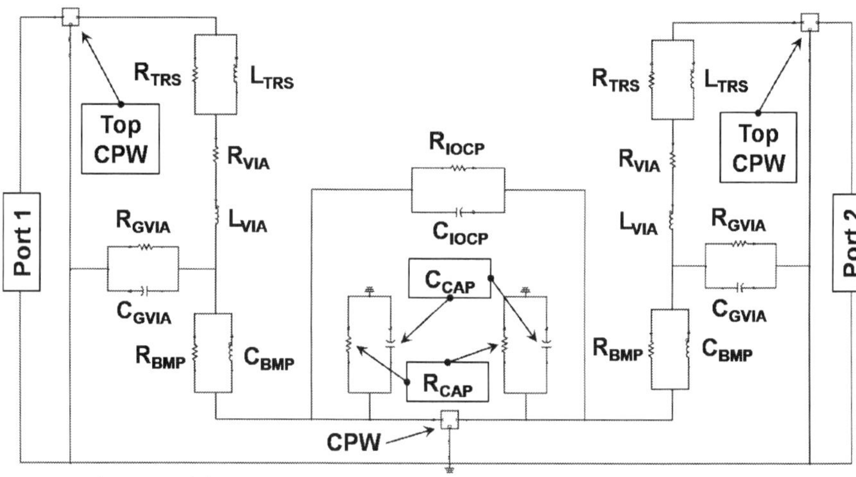

Fig. 17. Schematic of the lumped-element network describing the packaged CPW previously shown in Fig. 16-right.

The ground-signal-ground vertical vias are modelled according to the scheme of a standard CPW (Pozar, 2004) and the corresponding elements within the schematic of Fig. 17 are labelled as: R_{VIA}, L_{VIA}, R_{GVIA} and C_{GVIA}. The transitions between the top CPW and the vertical vias are modelled as a resistance and inductance in parallel (R_{TRS}, L_{TRS}) as well as the solder bumps connecting vertical vias with the capped CPW (R_{BMP}, L_{BMP}). Additional losses and capacitive coupling to ground, induced by the presence of the package above the CPW, are modelled with C_{CAP} and R_{CAP} and, finally, the direct input/output coupling through the cap is accounted for by R_{IOCP} and C_{IOCP}. As initial case, a package with a 500 μm thick silicon substrate, vertical vias diameter of 50 μm and lateral pitch of 250 μm (considered between the centre of the signal and of the ground vias) is taken into account. Starting from the HFSS simulation of such structure, the lumped elements value is extracted within ADS and reported in Table 6, thus validating the topology reported in Fig. 17 in the frequency range from 1 GHz up to 15 GHz.

R_{VIA}	L_{VIA}	R_{GVIA}	C_{GVIA}	R_{TRS}	L_{TRS}
110 mΩ	148 pH	630 MΩ	62.6 fF	331 mΩ	41 pH
R_{BMP}	L_{BMP}	C_{CAP}	R_{CAP}	R_{IOCP}	C_{IOCP}
9.07 Ω	55 pH	1 fF	200 GΩ	820 GΩ	17.4 fF

Table 6. Values extracted for the elements of the schematic reported in Fig. 17 for a 500 μm thick silicon package, with vias diameter of 50 μm and lateral pitch of 250 μm.

Fig. 18 reports the S11 and S21 parameters comparison between HFSS simulations of the packaged CPW and the network of Fig. 17 with the value reported in Table 6, showing a very good superposition of the curves.

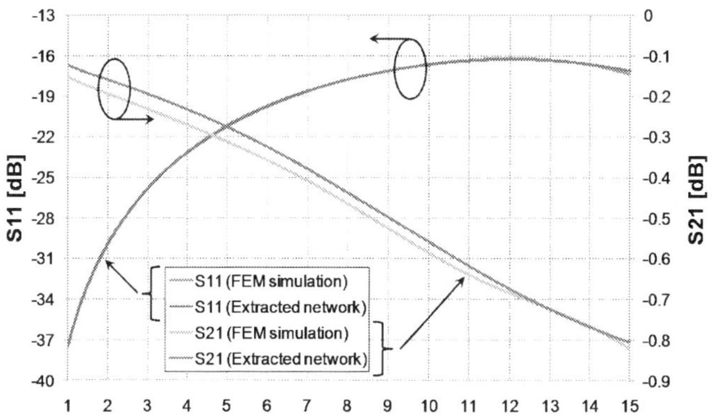

Fig. 18. Comparison of the simulated (in HFSS) and extracted network (see Fig. 17 and Table 6) S11 and S21 parameters in the frequency range from 1 GHz up to 15 GHz.

Subsequently, some critical technology degrees of freedom related to the package are alternatively modified in order to validate, on one side, the correctness of the topology reported in Fig. 17, and to analyze the influence of such variations on the network lumped components. Starting from the lateral via pitch, the whole structure is simulated in HFSS with a value of 200 μm and 300 μm, respectively, smaller and larger compared to the initial case discussed above. The ADS optimization is repeated for these cases and the only parameters allowed to change are $\mathbf{R_{GVIA}}$ and $\mathbf{C_{GVIA}}$. Their comparison concerning the three vias lateral pitch is reported in Table 7.

200 μm vias pitch		250 μm vias pitch		300 μm vias pitch	
R_{GVIA}	990 MΩ	R_{GVIA}	630 MΩ	R_{GVIA}	960 MΩ
C_{GVIA}	101 fF	C_{GVIA}	62.6 fF	C_{GVIA}	35.7 fF

Table 7. Values of the coupling capacitance and resistive loss between the signal and ground vias for different vias lateral pitches. The highlighted row corresponds to the most significant parameter exhibiting variations.

As expected, the coupling capacitance between the signal and ground vias increases when the lateral distance is smaller and decreased for a larger pitch. On the other hand, the resistive losses are so small that their variations can be neglected, as already mentioned in Subsection 3.1. However, such elements are kept in the network in order to extend its suitability to lossy substrates. Comparison of the S-parameters behaviour of the HFSS simulations and the network of Fig. 17 with the values reported in Table 7 (not reported here for sake of brevity) shows a good agreement as reported in Fig. 18. Another modified DOF is the via diameter. Starting from the capped CPW with lateral via pitch of 200 μm and silicon substrate thickness of 500 μm, via diameter is increased to 70 μm and 85 μm. In this case all the via parameters ($\mathbf{R_{VIA}}$, $\mathbf{L_{VIA}}$, $\mathbf{R_{GVIA}}$ and $\mathbf{C_{GVIA}}$) are allowed to change as well as the ones of the top CPW-to-via discontinuity ($\mathbf{R_{TRS}}$, $\mathbf{L_{TRS}}$) and via-to-solder bumps discontinuity ($\mathbf{R_{BMP}}$, $\mathbf{L_{BMP}}$). The extracted values are reported in Table 8.

50 μm via diameter		70 μm via diameter		85 μm via diameter	
R_{VIA}	110 mΩ	R_{VIA}	98 mΩ	R_{VIA}	56 mΩ
L_{VIA}	148 pH	L_{VIA}	82 pH	L_{VIA}	70 pH
R_{GVIA}	630 MΩ	R_{GVIA}	188 GΩ	R_{GVIA}	448 GΩ
C_{GVIA}	62.6 fF	C_{GVIA}	100 fF	C_{GVIA}	120 fF
R_{TRS}	331 mΩ	R_{TRS}	314 mΩ	R_{TRS}	100 mΩ
L_{TRS}	41 pH	L_{TRS}	10 pH	L_{TRS}	25 pH
R_{BMP}	9.07 Ω	R_{BMP}	3.15 Ω	R_{BMP}	2 Ω
L_{BMP}	55 pH	L_{BMP}	20 pH	L_{BMP}	20 pH

Table 8. Values of the via parameters, top CPW-to-via and via-to-solder bumps transitions for vertical vias diameter of 50 μm, 70 μm and 85 μm. The highlighted rows correspond to the most significant parameters exhibiting variations.

As final case, given the via diameter of 50 μm and the lateral pitch of 200 μm, the silicon package thickness is reduced to 400 μm and 300 μm. In this case all the via parameters ($\mathbf{R_{VIA}}$, $\mathbf{L_{VIA}}$, $\mathbf{R_{GVIA}}$ and $\mathbf{C_{GVIA}}$) are allowed to change as well as the additional coupling to ground and input/output elements ($\mathbf{C_{CAP}}$, $\mathbf{R_{CAP}}$, $\mathbf{R_{IOCP}}$ and $\mathbf{C_{IOCP}}$).

500 μm cap thickness		400 μm cap thickness		300 μm cap thickness	
R_{VIA}	110 mΩ	R_{VIA}	62 mΩ	R_{VIA}	20 mΩ
L_{VIA}	148 pH	L_{VIA}	54 pH	L_{VIA}	51 pH
R_{GVIA}	990 MΩ	R_{GVIA}	5 GΩ	R_{GVIA}	4.8 GΩ
C_{GVIA}	101 fF	C_{GVIA}	52 fF	C_{GVIA}	41 fF
C_{CAP}	1 fF	C_{CAP}	15 fF	C_{CAP}	1 fF
R_{CAP}	200 GΩ	R_{CAP}	225 GΩ	R_{CAP}	204 GΩ
R_{IOCP}	820 GΩ	R_{IOCP}	912 GΩ	R_{IOCP}	828 GΩ
C_{IOCP}	17.4 fF	C_{IOCP}	2.3 fF	C_{IOCP}	1 fF

Table 9. Values of the via parameters and additional coupling/loss due to the cap for a package thickness of 500 μm, 400 μm and 300 μm. The highlighted rows correspond to the most significant parameters exhibiting variations.

In conclusion, despite a few elements included in the network of Fig. 17 do not show significant changes, the most critical parameters (highlighted in Tables 7-9) change in compliance with physical consideration related to the package geometry variations in the FEM analyses. For example, the vias shunt (to ground) coupling capacitance decreases as the vias lateral pitch increases as well as when the cap thickness lowers. This proves the suitability of the chosen network (Fig. 17). Following the same approach, similar network topologies can be extracted referring to other frequency ranges, depending on the specific application the designer aims at.

7. Conclusion

In this chapter several aspects related to the mixed-domain electromechanical and electromagnetic simulation of RF-MEMS devices and network were reported. First of all, a fast simulation tool based on a lumped components MEMS model software library, previously developed by the author, was introduced and discussed. The elementary components, implemented in VerilogA programming language, within the Cadence IC development environment, are the flexible straight beam and the rigid suspended plate electromechanical transducer. Such elements, suitably connected together, allow the composition of complete RF-MEMS topologies and their fast simulation by means of the Spectre simulator.

Subsequently, the exploitation of the just mentioned software tool was discussed referring to an RF-MEMS variable capacitor (varactor), manufactured in the FBK surface micromachining technology. In particular, the model library was used in order to model the electromechanical behaviour (static pull-in/pull-out) of the mentioned varactor, also accounting for the most critical technology non-idealities, namely, residual stress within the electrodeposited gold and the surface roughness.

A methodology has been then discussed in details concerning the RF modelling of the variable capacitor. It is based on the extraction of a lumped-element network, accounting for the behaviour of the intrinsic device (shunt-to-ground tuneable capacitance), plus all the parasitic effects surrounding it, e.g. inductance, losses and coupling due to the input/output short CPW sections. Once the network arrangement is set, values of the lumped components are extracted by means of a commercial optimization tool, aiming at reproducing the S-parameters experimental characteristic of the tested device. The appropriateness of the defined network is validated both targeting several measured datasets, where only the intrinsic capacitance changes (collected for different applied bias levels), and comparing the corrective factors needed to account for the non-idealities in the electromechanical and electromagnetic modelling stages.

Furthermore, the fast simulation tool use was demonstrated also in the analysis of a hybrid RF-MEMS/CMOS voltage controlled oscillator (VCO).

Subsequently, the lumped element network approach was exploited also to simulate a complex RF-MEMS network, i.e. a reconfigurable RF/Microwave power attenuator, composed by multi-state resistive branches. In order to complete the overview on possible applications of the discussed modelling methodology, a lumped element network was extracted for a packaged CPW, based on FEM simulations of such structure (with and without cap).

By following the sequence suggested in this chapter, it is possible, stage after stage, to model all the critical aspects influencing the RF behaviour of the MEMS-based structures to be analyzed, like parasitic effects due to the device itself as well as introduced by the package, thus leading to a complete and accurate description of the real device that enables, at the same time, very fast simulations.

Application of such an approach eases the design phase that could be significantly speeded up by the definition of parameterized models, accounting for the parasitic effects plus package within a given technology. The just mentioned parametric models can be straightforwardly set up with the notions presented in this chapter. Moreover, the availability of the MEMS software library, developed by the author, would help in pursuing a complete, fast and accurate preliminary design of new MEMS-based RF simple component or networks. However, the method can be exploited even without such tool, as the main formulae describing the electromechanical behaviour of MEMS devices, as well as the non-idealities arising from the specific adopted technology process, were shown in details.

In conclusion, the material presented and discussed in this chapter might be of significant help for those who are involved in the design and performance optimization of RF-MEMS devices and networks. Indeed, the proposed methodology allows the inclusion of significant aspects of real devices, like technology non-idealities and RF parasitic effects, by keeping the simulation time and complexity very low.

Such method is very effective in the initial design optimization, when several degrees of freedom have to be studied, highlighting the trade-offs linking them. However, the method cannot completely replace the use of more accurate FEM tools, but can, in turn, reserve their use to the final optima definition, thus optimizing the time necessary to reach the desired final design, starting from a rough idea about the initial topology that could better suit the application requirements.

8. References

Chung, D.J.; Anagnostou, D.; Ponchak, G. & Tentzeris, M.M.; Papapolymerou, J. (2007). Light Weight MIMO Phased Arrays with Beam Steering Capabilities using RF MEMS, *Proceedings of the IEEE 18th International Symposium on Personal, Indoor and Mobile Radio Communications, PIMRC 2007*, pp. 1-3, ISBN 978-1-4244-1144-3, Athens, Greece, Sep. 2007, IEEE

Dambrine, G.; Cappy, A.; Heliodore, F. & Playez, E. (1988). A new method for determining the FET small-signal equivalent circuit. *IEEE Transactions on Microwave Theory and Techniques,* Vol. 36, No. 7, (Jul. 1988) page numbers (1151-1159), ISSN 0018-9480

Daneshmand, M. & Mansour, R. R. (2007). Redundancy RF MEMS Multi-Port Switches and Switch Matrices. *IEEE/ASME Journal of Microelectromechanical Systems,* Vol. 16, No. 2, (Apr. 2007) page numbers (296-303), ISSN 1057-7157

De Los Santos, H. J. (2002). *RF Mems Circuit Design for Wireless Communications,* Artech House, ISBN 1-58053-329-9, Boston, USA

Etxeberria, J.A. & Gracia, F.J. (2007). High Q factor RF MEMS Tunable Metallic Parallel Plate Capacitor, *Proceedings of the Spanish Conference on Electron Devices, 2007,* pp. 201-204, ISBN 1-4244-0868-7, Madrid, Spain, Jan. 2007, Piscataway, NJ

Fedder, G. (2003). Issues in MEMS macromodeling, *Proceedings of the 2003 IEEE/ACM Int. Workshop on Behavioral Modeling and Simulation (BMAS '03),* pp. 64-69, ISBN 0-780-38135-1, San Jose, CA, USA, Oct. 2003, IEEE

Gilleo, K. (2005). *MEMS/MOEM Packaging,* McGraw-Hill, ISBN 0-071-45556-6, Hoboken,NJ

Goldsmith, C.L.; Zhimin Yao; Eshelman, S. & Denniston, D. (1998). Performance of low-loss RF MEMS capacitive switches. *Microwave and Guided Wave Letters,* Vol. 8, No. 8, (Aug. 1998) page numbers (269-271), ISSN 1051-8207

Hyung, S.L.; Young, J. Y.; Dong-Hoon, C. & Jun-Bo, Y. (2008). High-Q, tunable-gap MEMS variable capacitor actuated with an electrically floating plate, *Proceedings of the IEEE 21st International Conference on Micro Electro Mechanical Systems,* pp. 180-183, ISBN 978-1-4244-1793-3, Tucson, Arizona, USA, Jan. 2008, IEEE

Iannacci, J.; Del Tin, L.; Gaddi, R.; Gnudi, A. & Rangra, K. J. (2005). Compact modeling of a MEMS toggle-switch based on modified nodal analysis, *Proceedings of the Symposium on Design, Test, Integration and Packaging of MEMS/MOEMS (DTIP 2005),* pp. 411-416, ISBN 2-84813-0357-1, Montreux, Switzerland, Jun. 2005

Iannacci, J.; Bartek, M.; Tian, J. & Sosin, S. (2006). Hybrid Wafer-Level Packaging for RF MEMS Applications, *Proceedings of the 39th International Symposium on Microelectronics (IMAPS 2006),* pp. 246-253, ISBN 0-930815-80-7, San Diego, CA, USA, Oct. 2006, IMAPS, Washington, DC, USA

Iannacci, J. (2007). *Mixed Domain Simulation and Hybrid Wafer Level Packaging of RF MEMS Devices for Wireless Applications,* Ph.D. Thesis Dissertation, University of Bologna, Italy, March 2007, ISBN, Available at: http://amsdottorato.cib.unibo.it/464/

Iannacci, J.; Gaddi, R. & Gnudi, A. (2007). Non-Linear Electromechanical RF Model of a MEMS Varactor Based on VerilogA© and Lumped Element Parasitic Network, *Proceedings of the 37th European Microwave Conference (EuMC),* pp. 544-547, ISBN 978-2-87487-000-2, Munich, Germany, Oct. 2007, Horizon House Publications Ltd, London, UK

Iannacci, J.; Bartek, M.; Tian, J.; Gaddi, R. & Gnudi, A. (2008). Electromagnetic Optimisation of an RF-MEMS Wafer-Level Package. *Sensors and Actuators A: Physical, Special Issue of Eurosensors XX 2006 Conference,* Vol. 142, No. 1, (Mar. 2008) page numbers (434-441), ISSN 0924-4247

Iannacci, J.; Giacomozzi, F.; Colpo, S.; Margesin, B. & Bartek, M. (2009). A General Purpose Reconfigurable MEMS-Based Attenuator for Radio Frequency and Microwave Applications, *Proceedings of the IEEE Region 8 EUROCON 2009 Conference,* pp. 1201-1209, ISBN 978-1-4244-3861-7, Saint Petersburg, Russia, May 2009, IEEE

Iannacci, J.; Repchankova, A.; Macii, D. & Niessner, M. (2009). A Measurement Procedure of Technology-related Model Parameters for Enhanced RF-MEMS Design, *Proceedings of the IEEE International Workshop on Advanced Methods for Uncertainty Estimation in Measurement AMUEM 2009*, pp. 44-49, ISBN 978-1-4244-3593-7, Bucharest, Romania, Jul. 2009, IEEE

Jing, Q.; Mukherjee, T. & Fedder, G. (2002). Schematic-Based Lumped Parameterized Behavioral Modeling for Suspended MEMS, *Proceedings of the ACM/IEEE International Conference on Computer Aided Design (ICCAD '02)*, pp. 367-373, ISBN 0-7803-7607-2, San Jose, CA, USA, Nov. 2002, ACM, New York, NY, USA

Larcher, L.; Brama, R.; Ganzerli, M.; Iannacci, J.; Margesin, B.; Bedani, M. & Gnudi, A. (2009). A MEMS Reconfigurable Quad-Band Class-E Power Amplifier for GSM Standard, *Proceedings of the 22nd IEEE International Conference on Micro Electro Mechanical Systems MEMS 2009*, pp. 864-867, ISBN 978-1-4244-2978-3, Sorrento, Italy, Jan. 2009, IEEE, Piscataway, NJ, USA

Maciel, J.J.; Slocum, J.F.; Smith, J.K. & Turtle, J. (2007). MEMS Electronically Steerable Antennas for Fire Control Radars, *Proceedings of the IEEE Radar Conference 2007*, pp. 677-682, ISBN 1-4244-0284-0, Boston, MA, USA, Apr. 2007, IEEE

Nguyen, C.T.-C. (2004). Vibrating RF MEMS for next generation wireless applications, *Proceedings of the IEEE Custom Integrated Circuits Conference*, pp. 257-264, ISBN 0-7803-8495-4, Orlando, FL, USA, Oct. 2004, IEEE

Novak, E.; Wan, D. S.; Unruh, P. & Schurig, M. (2003). MEMS Metrology Using a Strobed Interferometric System, *Proceedings of the XVII IMEKO World Congress*, pp. 178-182, ISBN 953-7124-00-2, Dubrovnik, Croatia, Jun. 2003, Kluwer Academic Publisher, Norwell, MA

Pozar, D. M. (2004). *Microwave Engineering*, J. Wiley & S., ISBN 0-471-44878-8, Hoboken, NJ

Przemieniecki, J. S. (1968). Theory of Matrix Structural Analysis, McGraw-Hill, ISBN 0-486-64948-2, New York, NJ

Rebeiz, G.M. & Muldavin, J.B. (2001). RF MEMS switches and switch circuits. *IEEE Microwave Magazine*, Vol. 2, No. 4, (December 2001) page numbers (59-71), ISSN 1527-3342

Tiebout, M. (2005). *Low Power Vco Design in CMOS*, Springer Technology & Industrial Arts, ISBN 3-540-24324-0, New York, NJ

Topalli, K.; Aydin Civi, O.; Demir, S.; Koc, S. & Akin, T. (2008). A Monolithic Phased Array using 3-bit DMTL RF MEMS Phase Shifters. *IEEE Transactions on Microwave Theory and Techniques*, Vol. 56, No. 2, (Feb. 2008) page numbers (270-277), ISSN 0018-926X

Varadan, V. K. (2003). *RF Mems & Their Applications*, John Wiley & Sons, ISBN 0-470-84308-X, Hoboken, NJ

Zine-El-Abidine, I.; Okoniewski, M. & McRory, J.G. (2003). A new class of tunable RF MEMS inductors, *Proceedings of the International Conference on MEMS, NANO and Smart Systems*, pp. 114-115, ISBN 0-7695-1947-4, Banff, Alberta, Canada, Jul. 2003, IEEE

16

Ultra Wideband Microwave Multi-Port Reflectometer in Microstrip-Slot Technology: Operation, Design and Applications

Marek E. Bialkowski and Norhudah Seman
The University of Queensland
Australia

1. Introduction

A microwave reflectometer is an instrument to measure a complex ratio between reflected and incident waves at an input port of a uniform transmission line terminated in a Device Under Test (DUT). The conventional reflectometer is formed by a four-port network with two ports connected to a microwave source and DUT, and the remaining ports coupled to a heterodyne receiver which acts as a Complex Ratio Detector (CRT). By using the heterodyne receiver technique, the two microwave signals are converted in the linear manner to an Intermediate Frequency (IF) of hundreds of kHz where they are processed using digital means. The use of the heterodyne technique enables a very large dynamic range of 100 dB or more for this type of reflectometer. However, as the ratio of two original microwave signals has to be preserved at IF, a very advanced electronic circuitry is required to accomplish the linear conversion process. This complicated electronics leads to a large size of the conventional reflectometer and its high price tag. Many applications require compact-size and low-cost reflectometers. They can be built using N-port networks, with N being greater than 5, equipped only in scalar (power) detectors. This chapter describes the concept of a multi-port reflectometer which employs scalar instead of complex ratio detector to determine the complex reflection coefficient of DUT. It is shown that such a device requires a suitable calibration and mathematical transformations of the measured power at selected ports of the N-port to obtain the complex reflection coefficient of DUT. Because of this requirement, the multi-port reflectometer uses a computer to perform calibrations and measurements. The use of a computer accelerates the calibration and measurement procedure and at the same time it does not create a considerable overhead to the total cost of this measurement instrument. The challenge is to obtain a low-cost fully integrated N-port network operating over an ultra wide frequency band, which can be used to develop a fully operational reflectometer. This challenge is addressed in the present chapter. Practical configurations of this measurement instrument are described and the design of a compact fully integrated N-port network in microstrip-slot technique to build a reflectometer operating over an ultra wide microwave frequency band of 3.1 to 10.6 GHz is given.

2. Multi-Port Reflectometer Concept

A multi-port reflectometer is a passive linear circuit with two input ports allocated for a power source and Device Under Test (DUT) and at least three output ports terminated in scalar power detectors to obtain the information about a complex reflection coefficient of DUT. A particular case of this device is a six-port reflectometer with four scalar detectors to determine in precise manner, the reflection coefficient of DUT. Having one more port with a power detector makes it less prone to power measurement errors than its five-port counter part. Being introduced in 1970s, a six-port, or in more general case, N-port reflectometer provides an alternative method to the conventional network analyser employing heterodyne receiver principle to measure impedance, phase or complex reflection coefficient of passive or active circuit (Hoer, 1975). For the six-port reflectometer, these parameters are obtained from the measured power at its four output ports.

Accuracy of six-port measurements is a function of linearity of the power detectors and the properties of the six-port network (Hoer, 1975). Because the six-port reflectometer can provide phase information by making only power (scalar) measurements of four different linear combinations of the two electromagnetic waves (incident and reflected at DUT), the requirement for phase information at the output ports of six-port is avoided. The other advantage of this technique is the reduced frequency sensitivity (Engen, 1977). Consequently, the phase locked source is no longer necessary in the design. As a result, the concept of six-port reflectometer can easily be extended to millimetre frequencies (Engen, 1977). The general block diagram of a six-port reflectometer is shown in Fig. 1.

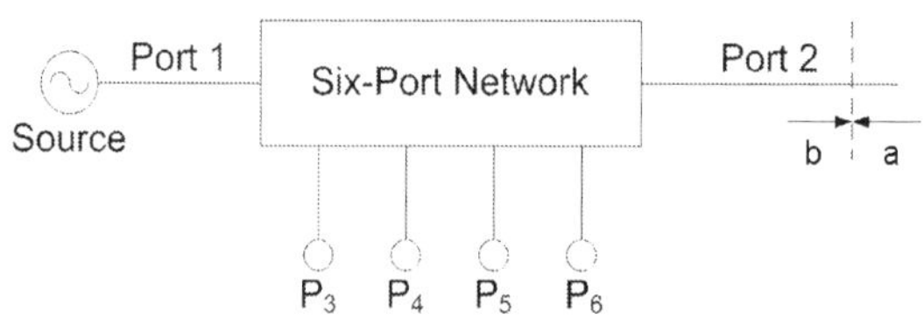

Fig. 1. General block diagram of a reflectometer employing six-port network.

As shown in Fig. 1, the microwave source is connected to Port 1, while Port 2 acts as the measurement port for Device Under Test (DUT). Here, variable b represents an incident signal while variable a, indicates a reflected signal. The other four ports (Port 3 to 6) are connected to scalar power detectors. The power reading from these 4 ports can be written as in (1) - (4) from the assumption that the network is arbitrary but linear (Engen, 1969; Engen, 1977):

$$P_3 = |b_3|^2 = |Aa + Bb|^2 \quad (1)$$

$$P_4 = |b_4|^2 = |Ca + Db|^2 \quad (2)$$

$$P_5 = |b_5|^2 = |Ea + Fb|^2 \quad (3)$$

$$P_6 = |b_6|^2 = |Ga + Hb|^2 \quad (4)$$

Evaluating the right sides of the above expressions gives real values. Alternatively, these expressions can be presented in the complex form by removing the "magnitude of" symbols. But, in this case only the magnitudes and not the phases of the resulting bilinear function are found from the measurements. Constants b_3 to b_6 are representing the signal voltages at the output ports. The unknown complex constants *A, B, C...H* in (5) - (8) can be obtained from the four sidearm power readings. The desired results are (Engen & Hoer, 1972; Hoer & Engen, 1973; Hoer, 1975):

$$|a|^2 = \Sigma_{i=3}^{6} \alpha_i P_i \tag{5}$$

$$|b|^2 = \Sigma_{i=3}^{6} \beta_i P_i \tag{6}$$

$$|ab| \cos\psi = \Sigma_{i=3}^{6} c_i P_i \tag{7}$$

$$|ab| \sin\psi = \Sigma_{i=3}^{6} s_i P_i \tag{8}$$

From these 4 unknowns, the general equation of reflection coefficient can be written as the ratio of reflected signal, *a* to incident signal, *b* (Engen & Hoer, 1972; Hoer & Engen, 1973; Hoer, 1975):

$$\Gamma = \frac{a}{b} = \frac{\Sigma_{i=3}^{6} (c_i + js_i) P_i}{\Sigma_{i=3}^{6} \beta_i P_i} \tag{9}$$

3. Geometrical Interpretation of Reflection Coefficient and Design Considerations

3.1 Geometrical Interpretation of Reflection Coefficient in Complex Plane

As presented in equation (9), the unknown reflection coefficient of measured load (DUT) is related to the power measurements by a set of complex constant *A - H*. These eight complex constants (*A - H*) and/or 12 real constants (c_i, s_i and β_i) can be determined from a suitable calibration procedure by applying 5 to 6 standards (Somlo & Hunter, 1982; Hunter & Somlo, 1985).

The principle of operation of a six-port reflectometer can be gathered by considering a simplified case of this device. The following representation can serve this purpose (Engen, 1977):

$$P_3 = |A|^2 |b|^2 |\Gamma - q_3|^2 \tag{10}$$

$$P_4 = |C|^2 |b|^2 |\Gamma - q_4|^2 \tag{11}$$

$$P_5 = |E|^2 |b|^2 |\Gamma - q_5|^2 \tag{12}$$

$$P_6 = |G|^2 |b|^2 |\Gamma - q_6|^2 \tag{13}$$

where q_3- q_6 are as follows:

$$q_3 = -\frac{B}{A}, \quad q_4 = -\frac{D}{C}, \quad q_5 = -\frac{F}{E}, \quad q_6 = -\frac{H}{G} \tag{14}$$

The above expressions (10) – (13) represent circles in the complex reflection coefficient plane which can be used as geometrical interpretation in determining the reflection coefficient. The circle centres are given by the unknowns q_3 to q_6, also branded as q-points, while the circle radii are given by the $|\Gamma - q_i|$ where i=3, 4, 5, 6.
The operation of the six-port reflectometer can also be described in terms of scattering parameters of a multi-port network. Complex constants A - H are first replaced by common complex constants m_i and n_i and then the incident signals at ports, b_i, (i=3, 4, 5, 6) can be rewritten as the following equation in terms of the incident and emergent signals at Port 2 (Somlo & Hunter, 1985):

$$b_i = m_i a + n_i b \tag{15}$$

Complex constants, m_i and n_i can then be expressed by the scattering parameters as follows (Somlo & Hunter, 1985):

$$m_i = S_{i2} - \frac{S_{i1} S_{22}}{S_{21}} \tag{16}$$

$$n_i = \frac{S_{i1}}{S_{21}} \tag{17}$$

The general equation of circle centre is given by the negative ratio of n_i and m_i which is analogous to the expression (14) (Somlo & Hunter, 1985):

$$q_i = -\frac{n_i}{m_i} = -\frac{S_{i1}}{S_{i2} S_{21} - S_{i1} S_{22}} \tag{18}$$

By assuming that approximately ideal components are used to construct the network, the parameter S_{22} is very close to zero. This simplifies the equation (18) to (Somlo & Hunter, 1985):

$$q_i = -\frac{S_{i1}}{S_{i2} S_{21}} \tag{19}$$

According to Probert and Carrol in (Probert & Carroll, 1982), the characterisation can be made more general for the multi-port network case. With the above assumption and the use of known input voltage, V_o at Port 1, the incident signal b_i (i=3....N) and reflection coefficient, Γ can be written as:

$$b_i = V_o \left(S_{i1} + S_{12} S_{i2} \Gamma \right) \tag{20}$$

$$\Gamma = \frac{b_i}{V_o S_{21} S_{i2}} - \frac{S_{i1}}{S_{i2} S_{21}} \tag{21}$$

Since q_i are given by (19), then the radius is $|\Gamma\text{-}q_i|$ and thus the circle radius can be calculated as:

$$radius = \left| \frac{b_i}{V_o S_{21} S_{i2}} \right| \tag{22}$$

where i=3, 4, 5,…., N (N = number of port) and b_i is proportional to $\sqrt{P_i}$.

An important characteristic of the properly selected six-port network is insensitivity of a reference port to the reflected signal from the DUT at Port 2. This port is in general assigned the special function of the incident signal power measurement. Since this port gives a good indication of the source power, the output of power detector connected to this port can be used to stabilize the signal source against power fluctuations or maintain the power at some set level (Woods, 1990). The inclusion of this special port does not therefore compromise the overall design of the six-port measurement system.

An alternative representation that can be used in the six-port analysis is through the use of the power ratio, p_i given by power at Port i to that at a reference port (Port 6 is selected in this analysis case) is (Somlo & Hunter, 1985):

$$p_i = r_i \left| \frac{\Gamma - q_i}{\Gamma - q_6} \right|^2, \qquad i = 3,\ 4,\ 5 \tag{23}$$

In (23) constant r_i is related to the scattering parameters and a real constant, K_i by the following equation (Somlo & Hunter, 1985):

$$r_i = K_i \left| \frac{S_{i1} S_{22} - S_{i2} S_{21}}{S_{61} S_{22} - S_{62} S_{21}} \right|^2 \tag{24}$$

The ratio of p_i to constant r_i can be written as:

$$e_i = \frac{p_i}{r_i} \cong \left| \frac{\Gamma - q_i}{q_6} \right|^2 \tag{25}$$

The reflection coefficient of DUT can be identified from the geometrical representation as the intersection of power circles. Fig. 2 illustrates this concept showing the intersection of the two circles that determine the reflection coefficient, Γ. The centres of the two circles are given by q_3 and q_5 (Engen, 1977).

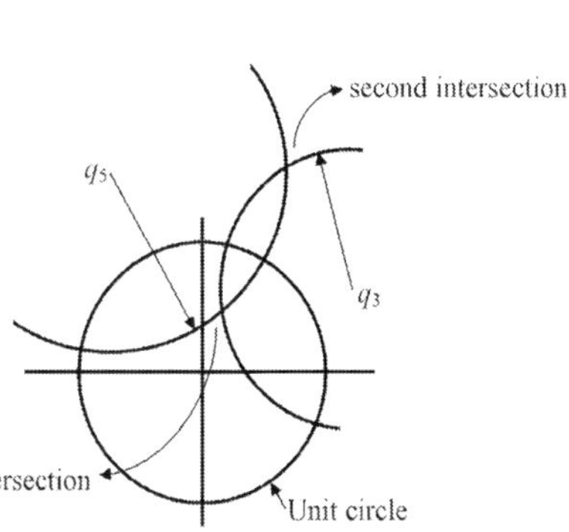

Fig. 2. Determination of the reflection coefficient, Γ from the intersection of two power circles.

This case is found for the five-port network configuration which does not make use of circle with centre q_4. The example presented in Fig. 2 shows that one intersection point falls within the region of reflection coefficient unit circle while the second point is outside it. In this case, the ambiguity in the proper choice of Γ is removed and a unique value is chosen on the basis that the reflection coefficient of a passive load is less than or equal to one. The passive load termination assumption has to be supported by the condition of a straight line connecting q_3 and q_5 that does not intersect the unit circle (Engen, 1977).
The close inspection of Fig. 2 indicates that solution offered by the five-port is prone to the power measurement errors. These power errors may result in a substantial error in the position of the reflection coefficient perpendicular to the line joining the circle centres of q_3 and q_5 (Woods, 1990). As explained in (Engen, 1977), a one percent error in the experimental measurement of $|\Gamma\text{-}q_3|$ and $|\Gamma\text{-}q_5|$ can cause the uncertainty of 10 percent in the measured reflection coefficient result.
The deficiency of the five-port reflectometer can be overcome by employing an extra power detector reading that is available in the six-port network. This is illustrated by introducing the third power circle, as shown in Fig. 3.

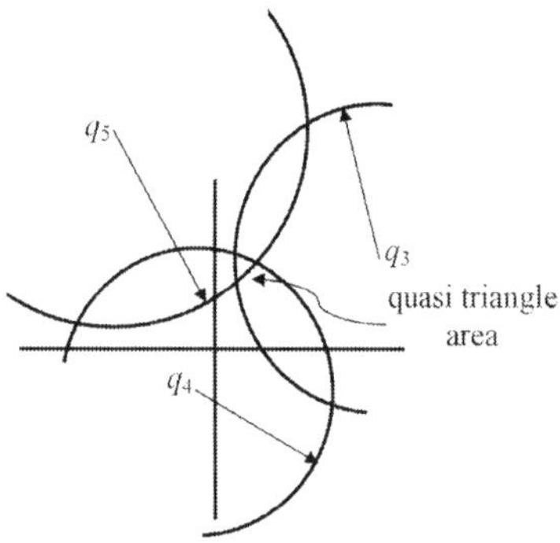

Fig. 3. Circle intersection failure when three circles are used to determine reflection coefficient, Γ.

From Fig. 3 it is apparent that the solutions for reflection coefficient are restricted more than in the case of five-port and a unique value can be determined without the assumption of the load being passive. This procedure can be interpreted as finding the intersection of three circles. Therefore, three circles solve the ambiguity when choosing between the two intersections given by two circles (Waterhouse, 1990). When the measured power values include errors, the three circles will not have a common point of intersection but will define a quasi-triangular area in the complex plane. Engen explained in (Engen, 1997) that this intersection failure is an indicator of the power meter error. Moreover, the measurement noise, nonlinearity in power measurement and imperfections in the calibration can also contribute to this phenomenon (Somlo & Hunter, 1985). Hence in practical cases, the multi-port measurement system being prone to power errors changes the ideal circles radii (Woods, 1990). A suitable configuration of multi-port has to be decided upon to counter this effect. The solution to this problem is related to the choice of locations of the q_i-points which characterize the multi-port. As can be observed in Fig. 3, locations of the q_i-points in the complex plane are important in keeping the area of the quasi-triangle to minimum. By making the proper choice of the q_i-points, the uncertainty of value for the Γ can be marked small (Somlo & Hunter, 1985).

Engen proposed that for the six-port reflectometer the q_i amplitudes should be in the range of 1.5 to 2.5 and their angular separation should be about 120°. The reasons for such conditions are explained in detail in the next section. When the multi-port with a larger number of ports is used more than three circles are available and the improved measurement accuracy is possible in situations where intersection failure occurs. The whole circle equation system can be solved simultaneously in a least-squares sense where statistical averaging or weighting can lead to the best solution (Engen, 1969; Engen, 1980).

It is apparent that the use of additional detectors can significantly improve the device performance and make it less sensitive to power measurement errors. Following this general concept, the system can be extended to seven or more ports. With the possible exception of a seven port, however, the accuracy improvement does not ordinarily warrant additional complexity (Engen, 1977).

3.2 Optimum Design Considerations

It has already been shown that the operation of six-port reflectometer is governed by the constants *A* - *H* which determine the coupling of the waves to the detectors (Woods, 1990). A set of the design rules for the six-port network can thus be formulated by establishing preferred values of these constants. A practical network can then be designed which conforms to these preferred values. The main parameter to be considered is the accuracy of the complex reflection coefficient measurement. However, as the detectors output voltages are processed by Analogue to Digital Converters, the other important factor which also needs be taken into account is the required voltage meters dynamic range.

The following are the considerations which lead to the guidelines for the six-port (or in a more general case, multi-port) reflectometer design.

From the graphical interpretation of operation of six-port reflectometer, the optimum design is related to selection of locations of the q-point circle centres, which correspond to the values of $-B/A$, $-D/C$, $-F/E$ and $-G/H$ in the complex plane. When the measurement accuracy of reflection coefficient is of concern, an optimum six-port reflectometer is the one that is least susceptible to detector power measurement errors. In the previous considerations, it has been pointed out that for the optimum design the q-points have to be separated evenly in phase and magnitudes. This six-port design strategy has been suggested by many researchers.

Somlo and Hunter explained in (Somlo & Hunter, 1985) that for the case of passive terminations with $|\Gamma| \leq 1$, the network has to be chosen in such a way that for the reference Port 6 $|q_6|$ has to be greater than 1. This geometrically means that q_6 is located outside the unit circle in the complex Γ plane. A similar choice they also suggested for the remaining q-points. This is to reduce the sensitivity of the power measurement to noise. If the opposite condition of $|q_i| \leq 1$, i=3, 4, 5 is chosen, then there are values of Γ which make the numerator in equation (23) and p_i small. In particular, the value of $\Gamma = q_i$ sets $p_i = 0$, which is greatly influenced by noise.

The restriction $|q_i|>1$ (i =3, 4, 5), also avoids the case $q_i = 0$ which has been argued against in detail by Engen in (Engen, 1977) on the basis of noise sensitivity when measuring a termination near a match, which is likely to be the one of the most important uses of the reflectometer. This condition can be explained using the example of having q_3=0, q_4=2 and q_5=j2 (Engen, 1977). In such a case, P_3 almost does not contribute to the determination of Γ when measuring $|\Gamma|$ with small magnitude such as 0.01. As a result, the most inaccurate power measurement (worst signal to noise ratio, SNR) occurs as the power incident on a detector approaches zero. Based on this argument the q values should be such that $|q_i| \neq 0$. However in contrast to the discussed $|q_i|>1$, Engen in (Engen, 1977; Engen, 1997) suggested the optimum value of $|q_i|$ to be chosen around 0.5. Their argument is valid if the measurement region is within $0 \leq |\Gamma| \leq 0.3$.

The choice of $|q_i|>1$ (i=3, 4, 5) postulated by Somlo and Hunter in (Somlo & Hunter, 1985), is also beneficial with regard to the voltage meters dynamic range. This range has to be not too large. If the conditions of $|q_6|>>1$ and $|q_i|>1$, i= 3, 4, 5 are implemented, the approximated dynamic range required for the power meters can be calculated as given by (Somlo & Hunter, 1985):

$$Dynamic \;\; range \;\; [dB] = 20\log_{10}\left[\frac{|q_i|+1}{|q_i|-1}\right] dB \tag{26}$$

With the condition of $|q_i|>1$ (i=3, 4, 5) and $|q_6| > 1$, one can pose the question whether the magnitudes of all the q_i,s have to be equal. If it is the case, complex constants, c_i and s_i are equal to zero. It is therefore essential that, geometrically, the q_i do not all lie on the circle with centre $\Gamma = 0$ on the complex Γ plane (Somlo & Hunter, 1985). This means that $|q_i|$ (i=3, 4, 5) have to be less than $|q_6|$ to meet the preferable design.

In addition to the above argument, the magnitude of q should not be too near to unity because p_i could be small for the fully reflecting terminations (Somlo & Hunter, 1985). Small values of p_i resulting from $|q_i| \cong 1$ decrease the measurement accuracy (Engen, 1977).

The remaining condition concerns the upper bound for the distance of the q-points with respect to the complex Γ plane origin. Since Γ is determined from its distances from q_3, q_4 and q_5 (Engen, 1977), it is proven that an ill conditioned situation will result if these distances become large in comparison with distances between q_3 and q_4, q_3 and q_5 or q_4 and q_5 (Engen, 1977). If the $|q_i|$ are too large, it can be seen from equation (25) that a small change to p_i represents a large change in Γ. Choosing $|q_i|$, i=3, 4, 5 to be large also places high resolving demands on the power meters (Somlo & Hunter, 1985).

Based on these argument, (Engen, 1977) postulated that magnitude of q_i should be in the range of $\sqrt{2}$ to 2. In turn, Yao in (Yao, 2008) made suggestion for using the range between 1 and 3. Additionally, Bilik in (Bilik, 2002) postulated the choice of magnitude of q-points approximately 2. It is worthwhile mentioning in the practical circuits these magnitudes of q-points fall to some extent short of the optimum design aims in (Engen, 1977). However, they are easier to achieve. Moreover, it appears that the theoretical loss in performance between such practical circuits and "ideal" ones may be small in comparison with the performance degradation which results from the use of non-ideal components (Engen, 1977).

With respect to the q-points spacing, the even spacing in the complex plane is postulated (Engen, 1977; Somlo & Hunter, 1985; Bilik, 2002). For the six-port reflectometer this requirement leads to 120° separation of q-points. For the more general case of multi-port network with N>6, the q-points are suggested to be separated by 360°/(N-3) (Probert & Carroll, 1982). Because practical circuits are unable to keep constant angular separation of q-points, Yao in (Yao, 2008) added the tolerance conditions. For the case of N=6 he suggested the phase separation range should fall between 100° and 140° with the ± 20° from the optimum 120°.

4. Integrated UWB Reflectometer

4.1 Reflectometer Design

The configuration of reflectometer chosen for practical development is shown in Fig. 4.

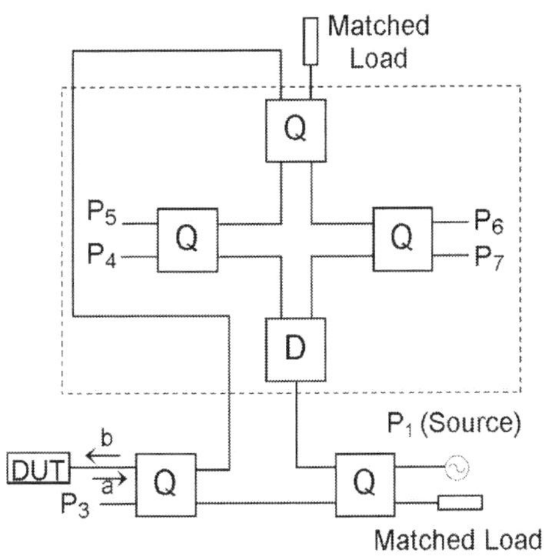

Fig. 4. Reflectometer configuration formed by five quadrature hybrids (Q) and one power divider (D).

The device is constructed using a seven-port network and includes five 3-dB couplers (Q) and one power divider (D). In this configuration, Port 1 is allocated for a microwave source while Device Under Test (DUT) is connected to Port 2. Five power detectors terminate Ports 3-7. Part of the reflectometer within the broken line is given the special name of Complex Measuring Ratio Unit (CMRU) or Correlator. It plays a similar role to the Complex Ratio Detector in the conventional four-port reflectometer based on the heterodyne receiver technique. The two couplers (Q) outside the CMRU are used to redirect the signals, *a* and *b* to measure the complex reflection coefficient of DUT. Note that in a more basic design, a single coupler is sufficient to perform this function. However, the use of two couplers provides a better signal balance which is of importance to achieving a better quality measurement of the reflection coefficient. A scalar detector terminating Port 3 of the divider D, outside the CRMU monitors the signal source power level.

The advantage of this seven-port configuration is that it allows for a real-time display of DUT complex reflection coefficient (Engen, 1977; Engen, 1977; Hoer & Roe, 1975; Hoer, 1977). In this case, the detector at Port 3 can be used in a feedback loop to maintain a constant power level from the source. The chosen configuration meets the condition of $|q_3|>1$ and $|q_i|<|q_3|$ where i=4, 5, 6, 7 and represents an optimal reflectometer configuration, as pointed by (Probert & Carroll, 1982), as its q_i (i=4, 5, 6, 7) points are spread by 90° in the complex reflection coefficient plane.

While undertaking a rough assessment of operation of the seven-port reflectometer of Fig. 4 it is important to find out by how much it diverges from the one using ideal components. The following mathematical expressions can be applied in this evaluation process.

Assuming an ideal operation of couplers and divider and the square-law operation of detectors (the measured voltages at detector outputs are proportional to power values at the detectors inputs) and by applying mathematical derivations similar to those in (Hoer, 1975), it can be shown that the reflection coefficient, Γ, of DUT for the configuration of Fig. 4 can be determined from (27):

$$\Gamma = \frac{a}{b} = \Gamma_1 + j\Gamma_2 = \frac{(P_4 - P_5) + j(P_6 - P_7)}{P_3} \tag{27}$$

where Γ_1 is the real component of complex reflection coefficient, Γ_2, the imaginary and $P_i = |V_i|^2$, (i=4, 5, 6, 7) are measured power at 4 ports.

It is apparent that the above expression can be used to obtain a real-time display of the DUT reflection coefficient as the difference operation can be achieved using analogue means and real and imaginary parts can be displayed in the polar form on an oscilloscope.

An equivalent representation of Γ can be obtained from knowing the scattering parameters of the seven-port constituting the reflectometer of Fig. 4. In this case, Γ can be determined using the following expression:

$$\Gamma = \frac{\left(|S_{41}|^2 - |S_{51}|^2\right) + j\left(|S_{61}|^2 - |S_{71}|^2\right)}{|S_{31}|^2} \tag{28}$$

Assuming ideal operation of couplers, dividers and square-law operation of detectors, the DUT reflection coefficient can also be obtained by geometrical means from an intersection of four circles defined by (29):

$$
\begin{aligned}
V_4 &= -\frac{b}{2\sqrt{2}}(\Gamma - q_4) \\
V_5 &= \frac{jb}{2\sqrt{2}}(\Gamma - q_5) \\
V_6 &= -\frac{b}{2\sqrt{2}}(\Gamma - q_6) \\
V_7 &= -\frac{jb}{2\sqrt{2}}(\Gamma - q_7)
\end{aligned}
\tag{29}
$$

where V_i represent the voltages measured at ports 4 to 7.

The four circles are defined here by the centres q_i and radii $|\Gamma - q_i|$ where i=4, 5, 6, 7.
In order to design the individual couplers (Q) and divider (D) constituting the reflectometer, CST Microwave Studio (CST MS) is used. Rogers RO4003C featuring a relative dielectric constant of 3.38 and a loss tangent of 0.0027 is chosen as a microwave substrate to manufacture these components. It has 0.508 mm thickness and 17 μm of conductive coating. The design of coupler and divider follows the initial guidelines explained in (Seman & Bialkowski, 2009) and (Seman et al., 2007), followed by the manual iterative process aided with CST MS.
In the present case, a three section coupler with rectangular shaped microstrip-slot lines is chosen. The microstrip-slot technique is also applied to a divider. A special configuration of divider proposed here makes it compatible with the coupler. Their design is accomplished using CST MS. Layouts of the coupler and the divider are generated with the use of CST MS as shown in Fig. 5(a) and (b), respectively.

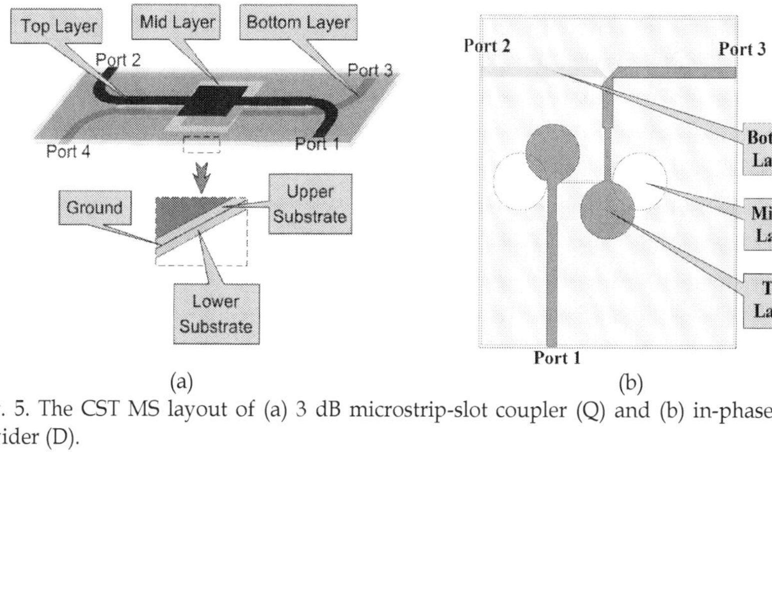

(a) (b)

Fig. 5. The CST MS layout of (a) 3 dB microstrip-slot coupler (Q) and (b) in-phase power divider (D).

The designed coupler has the simulated characteristic of return loss at its ports better than 20 dB whilst isolation between ports 1 and 4, and 2 and 3 is greater than 19 dB in the 3.1 to 10.6 GHz frequency band. In the same band, the coupling between ports 1 and 3 and 2 and 4 is 3 dB with a ±1 dB deviation. The phase difference between the primary and coupled ports is 90.5° ± 1.5°. The designed divider offers return losses greater than 12 dB at its input port and power division of -3 dB ± 1 dB between its output ports across the same band. The phase difference between the output ports is 0° ± 1° for 3 to 7 GHz and deteriorates to -1° to -3.5° for the frequency band between 7 and 11 GHz. These results indicate good performances of individual components. Therefore they can be integrated to form the reflectometer of Fig. 4.

The task of forming a reflectometer is accomplished in two stages. First, a Complex Measuring Ratio Unit (CMRU) in Fig. 6(a) is assembled. Then, two additional couplers are added to finalize the reflectometer design. Layout of the designed reflectometer providing the details of input and output ports, match terminated ports and screw holes is shown in Fig. 6(b).

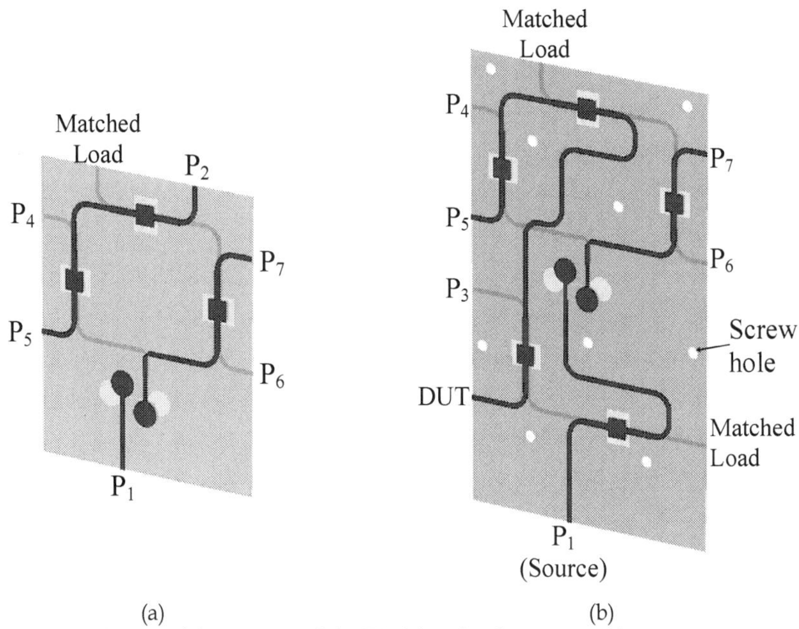

Fig. 6. CST MS layout of the integrated CMRU (a) and reflectometer (b).

4.2 Reflectometer Results

Fig. 7 presents a photograph of the fabricated reflectometer with the attached SMAs connectors but excluding power detectors. The device is formed by the CMRU and two additional couplers for rerouting signals to perform reflection coefficient measurements. The reflectometer uses two double-sided Rogers RO4003 PCBs.

In the fabricated prototype, the two substrates are affixed using plastic screws with diameter 3 mm to minimize air gaps between two dielectric layers. Sub-miniature A (SMA) connectors are included for detectors, a microwave source and DUT. They are also used for characterization of the seven-port using a Vector Network Analyser. The overall dimensions of this device excluding SMA connectors are 11.8 cm × 7 cm. These dimensions indicate the compact size of the developed reflectometer.

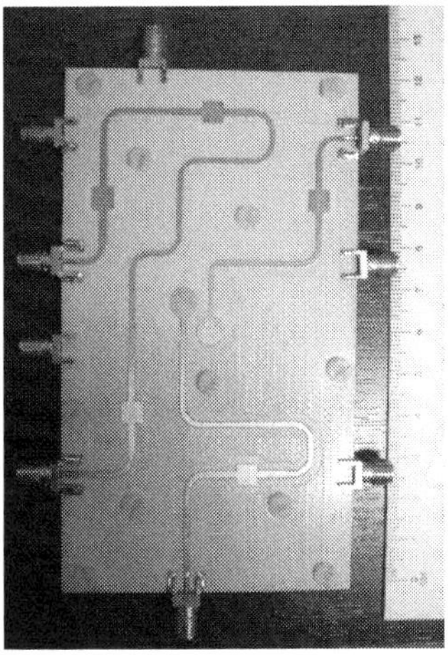

Fig. 7. Photograph of the fabricated reflectometer.

The CST MS simulated transmission coefficients at Port 4, 5, 6 and 7 referenced to Port 1 and 2 for this device are shown in Fig. 8.

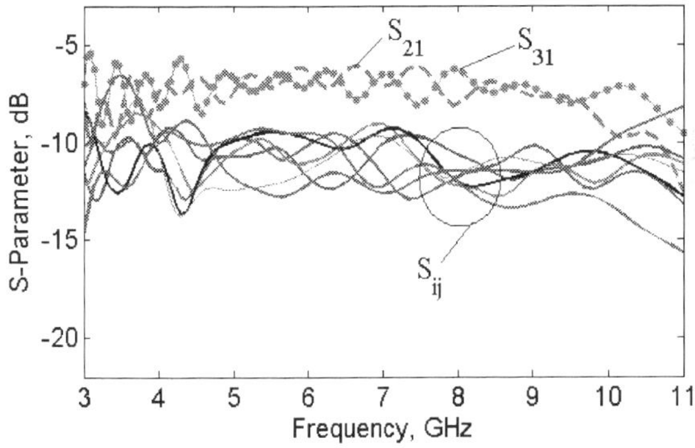

Fig. 8. Simulated transmission coefficients of designed reflectometer using CST MS where i=4, 5, 6, 7 and j=1, 2.

As observed in Fig. 8, magnitudes of the simulated parameters S_{21} and S_{31} are -7.3 dB ± 1.3 dB and -7.05 dB ± 1.35 dB for the frequency range of 3.5-9.8 GHz and 3.3-10.6 GHz, respectively. The simulated S-parameters (S_{ij}) at Port 4 to 7 with the reference to Port 1 and 2

show good performance of the seven-port network between 4 and 10 GHz. The worst case is for the parameter S_{72} which starts to deteriorate above 10 GHz.
Fig. 9 shows the measured results corresponding to the simulated ones of Fig. 8.

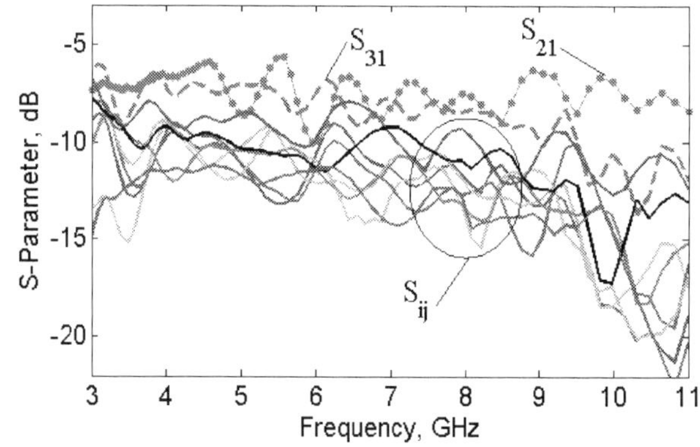

Fig. 9. Measured transmission coefficients of the fabricated reflectometer where i=4, 5, 6, 7 and j=1, 2.

There is similarity between the results shown in Fig. 8 and those of Fig. 9. However, the measured results exhibit larger ripples (±2 dB) between 3 and 9.5 GHz.

Fig. 10 presents the simulated and measured return loss characteristics at Port 1 and the simulated and measured transmission coefficients between port 1 and Port 8 and 9. Similarly, Fig. 11 presents the simulated and measured return loss at Port 2 and the simulated and measured transmission coefficients between Port 2 and selected ports of the seven-port reflectometer. Comparisons between the simulated and measured characteristics presented in Fig. 10 and 11 indicate a relatively good agreement.

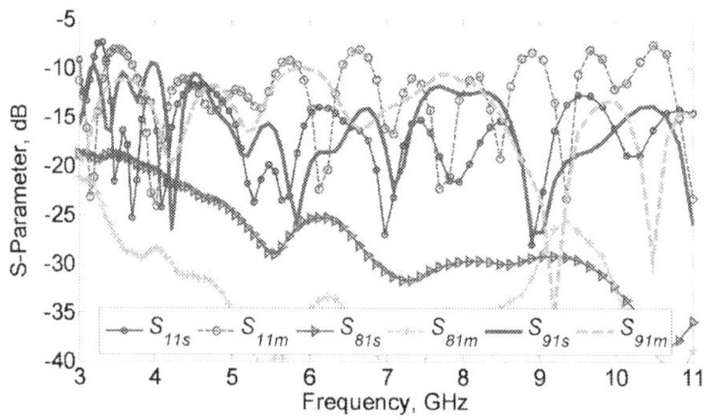

Fig. 10. Simulated and measured reflection coefficient at Port 1, and simulated and measured transmission coefficients between Port 1 to Port 8 and 9 of the reflectometer.

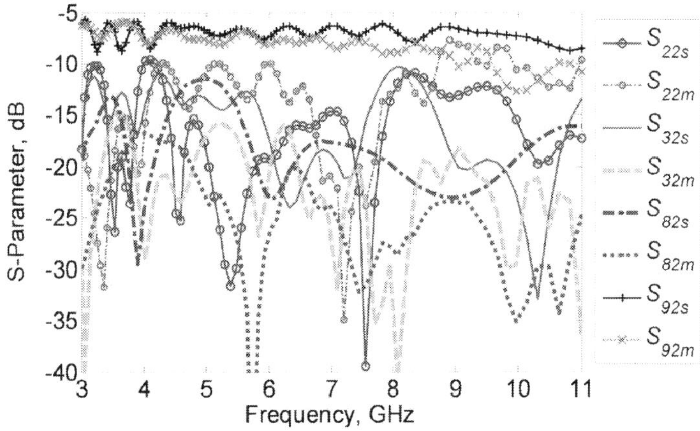

Fig. 11. Simulated and measured reflection coefficient at Port 2, and simulated and measured transmission coefficients between Port 2 and Port 3, 8 and 9.

The simulated or measured S-parameters can be used to assess the performance of the designed seven-port in terms of its q-points (i= 4, 5, 6, 7), which can be calculated using expression (18). For the ideal case, the chosen configuration of seven-port reflectometer offers the location of q_i at 2, j2, -2 and -j2. The location of these points with respect to the origin of the complex plane of 2 and the angular separation of 90° indicate the optimal design of this reflectometer.

Fig. 12 shows the simulated and measured locations of the q-points (i= 4, 5, 6, 7).

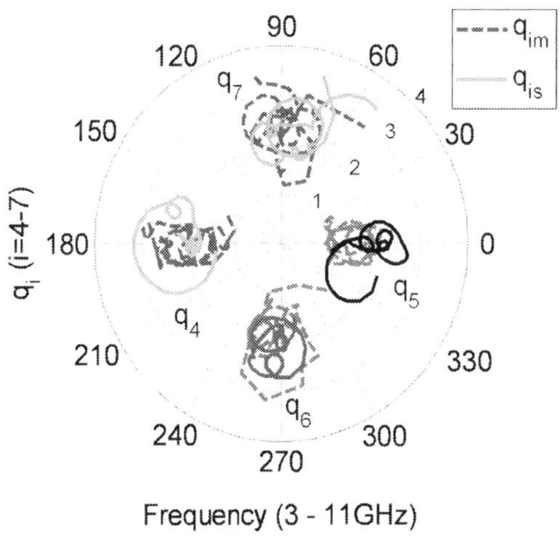

Fig. 12. Polar plot of the simulated (s) and measured (m) q_i - points (i=4, 5, 6, 7).

The simulated magnitudes of q_4, q_5, q_6 and q_7 are 2.3 ± 0.9, 1.9 ± 0.8, 2.1 ± 0.6 and 2.5 ± 0.9, while the measured ones are 2 ± 1, 1.6 ± 0.6, 2.1 ± 1.1 and 2.3 ± 1.1 in the frequency band between 3 and 11 GHz. Therefore there is a reasonable agreement between the two sets.
As observed from the polar plot in Fig. 12, the circle centres of q_i for this reflectometer deviate from the ideal separations of 90° (0°, 90°, 180° and 270°). The actual phase separation is given by $\pi/2+\O_0+k\Delta f$, where k and $\O_0$ are constants and Δf is the shift from the mid-frequency (Yao & Yeo, 2008). The measured phases of q_4, q_5, q_6 and q_7 are 180° ± 10°, 0° ± 20°, -90° ± 18° and 89° ± 19°, respectively from 3 to 10.6 GHz.
The measured phase characteristics q_i (i=5, 6, 7) can be referenced against q_4 by the following equation of (30):

$$\text{phase}(q_{\Delta i}) = \text{phase}(q_i) - \text{phase}(q_4) \qquad i = 5, 6, 7 \tag{30}$$

The measured phase ($q_{\Delta i}$) deviation compared to the ideal case is ± 20° for frequencies from 3 to 9.9 GHz.
Although Fig. 12 shows a good behaviour of q-point characteristics, better results could be obtained if the factors k, $\O_0$ and Δf were included in the design specifications. In the present case, the design of seven-port reflectometer was accomplished by just integrating individually designed Q and D components.
There is one remaining criterion of performance of the designed seven-port reflectometer and it concerns the magnitude of reference point q_3. The simulated and measured results for $|q_3|$ are shown in Fig. 13. They are dissimilar. However in the both cases the $|q_3|$ values are greater than 4.4. These results indicate that the reflectometer fulfils the optimum design specification of $|q_i|<|q_3|$.

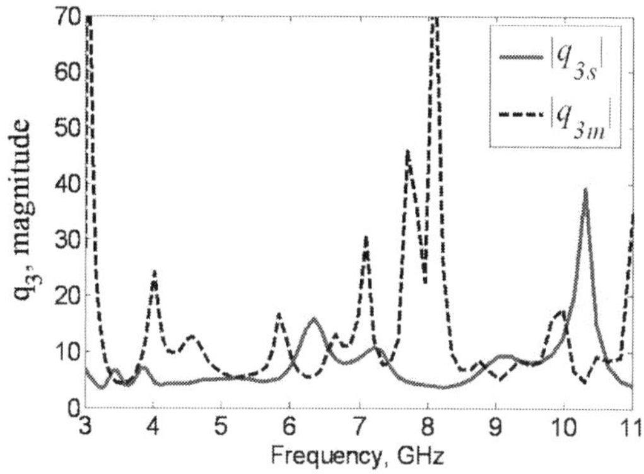

Fig. 13. Simulated and measured magnitude of q_3.

5. Calibration Procedure

Following its successful design and development, the reflectometer is calibrated prior to performing measurements. A suitable calibration procedure to the reflectometer offers high measurement accuracy that can be obtained with the error correction techniques. There are various methods for calibrating multi-port reflectometers. The differences between these methods include the number of standards, restrictions on the type of standards and the amount of computational effort needed to find the calibration constants (Hunter & Somlo, 1985). In (Hoer, 1975), Hoer suggested to calibrate a six-port network for the net power measurement. In this case, Port 2 (measurement port) is terminated with a power standard. The known power standard can be expressed as:

$$P_{std} = \sum_{i=3}^{6} u_i P_i \quad (31)$$

Then, the procedure is repeated with connecting three or more different offset shorts to replace power standard. The sliding short or variable lossless reactance also can be used. Therefore, the real net power at Port 2 is zero.

$$0 = \sum_{i=3}^{6} u_i P_i \quad (32)$$

The net power into unknown impedance can be measured with the known u_i real constants. P_i is also observed for two or more positions of a low reflection termination. This is an addition to the P_i for the three or more different positions of an offset or sliding short. After performing this set of measurements, all constants state which one requires to calculate reflection coefficient are determined (Hoer, 1975).

Calibration algorithms proposed in (Li & Bosisio, 1982) and (Riblet & Hanson, 1982) assume the use of ideal lossless standards having $|\Gamma|=1$. This notion was criticized by Hunter and Somlo which claimed that this would lead to measurement inaccuracies since practical standards are never lossless (Somlo & Hunter, 1982; Hunter & Somlo, 1985). Therefore, the information on the used non-ideal standards is important when high reflectometer accuracy is required. This information has to be used in the calibration algorithm. To perform the calibration process, Hunter and Somlo presented an explicit non-iterative calibration method requiring five standards. They suggested that one of the standards should be near match. This is to ensure the improvement of the performance of the calibrated reflectometer near the centre of the Smith chart (Somlo, 1983). The other four standards are short circuits offset by approximately 90° (Hunter & Somlo, 1985). These standards are convenient because of their ready availability. Also their use is beneficial in that their distribution is likely to avoid the accuracy degradation which can occur when measuring in areas of the Smith chart remote from a calibrating standard (Hunter & Somlo, 1985).

An alternative full calibration algorithm can be also obtained using 6 calibration standards (Somlo & Hunter, 1982). The proposed standards used in the procedure are four phased short-circuits (Γ_1, Γ_2, Γ_3, Γ_4), a matched load (Γ_5) and an intermediate termination ($0.3 \leq |\Gamma_6| \leq 0.7$). It is based on the general reflection coefficient six-port equation (9) and is separated into two equations of real, r and imaginary, x part as (Somlo & Hunter, 1982):

$$r = \frac{\sum_{i=3}^{6} c_i P_i}{\sum_{i=3}^{6} \beta_i P_i} \quad (33)$$

$$x = \frac{\sum_{i=3}^{6} s_i P_i}{\sum_{i=3}^{6} \beta_i P_i} \quad (34)$$

The constants are normalized by setting β_6 equal to 1. The other 11 real constants can be determined from the calibration (Somlo & Hunter, 1982). Then, equation (33) and (34) can be rewritten as:

$$\Sigma_{i=3}^{6} c_i P_i - r\Sigma_{i=3}^{5} \beta_i P_i = rP_6 \tag{35}$$

$$\Sigma_{i=3}^{6} s_i P_i - x\Sigma_{i=3}^{5} \beta_i P_i = xP_6 \tag{36}$$

These two equations are used to determine 11 real constants in the calibration procedure. The matrix to calculate the constants is given by (37) (Somlo & Hunter, 1982):

$$\begin{bmatrix} c_3 \\ : \\ c_6 \\ s_3 \\ : \\ s_6 \\ \beta_3 \\ : \\ \beta_6 \end{bmatrix} = \begin{bmatrix} P_{13} & .. & P_{16} & 0 & .. & 0 & -r_1P_{13} & .. & -r_1P_{15} \\ : & .. & & & & & & & \\ P_{43} & .. & P_{46} & 0 & .. & 0 & -r_4P_{43} & .. & -r_4P_{45} \\ 0 & .. & 0 & P_{13} & .. & P_{16} & -x_1P_{13} & .. & -x_1P_{15} \\ : & .. & & & & & & & \\ 0 & .. & 0 & P_{43} & .. & P_{46} & -x_4P_{43} & .. & -x_4P_{45} \\ P_{53} & .. & P_{56} & 0 & .. & & & & 0 \\ 0 & .. & 0 & P_{53} & .. & P_{56} & 0 & .. & 0 \\ P_{33} & .. & P_{66} & 0 & .. & 0 & -r_6P_{63} & .. & -r_6P_{65} \end{bmatrix}^{-1} \begin{bmatrix} r_1P_{16} \\ : \\ r_4P_{46} \\ x_1P_{16} \\ : \\ x_4P_{46} \\ 0 \\ 0 \\ r_6P_{66} \end{bmatrix} \tag{37}$$

where P_{ti} is a measured power at ith port when tth calibrating termination is connected to the measuring port.

From the above described alternative calibration techniques, it is apparent that the use of three broadband fixed standards such as open, short and match required in the conventional heterodyne based reflectometer is insufficient to calibrate a six-port reflectometer. To complete the calibration, at least two extra loads are required. To achieve the greatest possible spacing for the best calibration accuracy, it is beneficial to phase the offset shorts by 90° (Hunter & Somlo, 1985). Woods stated in (Woods, 1990) that to apply this ideal condition at many frequency points would require repeated tuning of standards. It may be time consuming and would rely on the expert operator (Woods, 1990). Because of these reasons, it may be appropriate to ease the ideal condition on 90° phasing of the sliding loads in favour of least adjustments to the standards (Woods, 1990). Assuming the standards are phased by at least 45° to obtain sufficient calibration accuracy, fixed positions of the short could be employed over a bandwidth of approximately 5:1 (Riblet & Hanson, 1982).

To calibrate the developed reflectometer, the method using six calibration standards, as proposed by Hunter and Somlo in (Somlo & Hunter, 1982), is chosen. This method offers a straight forward solution for the reflectometer constants and employs simple equations, which lead to the easy practical implementation of the calibration algorithm

In the chosen calibration procedure, three coaxial standard loads (matched load, open and short circuit), two phased-short circuits and an intermediate termination with magnitude of approximately 0.5 are used. For the last standard, a 3 dB coaxial attenuator open-circuited at its end is utilized. The information about the electrical characteristics of these standards in

the frequency band of 3 to 11 GHz is obtained from measurements performed with the conventional Vector Network Analyser (HP8510C). This information is used for the values r and x in equations (33) and (34). Knowing r and x, the calibration constants c_i, s_i and β_i are determined from solving the matrix equation similar to the one in (37).
The operation of the developed seven-port reflectometer is assessed by assuming an ideal operation of power detectors. To achieve this task in practice, the power values required in (33) and (34) are obtained from the measured S-parameters of the seven-port reflectometer with DUT present at Port 2. Therefore, $P_i = |S_{i1}|^2$ for i=4, 5, 6, 7, where S_{i1} is the transmission coefficient between port 1 and port i when port 2 is terminated with DUT.
The validity of the calibration method and measurement accuracy is verified by comparing the characteristics of three open-circuited coaxial attenuators of 3, 6 and 10 dB (Fig. 14) as measured by the seven-port reflectometer with those obtained using the conventional VNA (HP8510C). For the reflectometer, the complex reflection coefficient values are determined using equation (9).

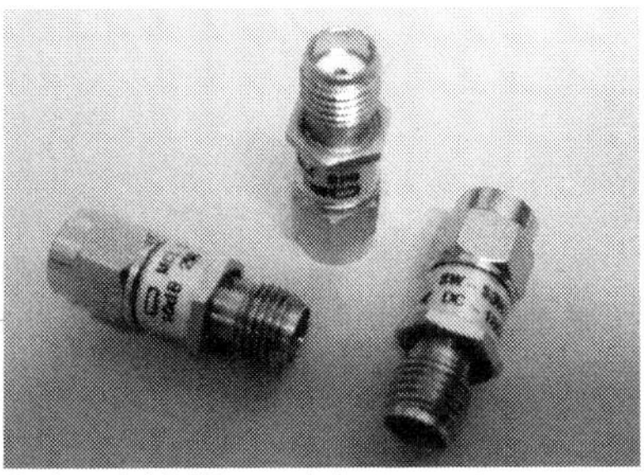

Fig. 14. Photograph of the 3, 6 and 10 dB coaxial attenuators.

The two sets of measured results for the magnitudes and phases of reflection coefficient are presented in Fig. 15 and Fig. 16.

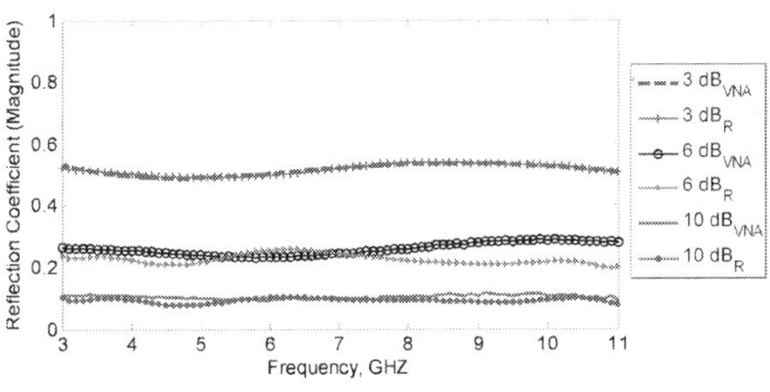

Fig. 15. Measured magnitude of reflection coefficient for three coaxial attenuators: 3, 6 and 10 dB obtained using the developed reflectometer (R) and VNA HP8510C (VNA).

As observed in Fig. 15, HP8510C provides the measured $|\Gamma|$ of 0.51 ± 0.02 for 3 dB, 0.25 ± 0.03 for 6 dB and 0.1 ± 0.05 for the 10 dB attenuator across the investigated frequency band. The calibrated seven-port reflectometer gives comparable results for $|\Gamma|$ which are 0.51 ± 0.02 for 3 dB, 0.22 ± 0.03 for 6 dB, and 0.1 ± 0.01 for the 10 dB attenuator.

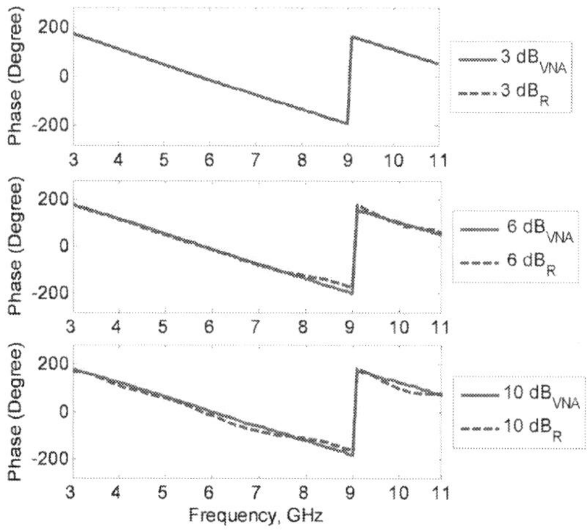

Fig. 16. Comparison of measured phase characteristic reflection coefficients of three coaxial attenuators of 3, 6 and 10 dB obtained using the developed reflectometer (R) and VNA HP8510C (VNA).

The best agreement occurs for the 3 dB attenuator, which was used in the calibration procedure. This agreement indicates validity of the calibration procedure as well as a very high measurement repeatability of the two instruments. The worst agreement between the reflectometer and the VNA measured results looks to be for the 6 dB attenuator, which is observed for the frequency range between 8 and 11 GHz. In all of the remaining cases the agreement is quite good. The observed discrepancies are due to the limited range of off-set shorts.

Because the attenuators have the same length, it is expected that they should have similar phase characteristics of reflection coefficient. This is confirmed by the phase results obtained by the reflectometer and the VNA, as shown in Fig. 16. An excellent agreement for the phase characteristic of 3 dB attenuator obtained with the reflectometer and the VNA again confirms excellent repeatability of the two instruments. For the remaining 6 and 10 dB attenuators there are slight differences of about ± 10° between the results obtained with the reflectometer and the VNA for some limited frequency ranges. Otherwise the overall agreement is very good indicating that the designed seven-port reflectometer operates quite well across the entire ultra wide frequency band of 3 to 11 GHz. Its special attributes are that it is very compact in size and low-cost to manufacture.

6. Applications

The designed seven-port reflectometer can be used in many applications requiring the measurement of a complex reflection coefficient. There is already an extensive literature on applications of multi-port reflectometers with the main focus on six-ports.

Initially, the six-port reflectometer was developed for metrological purposes (Bilik, 2002). The metrological applications benefit from the high stability of six-port reflectometer compared to other systems. Because of this reason, National Institute of Standards and Technology (NIST), USA has been using this type instrument from the 1970s (Engen, 1992), (Bilik, 2002).

Nowadays, six-port techniques find many more applications. For example, there are a number of works proposing six-port networks as communication receivers (Hentschel, 2005; Li et al., 1995; Visan et al., 2000). In this case, input to the six-port consists of two RF (radio frequency) of signals, one being a reference and the other one, an actual received signal. Different phase shifts and attenuations are used between the couplers, dividers or hybrids forming the six-port so that by the vector addition the two RF input signals generate different phases at four output ports of the six-port. The signal levels of the four baseband output signals are then detected using Schottky diode detectors. By applying an appropriate baseband signal processing algorithm, the magnitude and phase of the unknown received signal can thus be determined for a given modulation and coding scheme (Li et al., 1995; Visan et al., 2000). The six-port technique can also be applied to the transmitter with an appropriate modulation. Therefore, the six-port technique can be used to build a microwave transceiver. A particular use is foreseen in digital communication systems employing quadrature phase shift keying (QPSK), quadrature amplitude modulation (QAM) or code division multiple access (CDMA) (Xu et al., 2005).

Six-port techniques can be also used to build microwave locating systems, as explained in (Hunter & Somlo, 1985). This application requires and extra step to convert the frequency domain results to time- or space-domain. The required task can be accomplished using an Inverse Fast Fourier Transform (IFFT) to the data measured in the frequency-domain. The procedure leads to so-called step frequency pulse synthesis technique illustrated in Fig. 17. As seen in Fig. 17, a constant magnitude signal spanned from 3.5 to 9 GHz is equivalent to a sub-nanosecond pulse in the time domain.

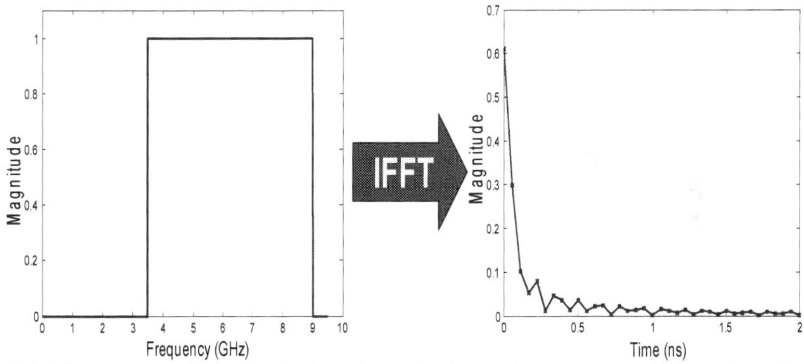

Fig. 17. Illustration of synthesized pulse technique: frequency and time domain pulse representation.

The locating reflectometer can be used to investigate waveguide discontinuities, as shown in (Hunter & Somlo, 1985), as well as to build a UWB radar system to measure distances in free space (Noon & Bialkowski, 1993) or perform internal imaging of objects (Bialkowski et al., 2006). The image of a scattering object in time/space domain can be constructed from the scattering signal measured at different viewing angles (Lu & Chu, 1999). Such monostatic radar systems (Edde, 1995) can be realized by connecting a UWB antenna to the port allocated for DUT in the developed seven-port reflectometer. The potential of using a reflectometer in a microwave imaging system is illustrated in Fig. 18.

In the presented setup, a UWB microwave source is connected to Port 1 while an antenna is connected to Port 2.

In the system illustrated in Fig. 18, the antenna transmits a step-frequency synthesized pulse signal to the object. The reflected signal from the object is received by the same antenna. The measured powers by scalar power detectors at Port 3-7 are converted to digital form by a precision Analog to Digital Converter (ADC). A PC included in this system provides control of the source, the reflectometer and ADC. Also it is used for data collection and post-processing. A UWB microwave system similar to the one shown in Fig. 18 aiming for an early detection of breast cancer is under development at the University of Queensland (Khor et al., 2007).

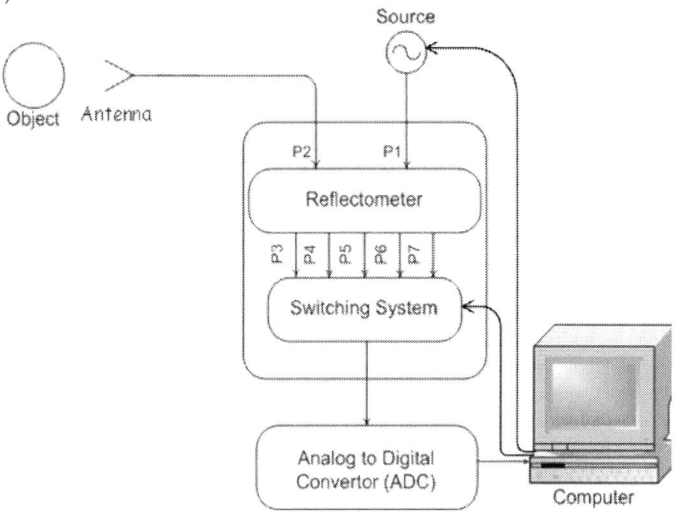

Fig. 18. Configuration of a microwave imaging system using a seven-port reflectometer.

7. Conclusion

This chapter has described a multi-port reflectometer which employs scalar instead of complex ratio detection techniques to determine the complex reflection coefficient of a given Device Under Test. The operation and optimum design principles of this type of microwave measurement instrument have been explained. Following that, the design of a seven-port reflectometer in microstrip-slot multilayer technology formed by five couplers and one in-phase power divider operating over an ultra wide frequency band of 3.1 to 10.6 GHz has been presented. It has been shown that the seven-port network forming this reflectometer fulfils optimum design requirements. The calibration procedure involving the use of six calibration standards of match load, open, short, two phased-shorts and an intermediate termination have been described for this reflectometer. The performance of the developed reflectometer has been evaluated for 3 different attenuators. The obtained results have shown that the designed device can be confidently used for UWB measurements. Possible applications of the developed device in communications, microwave imaging and metrology field have been pointed out and briefly explained.

8. References

Bialkowski, M. E.; Khor, W.C. & Crozier, S. (2006). A planar microwave imaging system with step-frequency synthesized pulse using different calibration methods. *Microwave and Optical Technology Letters*, Vol. 48, No 3, 2006, pp. 511-516, ISSN. 1098-2760.

Bilik, V. (2002). Six-Port Measurement Technique: Theory and Applications, *Proceeding of Radioelectronika 2002*, May 2002, ISBN. 80-227-1700-2.

Edde, B. (1995). *Radar: principles, technology, applications*, Prentice Hall, ISBN. 978-0-13-752346-7, Englewood Cliffs, New Jersey.

Engen, G. F. (1969). An introduction to the description and evaluation of microwave systems using terminal invariant parameters. *NBS Monograph 112*, October 1969.

Engen, G. F. & Hoer, C. A. (1972). Application of arbitrary six-port junction to power measurement problems. *IEEE Transactions on Instrument and Measurement*, Vol. IM-21, November 1972, pp. 470-474, ISSN. 0018-9456.

Engen, G.F. (1977). The six port reflectometer: an alternative network analyzer. *IEEE Transactions on Microwave Theory and Techniques*, Vol. 25, No. 12, December 1977, pp. 1075-1080, ISSN. 0018-9480.

Engen, G.F. (1977). An improved circuit for implementing the six-port technique of microwave measurements. *IEEE Transactions on Microwave Theory and Techniques*, Vol. MTT-25, No.12, December 1977, pp. 1080-1083, ISSN. 0018-9480.

Engen, G.F. (1980). A least squares solution for the use in the six-port measurement technique. *IEEE Transactions on Microwave Theory and Techniques*, Vol. MTT-28, No. 12, December 1980, pp. 1473-1477, ISSN. 0018-9480.

Engen, G.F. (1992). *Microwave circuit theory and foundation of microwave metrology*, IET, ISBN.0-86341-287-4, London, England.

Engen, G.F. (1997). A (historical) review of the six-port measurement technique. *IEEE Transactions on Microwave Theory and Techniques*, Vol. 45, No. 6, December 1997, pp. 2414-2417, ISSN. 0018-9480.

Hentschel, T. (2005). The six-port as a communications receiver. *IEEE Transactions on Microwave Theory and Techniques*, Vol. 53, No. 3, March 2005, pp. 1039-1047, ISSN. 0018-9480.

Hoer, C. A. & Engen, G. F. (1973). Analysis of a six-port junction for measuring *v, I, a, b, z*, Γ and phase. *Proceeding of IMEKO Symposium on Acquisition and Processing of Measuring Data for Automation*, ISBN. 9780444106858, Dresden, Germany, June 1973, North-Holland Pub Co.

Hoer, C.A. (1975). Using six-port and eight-port junctions to measure active and passive circuit parameters. *NBS Technical Note 673*, September 1975.

Hoer, C.A. & Roe, K.C. (1975). Using and arbitrary six-port junction to measure complex voltage ratios. *IEEE Transactions on Microwave Theory and Techniques*, Vol. 23, No. 12, December 1975, pp. 978–984, ISSN. 0018-9480.

Hoer, C.A. (1977). A network analyzer incorporating two six-port reflectometers. *IEEE Transactions on Microwave Theory and Techniques,* Vol. 25, No. 12, December 1977, pp. 1070–1074, ISSN. 0018-9480.

Hunter, J.D. & Somlo, P.I. (1985). An explicit six-port calibration method using 5 standards. *IEEE Transactions on Microwave Theory and Techniques,* Vol. MTT-31, No. 1, January 1985, pp. 69-72, ISSN. 0018-9480.

Khor, W.C.; Bialkowski, M. E.; Abbosh, A. M.; Seman, N., & Crozier, S. (2007). An ultra wideband microwave imaging system for breast cancer detection. *IEICE Transactions on Communications,* Vol. E85-A/B/C/D, No. 1, September 2007, pp. 2376 - 2381, ISSN. 0916-8516.

Li, J.; Bosisio, R. G. & Wu, K. (1995). Computer and measurement simulation of a new digital receiver operating directly at millimeter-wave frequencies. *IEEE Transactions on Microwave Theory and Techniques,* Vol. 43, No. 12, December 1995, pp. 2766-2772, ISSN. 0018-9480.

Li, S. & Bosisio, R. G. (1982). Calibration of multiport reflectometers by means of four open/short circuits. *IEEE Transactions on Microwave Theory and Techniques,* Vol. MTT-30, No. 12, July 1982, pp. 1085-1089, ISSN. 0018-9480.

Lu, H. C. & Chu, T. H. (1999). Microwave diversity imaging using six-port reflectometer. *IEEE Transactions on Microwave Theory and Techniques,* Vol. 47, No.1, January 1999, pp. 84-87, ISSN. 0018-9480.

Noon, D. A. & Bialkowski, M. E. (1993). An inexpensive microwave distance measuring system. *Microwave and Optical Technology Letters,* Vol. 6, No. 5, April 1993, pp. 287-292, ISSN. 1098-2760.

Probert, P. J. & Carroll, J. E. (1982). Design features of multi-port reflectometers. *IEE Proceedings. H, Microwaves, Antennas, and Propagation,* Vol. 129, No. 5, October 1982, pp. 245-252, ISSN. 0143-7097.

Riblet, G. P. & Hanson, E. R. B. (1982). Aspects of the calibration of a single six-port using a load and offset reflection standards. *IEEE Transactions on Microwave Theory and Techniques,* Vol. MTT-30, No. 12, Dec. 1982, pp. 2120-2124, ISSN. 0018-9480.

Seman, N.; Bialkowski M. E. & Khor, W. C. (2007). Ultra wideband vias and power dividers in microstrip-slot technology, *2007 Asia-Pacific Microwave Conference,* Vol. 3, pp. 1383 - 1386, ISBN: 978-1-4244-0748-4, Thailand, December 2007, IEEE, Bangkok.

Seman, N. & Bialkowski M. E. (2009). Design and analysis of an ultrawideband three-section microstrip-slot coupler. *Microwave and Optical Technology Letters,* Vol. 51, No. 8, August 2009, pp. 1889-1892, ISSN. 1098-2760.

Somlo, P. I. & Hunter, J. D. (1982). A six-port reflectometer and its complete characterisation by convenient calibration procedures. *IEEE Transactions on Microwave Theory and Techniques,* Vol. MTT-30, No. 2, February 1982, pp. 186-192, ISSN. 0018-9480.

Somlo, P.I (1983). The case for using a matched load standard for six-port calibration. *Electronic Letters,* Vol. 19, No. 23, November 1983, pp. 979-980, ISSN: 0013-5194.

Somlo, P. I. & Hunter, J. D. (1985). *Microwave impedance measurement,* Peter Peregrinus Ltd., ISBN. 0-86341-033-2, London.

Visan, T.; Beauvais, J. & Bosisio, R. G. (2000). New phase and gain imbalance correction algorithm for six port based direct digital millimetric receivers. *Microwave and Optical Technology Letters,* Vol. 27, No. 6, December 2000, pp. 432-438, ISSN. 1098-2760.

Waterhouse, R. D. (1990). *Millimeter-wave frequency-domain reflectometers using Schotty-Barrier Diode Detectors*. Ph.D. Dissertation, The University of Queensland, Australia.

Woods, G. S. (1990). *A computer controlled six-port network analyser*. Ph.D. Dissertation, James Cook University of North Queensland, Australia.

Xu, X.; Wu, K. & Bosisio, R. G. (2005). Six-Port Networks. *Wiley Encyclopaedia of RF and Microwave Engineering*, Vol. 5, February 2005, A John Wiley & Sons Inc., pp. 4641-4669, ISBN. 978-0-471-27053-9.

Yao, J. J. & Yeo, S. P. (2008). Six-port reflectometer based on modified hybrid couplers. *IEEE Transactions on Microwave Theory and Techniques,* Vol. MTT-56, No. 2, February 2008, pp. 493-498, ISSN. 0018-9480.

Yao, J. J. (2008). *Modifying design of four-port couplers for enhanced six-port reflectometer performance*. Ph.D. Dissertation, National University of Singapore, Singapore.

17

Broadband Complex Permittivity Determination for Biomedical Applications

Radim Zajíček and Jan Vrba
Czech Technical University in Prague, Dept. of Electromagnetic Field, FEE
Czech Republic

1. Introduction

Medicine has the essential profit from microwave technique such as not only a development of new devices but also an improvement of existing devices. Generally, we want to *Look and See* using microwaves in the medical diagnostics and imaging and to *Heat and Destroy* in the medical therapy. But also the non-thermal effects of electromagnetic fields have a serious part in studying the biological effects of electromagnetic fields.

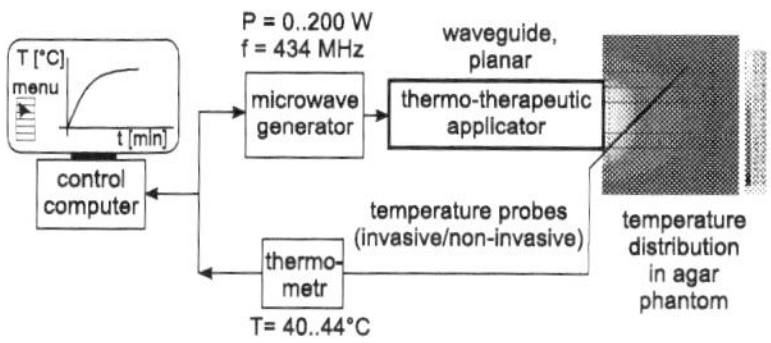

Fig. 1. Therapeutic Application of Microwave Technique: Microwave Hyperthermia

A knowledge of the dielectric parameters of materials is important for microwave or radio engineers involved in the analysis and synthesis of devices. Relative permittivity, loss factor and conductivity are the input parameters for electromagnetic field modelling and simulations. Although for many materials these parameters can be found in the tables, their experimental determination is very often necessary.

1.1 Applications of Microwaves in Medicine

The dielectric properties of biological tissues are the determining factors for the dissipation of electromagnetic energy in the human body and they are therefore the basic parameters for hyperthermia cancer treatment (Fig. 1). The measurement of the dielectric parameters of biological tissues is also a promising method in medical diagnostics and imaging. Knowledge of the complex permittivity[1] in an area under treatment, i.e. knowledge of the complex permittivity of healthy and tumor tissue, is extremely important for example in diagnosing tumor cell-nests in the human body or in the design of thermo-therapeutic applicators which transform electromagnetic energy into thermal energy in pathological tissue (Vrba, 2003).

[1] Complex permittivity is also known as a dielectric constant in literature.

Let's summarize the basic characteristics of microwaves, their advantages and limitations, and applications in the medicine:

General characteristic:

- from 100 MHz to 30 GHz frequency range
- diagnostic applications: a tumor detection based on differences in the tissue electrical properties
- therapeutic applications: a treatment based on the local heating or the regional (whole-body) heating - hyperthermia integrated with MRI, prostate hyperplasia, heart and other tissue ablation, angioplasty
- other applications: radiometry, telemetry, motion detection

Advantages of microwaves:

- offer a wide frequency range
- an ability to focus the energy
- a variety of simulation tools (methods for field solving[2])
- a relatively low cost of microwave components and devices
- a low if any health risk

Limitations of microwaves:

- a spatial resolution
- penetration depth of electromagnetic waves
- electromagnetic interferences

Summary of the human characteristics from microwave view point:

- differences in tissue properties (normal/tumor tissue, low/high water content)
- scattering of complex patterns of fields in the body
- individual anatomical differences

1.2 Complex permittivity

The complex permittivity is a quantity which desribes the electrical properties of materials. In case of non-conductors, dielectrics, the complex permittivity describes an interaction between the dielectric and the applied external electric field.

Polarization

The interaction of an electric field with a biological tissue has the origin in the response of the charge particles to the applied field. The displacement of these charge particles from their equilibrium positions gives rise to induced dipoles which respond to the applied field. Such induced polarization arises mainly from the displacement of electrons around nuclei (electronic polarization) or due to the relative displacement of atomic nuclei because of the

[2] FEM - Finite Element Method is utilized mostly in frequency domain, body parts are represented by surfaces and volumes are divided into tetrahedrons. FDTD - Finite Difference in Time Domain utilized voxel representation of body tissues.

unequal distribution of charge in molecule formation (atomic polarization). In addition to induced dipoles some dielectrics, known as polar dielectrics, contain permanent dipoles due to the asymmetric charge distribution of unlike charge partners in a molecule which tend to reorientation under the influence of a changing electric field, thus giving rise to orientation polarization. Finally, another source of polarization arises from charge build-up in interfaces

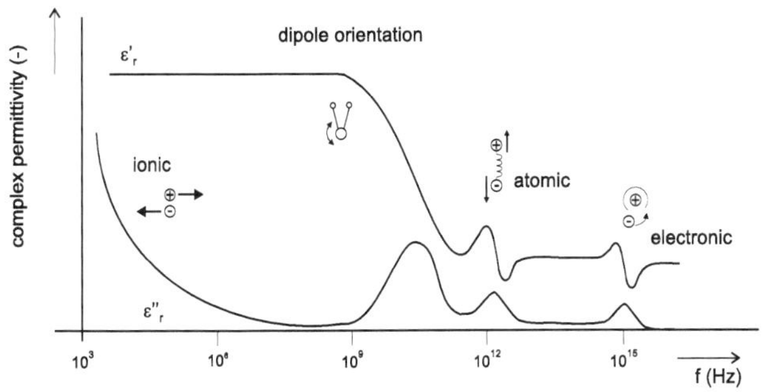

Fig. 2. Polarization effects at a broad frequency range

between components in heterogeneous systems, termed interfacial, space charge or Maxwell-Wagner polarization. The Maxwell-Wagner polarization and orientation polarization due to an alternating electric field together with d.c. conductivity are the basic of thermal effect of microwaves (Kittel, 1966).
Permittivity is known from the physics or theory of electromagnetic field as

$$\varepsilon = \varepsilon_0 \varepsilon_c \tag{1}$$

where ε_0 is free space permittivity and ε_c is complex relative permittivity (dielectrics are very often lossy). Complex relative permittivity can be given in turn as

$$\varepsilon_c = \varepsilon_r - j\varepsilon_r tan\ \delta \tag{2}$$

where ε_r is a real part of complex relative permittivity[3] and tan δ is the loss factor. For purely conductive losses

$$tan\ \delta = \frac{\sigma}{\omega\varepsilon_0\varepsilon_r} \tag{3}$$

applies, where σ is the medium conductivity.

Derivation of Complex Permittivity

It would be helpful if, through some elementary analysis, the complex nature of permittivity is demonstrated without having to assume this premise from the start. Amper's circuital law in its elementary form contains all the necessary components needed for this analysis. Maxwell modified Ampere's law by including a displacement current density term for sinusoidal electric field variations

$$rot\mathbf{H} = \mathbf{J} + j\omega\mathbf{D} \tag{4}$$

[3] The real part of complex relative permittivity is very often called only relative permittivity. One must carefully consider where it is possible (for example for the simplification of terms) to reduce complex relative permittivity to only relative permittivity.

where **H** is magnetic field strength, **J** is current density, **D** is electric flux density and t is time. The conduction current density is a function of the electric field vector (Ohm's law)

$$\mathbf{J} = \sigma \mathbf{E} \tag{5}$$

where σ is medium conductivity. Using material formula $\mathbf{D} = \varepsilon \mathbf{E}$ and $\varepsilon = \varepsilon_0 \varepsilon_r$ is

$$rot\mathbf{H} = \sigma \mathbf{E} + j\omega\varepsilon_0\varepsilon_r\mathbf{E} = j\omega\varepsilon_0\mathbf{E}\left(\varepsilon_r + \frac{\sigma}{j\omega\varepsilon_0}\right) = j\omega\varepsilon_0\mathbf{E}\left(\varepsilon_r - j\frac{\sigma}{\omega\varepsilon_0}\right) \tag{6}$$

Complex relative permittivity is finally expressed

$$\varepsilon_c = \varepsilon_r^{'} - j\varepsilon_r^{''} = \varepsilon_r - j\frac{\sigma}{\omega\varepsilon_0} \tag{7}$$

1.3 Complex permittivity determinaton

A more or less extensive tabulation of the dielectric properties of materials selected to illustrate the theoretical deliberations provided by the authors is common to all papers in the field of complex permittivity measurement. Nowadays, papers are instigated for example by the need for such information for electromagnetic dosimetry. To date there is no consensus on the dielectric data (Gabriel, 1996).

Systems using complex permittivity measurement exist, and some are even commercially available (e.g. the Agilent Technologies 85070 analyzer) but the Czech Technical University in Prague is not equipped with this type of material measurement system. Our work is motivated by the need for cheap[4] but accurate determination of the complex permittivity with the measurement technique which is also easy to manufacture and therefore would be interesting for many researchers.

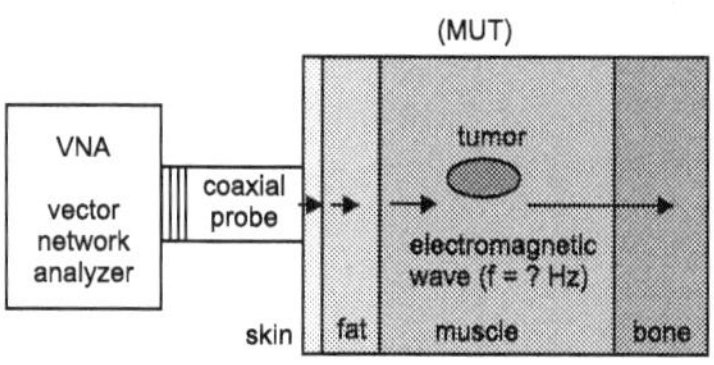

Fig. 3. This illustrates the principle of the reflection method - a medical diagnostics and imaging concept using microwaves. The different equivalent penetration depths of microwaves at different frequencies enable us to diagnose the tissues (layers) in the area under treatment.

2. Materials and Methods

There are several methods for measuring the complex permittivity - the resonance method, measurement in free space and the transmission line method (Fig. 4). If we want to use a broadband measurement method that is non-destructive and non-invasive, we should choose a reflection method on an open-ended coaxial line-a method based on the reflection coefficient measurement of an open-ended coaxial line attached to the sample of the measured dielectric.

[4] The cost of mentioned Agilent system is $5000 approximately, http://www.agilent.com.

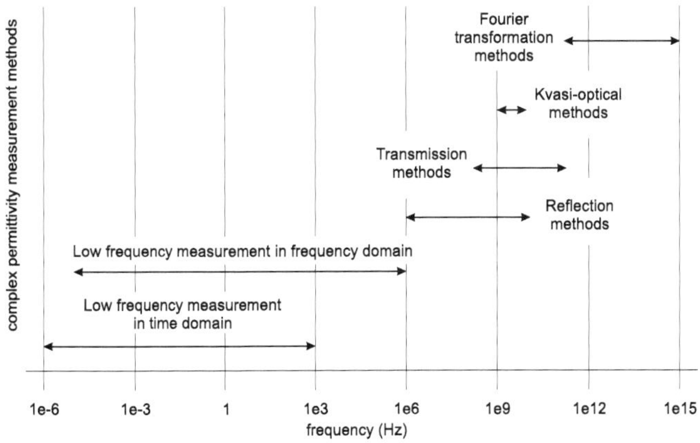

Fig. 4. Complex permittivity determination in a broad frequency band

2.1 Reflection Coefficient

Reflection coefficient relates the tangential components of the incident and the reflected electric fields and is a frequency dependent and complex quantity[5]

$$\Gamma = \frac{\mathbf{E}_r}{\mathbf{E}_i} \tag{8}$$

where $\mathbf{E}_r$ is the reflected wave and $\mathbf{E}_i$ is the incident wave (Novotný, 2005).

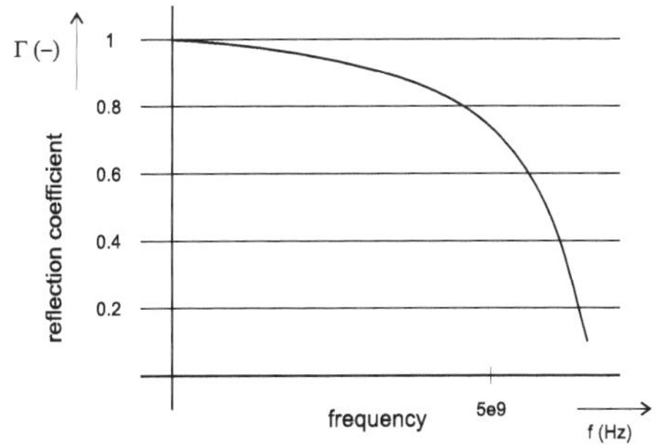

Fig. 5. Frequency behavior of the reflection coefficient

In case of interface between two materials with different impedances, the reflection coefficient is defined

$$\Gamma = \frac{\mathbf{Z}_1 - Z_0}{\mathbf{Z}_1 + Z_0} \tag{9}$$

[5] It is important to note the reflection coefficient is a voltage quantity and it is related to a power quantity with the term $\Gamma_P \approx \Gamma^2$. It is possible to find the reflection coefficient as R, Γ or S_{11} in technical literature.

where Z_0 and $\mathbf{Z}_1$ are impedances of materials-in described measurement technique the impedance of coaxial line is $Z_0 = 50\ \Omega$ and the impedance of a MUT sample is $\mathbf{Z}_1$ (generally complex impedance).

2.2 Relation between Reflection and Permittivity

The reflection method entails the measurement of the reflection coefficient on the interface between two materials, on the open end of the coaxial line (as a detector) and the material being tested (MUT, Fig. 6). It is a well-known method for determining dielectric parameters

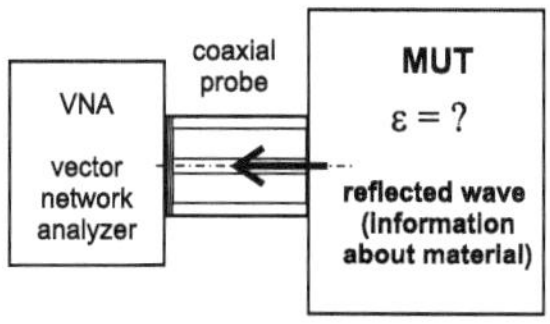

Fig. 6. Dielectric measurement from the viewpoint of electromagnetic field theory and wave propagation on the interface between the coaxial probe and the MUT

(Stuchly et al., 1982). This method is based on the fact that the reflection coefficient of an open-ended coaxial line depends on the dielectric parameters of the MUT that is attached to it. To calculate the complex permittivity from the measured reflection coefficient, it is useful to use an equivalent circuit of an open-ended coaxial line. Fig. 6 shows the complex permittivity measurement from the point of view of electromagnetic field theory and the propagation of the electromagnetic wave on the interface between two materials with different impedances (Novotný, 2001). The probe translates changes in the permittivity of a MUT into changes in the input reflection coefficient of the probe.

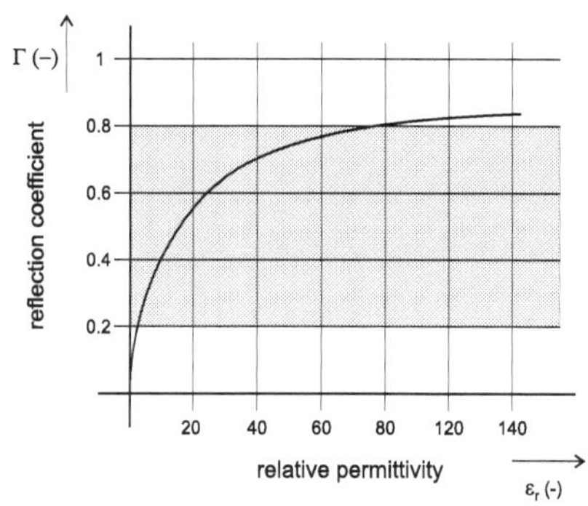

Fig. 7. Reflection coefficient versus relative permittivity

The reflection coefficient is varying a lot for the small values of relative permittivity (approximately less then 20, Fig. 7). Complex permittivity measurement is more sensitive and hence precise. Conversely, for high values of relative permittivity (for example between 70 and 90) there is a little change of the reflection coefficient and the measurement has more uncertainties.

The interface between the measurement probe and a MUT sample represents an impedance jump because of biological tissues have extremely high permittivity values. At low frequencies their permittivity can be more than 100 and the value of the loss factor more than 0.1. An accurate evaluation is very difficult because the reflection coefficient is close to 1 (Fig. 5). This means that only a very small part of the incident energy penetrates into the sample.
The surface of the sample of MUT must be in perfect contact with the probe. The thickness of a measured sample must be at least twice an equivalent penetration depth of the electromagnetic wave d. This ensures that the waves reflected from the far MUT interface are attenuated by approximately -35 dB, which assures that their effect on the measured reflection coefficient is insignificant.

$$d = \sqrt{\frac{2}{\omega\mu\sigma}} = \frac{1}{\omega}\sqrt{\frac{2}{\mu\varepsilon_0\varepsilon_r tan\,\delta}} \tag{10}$$

The dependence of equivalent penetration depth d on dielectric parameters ε_r and tan δ and also frequency f denotes Eq. 10. Typical values of d for distilled water and biological muscle tissue at different frequencies are summarized in Tab. 1.

Materials	f (MHz)	$2d$ (mm)
distilled water	434	71.2
	915	34.0
	2450	12.8
muscle tissue	434	25.9
	915	24.2
	2450	15.6

Table 1. Equivalent penetration depth of the electromagnetic wave into different materials: distilled water and muscle tissue

2.3 Coaxial Probes

When a wave has only transverse field components for both the electric and magnetic field and no longitudinal components ($E_z = 0$ and $H_z = 0$), the wave is transverse electromagnetic (TEM). Transverse electromagnetic waves are very much appreciated in practice because they

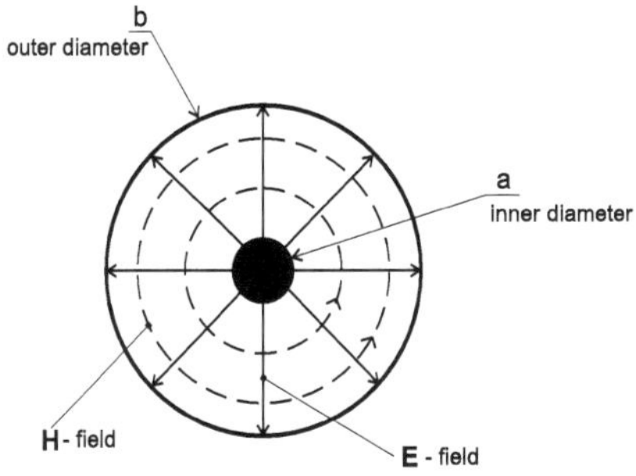

Fig. 8. Coaxial transmission line.

have only four components, with no longitudinal components. On the other hand, uniform plane waves also characterize a very simple structure. As an example, the very well known coaxial cable has a TEM wave as the main propagating mode (Fig. 8). The surface of constant phase is indeed a plane, perpendicular to the direction of propagation, hence the wave is planar. The field, however, is not constant over the plane because it varies between the two conductors according to a $1/r$ law. Hence, the TEM wave in a coaxial cable is a nonuniform plane wave.

N&SMA Connector Type Probe

N type coaxial connectors are medium size units which have constant 50 Ω impedance and provide radio frequency performance up to 11 GHz. SMA (SubMiniature, version A) connectors are coaxial RF connectors developed as a minimal connector interface for coaxial cable with a screw type coupling mechanism. The connector has a 50 Ω impedance. It offers excellent electrical performance from DC to 18 GHz.

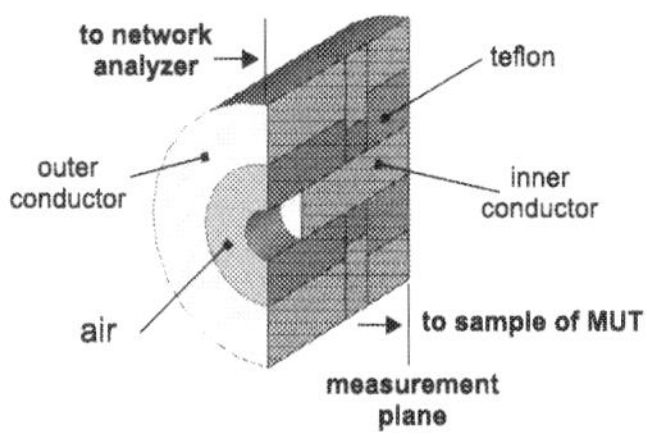

Fig. 9. Measurement coaxial probe: description and dimensions - inner/outer diameter: N connector type probe 3/7 mm, SMA connector type probe 1.3/4 mm

We have adapted the standard N and SMA connectors for this measurement method from which the parts for connecting to a panel were removed (Fig. 9). The measurement probe can be described by the equivalent circuit consisting of fringing capacitance between inner and outer conductor out of the coaxial structure and radiating conductance which represents propagation losses (Fig. 10). The capacitance and conductance are frequency and permittivity dependent and are also dependent on the dimensions (inner and outer diameters) of the probe.

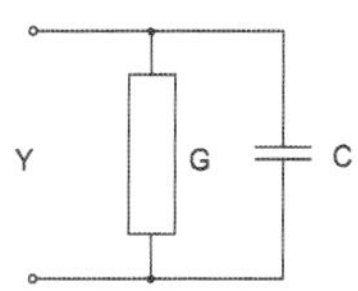

Fig. 10. An equivalent circuit of the open end of coaxial line: admittance of the probe Y has two components - fringing capacitance C and radiating conductance G

A measurement probe can be described as an antenna with the input admittance ($Y = G + jB$, where G is conductance and B is susceptance) expressed as

$$Y_0 = G_0(\varepsilon_c, \omega) + j\omega C_0(\varepsilon_c, \omega) \tag{11}$$

where C_0 and G_0 are constants if the antenna radiates in free space ($\varepsilon_c = 1$).
If we change the free space surrounding the medium for a lossy biological tissue medium, the dependence of this antenna admittance on the properties of surrounding medium can be derived using Deschamps' theorem (Deschamps, 1962). Admittance changes - Y (in Eq. 12) is now an admittance of the antenna immersed in a medium of permittivity ε_c (Fig. 11)

$$Y(\varepsilon_c, \omega) = \sqrt{\varepsilon_c} Y_0 \left(1, \omega\sqrt{\varepsilon_c}\right) \tag{12}$$

This means that the admittance of medium of permittivity ε_c at angular frequency ω is the same as the admittance measured in the free space at frequency $\sqrt{\varepsilon_c}$ -times higher furthermore multiplied by $\sqrt{\varepsilon_c}$.

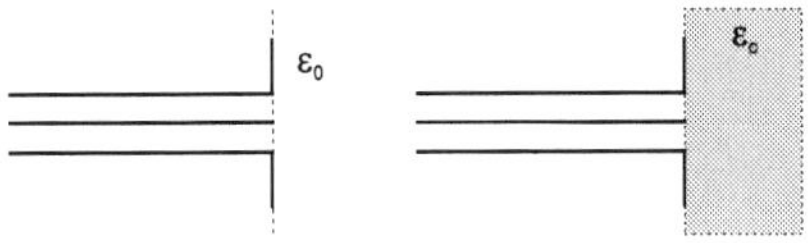

Fig. 11. Illustration to the principle of Deschamps' Theorem: a coaxial probe radiates into free space and then into the lossy medium

The capacitance C_0 can be considered to be a constant (in free space). The radiation into the dielectric surrounding the antenna can be calculated when the dielectric is air (Liu, 1986). Expanding the Bessel function into a Maclaurin series yields

$$G_0 = \frac{Y_0}{ln\frac{a}{b}} \int_{\pi/2}^{0} \left[J_0(ka\sin\Theta) - J_0(kb\sin\Theta)\right]^2 \frac{d\Theta}{\sin\Theta} \tag{13}$$

$$\approx \frac{Y_0}{ln\frac{a}{b}} \left[G_1\left(f^4\right) + G_2\left(f^6\right) + ...\right] \tag{14}$$

where $J_0(x)$ is the Bessel function and G_i are parameters dependent on the antenna dimensions (radius a and b) and wavelength. An investigation of coefficients G_i showed that $G \approx G_1$ for the N and SMA connector type probes. Therefore the radiation effect of coaxial probe, which is represented as radiation conductance in the probe's equivalent circuit, can be approximate as varing with frequency f^4

$$G_1\left(f^4\right) = \frac{2}{3}\left(a^2 - b^2\right)\frac{\pi^4}{c^4} f^4 \approx f^4 \tag{15}$$

Let us apply Deschamps' theorem - Eq. 12 to Eq. 11

$$Y_0 = G_0 + j\omega C_0 \Leftrightarrow Y(\omega, \varepsilon_c) = \sqrt{\varepsilon_c} G_0 \left(\sqrt{\varepsilon_c}\right)^4 + \sqrt{\varepsilon_c} j\omega C_0 \sqrt{\varepsilon_c} \tag{16}$$

The input admittance of the coaxial transmission line as the antenna in a lossy dielectric is then

$$Y = \sqrt{\varepsilon_c^5} G_0 + j\omega\varepsilon_c C_0 \tag{17}$$

where, with respect to the measurement, Y is the measured admittance of the probe, C_0 and G_0 are constants given by the equivalent circuit of the probe in free space (Fig. 10).

Admittance Y is related to the measured reflection coefficient S_{11}

$$Y = Y_0 \frac{1 - S_{11}}{1 + S_{11}} \tag{18}$$

where $Y_0 = 1/(50)\ \Omega = 0.02$ S is the characteristic admittance of the probe.
The usable frequency range of the Eq. 17 results from the theory of wave propagation on the microwave transmission line. The incident wave is in TEM mode. The equivalent circuit of the probe consisting of the fringing capacitor is useful for a low frequency range. The higher order modes excited at the discontinuity can be various TM_{0n} modes. The equivalent circuit of the probe consisting of fringing capacitance and radiating conductance is useful in the frequency range above approximately 1 GHz. The range of the operation can be better defined and extended by a field analysis summing a series of TM modes needed to meet boundary conditions at the probe face. TM_{01} is the lowest circular symmetric mode which sharply increases the loss by radiation. Measurement probes are under investigation in the frequency range from 300 kHz (TEM, fringing field) to 3 GHz (TM_{01}, also radiating conductance).
The equivalent circuit has an imperfectly defined range of frequency and permittivity in which a specified uncertainty can be achieved. Equivalent circuits of the probe never take account of the finite MUT sample and finite ground plane flange which give rise to resonances in the response.
The method of solving the complex Eq. 17 consists of splitting it into real and imaginary parts, thereby obtaining a set of two real nonlinear equations for the two real unknowns, which are either C_0 and G_0 (when calibrating the probe) or the real and imaginary part of the complex permittivity ε_c of the MUT. To break this down into stages:

- splitting of Eq. 17 into real and imaginary parts
- to obtain C_0 and G_0, admittance Y for a material with a known complex permittivity ε_c (e.g. distilled water) is measured and the set of two equations is solved for the unknowns C_0 and G_0
- to measure of complex permittivity, the admittance Y of a MUT is measured and the set of the two equations in solved for the unknown real and imaginary parts of the ε_c

2.4 Modeling and Simulations

An evaluation of this measurement method involves numerical calculation and modeling (Fig. 13). A numerical simulation based on a Finite Integration Technique (FIT, (CST, 2009)) is used to calculate the reflection coefficient on the interface between the probe and the sample of a MUT (Hudlička, 2006). The system that we modeled consists of two parts, i.e. the sensor and the sample of MUT. The modeling is focused on a model with distilled water - the Debye model of distilled water is implemented (Eq. 19). This model was also used for the evaluation of measurement uncertainty.

Choice of Reference Liquid

The material model of distilled water was recognized by Debye in 1926 and describes the complex permittivity of distilled water as a function of frequency.

$$\varepsilon_c = \varepsilon_r^{'} - j\varepsilon_r^{''} = \varepsilon_\infty + \frac{\varepsilon_s - \varepsilon_\infty}{1 + j\omega\tau} \tag{19}$$

where ε_∞ is optical permittivity at high field frequencies, ε_S is static permittivity at low field frequencies and τ is electrical relaxation time (a measure of molecules and dipoles mobility). The values of these parameters for distilled water are: ε_∞= 4.6, ε_S = 78.3 and τ = 8.07 ps.
The condition that *a reference material should have similar dielectric properties as the MUT-Űbiological tissue*, could not be totally fulfilled because of the very high values of the real part of complex permittivity of biological tissues and their lossy character.

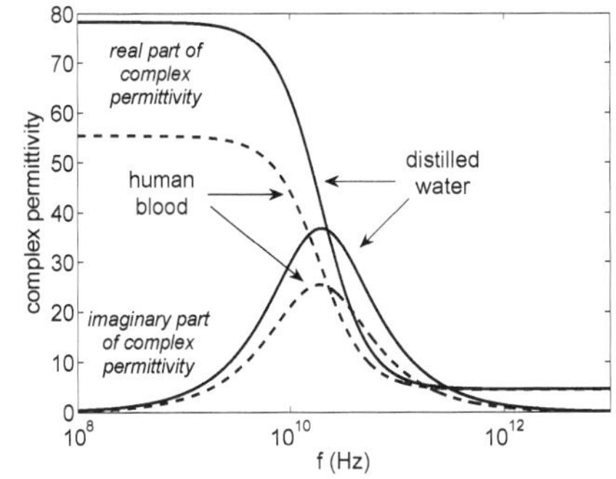

Fig. 12. Dielectric behavior of distilled water and human blood

Fig. 12 shows dielectric properties of the distilled water and the human blood. Distilled water has high values of the real part of complex permittivity and very low values of the imaginary part. Therefore the distilled water is useful reference liquid mainly for the determination of the real part of complex permittvity but neither this condition cannot be satisfied in the whole expected frequency range. Another used reference liquids are alcohols for the dielectric measurements. Unfortunately, an alcohol as a hygroscopic liquid is a time unsteady caliber. Additionaly, the dielectric parameters are known from the tables for only 100 % virgin alcohols. Their use in the calibration makes measurements not very precise.

De-embedding the Measurement System

EM field simulation provides information about the reflection coefficient Γ at a reference plane of the excitation port but in order to calculate complex permittivity, the measurement technique needs to obtain the reflection coefficient (Γ_x = S_{11}) on the interface between the probe and the MUT sample (Fig. 13). In addition, the measurement probe represents an inhomoge-

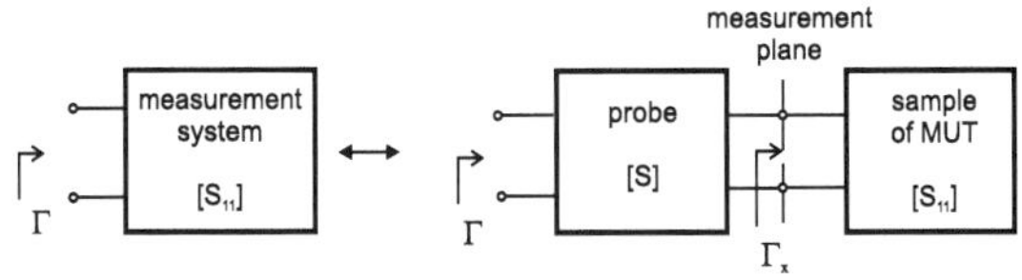

Fig. 13. A model of a reflection coefficient measurement: one-port network (reflection measurement) and the de-embedding of the measurement system

neous transmission line. It is not possible to shift simply the reference plane because of the different dielectric materials (air and teflon) inside the N connector. It is therefore necessary to perform a de-embedding procedure for the measurement system in order to obtain the correct phase information of the reflection coefficient. Fig. 14 shows the comparison of original data from the numerical modelling and also the measurement with the de-embedded data in the Smith chart. The difference is obvious.

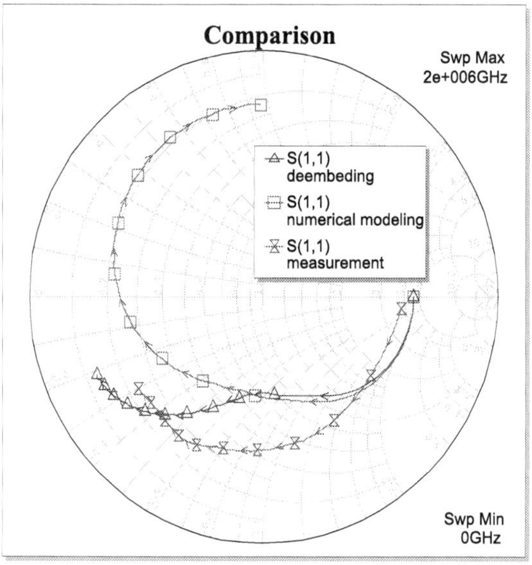

Fig. 14. Comparison of original and de-embedded data: values valid for the agar phantom material

2.5 Measurement System

A typical measurement system using the reflection method on an open-ended coaxial line consists of the network analyzer, the coaxial probe and software. Our measurements (Zajíček, 2009) were done with the aid of an Agilent 6052 network analyzer in the 300 kHz to 3 GHz frequency range. The coaxial probe is placed in contact with a MUT (Fig. 15). Complex permittivity measurement is very fast and proceeds through the three stages. First the calibration of vector network analyzer (VNA) is performed. Then the calibration is carried out using a reference material (with the known dielectric constant ε_C). And, finally, the reflection coefficient of MUT is measured. The complex permittivity of MUT is evaluated with the aid of a PC (in our case using MatLab).

2.6 Calibration Kit for Vector Measurement

A calibration kit is a set of physical devices called standards. Each standard has a precisely known or predictable magnitude and phase response as a function of frequency. To be able to

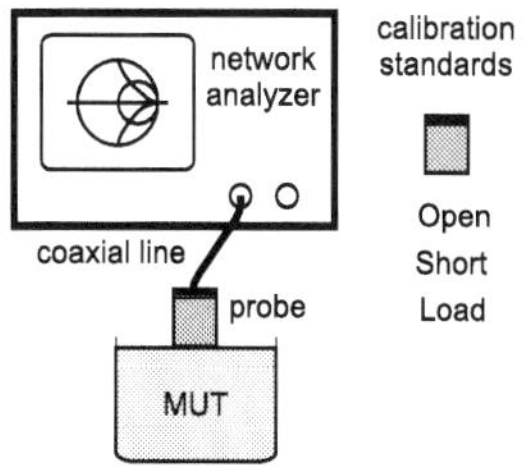

Fig. 15. Measurement system: vector network analyzer, measurement probe and calibration standards for vector measurement

use a particular calibration kit, the known characteristics from each standard in the kit must be entered into the network analyzer memory. The electrical properties of calibration standards

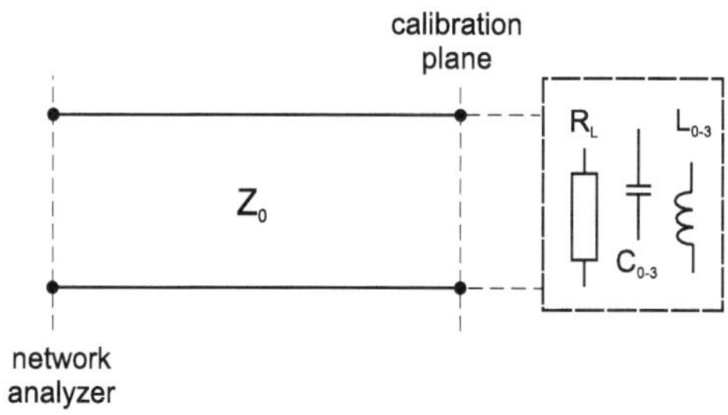

Fig. 16. Specifying the calibration standards

are determined using the modeling in AWR MicroWave Office software. If the actual response of each calibration standard is known (from electromagnetic field solver), then it is possible to determine the value of element in its equivalent circuit (Fig. 16) by tuning this element in order to achieve right (actual) response (Zajíček, 2009).

Fig. 17. Measurement kit: a panel N connector, a measurement probe and open and short calibration standards

The measurement system is a one-port network so the measurement is reduced to measuring the input reflection coefficient S_{11} alone. The OSL (Open, Short and Load) calibration method is performed on the interface between the probe and the MUT sample (reference plane). A coaxial calibration kit was mechanically developed by adapting the panel N and

Fig. 18. Measurement kit: a panel SMA connector, a measurement probe, open, short and matched (50 Ω) load calibration standards

SMA connectors in the same way as the measurement probe (Fig. 17 and Fig. 18). The short standard is made by the connector which is shorted in the measurement plane by a metal plate. The coaxial open standard is created by two connectors (second one represents the air cavity) and the load standard is a commonly used 50 Ω termination.

3. Step-by-Step Measurement Procedure

The measurement stages for the MUT are:

- to calibrate the network analyzer
- to measure S_{11} for a substance with a known ε_c (distilled water), to compute Y using Eq. 18 and, solving Eq. 17 as outlined above, to determine the constants C_0 and G_0
- to measure S_{11} for any desired MUT, to compute Y (Eq. 18) and to solve Eq. 17 for the unknown real and imaginary part of complex permittivity ε_c. Since the equations are of $5th$ order in terms of ε_c, care must be taken to select the physically correct solution (positive real part and negative imaginary part of the ε_c)
- if needed, to derive any quantities of interest, such as relative permittivity ε_r, loss factor $\tan\delta$ or conductivity σ from the ε_c

4. Uncertainty Analysis

The result of dielectric measurement is only an approximation or estimate of the value of the complex permittivity and thus the result is only complete when it is accompanied by a quantitative statement of its uncertainty.

If the measurement device taken to include measurement standards and reference materials is tested through a comparison with a known reference standard and the uncertainties associated with the standard are assumed to be negligible relatively to the required uncertainty of the test, the comparison is viewed as determining the error of the device. The reference standard is the Debye model of distilled water and the standard and combined standard uncertainty of the complex permittivity determination of distilled water is evaluated in the following section.

Sources of an uncertainty are distinguished from the view point of measurement technique. The measurement technique is based on the measurement of the reflection coefficient with the aid of a vector network analyzer. Generally, the calibration of network analyzer and the calibration of measurement probe by means of reference liquid is considered. Last but not least, the condition as a location of coaxial cable which connects the network analyzer with the probe is assumed as a possible contribution to the uncertainty.

This uncertainty evaluation is also the verification of self-consistency of the developed relation between the measured reflection coefficient and the calculated complex permittivity (Eq. 17). Uncertainty evaluation is based on the relevant information available from previous measurement data and experience and knowledge of the behavior and property of the distilled water, and the measurement instruments used (referred to as Type B uncertainty evaluation). Sources of uncertainties and related standard and combined standard uncertainties (Tab. 2 and 3) are evaluated with the aid of guidelines (NIST, 1999).

source of uncertainty	f (MHz)	standard uncertainty (%)
repeated observations	30	0.26
	434	0.29
	915	0.34
random effects	30	4.93
	434	2.16
	915	3.67
systematic effects	30	1.01
	434	0.58
	915	4.50
combined standard uncertainty	30	5.12
	434	2.29
	915	5.90

Table 2. The uncertainty budget, N-type probe: sources of measurement uncertainty are evaluated at important frequencies from a medical point of view

source of uncertainty	f (MHz)	standard uncertainty (%)
repeated observations	434	0.33
	915	0.29
	1800	0.19
	2450	0.37
random effects	434	3.06
	915	2.93
	1800	2.81
	2450	2.14
systematic effects	434	0.58
	915	0.67
	1800	3.69
	2450	20.9
combined standard uncertainty	434	3.19
	915	3.07
	1800	4.72
	2450	21.3

Table 3. The uncertainty budget, SMA-type probe: sources of measurement uncertainty are evaluated at important frequencies from a medical point of view

The applied recommendations given in (NIST, 1999) and mentioned also above are following:

- *Repeatability*
  The measurement procedure was performed twenty times over a short period of time (minutes) in a single location with the one-off application of measuring instruments in order to observe the same results.
- *Random Effects*
  The measurement procedure was performed ten times over a long period of time (days and months) at a single location with the different application of measuring instruments in order to observe different results. The conditions are generally changed by locating the coaxial cable in a different position between the measurement probe and the network analyzer. The calibration of the network analyzer was performed before each measurement as well as the calibration of the measurement probe by means of distilled water.

It is important to note that complex permittivity is a variable quantity - it changes with frequency, temperature, mixture, pressure and the molecular structure of the MUT. Frequency has a significant influence on changes in the complex permittivity of biological substances. This is the reason for evaluating the uncertainties separately at each frequency of interest for microwave applications.

5. Results

The relative permittivity of lossy materials is a heavily frequency-dependent quantity. Because of the decreasing ability of particles to follow rapid changes of electrical field, the relative permittivity decreases with increasing frequency. The frequencies in the following tables have been selected because of their interest from an industrial, scientific and medical point of view.

5.1 Home-made phantom material

Human tissues can be classified into those with high water content such as muscle, brain, and the internal organs and those with low water content such as fat and bone. The present biological tissue-equivalent phantom[6] simulates the characteristics of the high-water-content tissues.

ingredients	weight (g)
de-ionized water	3375
agar gelatine	104.6
solidum chloride (NaCl)	37.60
sodium azide (NaN_3)	2.000
TX-151	84.40
polyethylene powder	337.5

Table 4. The composition of the agar phantom - ingredients for developing a biological muscle tissue-equivalent

[6] This phantom material was manufactured by Tomáš Dřížďal at the Department of Radiation Oncology, Erasmus MC - Daniel den Hoed Cancer Center, Rotterdam.

The tissue-equivalent phantom can be made of agar, deionized water, polyethylene powder, sodium chloride (NaCl), TX-151, and sodium azide (NaN_3) (Tab. 4). The polyethylene powder is used to adjust the relative permittivity while the conductivity is mainly adjusted by the sodium chloride concentration. Since the agar solution and the polyethylene powder cannot be mixed directly, TX-151 is used to increase the viscosity. Sodium azide is added as a preservative. The advantages of this particular phantom are the ease of use of the original materials and the possibility of manual processing with no need for special production equipment. It is also easy to machine and to cut into arbitrary shapes. The phantom maintains its shape and is mechanically strong. By manipulating the agar, a certain amount of adjustment of the mechanical strength is possible. Hence, this phantom is useful for splitting the phantoms.

f (MHz)	ε_r (-)	$\tan\delta$ (-)	σ (S/m)
434	60.9	0.86	1.26
915	58.6	0.54	1.60
1800	53.1	0.47	2.50
2450	48.5	0.48	3.14

Table 5. Dielectric parameters of a home-made muscle tissue phantom

The electrical parameters of the muscle tissue equivalent are described in Tab. 5. Different values of these biological parameters may be required for experimental work. For this reason, it is desirable that the electrical characteristics of the phantom be adjustable within a certain range. In this phantom, the electrical characteristics can be adjusted to a certain extent by modifying the composition shown in Tab. 4. Hence, phantoms are fabricated with varying amounts of polyethylene powder and sodium chloride in order to adjust their permittivity characteristics. To facilitate mixing the polyethylene powder into the agar solution in order to enable the smooth fabrication of the phantom, the amount of TX-151 is dependent on the amount of polyethylene powder. The conductivity is affected by both the polyethylene and sodium chloride whereas the relative permittivity is mainly determined by the polyethylene. Hence, the composition of the phantom with a desired characteristic can be determined first by deriving the amount of polyethylene needed for the desired relative permittivity and then adjusting the conductivity by means of sodium chloride. More details can be found in (Koichi, 2001).

5.2 Commercially available phantom material

This phantom is a tissue-equivalent material, in this case an equivalent of biological muscle tissue. An agar phantom (agar gelatine) is the most commonly used phantom in the testing of thermotherapy applicators, and the use of the phantoms is significant in the measurement of impedance matching and Specific Absorption Rate (SAR).

f (MHz)	ε_r (-)	$\tan\delta$ (-)	σ (S/m)
434	57.5 (56.9)	0.31 (0.59)	0.45 (0.81)
915	55.9 (54.9)	0.29 (0.34)	0.85 (0.95)
1800	52.7 53.5)	0.27 (0.25)	1.42 (1.34)
2450	51.1 (53.5)	0.23 (0.24)	1.63 (1.74)

Table 6. The dielectric parameters of the agar phantom: values of a commercially available phantom DUBLAGA (Zajíček, 2008) and for comparison the values in brackets for muscle tissue (Gabriel, 1999)

The agar phantom is a relatively good equivalent of biological muscle tissue. There is good agreement in the relative permittivity but in the loss factor or conductivity the difference is mainly revealed at lower frequencies (Tab. 6). It appears that the agar has lower water content and is not as lossy as the muscle tissue. If the agar phantom is used, the age of the phantom must be considered.

5.3 Saline phantom

Another type of phantom is a saline liquid (3g of NaCl in 1l of distilled water), simulating biological tissue. A hyperthermia system with the liquid phantom can be used for the evaluation of microwave applicators. This phantom offers 3-D electromagnetic field distribution measurements e.g. the distribution of SAR can be easily calculated with the aid of dipole antennas.

f (MHz)	ε_r (-)	$\tan\delta$ (-)	σ (S/m)
434	84.9	0.21	0.57
915	74.4	0.31	1.18
1800	66.4	0.36	2.38
2450	60.2	0.41	3.38

Table 7. Dielectric parameters of saline phantom used in 3-D electomagnetic field distribution measurements

5.4 Melanoma tumor

Another dielectric measurement was performed on an experimental animal with an implanted malignant melanoma. The task was to describe the dielectric properties of the tumor after microwave hyperthermia. In the case of the design of a planar applicator for this experiment

Tissue (measurement)	ε_r (-)	$\tan\delta$ (-)	σ (S/m)
tumor (non-invasive)	51.2	0.174	1.21
tumor (invasive)	53.0	0.145	1.05
skin (ex-vivo)	11.1	0.035	0.053

Table 8. Analysis of the dielectric parameters of a melanoma implanted in a mouse

where the dimensions of both animal and tumor are small and care must be taken regarding the effective depth of tissue heating, the dielectric parameters are very important. Tab.

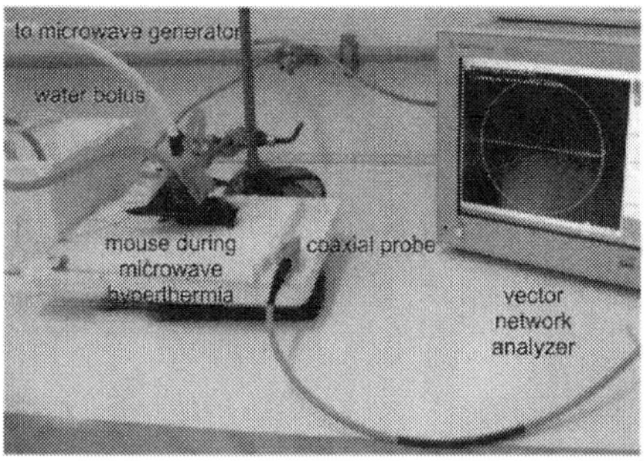

Fig. 19. Microwave hyperthermia: experimental therapy on a laboratory mouse, measurement of the dielectric parameters of an implanted tumor

8 summarizes results measured at a frequency f = 2.45 GHz. The tumor dimensions were 30x18 mm. Hyperthermia was applied for a period of 15 minutes with a continual power of 30 W and the achieved temperature in the tumor was 45 ^{0}C.

5.5 Biological tissue

Fig. 20 summarizes the values measured on the author's arm and values modeled using a four-layered model of biological tissue. The simulation and the measurement values are based on a model and in vivo sample respectively, these being inhomogeneous (layered). Microwave applicators are usually designed and tested on the agar phantom described in the section below as a homogeneous equivalent of biological muscle tissue. This disparity may affect the impedance matching of microwave applicators. If the complex permittivity of the layered treated area is considered in the design of applicators, a more realistic impedance matching could be achieved.

6. Limitations

Complex relative permittivity is used in calculations of electromagnetic field distribution and is inversely related to the square root in these calculations. This means that the measurement uncertainties from Tab. 2 and 3 are further suppressed. No evaluation of the measurement uncertainties in the case of determining the imaginary part of complex permittivity is presented. For distilled water, this imaginary part has extremely low values (lower than 1) and any evaluation is difficult - there is a high level of uncertainty when is only a small difference between the measured and Debye values.

7. Conclusion

The complex permittivity determination based on reflection coefficient measurement is suitable for the determination of the dielectric parameters of materials in wide bands. This method was described from the viewpoint of electromagnetic field theory and the coaxial probes were described with the equivalent circuit as an antenna in a lossy medium respecting radiation effects at higher frequencies. Some materials were measured and where possible the comparison between measurement (modeling) and values from tables was carried out.

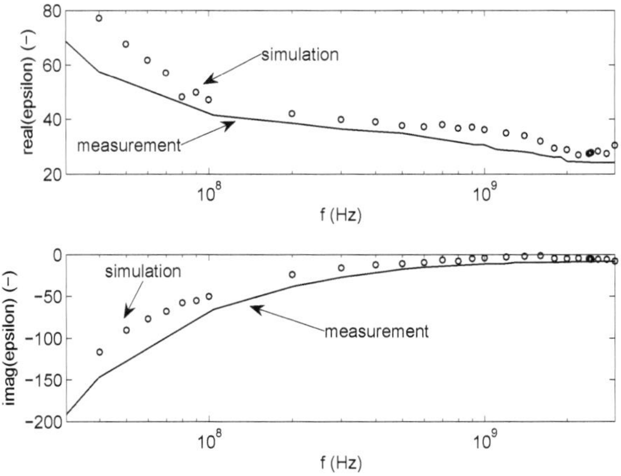

Fig. 20. Complex permittivity measured on author's arm and its comparison with the simulation

Where the evaluation is required to be complete, the uncertainty of measurement has to be specified. The results obtained indicate that the accuracy may be sufficient for most practical applications (2.3-6 % depending on working frequency).
Future perspective of studied method for determining the complex permittivity is in an investigation of layered tissues. Experimental measurement on the layered tissue showed that method yields reasonable approximation of complex permittivity. It could be quatified which range of tissue thisknesses can be considered as sufficient. This would require e.g. statistical analysis of the distribution of the tissue thickness and how they affect the final outcome of the measurement.

8. References

Deschamps, G., A. (1962). Impedance of an antenna in a conducting medium. *IRE Transactions on antennas and propagation*, p. 648-650.

Gabriel, C. et al. (1996). The dielectric properties of biological tissues: I. Literature Survey. *Physics in Medicine and Biology*, Vol. 41, p. 2231-2249.

Hudlička, M., Hazdra, P.: *Finite Integration Technique*. Modeling of Fields. IEEE Czechoslovakia Section, p. 58-77.

Internet website address: http://www.p2pays.org/ref/18/17627.pdf Guidelines for Evaluating and Expressing the Uncertainty of Measurement Results by NIST.

Internet website address: http://niremf.ifac.cnr.it/tissprop/ Dielectric properties of body tissues developed by C. Gabriel and colleagues.

Internet website address: http://www.cst.com/ CST MW Studio software.

Kittel, C. (1966). *Introduction to Solid State Physic*, John Wiley&Son, 2_{nd} edition., p. 157-181.

Koichi I. et al. Development and Characteristics of a Biological Tissue-Equivalent Phantom for Microwaves. *Electronics and Communications in Japan*, Part 1, Vol. 84, No. 4.

Liu L., X. et al. (1986). Improvement in Dielectric Measurement Technique of Open-ended Coaxial Line Resonator Method. *Electronics Letters*, Vol. 22, No. 7, p. 373-375.

Novotný, K. (2005). *Theory of Electromagnetic Field*, Press CTU in Prague.

Novotný, K. (2001). *Theory of Electromagnetic Field II: Field and Waves*, Press CTU in Prague.

Stuchly M., A. et al. (1982). Measurement of RF permittivity of biological tissue with an opend-ended coaxial line: Part II-Experimental results. *IEEE transactions on MTT*, Vol. 30, no.1, p. 82-92.

Vrba, J. (2003). *Medical applications of microwave technique*, Press CTU in Prague, p. 46-61.

Zajíček, R. et al. (2008). Broadband Measurement of Complex Permittivity Using Reflection Method and Coaxial Probes. *Radioengineering*, Vol. 17, No. 1, p. 14-19, ISSN 1210-2512.

Zajíček, R. (2009). *Application of Complex Permittivity in Medical Diagnostics and Imaging*, Doctoral Thesis, CTU in Prague.

18

Microwave Dielectric Behavior of Ayurvedic Medicines

S.R.Chaudhari[#1] ,R.D.Chaudhari[*2], and J.B.Shinde[#3]
[#1] *Dept of Physics,Baburaoji Gholap College, Pune,M.S.,India*
[*2] *Engg Dept,College of Agriculture,Pune, M.S.,India*
[#3] *Dept of Physics,Deogiri College, Aurangabad,M.S.,India*

1. INTRODUCTION

In Material Science, characterization of materials is a significant activity. Chemical composition and structural features decides the properties of material. The properties of material also depend on the degree of molecular order.

The basics of molecular interaction are the hydrogen bonding. Hydrogen bonds occur between hydrogen containing dipoles and an electronegative element. Electro-negativity provides us a relative activity of atom in molecule to attract bonding electrons. In the present work interaction of Hydroxyl -OH group in Ethanol and Methanol at 15^0C, 25^0C ,35^0C and 45^0C is studied. In the present work interaction of Ayurvedic Medicines (Arishta group) such as Ashokarishta, Punarnvarishta and Dashmularishta from Arishta group are taken with Ethanol and Methanol.

Time Domain Spectroscopy (TDS) technique gives information in a wide frequency range from 10 MHz to 20 GHz. In the present work, reflected part of the pulse is used to obtain dielectric relaxation data. Prof. Cole developed this technique. It is very useful, economic and fast as compared to other techniques. It requires very small amount of sample and in single measurement we get permittivity and dielectric loss over wide range of frequency.

The Hewlett Packard HP 54750 sampling oscilloscope with HP 54754A TDR plug in module has been used. The TDR setup consists of step generator, sampling head, sample cell and broadband storage oscilloscope. A fast rising step pulse from generator propagates through coaxial transmission line and reaches dielectric sample placed in sample cell connected as open-ended load. It is partly transmitted and partly reflected at air dielectric interface. Both reflected as well as transmitted step from sample contains information about dielectric behavior of sample. In the present work reflected step with and without sample is recorded in the oscilloscope. This time domain data is transformed into frequency domain data using Fourier transformation. Frequency domain data is used to obtain complex reflection coefficient $\rho^*(\omega)$ over frequency range of 10 MHz to 20 GHz. Complex reflection coefficient gives permittivity and dielectric loss in selected frequency range. But normally there occurs error in this data at higher frequency due to fringing field, multiple reflections or due to quarter wave resonance in case of high lossy liquids. The complex reflection data is called 'RAW' data. An error in 'RAW' data is corrected by bilinear calibration process.

The corrected data is called 'COR' data. The dielectric parameters of the Ayurvedic Medicines are obtained by fitting 'COR' data to Harviliak Negami equation-

$$\varepsilon^*(\omega) = \varepsilon_\infty + \frac{\varepsilon_0 - \varepsilon_\infty}{\left[1 + (j\omega\tau)^{(1-\alpha)}\right]^\beta} \tag{1}$$

If we put α =0, β =1 then equation represents simple Debye model. A Least Square Fit method is used to obtain dielectric parameters.

The dielectric relaxation time (τ) of biological material is related to nature of intermolecular bonding, size of molecule, mobility of molecules in solutions, molecular volume, viscosity and temperatures similarly, the permittivity(ε) is related to square of molecular dipole moment, and the value of permittivity(ε) is related to size of molecule in solution and temperatures. Thus information at molecular level can be gained from a study of dielectric behavior. To understand structural change in system Excess Permittivity (ε^E), Excess inverse relaxation time $(1/\tau)^E$ and Bruggeman factor (f_B) are obtained. Thermodynamic parameters i.e. activation energy in KJ/mole, change in enthalpy (ΔH) and change in entropy (ΔS) are calculated using Eyring's equation to understand molecular dynamic of the system.

2. AYURVEDA

Ayurveda[1] is a holistic healing science, which comprises of two words, Ayu and Veda. Ayu means life and Veda means knowledge or science. So the literal meaning of the word Ayurveda is the science of life. One of the basic principles of Ayurveda, the recognition of the inner reflecting the outer.

By using ayurvedic and herbal medicines you ensure physical and mental health without side effects. The natural ingredients of herbs help bring "arogya" to human body and mind. ("Arogya" means free from diseases). The chemicals used in preparing allopathy medicines have impact on mind as well. One should have allopathy medicine only when it is very necessary. According to the original texts, the goal of Ayurveda is prevention as well as promotion of the body's own capacity for maintenance and balance. Ayurvedic treatment is non-invasive and non-toxic, so it can be used safely as an alternative therapy or alongside conventional therapies. Ayurvedic physicians claim that their methods can also help stress-related, metabolic, and chronic conditions. Ayurveda has been used to treat a acne, allergies, asthma, anxiety, arthritis, chronic fatigue syndrome, colds, colitis, constipation, depression, diabetes, flu, heart disease, hypertension, immune problems, inflammation, insomnia, nervous disorders, obesity, skin problems, and ulcers.

3. DIELECTRIC STUDY IN MEDICINE

The recent application of dielectric studies on biological molecules has shown the possibility of treating malignant diseases by studying the responses of diseased cells of the body to the radiation used and then localizing the heat in the area of interest. The energy necessary for hyperthermia of a 20 mm diameter tumor by 5°C in one minute, if energy is efficiently transmitted to the tumor. Neoplastic tissue has a higher dielectric loss factor than healthy tissue because of its higher water content. For example, skin carcinoma contains 81.6% as compared with 60.9% in normal epidermis; hepatoma contains 81.9% as compared with 71.4% in liver. These differences are sufficient to enable us to discriminate between different types of carcinoma by diagnostic radio - frequency imaging.

The information such as molecular flexibility or rigidity, shape and size etc. obtained using dielectric relaxation as the probe which serves the basis for determining its carcinogenic or anti-carcinogenic action, is of vital importance for investigation of cancer.

The values of dielectric constant and dielectric relaxation time for mouth cancer patient's saliva are ($\varepsilon_0 = 81.68,\ \tau = 13.29$ ps) larger than normal person's saliva ($\varepsilon_0 = 76.57,\ \tau = 11.38$ ps).Now a days, popularity of Ayurvedic Medicines becomes worldwide. Ayurvedic Medicines used in Gyneac problems in human body are considered in the present work.

Dielectric spectroscopy is a branch of spectroscopy where one gets information about structural changes and molecular interactions through dielectric relaxation data. In the present work interaction of Ayurvedic Medicine such as Ashokarishta, Punarnvarishta, and Dashmulrishta from Arishta group are taken with Ethanol and Methanol

4. DIELECTIRC RELAXATION SPECTROSCOPY

Dielectric Relaxation Spectroscopy (DRS) probes the interaction of a macroscopic sample with a time-dependent electric field[2]. The resulting polarization either expressed by the frequency-dependent complex permittivity and conductivity or as an impedance spectrum, characterizes amplitude and timescale (via the relaxation time) of the charge-density fluctuations within the sample. Such fluctuations generally arise from the reorientation of the permanent dipole moments of individual molecules or from the rotation of dipolar movements in flexible molecules, like polymers. Other possible mechanisms include the transport of ions or the appearance of interfacial charges in heterogeneous systems. The timescale of these fluctuations depends on the sample and on the relevant relaxation mechanism. Relaxation times range from several picoseconds in low-viscosity liquids to hours in glasses, probably marking DRS as the technique with the most extensive coverage of dynamical processes. The corresponding measurement frequencies range from 10^{-4} Hz to 10^{12} Hz, which requires a series instruments for complete coverage. However, it is generally sufficient to concentrate on a smaller frequency range adapted to the sample properties. In contrast to conventional spectroscopic methods, like NMR or vibration spectroscopy, DRS is especially sensitive to intermolecular interactions. DRS is able to monitor cooperative processes and thus provides a link between molecular spectroscopy, which monitors the properties of the individual constituents, and techniques characterizing the bulk properties of the sample, especially the viscoelastic and archeological behavior. The decomposition of the dielectric spectrum into its individual relaxation processes informs on the relative amplitudes and characteristic times of the underlying molecular motions. Dielectric relaxation studies on binary mixture are important for understanding the hydrogen bonding and intermolecular interaction in the mixture. The dielectric relaxation study of solute - solvent mixture of microwave frequency gives information about molecular interaction in the system, formation of monomers and multimers. Dielectric Spectroscopy is being successfully used to determine the time of relaxation of electrolytes in solution.

DRS is widely applied in the characterization of ion-conducting solids, polymers and mesophases. But it is also of large potential interest for the investigations of liquid and colloidal systems. Additionally, the effects studied by DRS are of increasing importance for technical applications like dielectric heating or remote sensing.

5. DIELECTRIC POLARIZATION

When a dielectric is placed between charged plates, the polarization of the medium produces an electric field opposing the field of the charges on the plate. The dielectric constant k is defined to reflect the amount of reduction of effective electric field . The permittivity is a characteristic of space, and the relative permittivity or "dielectric constant" is a way to characterize the reduction in effective field. Because of the polarization[1] of the dielectric. The capacitance of the parallel plate arrangement is increased by factor k.
According to spatial arrangement of charges in a molecule the molecules are classified as polar and non-polar[6]. A polar molecule has permanent dipole moment. The dipole moment depends on the size and symmetry of the molecule. Although the total number of positive and negative charges is equal to the distribution of two kinds of charges is different. Non-polar dielectric consists of molecules with positive and negative charges such that their effective center of charge distribution coincides. Thus dipole moment of non-polar dielectric material is zero in absence of electric field.
If a distance 'd' separates the charge +q and -q, it forms a dipole moment given by - qd. In a molecule 'q' is of the order of electronic charge, 10^{-10} e.s.u., while the 'd' is of the order of 10^{-8} e.s.u. Therefore unit of dipole moment is 10^{-18} e.s.u., and is called a 'Debye', abbreviated as 'D'. In the case of non-polar molecules, the centers of positive and negative charges coincide with the centers of symmetry of the molecule, therefore they have zero dipole moment. e.g. Benzene, Methane. Polar molecules always have a permanent dipole moment, even in the absence of an external electric field.
When the electric field is applied to dielectric, the molecular charges get displaced. The total charge passing through unit area within the dielectric, perpendicular to the direction of applied field is called polarization. The polarizations are of three types, Electronic Polarization (P_e), Atomic Polarization (P_a), Orientation Polarization (P_o).
Orientation polarization is property of the polar molecules. It is due to rotation of permanent dipoles of dielectric medium. The molecular dipoles orient in the direction of the applied field. It is function of molecule size, viscosity, temperature, and frequency of applied field. Orientation polarization takes a time of the order of 10^{-12} to 10^{-10} sec., corresponding to period of microwave frequency region. The total polarization is,

$$P_t = P_e + P_a + P_o \tag{2}$$

Thus the polar materials have greater permittivity than the non-polar, because of additional polarization due to orientation The insulator whose behavior gets modified in the electric field are called as dielectric. Dielectric materials are bad conductors of electricity. When these materials are placed in dielectric field, displacement of positive and negative charges in molecule takes place. When the change in the behavior of dielectric is independent of the direction of the applied field, the dielectric is called Isotropic. On the other hand if the change in behavior of dielectric depends on the direction of applied field the dielectric is called anisotropic. The positive and negative charge distribution separated by some distance can be treated as dipole. Applied electric field forces this molecular dipole to align in the direction of field. This alignment of molecular dipoles in the direction of the field is called polarization dielectric constant . It is a measure of ability of material to get polarized in the direction of applied electric fields. Hence dielectric material store applied electrical energy in the form of polarization.

6. STATIC AND DYNAMIC PERMITTIVITY

The theories of dielectric relaxation can be broadly divided into two parts as theories of static permittivity and theories of dynamic permittivity. The polar dielectric materials having a permanent dipole moment, when placed in steady electric field so that all types of polarization can maintain equilibrium with it, the permittivity of material under these conditions is called as static permittivity (ε_o), when dielectric material is placed in electric field varying with frequency, then permittivity of material changes with change in frequency of applied field. This is so because with increasing frequency molecular dipoles cannot orient faster to come up with applied field. Thus permittivity of material falls off with frequency of applied field, the frequency dependent permittivity of material is called as dynamic permittivity The different theories of static and dynamic permittivity like Clausius Mossotti Equation, Debye Theory of Static Permittivity, Onsager Theory of Static Permittivity, Frohlich's theory are used as well as, the Debye Model ,the Cole-Cole Model,the Davidson-Cole Relaxation Model,the Havriliak-Negami Model are used.

Dielectric relaxation occurs when; the externally applied alternating field polarizes a dielectric material. The decay in polarization is observed on removal of the field. The decay in polarization occurs due to orientation of electric dipoles in an electric field. This depends on the internal structure of a molecule and on molecular arrangement. The orientation polarization decays exponentially with time; the characteristic time of this exponential decay is called relaxation time. It is defined as the time in which this polarization reduces to $(1/\tau)^{th}$ times the original value. Dielectric relaxation is the cause of anomalous dispersion in which permittivity decreases with increasing frequency.

Under the influence of an ac electric field, the polar molecules of a material orient themselves and attain an equilibrium distribution in molecular orientation. When the polar molecules are of large size or frequency of ac field is very high or the viscosity of the medium is very large, the orientation of molecules is not fast enough for the attainment of equilibrium with the applied field. The polarization then acquires a component out of phase with the field and the displacement current acquires a conductance component in phase with field, resulting in thermal dissipation of energy. The permittivity thus acquires a complex characteristic.

In such cases it is used to relate the displacement $\vec{D} = \varepsilon^* \vec{E}$. The complex permittivity ε^* can be written as $\varepsilon' - i\varepsilon''$, where ε' is real part proportional to stored energy and ε'' is imaginary part and it is dielectric loss.

7. TDR TECHNIQUE

Time Domain Spectroscopy has become a widespread method of investigation for variety of substances. Dielectric study provides information about charge distribution in a molecular system. The recently developed Time Domain Reflectometry has proved to be very effective and efficient for determination of dielectric constant and loss in the frequency range of 10 MHz to 10 GHz. Time Domain Reflectometry (TDR) method is the most suitable method for determination of frequency dependent dielectric parameters of material. This method was first introduced by Hugo - Fellner -Feldegg[5] (et al in 1969) and developed by many workers in field of dielectric spectrometry.. TDR technique is being adopted to measure static conductivity of electrolytic solution.

In TDR method, a fast rising step pulse is allowed to incident on sample under investigation. The reflected pulse from sample contains the information regarding dielectric behavior of sample. The Fourier transformation of step pulse gives us frequency components contained in step pulse. Thus incident step pulse is treated as mixture of waves with different frequencies. The lower limit of frequency spectrum contained in step pulse depends on time window used, while upper limit depends on rise time of pulse. Frequency dependent permittivity parameters of sample can be determined from its response to incident step pulse.

In order to obtain frequency dependent dielectric parameters one needs a step generator, a sampling oscilloscope, a sample holder and mathematical expression with computer software. The step generator must be capable of generating step pulse with rise time adequate enough to give the highest frequency components of interest with considerable magnitude. The broadband oscilloscope is required to handle broad frequency spectrum contained in step pulse with sufficient accuracy. A transmission line is needed to carry signal from step generator to sample holder. The transmission line as well as sample holder must be capable of holding high frequency signals.

For faithful transmission characteristics impedance of components must be matched. Any impedance mismatching in this signal path carries multiple reflection, which can disturb signal of our interest. Practically multiple reflection of signals, when it passes from one component to other cannot be avoided totally, but can be minimized to acceptable level by making same precautions. The time domain data is converted into frequency domain data using Fourier transformation.

Experimental setup consists of sampling oscilloscope HP 54750A, TDR module HP 54754A, a transmission line and sample cell. The HP 54750A-sampling oscilloscope is very precise instrument for digital data acquisition of instantaneous signals. The working of instrument depends on front panel keys as well as menus invoked after pressing any front panel key. The menus of function are displayed along the right side of display screen. These menus are called soft key menus. Soft key menus list functions other than those accessed directly by the front panel keys. To activate a function on soft key menu can be accessed by pressing unlabeled key immediately next to the annotation on the screen. The unlabeled keys next to the annotation on display are called soft keys. Front panel of the instrument includes a display area and several functional areas, which includes control, storage, autoscale, entry devices, setup, and system. Control section includes three keys clear display, run and stop signal. These keys are used to clear screen, start data acquisition and stop data acquisition respectively. Storage section includes four keys disk, waveform, setup and print. Disk key is used to access information from 1.44 MB Floppy Disk Drive. We can store the waveforms on disk or load waveform from disk. Waveform key is used to store current waveform in memory of oscilloscope. Four waveforms can be stored at a time in oscilloscope memory. Setup is used for setting waveform. Print key is used to print current waveform or waveform in memory. Autoscale section contains only single key Autoscale. This Autoscale key causes the instrument to quickly analyze the signal. Then, it sets up vertical, horizontal and triggers to best display that signal. Entry devices are the keypad, the arrow keys and the knob. Entry devices can change the numeric settings of some soft-keys, such as trigger level, or to select an item from the list of choices. The setup section includes seven keys. Time base, Trigger, Acquisition, Display, marker, Define meas. and Math. With time base key we can change horizontal position of waveform and also the time window. Trigger can be used to change trigger level of signal. Acquisition key is used to set number of data acquisition points and also number of times the averaging is done. Marker key can be used for setting markers on waveform during measurement of specific parameters. One can also put meas. (measurement marker lines) during measurement. Math function key is used to perform few mathematical operations such as addition and subtraction of two waveforms or even Fourier transform of waveform.

The HP 54754A TDR plug in module is capable of performing both, single ended TDR measurements as well as differential TDR measurements. These measurements include characterizing micro strip lines, PC board traces and coaxial cables. The plug in module takes up two, out of four mainframe slots. In single ended TDR measurement, a positive going step (a fast rising step voltage pulse of 200 mv with 39 ps rise time) is launched on one of the channels while the other channel is terminated using short. In differential TDR measurement, a positive going step is launched on channel 1 and an effective negative going is launched on channel 2. The response controls are provided which shows the single ended or differential mode response of a TDR system. A fast rising step voltage pulse of about 200 mV amplitude and 43.8486 ns rise time with repetition frequency of 12.4 GHz is generated and is propagated through a coaxial transmission line. The sample is placed at the end of the coaxial transmission line in a standard Military application (SMA) coaxial cell. The SMA cell used for this work had 3.5 mm outer diameter and 1.52 mm effective pin length. The step pulse generated by tunnel diode and the pulse which is reflected from the sample cell were sampled by a sampling oscilloscope in the time window of 1.3 ns. The reflected pulse without sample $R_1(t)$ and with sample R_x (t) averaged 64 times and digitized with 1024 points in oscilloscope memory and transferred to PC through a 1.44 floppy diskette drive.

The temperature controller system with water bath and a thermostat has been used to maintain the constant temperature within the accuracy limit of ±1 ^{0}C. The sample cell is surrounded by a heat-insulating container through which the water of constant temperature using temperature controller system is circulated. The temperature at the cell is checked using the thermometer. The combination of Ayurvedic Medicines with Ethanol and Methanol are taken at different temperatures such as 15^{0}C, 25^{0}C, 35^{0}C and 45^{0}C. No work has been reported on this systems using Time Domain Reflectometry (TDR) technique. All the above systems are studied by preparing 11 concentrations by volume fraction 'X' of solutions such as 0%, 10%, 20%,…….100% with the two pure liquids. Temperature dependent variations in dielectric parameters and thermodynamic parameters for four different temperatures are reported.

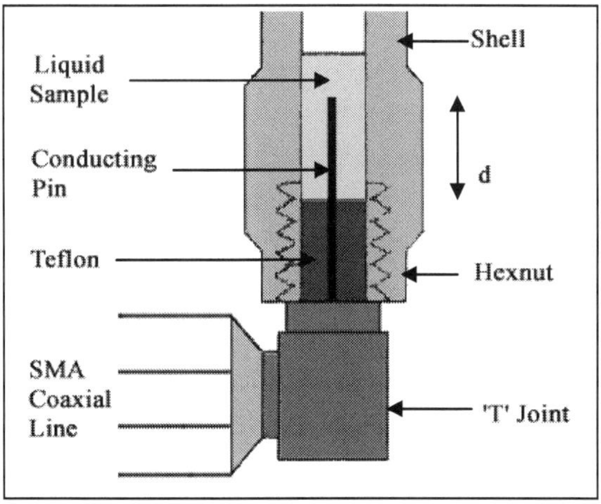

Fig.1. Geometrical construction of SMA cell

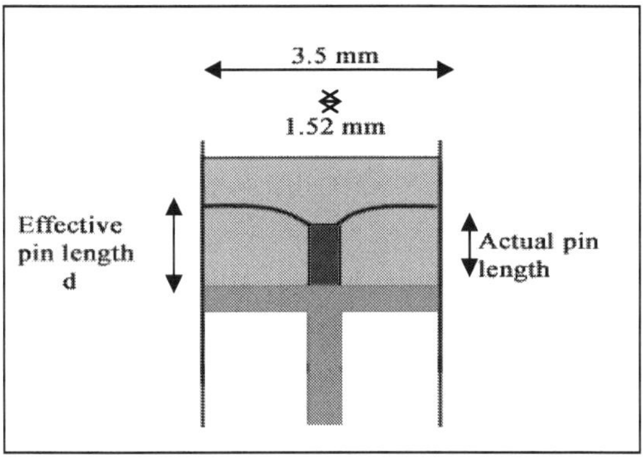

Fig. 2. Fringing field and SMA cell dimension

8. DIELECTRIC RELAXATION BEHAVIOUR OF AYURVEDIC MEDICINES

The dielectric relaxation study at microwave frequency gives information about solute - solvent interaction and liquid structure of mixture. By using Time Domain Reflectometry technique different types of liquids such as pharmacologically important drugs, n nitrites, glucose water mixtures, Binary mixtures, Biological samples (Methanol - algae), amides, Alcohols, electrolytes, liquid crystals were studied. Ayurvedic medicines from 'Arishta' group are selected to observe velatine structural changes as well as changes in interaction of these molecules with ethanol and methanol. Static permittivity, relaxation time, Bruggeman factor, Activation energy, enthalpy, entropy are reported for various systems and for different temperatures. The preparation of Ayurvedic medicine is always complex in nature, as it contains alcohol, sugar, variety of medicinal herbs, their leaves, flowers, fruits, peels, roots, sap (gum), resins. It also contains shells, conches cowries coral and pearls found in the sea and metals like gold, silver, lead, mercury, copper or iron..

'Arishta' is prepared with the help of extract of medicinal material or juice and it is mixed with jaggery, sugar, honey, or other sweeteners. According to Ayurvedic science it is fermented, brewed for a period for 2-3 months. The process (Kinwa) of fermentation occurs at certain temperature near about 30°C to 35°C After formation of 'Arishta' percentage of alcohol is there e.g. Ashokarishta contains 7.4% of alcohol, Dashmularishta contains 8.8% of alcohol, Punarnarishta contain 6.4% of alcohol etc. The Sanskrit name of medicinal plants and metals has been used to indicate the standard names of 'Arishta' group. Eg. Ashokarishta contain 'Ashoka' as medicinal plant as well as another 14 different plants are used in a minor portion.

Dashmularishta contain 10 different roots hence the name Dashmularishta. Basically Dashmularishta is used to increase the immune system of human being. Punarnvarishta contain 'Punarna' as a medicinal plant. It is used to improve the working of heart, liver, pancreas, kidney etc. These Ayurvedic Medicines also consist of water, carbohydrates, protein, fats, alkaloid and alcohol molecules. The functional groups commonly present in these molecules are hydroxyl (-OH), aldehyde (-CHO), carbonyl (>C=O), Caboxylic (-COOH), amine (-NH-), methane (C-H) and cyanide (C-N).

Hydrogen bonding is the basics of all molecular interactions. The distinguishing feature of hydrogen bonding is the involvement of a specific H atom of a proton donar group with a localized site of high electron density in the same or another molecule. Another important feature of hydrogen bonding and of other weak attractive interactions in solution is that, at ordinary temperature, only a fraction of the molecules are generally associated. At equilibrium while a certain number of new complexes are continually formed, an equal number of complexes are continually broken due to the kinetic energy of motion of the interacting molecules. Basically hydrogen bonding occurs between a proton donar and proton accepter group. The hydroxyl (-OH), carboxyl (-COOH), amine (-NH-) and cyanide (C-N) are proton donar as well as proton accepter group. Hydrogen bonds occur between hydrogen containing dipoles and an electromagnetic element. The carboxyl group and oxygen atom have more electronegativity. Electronegativity provides us a relative ability of atom or functional group in molecule to attract bonding electrons.

The enthalpy of hydrogen bonds generally falls in the range of 1 to 10 Kcal/mole. Oxygen is a good proton accepter whether it is attached to phosphorous, to sulphur, to carbon or to nitrogen. The anions of electronegative atom from strong hydrogen bonds. Time Domain Reflectometry (TDR) is an effective approach to understand molecular interactions in liquid. Time Domain Reflectometry in reflection made is used as technique. It is very interesting to correlate dielectric parameters to molecular dynamics in aqueous solutions, hydrogen bonding change in size of molecular entities as well as their speed of rotation, in presence of different types of solutes was carried out by many research groups in the field of dielectric spectroscopy.

9. DATA ANALYSIS

The time dependent data were processed to obtain complex reflection coefficient $\rho^*(\omega)$ over the frequency range from 10 MHz to 20 GHz. Using Fourier Transformation as,

$$\rho^*(\omega) = (c/j\omega d)^* [p(\omega)/ q(\omega)] \tag{3}$$

Where p(ω) and q(ω) are Fourier transforms of $[R_1(t)- R_x(t)]$ and $[R_1(t) + R_x(t)]$ respectively, c is velocity of light, ω is angular frequency, $j = \sqrt{-1}$, d is effective pin length and the complex permittivity spectra $\varepsilon^*(\omega)$ were obtained from reflection coefficient $\rho^*(\omega)$ spectra by applying a bilinear calibration method[8]. The example of $\rho^*(\omega)$ and $\varepsilon^*(\omega)$ spectra are shown in Fig. 3 and 4 respectively.

The experimental value of $\varepsilon^*$are fitted with the Debye equation,

$$\varepsilon^* = \varepsilon_\infty + \frac{(\varepsilon_0 - \varepsilon_\infty)}{(1 + j\omega\tau)} \tag{4}$$

With ε_0 and τ as fitting parameters. In Eq. 4, ε_0 is the static permittivity, τ is the relaxation time and ε_∞ is the permittivity at high frequency. The value of ε_∞ is taken to be 3.0 for all the system studied, as for the frequency range considered here, the ε^* is not sensitive with respect to ε_∞. A non-linear Least-Squares fit method used to determine the values of dielectric parameters.

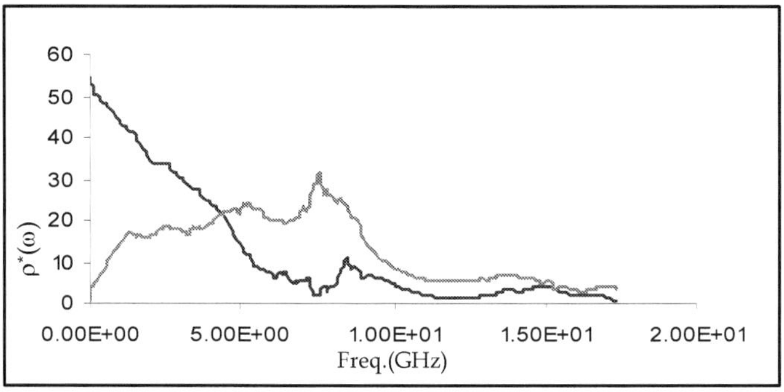

Fig. 3 $\rho^*(\omega)$ spectra for 80% Ashokarishta + 20% Ethanol at 15°C

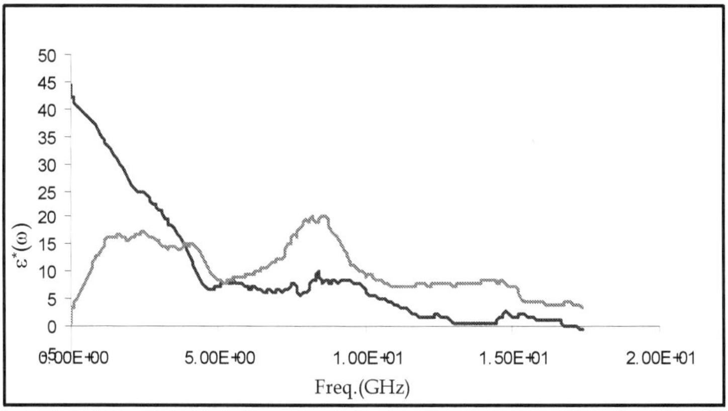

Fig. 4. $\varepsilon^*(\omega)$ spectra for 80% Ashokarishta + 20% Ethanol at 15°C

10. AYURVEDIC MEDICINE 'ASHOKARISHTA'AND ETHANOL

The Frequency dependent complex permittivity spectra $\varepsilon^*(\omega)=\varepsilon'(\omega)-j\varepsilon''(\omega)$, in frequency range of 10 MHz to 20 GHz, for mixture of Ashokarishta and Ethanol is presented in this section.

Ashokarishta (Shri. Baidyanath pharmaceutical ltd., Kolkata) and Ethanol (Changshu yang Yuan Chemical, China) were obtained commercially and used without further purification. The solutions were prepared at different volume percentage of Ashokarishta in Ethanol in steps of 10% vol., within 0.01% error limit. The experiment is performed for 11 concentrations, at 15°C, 25°C, 35°C and 45°C temperatures. The relative change in dielectric parameters with increasing concentrations of solute and temperature are presented in this section. The bilinear calibration method is used to obtain complex permittivity $\varepsilon^*(\omega)$ from complex reflextion coefficient $\rho^*(\omega)$. Reflection coefficient spectra $\rho^*(\omega)$ for pure Ethanol is used to correct $\rho^*(\omega)$ of entire concentrations range at each temperature. The corrected data for each mixture is fitted to Debye equation to obtain static permittivity (ε_0) and relaxation time (τ). As the frequency range of dielectric investigation in present work is from 10 MHz to 20 GHz, the value of (ε_0) obtained is just fitting parameter. This value does not correspond to real value of permittivity which one gets after completion of dispersion processes related to vibration and electronic motions in liquid. It is found reasonably satisfactory procedure to keep value (ε_∞) fix. In the present work value of (ε_∞) is kept fix to 3.00. The corrected spectra $\varepsilon^*(\omega)=\varepsilon'(\omega)-j\varepsilon''(\omega)$ for mixtures of Ashokarishta - Ethanol system for different concentrations at 35°C is shown in fig. 5. Gradual decrease in permittivity and dielectric loss with increasing volume fraction of Ethanol can be observed from this fig. The fall in permittivity starts at low frequency for Ashokarishta rich region but this point shifts to higher frequency in Ethanol rich region. This indicates small change in alignment of molecular dipoles with addition of Ethanol in Ashokarishta. The Cole- Cole plot for Ashokarishta and Ethanol system at 35°C is shown in fig. 6. The Cole Cole plot follows the Debye semicircle, which indicates that relaxation process in Ashokarishta-Ethanol system can be explained with single relaxation time.

10.1 Static Permittivity and Relaxation Time

The Values of static permittivity (ε_0) and relaxation time (τ) of Ashokarishta - Ethanol system for 11 different concentrations at four temperatures are listed in table 1. The Variation in static permittivity (ε_0) with increasing volume fraction of Ethanol is shown in fig 7. Static permittivity of mixture decreases linearly with increasing volume fraction of Ethanol. Decrease in permittivity with increasing temperature is also observed for all concentrations. This fall in permittivity with increasing amount of solute indicates linear increase in effective dipole moment of the system.

The variation in relaxation time (τ) with increasing volume fraction of Ethanol is shown in fig. 8. Relaxation time increases with increasing volume of Ethanol in solution. Raise in relaxation time (τ) is slow up to 50% Ashokarishta - Ethanol solution. With increase in volume fraction of Ethanol above 50%, relaxation time rises fast. Gradual decrease in relaxation time with increasing temperature is also observed for all concentrations. Fast increase in relaxation time in Ethanol rich region shows that there is significant change in size of molecular structures in this region. The slow increase in relaxation time indicates decrease in density of larger molecular structures in solutions. This increase in relaxation time indicates increase in amount of hydrogen bonding between solute and solvent molecules, which leads to smaller molecular structures rotating fast.

10.2 Excess permittivity (ε^E) and excess inverse relaxation time $(1/\tau)^E$

The structural changes in binary mixture can explored by determining excess properties. The plot of excess permittivity (ε^E) and excess inverse relaxation time $(1/\tau)^E$ with change in volume. fraction of Ethanol is shown in fig 9 & 10. The values of excess permittivity are positive for all concentrations and at all temperature. The positive values of excess permittivity in mixture indicate increase in effective dipole moments, in proportion to their volume fraction, in pure liquids. This increase in effective dipole moment can be attributed to formation of new smaller structures, may be due to hydrogen bonding, with dipole moment more than addition of dipole moments of constituting molecules. The decrease in relaxation time from value of pure ethanol to 100% Ashokarishta can be explained with breaking of hydrogen bonds in mixture. Further increase in volume fraction of Ashokarishta increases density of comparatively smaller molecules, which leads to decrease in relaxation time. It must be noted that this decrease in relaxation time is rapid which shows formation of smaller structures.

The excess inverse relaxation time $(1/\tau)^E$ reveals speed of rotation of molecular structure. The value of $(1/\tau)^E$ gives us frequency at which dielectric loss is maximum. Positive value of $(1/\tau)^E$ shows increase in frequency at which peak value of dielectric loss occurs.

10.3 Bruggeman Factor

The experimental values together with ideal and theoretical values of Bruggeman of (f_B), plotted against change in volume fraction of Ethanol are shown in fig.11. The values of Bruggeman factor (f_B) of Ashokarishta-Ethanol system for 11 different concentrations at four-temperature are listed in table 2. The Bruggeman mixture formulae state linear relationship between (f_B) and volume fraction of solvent by assuming that there is no interaction between solute and solvent. Modified Bruggeman mixture formula can be used if two components in binary mixture interact. The experimental value of (f_B) for Ashokarishta - Ethanol are fitted to modified Bruggeman mixture formula. When value of numerical fitting parameter "a" is unity, modified Bruggeman mixture formula reduces to original Bruggeman mixture formula. Decrease in value of "a" below unity shows increase in effective volume fraction of solvent in mixture. The small values of "a" indicates significant expansion in effective volume of solvent as well as weak interaction between solute and solvent. Furthermore values of "a" changes remarkably with change in temperature, which shows temperature dependent nature of molecular interactions.

10.4 Thermodynamic Parameters

The variation of conductivity for Ashokarishta with volume fraction of ethanol is shown in Fig. 12. The values of molar enthalpy of activation (ΔH) in KJ/Mole, and Entropy (ΔS) in J/^{0}Kmole, obtained from Eyring's equation are given in table no. 3.

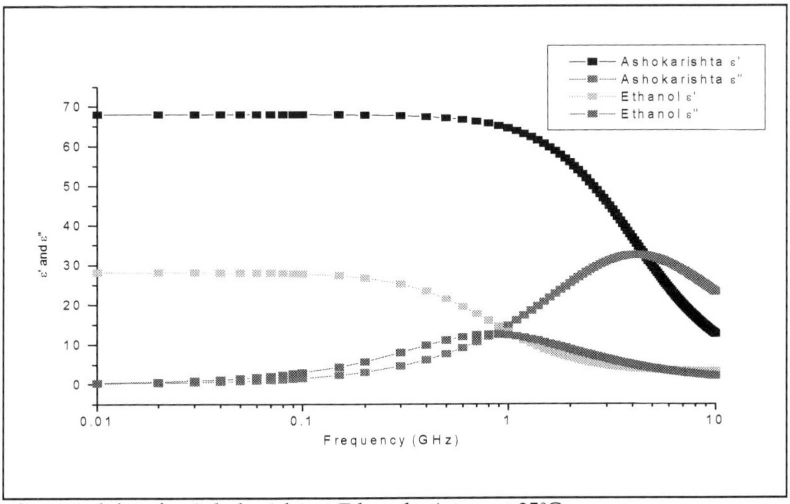

Fig.5. Corrected data for Ashokarishta + Ethanol mixture at 35^{0}C

The values of (ΔH) and (ΔS) decreases with increases in a volume fraction of Ashokarishta in Ethanol. The value of activation enthalpy gives an idea about nature of compactness in molecules of liquid. The smaller value of (ΔH) shows weaker hydrogen bonding in solute and solvent with decrease in Ashokarishta concentration in mixture. The plot of change in enthalpy and entropy in variation with volume fraction of Ethanol is shown in Fig. 13 and Fig. 14. The variation in free energy of activation with volume fraction of Ethanol in solution is shown in Fig. 15. Arrhenius plot i.e. plot of log (τT) verses 1000/T for Ashokarishta - Ethanol system is shown in the Fig. 16. The nature of plot is almost linear.

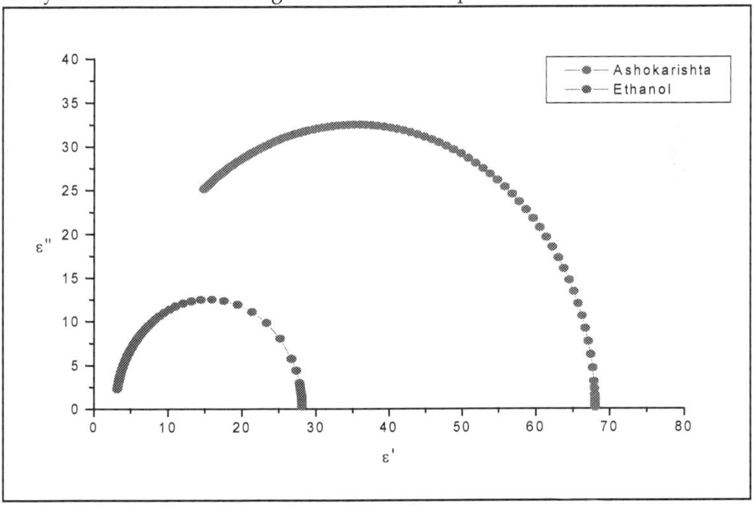

Fig.6. Cole -Cole plot for Ashokarishta + Ethanol

Vol. fraction of Ethanol	Static Dielectric constant (εo)				Relaxation Time (τ)			
	15 ⁰C	25 ⁰C	35 ⁰C	45 ⁰C	15 ⁰C	25 ⁰C	35 ⁰C	45 ⁰C
0.0	68.02	65.74	61.18	57.14	38.18	32.18	27.58	25.99
0.1	65.37	62.28	60.82	58.01	43.51	35.62	31.33	26.33
0.2	61.26	58.22	56.74	55.47	49.08	41.66	35.18	31.28
0.3	56.45	54.6	53.2	51.31	53.35	47.54	42.49	31.86
0.4	51.91	50.77	48.95	47.21	58.66	52.98	44.45	36.63
0.5	47.96	46.47	44.54	42.82	64.47	63.67	48.17	37.58
0.6	43.15	41.53	39.29	37.75	73.01	71.45	53.05	46.08
0.7	38.48	39.16	35.66	34.84	83.65	72.55	63.97	57.59
0.8	33.75	32.45	30.98	30.39	98.19	90.30	72.10	66.52
0.9	27.42	27.03	25.61	23.63	149.68	140.89	121.41	90.52
1.0	28.12	26.24	24.94	22.30	194.45	150.75	125.91	108.48

Table 1. Dielectric parameters for Ashokarishta at different temperatures

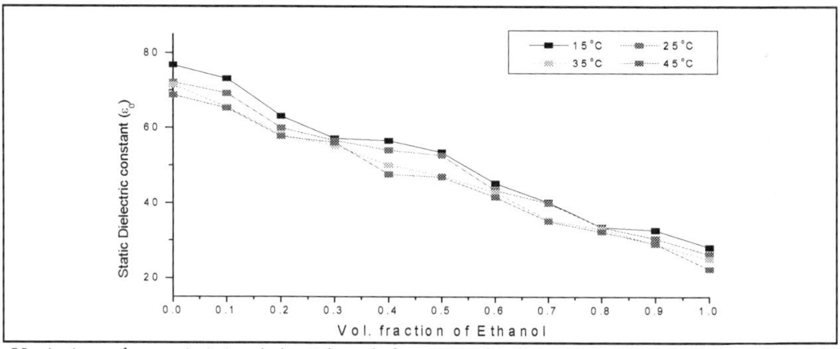

Fig. 7. Variation of permittivity (ε_0) with vol. fraction of Ethanol at various temperatures for Ashokarishta.

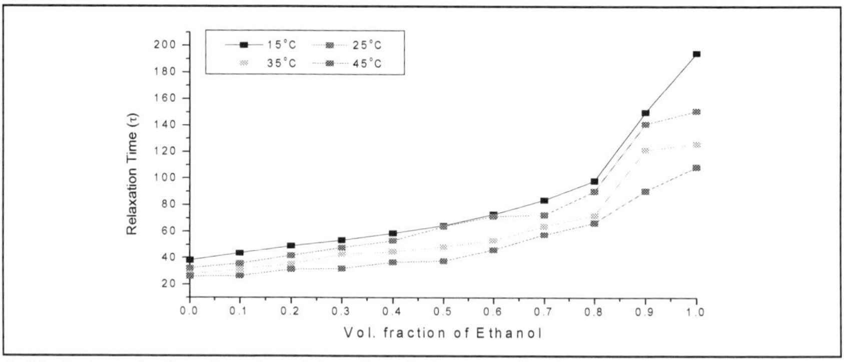

Fig. 8. Variation of relaxation time (τ) with vol. fraction of Ethanol at various temperature for Ashokarishta.

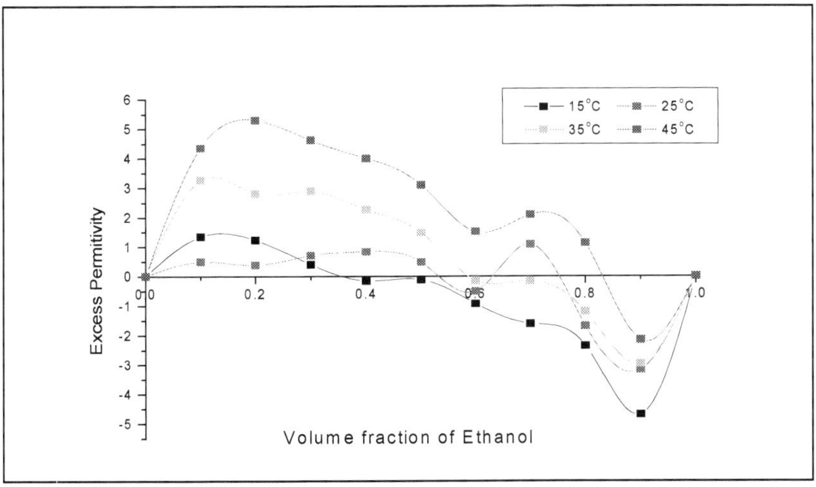

Fig. 9. Variation of excess permittivity (ε^E) with volume fraction of Ethanol in Ashokarishta at various temperatures

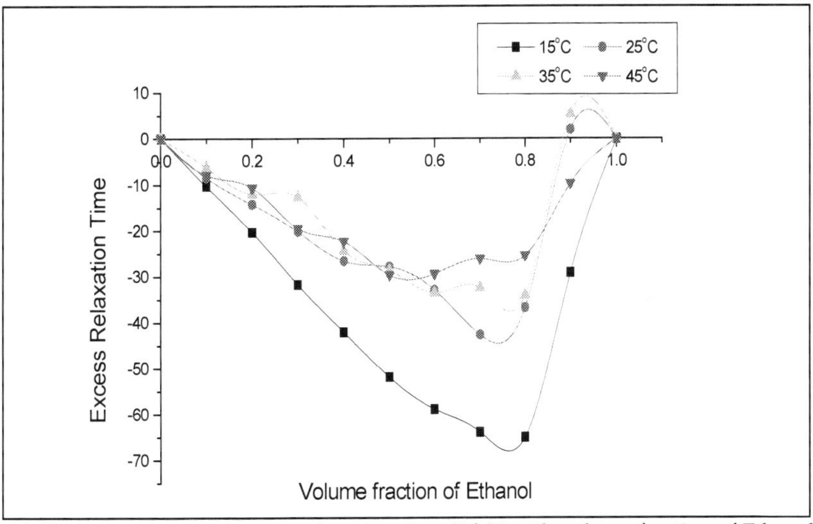

Fig. 10. Variation of excess inverse relaxation time $(1/\tau)^E$ with volume fraction of Ethanol at various temperatures for Ashokarishta.

Vol.fraction of Ethanol	Ideal value for F_b	15 ⁰C,a=0.933		25 ⁰C,a=1.0068		35 ⁰C,a=0.943		45 ⁰C,a=1.090	
		Expt.	Theor	Expt.	Theor	Expt.	Theor	Expt.	Theor
0.0	1.0	1.000	1.000	1.000	1.000	1.000	1.000	1.000	1.000
0.1	0.9	0.946	0.906	0.929	0.899	0.992	0.905	1.019	0.891
0.2	0.8	0.860	0.810	0.843	0.798	0.899	0.809	0.961	0.785
0.3	0.7	0.755	0.714	0.763	0.698	0.817	0.711	0.863	0.681
0.4	0.6	0.652	0.616	0.676	0.598	0.713	0.613	0.762	0.578
0.5	0.5	0.558	0.516	0.574	0.498	0.601	0.514	0.648	0.477
0.6	0.4	0.438	0.416	0.451	0.398	0.459	0.413	0.509	0.378
0.7	0.3	0.313	0.314	0.388	0.299	0.354	0.311	0.424	0.281
0.8	0.2	0.178	0.210	0.198	0.199	0.209	0.209	0.286	0.185
0.9	0.1	0.023	0.106	0.026	0.100	0.024	0.105	0.051	0.091
1.0	0.0	0.000	0.000	0.000	0.000	0.000	0.000	0.000	0.000

Table 2. Bruggeman factor for Ashokarishta-Ethanol mixture

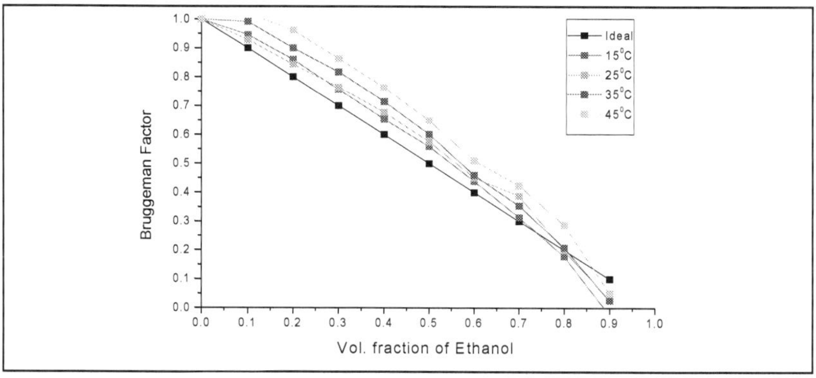

Fig. 11. Variation of Bruggeman factor (F_b) with vol.fraction of Ethanol in Ashokarishta at various temperatures

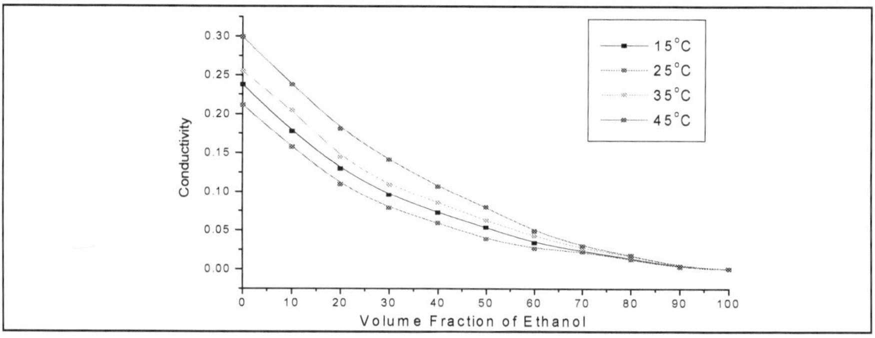

Fig. 12. Variation of conductivity for Ashokarishta with vol.fraction of Ethanol at various temperatures.

Vol. frac. of Ethanol	Entropy (ΔS) J/°Kmole	Enthalpy(ΔH) KJ/Kmole
0.0	-0.019	7.49478
0.1	-0.0116	9.942
0.2	-0.0157	9.07856
0.3	-0.0135	10.0227
0.4	-0.00534	12.64
0.5	-0.00917	11.809
0.6	-0.0156	10.195
0.7	-0.0274	6.9836
0.8	-0.0251	8.08786
0.9	-0.0224	9.9656
1.0	-0.016	12.2479

Table 3. Activation Enthalpy and Entropy of Ashokarishta -Ethanol

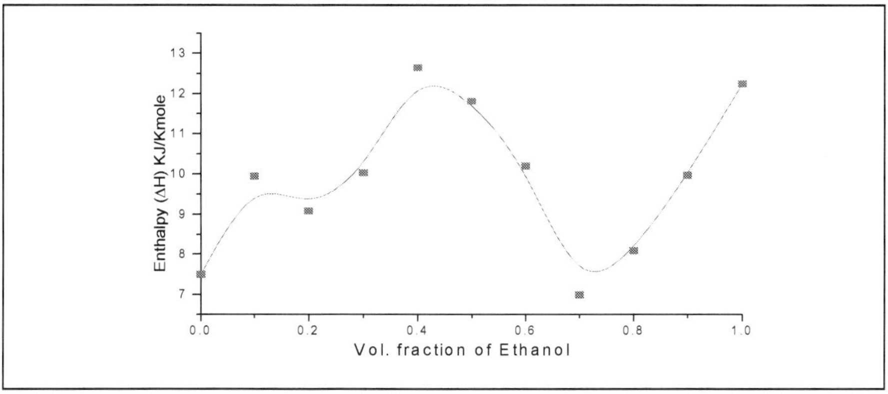

Fig. 13. Variation of Enthalpy for Ashokarishta + Ethanol

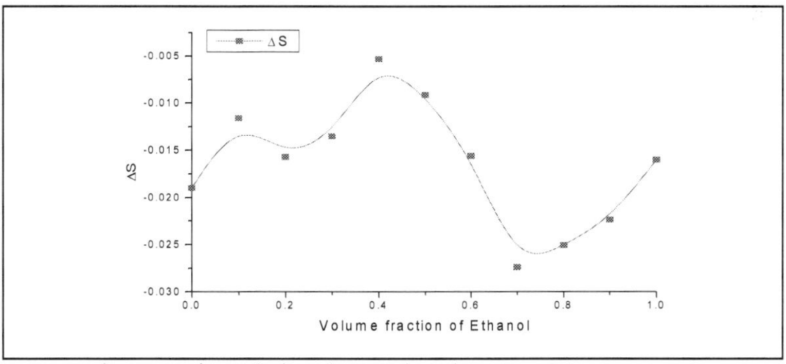

Fig. 14. Variation of Entropy for Ashokarishta + Ethanol

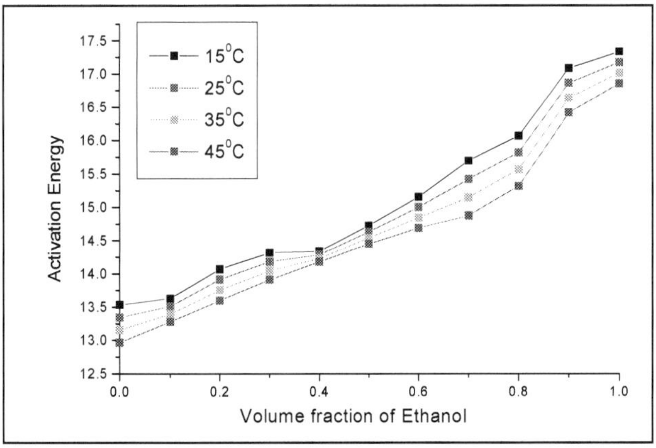

Fig.15. Variation of free energy of activation for Ashokarishta + Ethanol

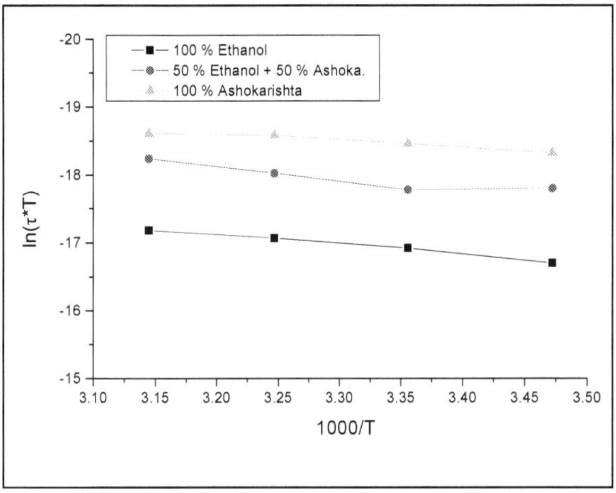

Fig.16. Arrhenius plot of Ashokarishta + Ethanol mixture

11. RESULTS AND DISCUSSION

The temperature dependent dielectric Relaxation as well as frequency dependent dielectric Relaxation has been used to investigate the information of dielectric properties of biological materials. The study of dielectric properties of these materials are of great assistance in exploring the molecular structure and dynamics of condensed matter. We can investigate the information such as molecular flexibility or rigidity, shape and size etc.

When a molecular system is placed in an electric field, there is always the tendency for the electrically charged species to move along the appropriate direction, causing the atom to develop an induced dipole moment. Permittivity of material reflects materials ability to get polarized with applied electric field. The amount of polarization depends on factors such as size of molecule, effective dipole moment and temperature. In microwave region major contribution to total polarization is orientation polarization. As frequency of applied field increases, it is expected that permittivity should decrease, since molecular orientation cannot cope up with speed with which applied field changes. Hence increase in frequency of applied field decreases alignment of molecular dipoles, which ultimately decrease permittivity. It is very interesting to observe frequency of point from which fall in permittivity begins. This point indicates the beginning of dispersion process. The shift in this point with change in temperature for biological sample gives us an idea about change in induced polarization. Relaxation time of biological material can be related to the size of molecule and mobility of molecules in liquid.

If relaxation time decreases it is correlated that due to decrease in size of molecules as well as to increase in mobility of molecules in liquid. If polar solute molecules are spherical and large by comparison with the solvent molecules then the orientation relaxation of the solute molecules can usefully be described using Debye's model. In this model the dipolar solute molecules are considered as spheres whose rotation is opposed by the viscosity of the surrounding solvent medium.

12. REFERENCES

A. Singh, V. K. Kaul, V. P. Mahajan, A. Singh, L. N. Mishra and R. S. Thakur, *Indian J. Pharm. Sci.*, 48(1986) 137.

A. Surowiec, S. Stuchly, R. Barr and A. Swarup, Dielectric properties of breast carcinoma and surrounding tissues, *IEEE Trans Biomed.Eng.*,35(1988)257.

Cole K. S. and Cole R. H., *J. Chem. Phys.*, 9, 341 (1941).

Collie C. H., Ritson D. M. and Hasted J. B., *J. Chem. Phys.*, 16, 1 (1948).

Davidson D. W. and Cole R. H., *J. Chem. Phys.*, 19, 1484 (1951).

Debye P. and Falkenhagen H., *Phys. Zeit.*, 29, 121, 401 (1928).

Debye P., *Polar molecules*, Chemical Catalog Company, New York (1929).

Debye, P. and Huckel E., *Phys. Zeit*, 24, 305 (1923).

H. Fellner-Feldegg, J. *Phys. Chem.*, 73, 613 (1969).

HP 54750A *Oscilloscope user's guide.*

HP 54754A *TDR Plug-in Modules* user's and programmer's guide.

J. F. S. Ferreira, J. E. Simon and J. Janick, *Plant Med.*, 61 (1999) 167.

K. S. Cole and R. S. Cole, Dispersion and absorption in dielectrics, *1.*

P. Debye , *Polar Molecules*. The Chemical Catalog Co.New York,(1929).

R. H. Linnell And S. N. Vinogradov, *Hydrogen Bonding*, Van Nostrand Reinhold Company, New York (1971).

V. Lad, Ayurveda, *The Science of Self-Healing*, A Pratical Guide, Lotus Press, USA (1984).

19

Analysis of Power Absorption by Human Tissue in Deeply Implantable Medical Sensor Transponders

Andreas Hennig, Gerd vom Bögel
Fraunhofer IMS, Duisburg
Germany

Abstract

The use of sensor transponder systems in medicine opens valuable possibilities in therapy and diagnostics. This chapter is about physical effects by the use of sensor transponder technology for medicine applications, especially for deeply implanted passive powered sensor transponders. This chapter will inform about present and future applications. The influence of human body on the energy transmission in a sensor transponder system is shown. The dielectrical properties of human tissue are discussed. A way how to estimate losses in an analytical way and with the use of a 3D FTDT method is presented. Finally, a design example of an energy transmission for a sensor transponder is shown with a calculation of the optimal frequency and experimental results.

1. Introduction

The treatment of cardiovascular diseases can be significantly improved by continuous monitoring of parameters such as blood pressure, temperature, etc. Miniaturized sensor transponders, implanted into the human body, can improve a therapy considerably. These transponders can be located in different places in the body, monitoring the performance of the heart circulation system. Such transponders make a cabling of the whole body unnecessary. Especially so-called passive transponder systems are of interest, because such implants normally stay inside the body for a long period. Thus, a supply by a local battery is not possible. In such systems, the transponder is contactlessly supplied by a field from the reading device located outside the body. The maximum possible distance between the reading device and the implanted transponder is of high interest, e.g. to make such a system suitable for corpulent patients. Because of that, it is essential to estimate the power loss caused by the heat capacity produced in human tissue. Moreover, human tissue has frequency depending properties. One of the most important tasks is to find the best carrier frequency for the wireless energy transmission. Figure 1 shows an example of a model for an implanted sensor transponder near to the heart and a corresponding reader. The coil of the reader produces an alternating magnetic field. A small part of the magnetic flux couples with the transponder coil. In consequence, a voltage is induced in this coil. With this voltage the electronic of the transponder is supplied with power. On the right hand there can be seen a sensor transponder in shape of a stick developed at Fraunhofer IMS. On the chip photo the front end, sensor and digital part are visible.

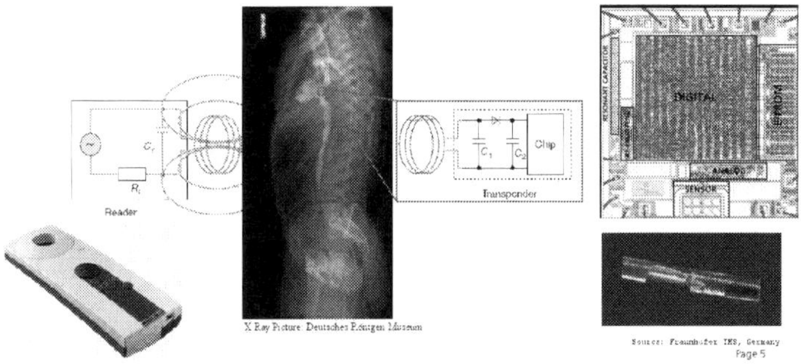

Fig. 1. Model of an inductive sensor transponder system for medical

2. Preliminary Considerations

2.1 Limitations and Requirements

The dimensions of an implantable transponder should not be more than several millimetres. Otherwise an implantation by a catheter is not possible. From this it follows, that only small antennas in the shape of a stick are supposed. The induced voltage is proportional to the size of the area encircled from the windings. Losses in the energy transmission through human tissue determine the available energy. Transponders with additional sensors consume significantly more energy than simple id transponders. These facts reduces the maximum possible distance. For this application, the distance can reach half a meter.

2.2 Frequencies and Antennas

Normally transponder systems work with ISM frequencies. Those are 100 to 150 kHz, in high frequency (HF) 6.78 MHz, 13.56 MHz, 27.125 MHz and 40.68 MHz, as well as 433.92 MHz, 869 MHz and 2.4 GHz in ultra high frequency band (UHF). In LF and HF systems, the distance between reading device and transponder is significantly smaller than the wavelength. This is why it is usually called near-field area. For UHF frequencies the distance is usually bigger than the wavelengths and is thus called far-field. In near-field areas field components are not related to one another and can thus be observed separately. In far-field areas, on the other hand, there are electromagnetic waves. The dimensions of transponder antennas are restricted to a few millimetres. A dipole-antenna to receive electromagnetic waves in ultra high frequency would be too big. Furthermore, an absorption effect is expected, which makes a passive use of the transponder impossible. This is why only low and high frequencies are of interest. For passively working transponder systems usually only the magnetic component is used to transfer energy and data. Once constructed, those antennas can be very small. A transfer via electric components would be very inefficient over such a distance. Antennas for magnetic transfer consist of a coil. There, two different types of concepts can be distinguished: air coils and ferrite coils. The advantage of air coils is that they can also easily be realised with bigger cross sections surfaces. Ferrite coils, on the other hand, show high inductance results on relatively small space. Moreover, there is a loss of ferrite material depending on the frequency.

2.3 Regulations for the Emission of Magnetic Fields

The EN regulation 300 330 specifies a maximum strength which is allowed to be created by a device within 10 metres. The exact strength in immediate proximity of the device is not given.

For the dimensioning of reader device antennas it has to be noted that the maximum allowed field strength should not be exceeded but at the same time there has to be enough strength for the transponder. Important parameters that define the strength are the amplitude of voltage over antennas and their geometry. Figure 2 shows a small part of this norm.

Frequency range (MHz)	**H-field strength limit (H_f) dBμA/m at 10 m**
$0{,}009 \leq f < 0{,}315$	30
$0{,}009 \leq f < 0{,}03$	72 or according to note 1
$0{,}03 \leq f < 0{,}05975$ $0{,}06025 \leq f < 0{,}07$ $0{,}119 \leq f < 0{,}135$	72 at 0,03 MHz descending 3 dB/oct or according to note 1
$0{,}05975 \leq f < 0{,}06025$ $0{,}07 \leq f < 0{,}119$ $0{,}135 \leq f < 0{,}140$	42
$0{,}140 \leq f < 0{,}1485$	37,7
$0{,}1485 \leq f < 30$	-5 (see note 4)
$0{,}315 \leq f < 0{,}600$	-5
$3{,}155 \leq f < 3{,}400$	13,5
$7{,}400 \leq f < 8{,}800$	9
$10{,}2 \leq f < 11{,}00$	9
$6{,}765 \leq f \leq 6{,}795$ $13{,}553 \leq f \leq 13{,}567$ $26{,}957 \leq f \leq 27{,}283$	42 (see note 3)
$13{,}553 \leq f \leq 13{,}567$	60 (see notes 2 and 3)

NOTE 1: For the frequency ranges 9 to 70 kHz and 119 to 135 kHz, the following additional restrictions apply to limits above 42 dBμA/m:

- for loop coil antennas with an area $\geq 0{,}16$ m^2 table 4 applies directly;
- for loop coil antennas with an area between 0,05 m^2 and 0,16 m^2 table 4 applies with a correction factor. The limit is: table value + 10 × log (area/0,16 m^2);
- for loop coil antennas with an area < 0,05 m^2 the limit is 10 dB below table 4.

NOTE 2: For RFID and EAS applications only.
NOTE 3: Spectrum mask limit, see annex G.
NOTE 4: For further information see annex H.

Fig. 2. Limitation of field strength at 10 m meeting EN 300 330

The maximum allowed strength at 10 m is different for every frequency. For example, at 133 KHz there is a strength of 66 $dB\mu A/\text{m}$ allowed. At 6.78 MHz 42 $dB\mu A/\text{m}$ are valid and at 13.56 MHz, 60 $dB\mu A/\text{m}$. Figure 3 showes a summary of the existing frequency bands. Shaded bars show frequencies for the use in industrial and medical applications (ISM).

3. Analysis of Power Absorption in Human Body

In order to operate the passive implant, a transfer of energy from the reading device is necessary. This shall be realised wire-free through the human body directly to the implant. For this, distances of up to half a metre have to be covered within the human tissue. Losses are a normal part of the process. Part of the energy is changed into heat in the body. The aim is to keep losses small in order to enable an ideal energy transfer to a passive transponder. Apart from undesired warming of tissues which, from a medical perspective, could be dangerous, it is also necessary to keep losses small in view of the whole system. Bigger losses would lead

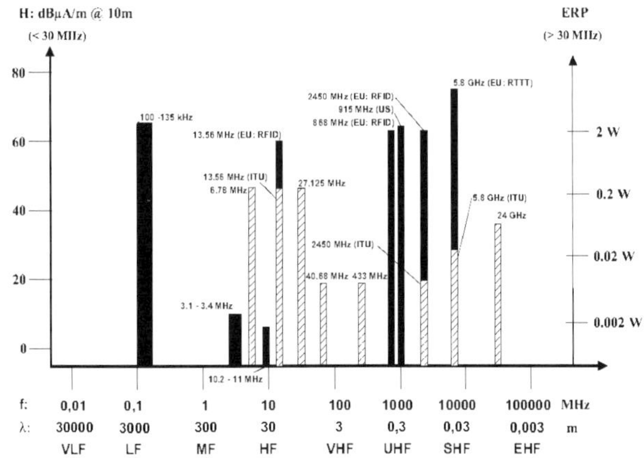

Fig. 3. Frequency areas [4]

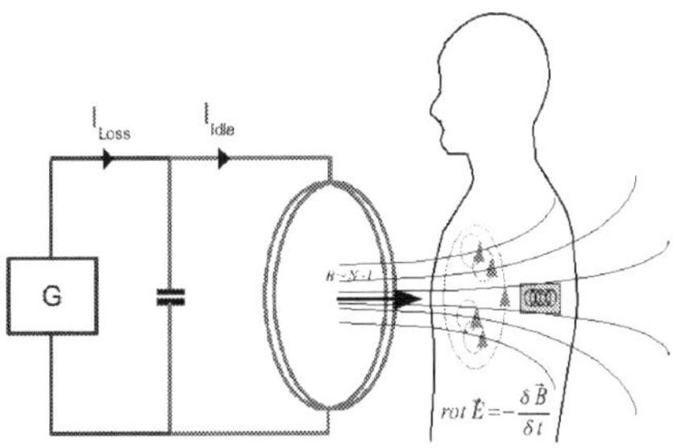

Fig. 4. Induced eddy currend caused by magnetic field changing in time

to a reduction of coverage and thus eventually make the system useless for its application. To minimise losses, it is necessary to find frequencies at which the transfer is at its peak.

The antenna coil of a reading device creates an alternating magnetic field. A small part of the magnetic flow which is produced goes through the coil of the transponder. As a consequence there is an induction of voltage in the coil. With the help of this voltage the electronic circuits are operated. If the medium surrounding and in between the coils is not conductive, energy transformation cannot take place in this medium. Only a small part of the energy provided by the generator is used to supply the transponder. A big part is transformed into heat in the antennas and is lost.

When non-magnetic conductive material lies between the antenna coil of the reading device and the transponder, it is permeated by a magnetic flow. Each alternating magnetic field with electricity-conductive material results in electric eddies if it is set vertical to the direction of the magnetic field. These are electric fields which are adjusted to circular paths. They can be described with the 2nd Maxwell equation.

$$rot\,\vec{E} = -\frac{d\,\vec{B}}{dt} \tag{1}$$

The exact electric field strength depends on temporal change of the magnetic flow. The rotation mathematically describes the development of eddies. Each electric field, which exists in conductive materials, results in current density J.

$$\vec{J} = \kappa\,\vec{E} \tag{2}$$

The current flow now takes place along circuit paths. This is why it is also called eddy current. This current flow now also results in an alternating magnetic field. The orientation of the field is opposed to that of the reading device. This leads to a partial loss. Thus, one could say that magnetic fields can permeate through the material without any problems but interfere with another field, which then ultimately leads to a destructive heterodyne. The magnetic field strength at the transponder is thus weakened by eddy currents. Less voltage is induced in the antenna of the transponder. This is why the transponder has less energy at its disposal.
A big part of the energy in conductive materials is transformed into heat. The power dissipation can be described as follows:

$$P = \frac{1}{2}\int_V \frac{\|J^2\|}{\sigma} dV \tag{3}$$

The heat capacity is dependent on the conductivity of the material and its volume. In order to specify losses in the human body very precisely, both conductivity as well as the volume of each tissue, which lies between reading device and implant, has to be known. In the following chapters the conductivity of human tissues is discussed. Afterwards, the losses with various frequencies are assessed.

3.1 Characterisation of Human Tissue

Human tissue does not have magnetic properties. The permeability of tissues is tantamount to that of a vacuum. This means that a magnetic field is not directly influenced by it. However, if it is an alternating magnetic field, electrically induced eddy currents have an impact on it. The following chapter thus concentrates on dielectric properties of tissues.
The human body is an inhomogeneous medium. It consists of different types of tissues and liquids. These are, among others, skin, muscles, fat, blood or organs such as lung, liver or heart. Each of these tissues and liquids shows different dielectric properties. Furthermore, these properties also vary in frequency.
In a simplified view, molecules of heterogeneous materials, such as water, can be treated as electric dipoles. If a magnetic field is applied from the outside, they adjust to this field. This process is called dipole polarisation. The impact of the field of adjusted dipoles on the field strength as a whole is expressed with the dielectric constant ε. The resulting electric field strength is called electric induction density.

$$\vec{D} = \varepsilon\,\vec{E} \tag{4}$$

With temporal-variant electric fields, the adjustment is only reached after a certain period of time. This time is described with the constant τ in mathematics. It is also referred to as relaxation time. In a sinusoidal alternating field this relaxation effect is shown by a distortion of phase between electric induction density and the electric field. This can be expressed with a complex permittivity.

$$\varepsilon = |\varepsilon| \cdot e^{j\delta} = \varepsilon' + j\varepsilon'' \quad (5)$$

If the frequencies are low, dipoles can follow the constructed field. For higher frequencies the re-orientation of dipoles is slowed down and the differences between the phases become bigger. This is shown in the course of real- and imaginary parts via the frequency. Figure 5 shows the distribution qualitatively. ε'_{r0} sets the permittivity at frequency zero. The imaginary

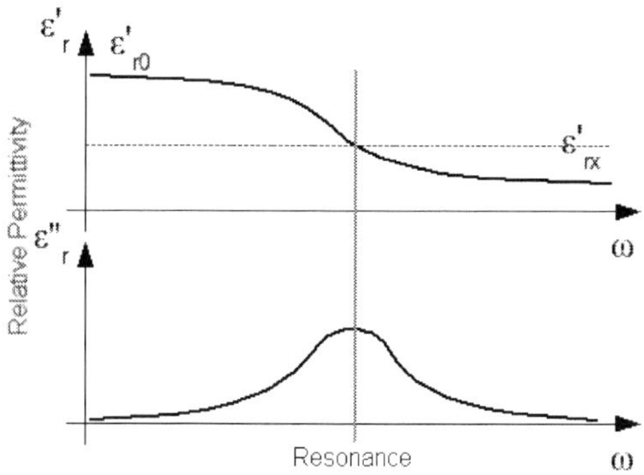

Fig. 5. Real and imaginary part of complex permittivity

part is distinguished from zero as soon as the real part is changing over the frequency. If the imaginary part increases, the phase difference between the outer field and the resulting induction density increases as well. In this case energy is transformed into heat. Thus, if the imaginary part is at its maximum size, the loss is at its maximum, too. In order to specify losses, the electric conductance is needed. The relation between permittivity and conductivity can be derived from Maxwell's equation.

$$\vec{J} = (j\omega\epsilon + \sigma)\vec{E} \quad (6)$$

After including permittivity in a complex form, conductivity can be divided into real and imaginary parts.

$$\vec{J} = j\omega(\epsilon' - j\epsilon'')\vec{E} + \sigma\vec{E} \quad (7)$$

$$= (\sigma + \omega\epsilon'')\vec{E} + j\omega\epsilon'\vec{E} \quad (8)$$

$$Real \quad : \sigma' = \sigma_{stat} + \omega\epsilon'' \quad (9)$$

$$Imaginary \quad : \sigma'' = \omega\epsilon' \quad (10)$$

The real part, which is also called effective conductance value, can be used to characterise losses in conductive materials. The conductivity for low frequencies is tantamount to the

static part σ_{stat}. For higher frequencies, imaginary permittivity exceeds. Thus, if the spectrum of a material's complex permittivity is known, the complex conductance can be determined. Together with geometry, losses for each frequency can be calculated.
Human tissues are heterogeneous materials. A variety of relaxation times exists. For the mathematical description of permittivity, in contrast to water for instance, higher terms are necessary. The frequency behaviour can be foretold with the help of the so-called Cole-Cole equation. This equation is based on the mathematical description of the above described model (1).

$$\hat{\epsilon}(\omega) = \epsilon_{inv} + \sum_n \frac{\Delta\epsilon_n}{1 + (j\omega\tau_n)^{(1-\alpha_n)}} + \frac{\sigma_i}{j\omega\epsilon_0} \tag{11}$$

σ_i is the static conductivity and ε_0 the permittivity of the vacuum. τ describes the relaxation time. Further concrete parameters for n=1...4 can be seen in (1). Parameters for all tissues, which are of interest here, are available, for instance for fat, heart tissue, lung, muscles and skin. Through the evaluation of this equation, dielectric properties of all interesting frequencies can be measured.

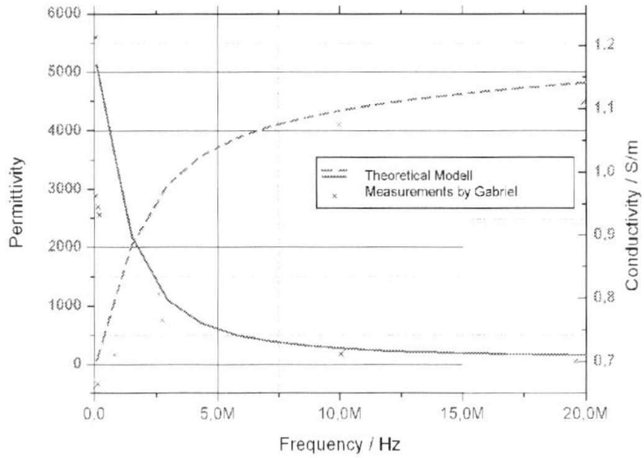

Fig. 6. Permittivity and conductivity of blood

Figure 6 exemplifies the properties of blood. The normal line shows the distribution of permittivity and the dotted line the distribution of conductivity for certain frequencies. Dielectric properties of human tissue were first characterised through an experiment carried out by (1). The crosses markes the measurement values investigated by Gabriel. Data are available for a frequency of 10 Hz to 20 GHz.
The conductance value includes both the static conductance value as well as the above described frequency-dependent value. First, the conductance value is important as the absorption of the field created by the reading device increases with increasing conductivity. According to figure 6 it can be expected that the absorbability in the frequency area shown increases with increasing frequency. This will be analytically evaluated in the next chapter.

3.2 Analytical Estimation of Power Absorption

In order to estimate losses for different frequencies, an analytical calculation has to be carried out. For this, a simplified model of the human body is constructed. Afterwards the power absorption which is transformed into heat is calculated for different frequencies. In order to classify the impact of energy transfer to the transponder, the available energy is set in relation to transfer in air in a FDTD simulation.

3.2.1 Simplified Body Model

If a transponder is directly attached to the heart and the reading device on the chest of a patient, there is more than one type of tissue between the antennas. Table 1 shows a list of tissues and their conductance values. In order to calculate the losses, the volume of each tissue, their conductance values and the distribution of current density induced have to be taken into consideration. However, here the aim is to have a simplified estimation, this is why a homogeneous model of the body is used in the following. It only contains one conductance value and consists of a simple geometry. For a "worst case scenario" conductance values of blood can be used because it has the highest conductivity in comparison to other types of tissues. For a second calculation an average conductance value is used. For the analytic calculation of induced eddy currents in bodies, a cylinder is especially useful because of its simple geometric form. Thus, a cylinder is assumed which has a field-creating coil at its front surface. The length is chosen so that the transponder is included and the distance to the field-creating coil is approximately 50cm. The diameter can be chosen accordingly. Figure 7 shows the model.

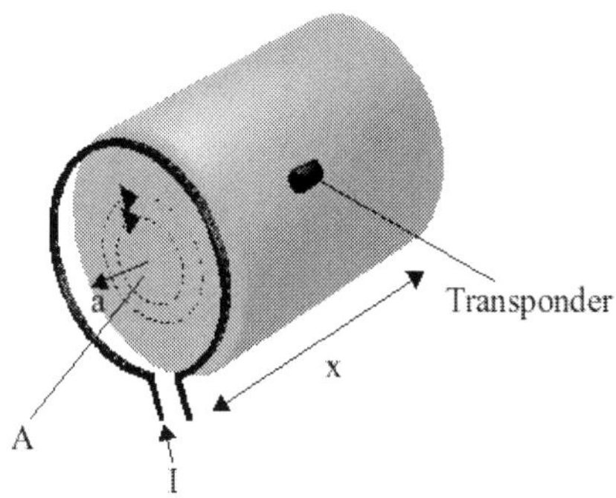

Fig. 7. Modell for estimation of power absoption

A length of l = 80 cm and a radiant of a = 15 cm were choosen.

3.2.2 Estimation of Power Loss

As described at the beginning, an alternating magnetic field, which occurs within a conductive medium, induces eddy currents. These ultimately lead to heating the medium. The heat capacity which is transformed in a cylinder - figure 7 - needs to be estimated. The heat capacity can be assessed as follows (3):

$$W = \frac{1}{2} f \mu H^2 A x K(a/p) \qquad ,with \tag{12}$$

$$K(a/p) = \frac{ber(\sqrt{2}\frac{a}{p})ber'(\sqrt{2}\frac{a}{p}) + bei(\sqrt{2}\frac{a}{p})bei'(\sqrt{2}\frac{a}{p})}{ber^2(\sqrt{2}\frac{a}{p}) + bei^2(\sqrt{2}\frac{a}{p})} \sqrt{2}\frac{a}{p} \tag{13}$$

- H is the magnetic field strength on the axis of the cylinder, μ is the permittivity, σ the conductance value, x the length of the cylinder and K a correction factor. The correction factor K describes the inhomogeneous distribution of current density an depends on the radius of the cylinder and penetration of the skin. The magnetic field within the cylinder is not homogeneous. Furthermore, the current density of eddy currents over the radius is not homogeneous due to energy-displacement effects.
The magnetic field strength H decreases with increasing distance x to the field-creating coil. This can be described with the Biot-Savart law.

$$H(x) = I \cdot \frac{a^2}{2(x^2 + a^2)^{\frac{3}{2}}} \tag{14}$$

Here, I is the intensity of current through the field-creating conductor loop. For an infinitely small part of the cylinder parts of the heating capacity can be described as

$$dW = \frac{1}{2} f \mu A K(a/p) \cdot I^2 \cdot \frac{x \cdot a^4}{4(x^2 + a^2)^{\frac{6}{2}}} dx \tag{15}$$

For a homogeneous cylinder model the heating capacity through integration over the whole length is:

$$\int_0^l \frac{1}{2} f \mu A K(a/p) \cdot I^2 \cdot \frac{x \cdot a^4}{4(x^2 + a^2)^{\frac{6}{2}}} x dx \tag{16}$$

With the conductivity of blood an heart the following results could be achieved. The conductivity of heart was choosen, because it has conductivity values near to the mean value of all tissues.

Frequency	133 kHz	3 MHz	6,78 MHz	13,56 MHz	27 MHz	40 MHz
Blood	2.27 μW	1.17 mW	4.55 mW	9.99 mW	17.91 mW	40.34 mW
Heart	1.80 μW	0.53 mW	2.81 mW	9.06 mW	19.75 mW	26.38 mW

In comparison, the electronik of the transponder consumes $90\mu W$. The volume of the cylinder in which the heat capacity is transformed is about $56{,}5dm^2$. The resulting heat capacity per unit volume is about 80nW. This value is quite safe for medical accounting purposes. In order to analyse the influence to the transponder-system, it is necessary to see the energy, which the transponder has at its disposal, in relation to a transfer via air. This is how the impact of absorption of the human body is visible.

3.3 Estimation by FDTD Method

In order to be able to analyse the frequency response in more detail and assess the strength of energy reduction of the transponder, a 3D field simulation was carried out. With a software it is possible to calculate the components of electric and magnetic fields. Furthermore, currents and voltage can be detected. The software is based on the FDTD (Finite Differences Time Domain) method which is based on the rough discretization of Maxwell's equations. A simple 3D model of the human body was constructed which contains all types of tissues that can be found between a reading device and the transponder. In the following this model will be called "inhomogeneous model". For each type of tissue the corresponding permittivity and conductance values were typed in (cf. table 1). In order to make the simulation more realistic, information about the volume of the tissues were extracted from a 2D MRT cross section. Figure 8 shows this process.

Tissue/Freq.	133 KHz	3 MHz	6.78 MHz	13.56 MHz	27 MHz	40 MHz
Skin	0.085347	0.06314	0.14712	0.23802	0.42748	0.45401
Fat	0.024484	0.02595	0.027775	0.030354	0.032909	0.03409
Muscle	0.36889	0.56805	0.6021	0.62818	0.654	0.6692
Lung	0.27613	0.3855	0.42109	0.45158	0.48429	0.50462
Bone	0.084146	0.10256	0.11585	0.12845	0.14185	0.15009
Heart	0.22405	0.41127	0.47134	0.52617	0.58769	0.62687
Blood	0.70494	0.98268	1.0673	1.117	1.158	1.1801

Table 1. Conductivities in S/m of different tissues at different frequencies (2)

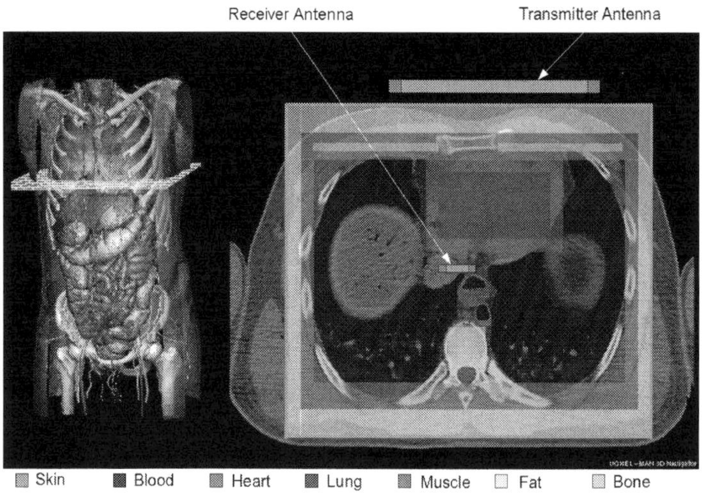

Fig. 8. Approximation der Volumen-Information (MRT-picture: Deutsches Röntgen-Museum)

A cross section of the human body at the level of the heart can be seen. In order to create a field and measure the strength close to the transponder, two types of antenna were used. With this simulation absorption and frequency behaviour can be analysed quickly.

In order to assess the absorption strength of the human body, it is necessary to eliminate factors from the antenna which might have an impact. For this matter, a type of reference simulation was carried out. In this simulation the human body was replaced with air. The measured voltage values at the transponder antenna will then be offset with the results of following simulations.

Simulations with a further model were used to assess absorption effects realistically. In the following this model is referred to as "homogeneous model". It reflects a worst-case scenario. For this purpose dielectric parameters of blood were used for all tissues since blood has a higher conductivity than other tissues.

In order to extract information about frequency-dependent absorption from the results of the three simulations, quotients were made from conductance values of the homogeneous model and the inhomogeneous model based on the reference model.

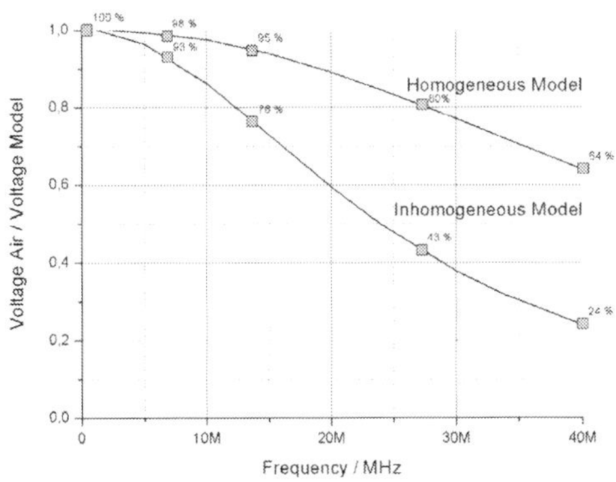

Fig. 9. Frequency depending attenuation

Figure 9 shows the voltage which can be induced in the transponder in comparison to a transfer via air. If there is only air in the transfer system, the quotient is one for all considered frequencies. First, it is clearly visible that the absorption capacity generally increases with higher frequencies and thus induced voltage decreases. For a frequency of 40 MHz the voltage decreases to 24 and 64 per cent respectively in the homogeneous model. In a low-frequency area, on the other hand, absorption is hardly detectable. However, which frequency is best for transferring a maximum of energy does not only depend on the absorption but also on further characteristics of the transmission. According to the induction-law, for instance, the induced voltage stands in proportion to frequency. Thus, it can be expected that there is a frequency at which the induced voltage is at its maximum. Furthermore, the characteristics of the antennas used have to be analysed. The following chapters deal with this topic. In chapter 4.2 the ideal frequency will be established with regard to all findings.

4. Example of Energy Transmission in a Sensor Transponder System

4.1 Frequency Behaviour of Induced Voltage at the Transponder

The voltage induced in the transponder coil is used to provide the power supply to the transponder electronic. To improve the efficiently, an parallel resonant circuit is formed by an additional capacitor connected in parallel with the transponder coil. Figure 10 shows the equivalent circuit of the transponder.

The resistor Ri represents the natural resistance of the transponder coil L1 and the current consumption of the transponder electronic is represented by the load resistor RL. If a voltage Ui is induced in the coil L1, the voltage Ul can be measured at the load resistor RL. It is a result of the voltage Ui minus the current i multiplied with the coil impedance and Ri. The so called quality factor represents the relationship between the induced voltage at L1 and the voltage at the transponder electronic. A higher quality factor causes a higher voltage ul and a higher

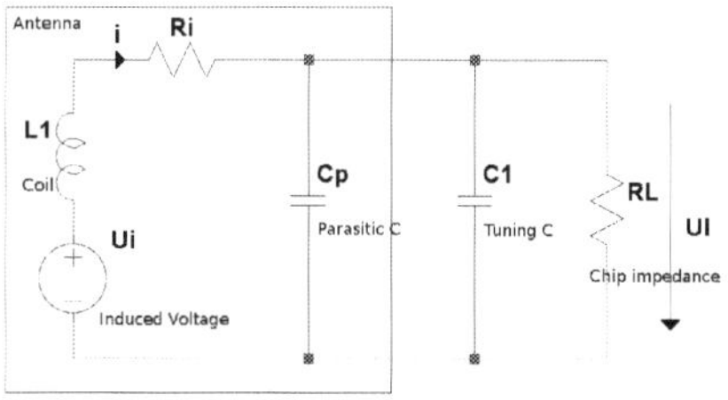

Fig. 10. Equivalent Circuit of a Transponder

maximum possible distance between reader and transponder. It can be calculated with the following formula relating to the equivalent circuit (4):

$$Q = \frac{1}{\frac{R_i}{\omega L_1} + \frac{\omega L_1}{R_L}} \tag{17}$$

By analysis of this formula it can be seen, that for every pair of Ri and RL there is a L1 at which the quality factor is at its maximum. And this maximum value of the quality factor is different for every frequency. So if the optimal L1 is calculated for every frequency, the maximum possible quality factor versus frequency could be calculated.

4.2 Optimal Frequency

The induced voltage Ui is reduced by the loss effects described in chapter 4.1. Because Ui is proportional to the quality factor, it is allowed to multiply the quality factor calculated with 17 together with the results of the graph's in figure 9. Figure 11 shows the evaluation of equation ?? considering the effects described before.

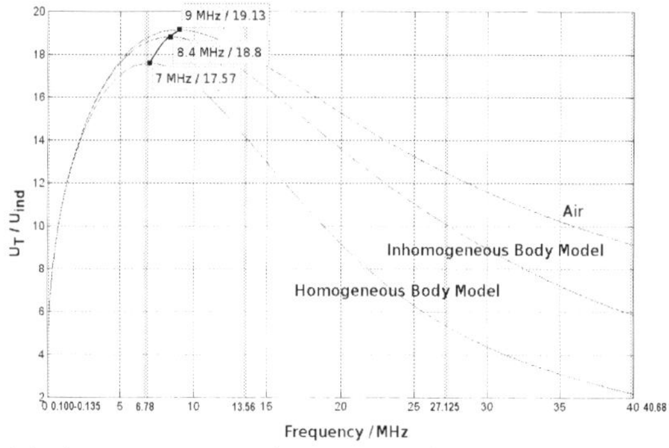

Fig. 11. Influence of the human tissue to the optimum frequency

First of all, a great difference in induceable voltages between LF and HF frequencies can be seen. For low frequencies, the quality factor is much smaller than for the HF case. The simulation shows a maximum quality factor for all simulations between 7 MHz and 9 MHz. If the coils are sourrounded by air, there will be an optimal frequency of about 9 MHz. This optimal frequency becomes lower, when human tissue is between the coils. For the homogeneous model, in worst case, an optimal frequency is about 7 MHz. In the inhomogeneous model, that is more realistic, the highest quality factor could be optained with 8.4 MHz. It can be said, that the human tissue reduces the optimal frequency value, at which the most voltage can be induced respectively the highest transmission range could be achieved. The optimal frequency can be observed near to the 6,78 MHz ISM band. In comparison to LF ISM Band the amount of induced voltage is about 4 times higher. In comparison to 13.56 MHz a power of maximum 20 % less is necessary to get the same transmission range.

4.2.1 Practical Measurement

An experimental measurement shall determine the maximum achievable distance. For this experiment, a circular coil with a single winding and an aperture of 26 cm was used to produce the magnetic field. A frequency of 13,56 MHz was chosen. A test transponder was developed to measure the energy that can be provided to an implanted transponder. It consists of a ferrite rod coil, an HF front-end and a load resistor that simulates the impedance of a transponder circuit. To create a substitute that simulates the electric properties of the human body, a phantom substance was prepared following a recipe described in [2]. The main goal of the experiment is to measure the voltage induced at the transponder coil when it is placed inside this substance at different distances from the reader coil. 50 L of the phantom substance is obtained. It was placed in a container large enough to allow the transponder to be placed in a similar position as in a human body. Following the specifications in the article, the container should be made of an electrical insulator and non-magnetic material. In our case, the container has a cylindrical form, which is sufficiently similar to a human body. Another requirement is a minimum volume of substance. It is specified that a mass of at least 30 kg of phantom material is necessary. Generally, a homogeneous phantom is accurate enough to simulate a human body, in this way it is not necessary to incorporate materials of different conductivity inside the container. Figure 12 (a) shows the measurement setup.

With this measurement it is possible to determine in how much surrounding tissue a transponder can work. To measure the provided energy for different distances from the reader, the voltage at the load resistor in the test-transponder was measured versus the distance. The chip used in sensor transponders usually works with voltages greater than 3 V. Therefore, the transponder would be provided with enough energy at a distance where the voltage is still higher than this voltage. Figure 12 (b) shows the measurement results.

The measurement was done with a voltage amplitude at the reader coil of 300 V and a load resistor in the test transponder of 60 kOhm and 100 kOhm. These values were chosen empirically. The diagram shows the voltage that would be available for a chip in different distances. The voltage is grater than 3 V for distances up to 43 cm.

The experimental measurement shows, that a sensor transponder can work inside a human tissue up to a distance of 40 cm.

5. Conclusion

The influence by the human tissue on the inductive energy transmission was considered for the design of a sensor transponder system. For the given constraints to the transponder antenna an optimal frequency could be found. The loss effects decrease this optimum frequency.

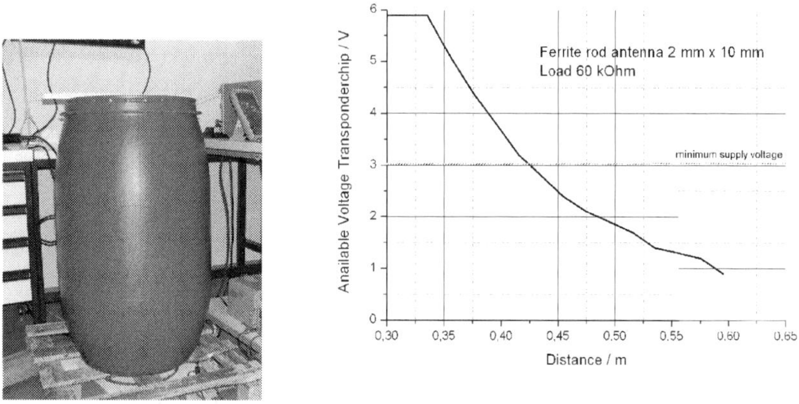

(a) Measurement setup (b) Results

Fig. 12. Practical measurement

A carrier frequency around 6,78 MHz is an optimal choice for our constraints. Measurements have determined the achievable transmission distance through human body.

6. References

[1] S Gabriel, R W Lau und C Gabriel; The dielectric properties of biological tissue; Phys. Med. Biol. 41 (1996) PP 2271-2293

[2] IFAC ;Dielectric Properties of Body Tissues in the frequency range 10 Hz - 100 GHz;http://niremf.ifac.cnr.it

[3] A. V. Vorst, A. Rosen, Y. Kotsuka; RF/Microwave Interaction with biological Tissues; John Wiley & Sons Inc.; Canada USA; 2006

[4] Klaus Finkenzeller; RFID-Handbook; Hanser; München Wien; 2006

[5] A. Hennig; RF Energy Transmission for Sensor Transponders Deeply Implanted in Human Bodies; EmuW IEEE 2008

20

UHF Power Transmission for Passive Sensor Transponders

Tobias Feldengut, Stephan Kolnsberg and Rainer Kokozinski
Fraunhofer Institute for Microelectronic Circuits and Systems (IMS)
Germany

1. Introduction

The importance of wireless sensors in medical systems, automotive applications, and environmental monitoring is growing continuously. A sensor node converts physical values such as pressure, temperature, or mechanical stress to digital values. The wireless interface connects it to a base station or a network for further data processing. Most of these products are required to be light, cheap, long lived, and maintenance free. Remote powering of transponder tags is a key technology to meet these demands, because it obviates the need for a battery. Near field systems usually operate in the low frequency range, typically between the 133 kHz (LF) and the 13.56 MHz (HF) ISM bands. While LF and HF systems operate in the magnetic near field via inductive coupling between two coils, UHF systems use electromagnetic waves in the far field of the base station. The range of the available inductive systems is typically limited to less than one meter, which motivates the use of far field energy transmission at ultra high frequencies. This chapter presents the design of a passive long range transponder with temperature sensor. The system is shown in figure 1.

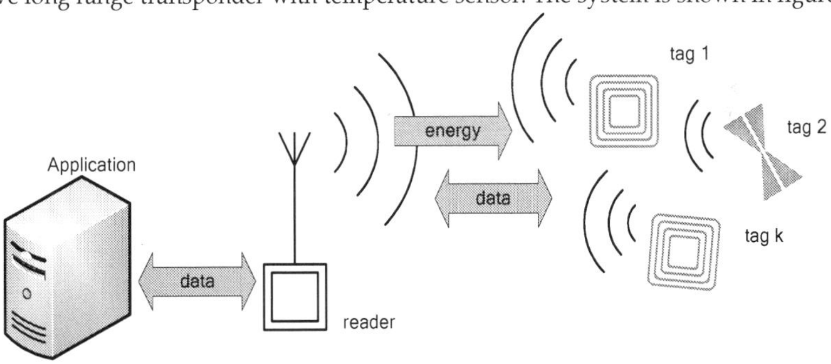

Fig. 1.passive far-field transponder system

A base station transmits an 868 MHz carrier wave that is modulated with the forward link data. In the transponder chip, the antenna voltage is rectified and multiplied to serve as the supply voltage for the integrated circuits including the sensor and a digital part. When the tag is transmitting data to the reader (backward link), it switches its input impedance between two different states to modulate its own radar cross section. The transponder is shown in figure 2. It consists of an integrated circuit and an antenna. The ASIC comprises an analog front-end as an air interface, a digital part for protocol handling, as well as non-volatile memory. The temperature sensor and the readout circuit are integrated on the same chip.

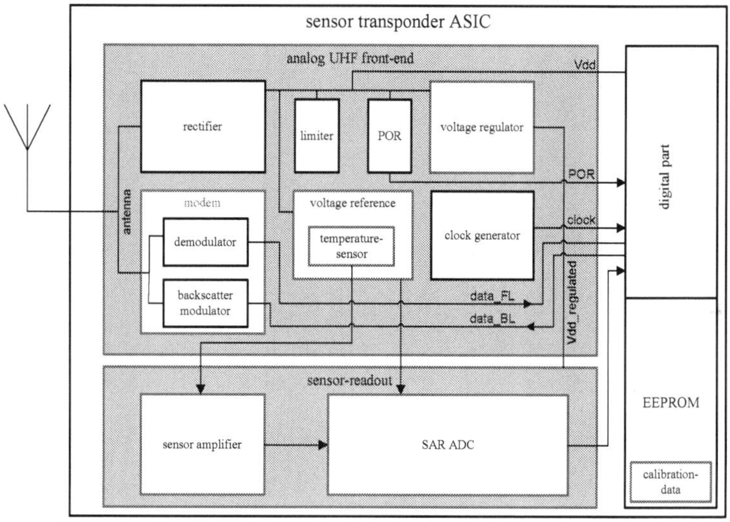

Fig. 2. sensor transponder architecture

The power supply block generates a stable 1.5 V voltage for the other circuit blocks by rectifying and regulating the incoming RF signal. The modem contains a simple low-power ASK demodulation circuit and a modulation switch. The carrier frequency from the reader is far too high to serve as a clock for the digital part, so that a local oscillator circuit is required. A bandgap circuit generates supply independent reference voltages and bias currents. It also generates a temperature-dependent voltage that is amplified to serve as the temperature sensor. This chapter is focused on the design of the analog front-end

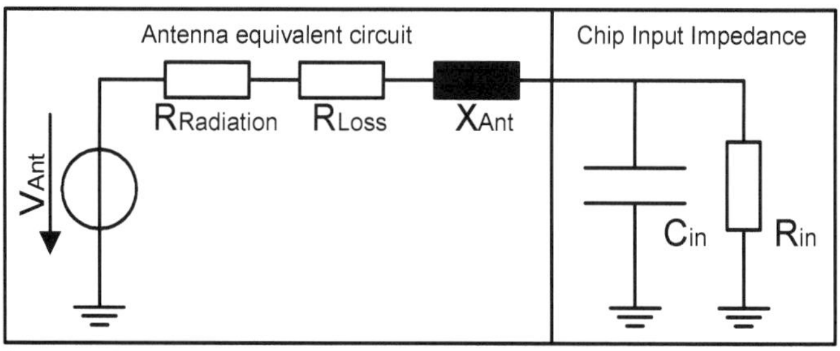

Fig. 3. simple equivalent circuit of transponder input

According to the well known Friis relation

$$P = P_{EIRP} G \frac{\lambda_{RF}^2}{(4\pi d)^2} \tag{1}$$

the power P that is available at the location of the transponder tag is related to the antenna gain G, the distance from the base station d and the wavelength λ_{RF}. The available power is sufficient to power the integrated circuits even in a far distance, but the high frequency antenna voltage is critically low. Figure 3 shows a simplified equivalent circuit of the tag input and the antenna. The antenna can be modelled as a radiation resistance $R_{Radiation}$, a loss resistance R_{Loss} and a reactive part X_{Ant}. The input of the transponder is modelled as a resistor and a capacitor as a linear approximation of the actual rectifier input impedance [Curty et. al, 2005]. Antenna matching is used to achieve high input voltage amplitude as well as power matching. The amplitude of the incoming signal is often as low as the threshold voltage of the rectifying devices, and sufficient rectifier efficiency is therefore difficult to achieve. Chapter 2.1 is focused on the rectifier optimisation.

2. Front-End Design

The analog front-end is mainly used for supply voltage generation, modulation and demodulation of data, clock synthesis, and reference voltage generation. In order to achieve a long range operation, all circuit blocks need to be optimised for ultra low power consumption. The main circuit blocks, namely the rectifier, the bandgap reference, the modem and the clock generator will be presented.

2.1 Rectifier

The rectifier is the most critical circuit for efficient energy transmission. The input from the antenna is a high frequency (868 MHz) signal with amplitude of less than 500 mV at large distance from the base station. Rectifying diodes are required to operate at (or slightly below) the threshold voltage. Recent research efforts have focused on the modelling and the optimisation of the typically used multi-stage Dickson charge pump [Curty et. al., 2005]; [Karthaus & Fischer, 2003]. This circuit is shown in figure 4.

Ideally, diode D1 and capacitor C1 lift the AC input voltage up by its peak value. Diode D2 and capacitor C2 create a peak detector, so that the output voltage of the first stage is set to twice the input amplitude. Several stages are cascaded to reach an output voltage that is high enough for the reliable operation of all circuits. At high frequencies and at low input voltage levels, the behavior of actual implementations differs significantly from the predictions of this simplified explanation [Karthaus & Fischer, 2003]. This fact results from the parasitics of real world devices, especially in cheap standard CMOS solutions. The following effects are detrimental to the rectifier performance. The diodes exhibit

- forward voltage drop,
- significant substrate and junction capacitances,
- reverse leakage current, and substrate leakage.

These values depend not only on the diode area, but also on the output current of the rectifier, on the temperature, and on random process variations. In addition to the diode parasitics, integrated capacitors exhibit parasitic substrate capacitances, series resistance, limited capacitance values, and leakage current. Finally, package parasitics, pad capacitance, and metal line parasitics can not be neglected. All of the above mentioned effects need to be considered and make the rectifier design a challenging task.

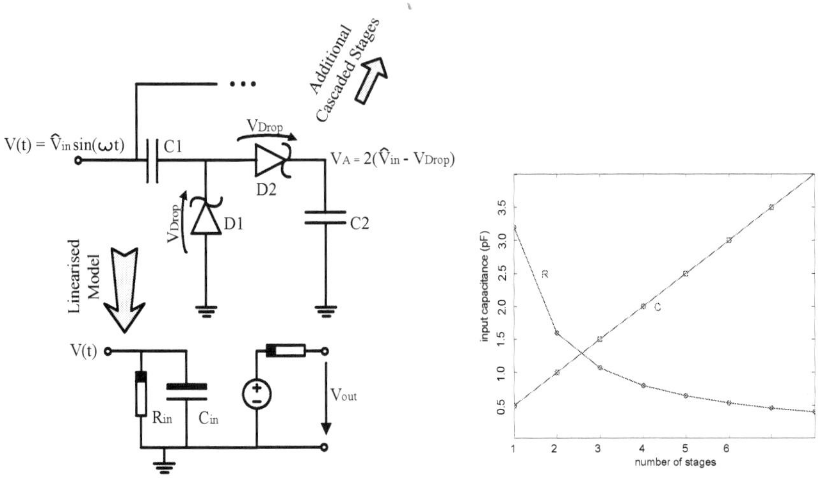

Fig. 4. basic rectifier stage, linearised model, and input impedance [Curty et. al., 1005]

Figure 4 also shows a linear model of the rectifier circuit [1]. It consists of an input resistance and an input capacitance, as well as an output voltage source and an output resistance. The antenna should be inductively matched to this input impedance of the rectifier. The tag also exhibits an input capacitance due to the above mentioned parasitic capacitances.

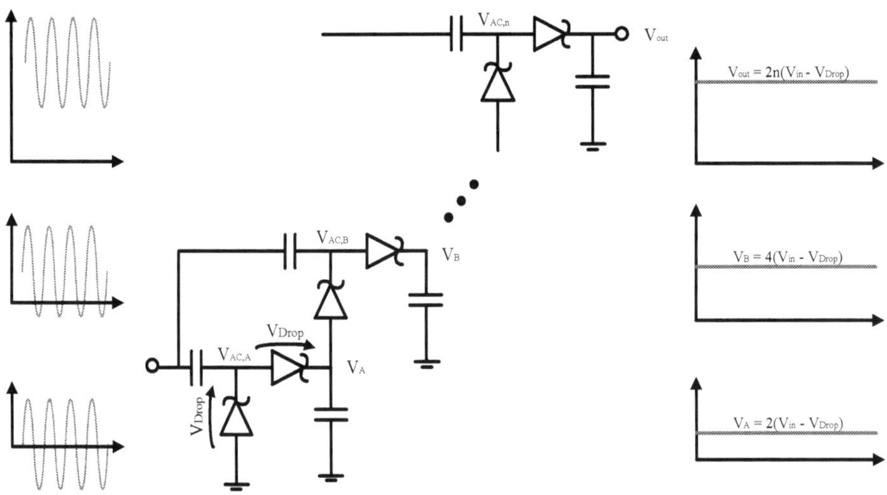

Fig. 5. multi stage voltage multiplier

At high quality factors, the bandwidth of the system is reduced, and inductive antenna matching is more difficult to achieve, so the input capacitance should be minimised. In summary, the goal of the rectifier design is to minimise the required input voltage and to achieve a large input impedance and low input capacitance for a given output current consumption. The major design concerns are therefore minimising the capacitance between AC and DC nodes as well as reducing the voltage drop

Figure 5 shows a multiplier stack with several stages, as well as the resulting waveforms at the AC terminal and the DC nodes. Each diode reduces the achievable output voltage by one threshold voltage drop. This voltage drop is a significant issue, as 12 diodes create a voltage drop of more than 2 V. One way to reduce this voltage drop is to use transistor diodes with threshold voltage compensation. This approach is depicted in figure 6 [Nakamoto et. al, 2007]. The gate of the transistor diode is forward biased by one threshold voltage, so that the device effectively acts as a diode with zero threshold voltage. The threshold voltage of transistors varies with temperature and process fluctuations, and the compensation voltage needs to track this variation. A compensation voltage that is too large would significantly increase reverse leakage currents, because the transistor could not close properly (see figure 7).

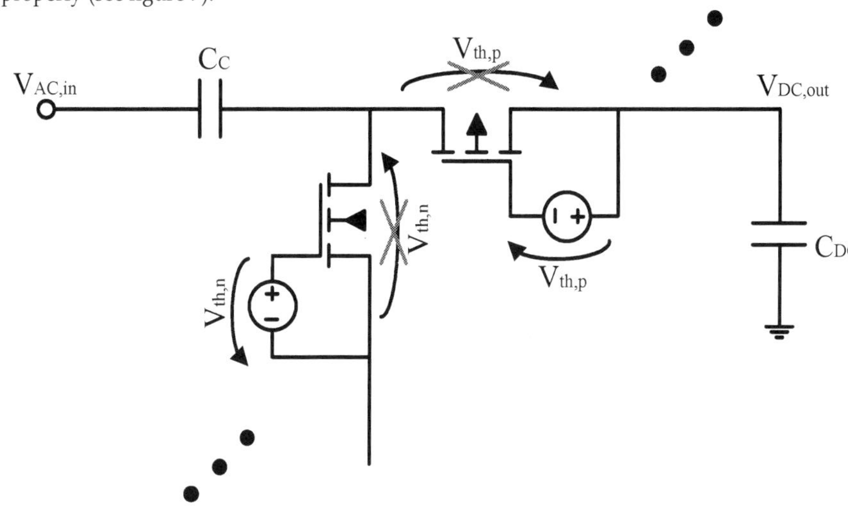

Fig. 6. rectifier stage with threshold voltage cancellation

Figure 8 shows the compensation voltage generator for NMOS transistors. It consists of a large resistor and a transistor diode M1 that is matched to the rectifying diode M2. Both diodes are designed to produce the same threshold voltage drop Vth.

The DC voltage at the anode of M2 is V_S. When the voltage at terminal V_B is larger than V_S+V_{th}, the compensation voltage at the gate of the rectifying diode equals one threshold voltage in a first order approximation, independent of V_B. When the diodes are matched, the compensation works independent of the temperature, the bulk-source voltage and process variations. The compensation of PMOS transistors works in an analog fashion.

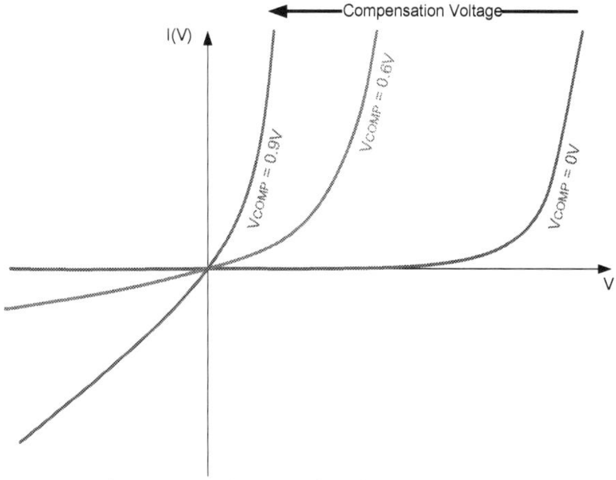

Fig. 7. I(V) characteristic of transistor diode with Vth compensation voltage

Figure 9 shows the complete rectifier circuit [Feldengut et. al., 2008]. It consists of two separate charge pumps, the first of which serves only as a compensation voltage generator (B) for the NMOS transistors in the main rectifier (A). The second rectifier (B) consists of eight stages with minimum area standard CMOS Schottky diodes. It can generate a large output voltage because it is almost unloaded. The output of this second rectifier is applied to the VB terminal of the compensation circuit in figure 8.

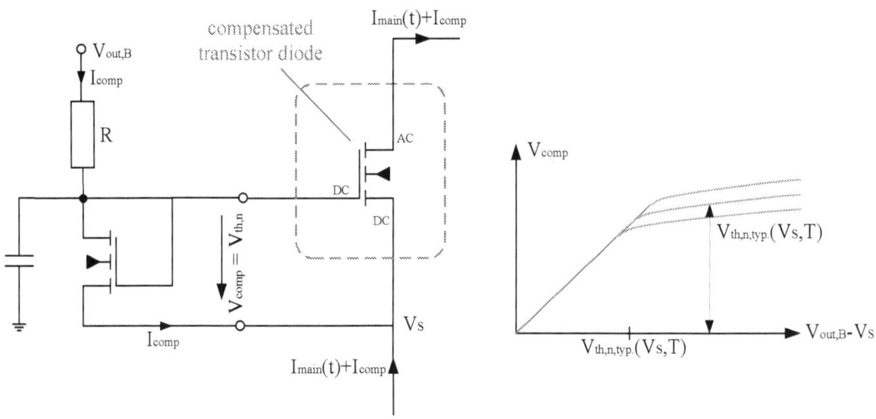

Fig. 8. compensation voltage generator

The total current consumption of the voltage dividers for the threshold voltage generation is less than 150 nA, while the output current of the six-stage main rectifier is typically several micro Amperes.

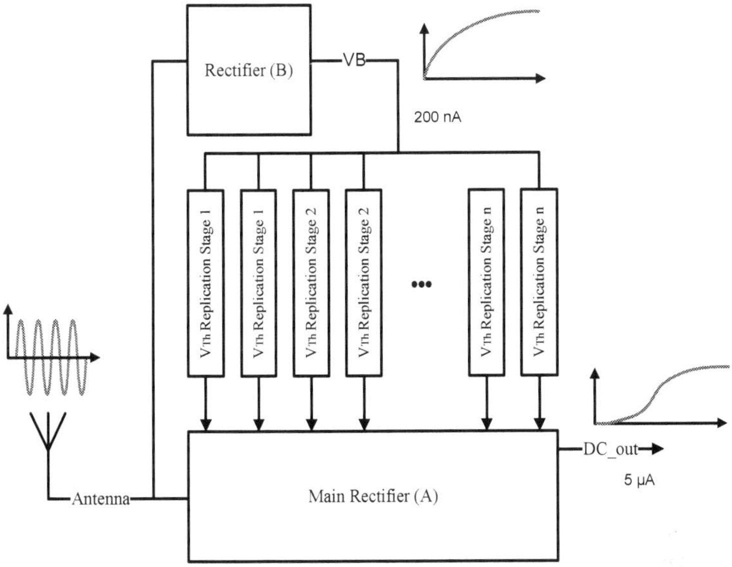

Fig. 9. proposed rectifier approach

The second rectifier (B) presents an additional load to the antenna. Its input capacitance and resistance appear in parallel to the main rectifier's input. The negative impact of this additional load is limited, because the input resistance is very large and the input capacitance is very small due to the minimum diode area and the low output current of rectifier (B). The circuit was implemented in a standard 0.35 µm CMOS process with Schottky diodes, double poly layers and high res poly resistors. The complete topology is shown in figure 10. Not all the voltage dividers for the threshold voltage reproduction are connected to the output of rectifier (B).

To reduce the current load and the required resistor size, the voltage dividers that compensate the lower stages of rectifier (A) are connected to intermediate output stages of rectifier (B). The voltage across the resistors is therefore very small. The compensation transistors have a much smaller aspect ratio, so that the voltage drop equals the threshold voltage across the rectification transistors, even when the current through the voltage dividers is several times smaller than the current flowing through the main rectifier stack (A).

Figure 11 shows the results for the conventional Schottky diode rectifier as well as for the proposed circuit. The load resistance is 300 kOhm and the output capacitor is 100 pF in both cases. The antenna resistance is 300 Ohm. At a distance of 4.5 m between the base station and the transponder, the input power is -11.3 dBm at a transmitted power of 2 W and a carrier frequency of 868 MHz. At this distance, the proposed rectifier (fig. 10) can power a transponder chip with 1.5 V supply voltage and 5 µA DC current, while the output voltage of the conventional circuit (fig. 5) is close to zero in the chosen process technology.

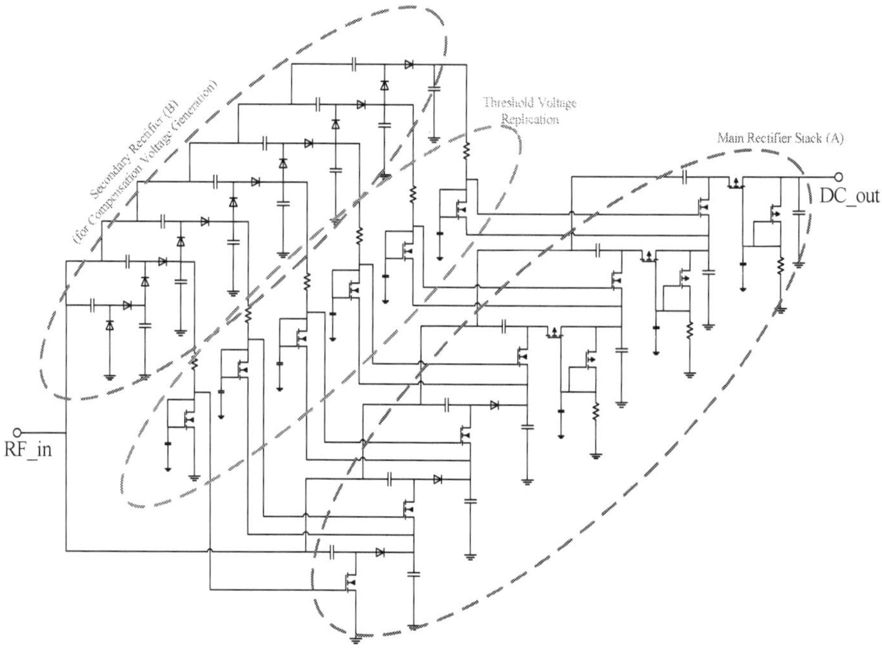

Fig. 10. complete circuit implementation

The transient start-up behaviour of the circuit is shown for different input voltages in figure 12. The output capacitor was reduced to only 20 pF in order to reduce the simulation time. The start-up waveform differs significantly from the typical capacitor loading waveform of a conventional diode multiplier. Once the compensation voltage has been build up, the rectifying transistors' efficiency is significantly increased, which changes the conversion efficiency. For very high output voltages of more than 2.5 V, some transistors are over compensated and begin to exhibit reverse leakage current. This leads to an abrupt stop of the output voltage increase.

Depending on the process technology, the number of stages may have to be reduced by one or two in each of the two rectifier stacks in order to reduce the input capacitance. The input capacitance has two negative effects: the first is that the bandwidth of the system is significantly reduced when the quality factor Q is high. The second issue with a large input capacitance is that it may also reduce the real part of the input impedance. When a large parasitic resistor lies in series to a large parasitic capacitance (this is often the case for substrate parasitics of diodes and capacitors in bulk CMOS), the equivalent parallel RC tank has a reduced resistance at the frequency of interest.

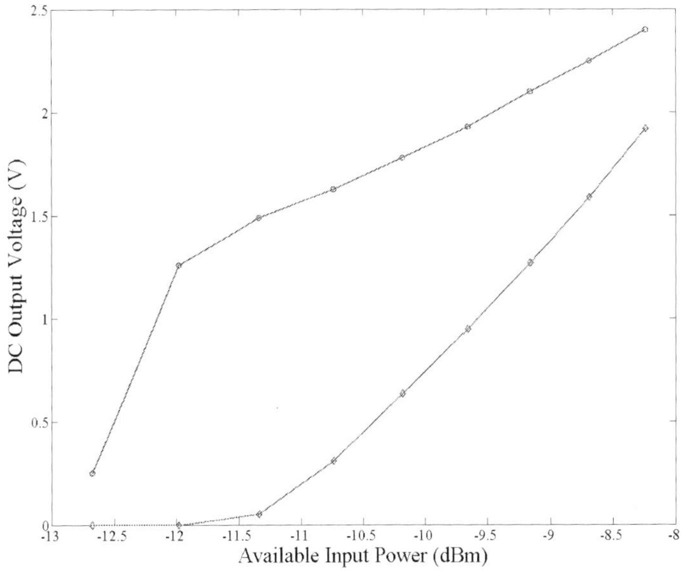

Fig. 11. Simulated output voltage as a function of input power under ideal matching conditions

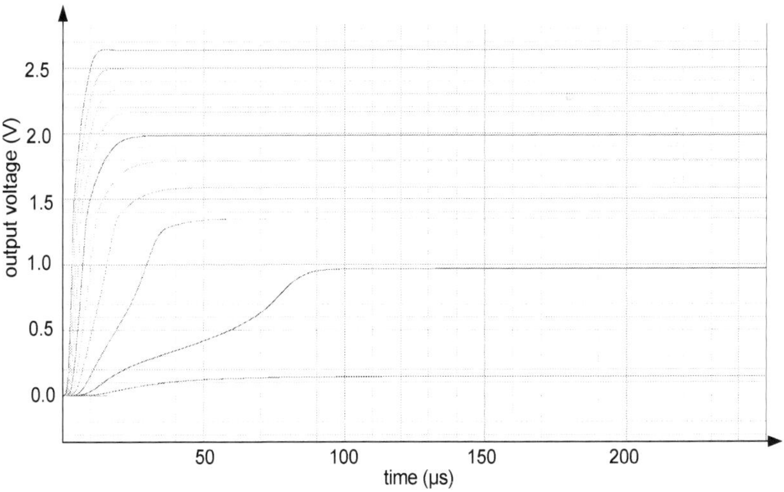

Fig. 12. simulation of transient start-up for different input voltages (reduced output capacitance to limit simulation time)

2.2 Reference and Temperature Sensor

The ADC and several other analog circuit blocks of a sensor transponder require precise reference voltages and currents that are insensitive to variations in the supply voltage, the temperature, and process variations. A low voltage bandgap reference circuit with low current consumption is implemented for this purpose [Razavi, 2001]; [Lee, 1998]. The topology is depicted in figure 13. The bandgap voltage reference core (middle) uses large high resistance poly resistors to limit the current in both branches to less than 1μA. These resistors have a negative temperature coefficient, so that the branch currents have a positive temperature coefficient (PTAT- Proportional To Absolute Temperature). A separate current reference circuit is required in order to obtain a temperature independent current. The V_{BE} of one of the PNP transistors is buffered and applied across another resistor with negative temperature coefficient, resulting in a nearly temperature-independent current.

The implementation of this circuit on a passive UHF transponder can be an issue, because the unregulated supply voltage from the rectifier (the input voltage for the reference) has a large dynamic range. The unregulated voltage is also very unstable due to the large output resistance and limited output capacitance of the rectifier. Furthermore, the required DC voltage level and the current consumption have to be reduced as far as possible in order to achieve long operating range. However, large resistor values and currents of less than 1μA lead to more mismatch, increased noise, and possibly stability issues. The operational transconductance amplifier provides limited gain and speed.

The reference circuit also serves as the temperature sensor because the voltage across the bipolar transistor is temperature dependant. The output signal is amplified and then converted by the ADC according to figure 2. Figure 14 shows the ADC input signal as a function of the temperature.

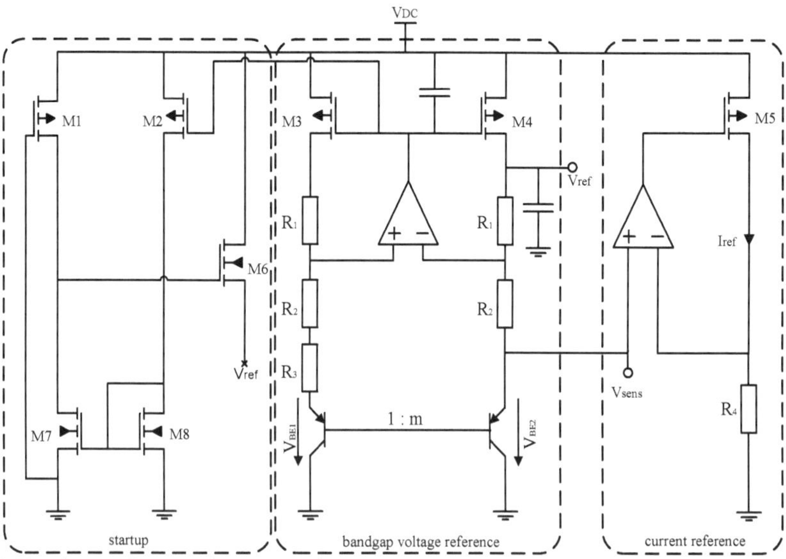

Fig. 13. Bandgap voltage and current reference

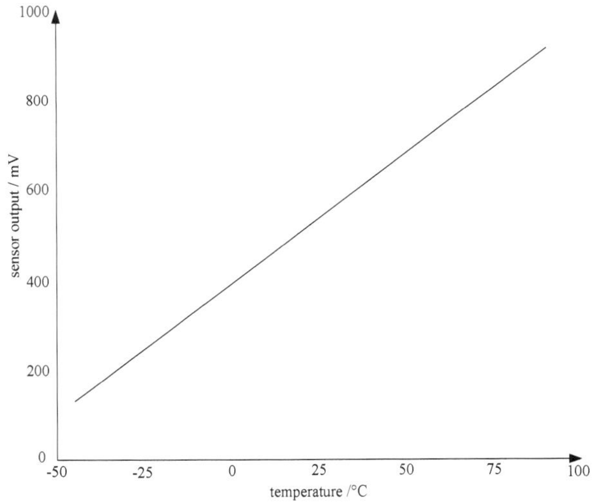

Fig. 14. output voltage of the temperature sensor after amplification

A low power successive approximation ADC is used to convert the sensor output to digital data. The operation principle is shown in figure 15. A digital value from the SAR register is converted to an analog value to be compared to the sampled input signal $V_{S/H}$. depending on which value is larger, the digital value in the register is either increased or reduced. After several cycles, the value of the SAR register is a digital representation of the analog input [Bechen, 2008].

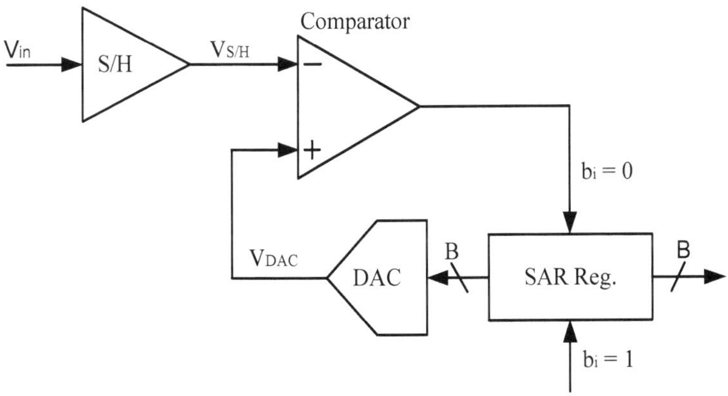

Fig. 15. SAR ADC architecture

Figure 16 shows the circuit implementation. The comparator is implemented using CMOS inverters. The first inverter is shortened from input to output to generate the middle voltage that determines the toggle threshold of the inverter. During the comparison phase, this voltage is either increased or reduced, depending on the input voltage level. The DAC is implemented with a capacitive array to reduce the static current consumption compared to a resistive voltage divider. A scaling capacitor is connected between the MSB and the LSB to limit the total capacitor area [Bechen, 2008].

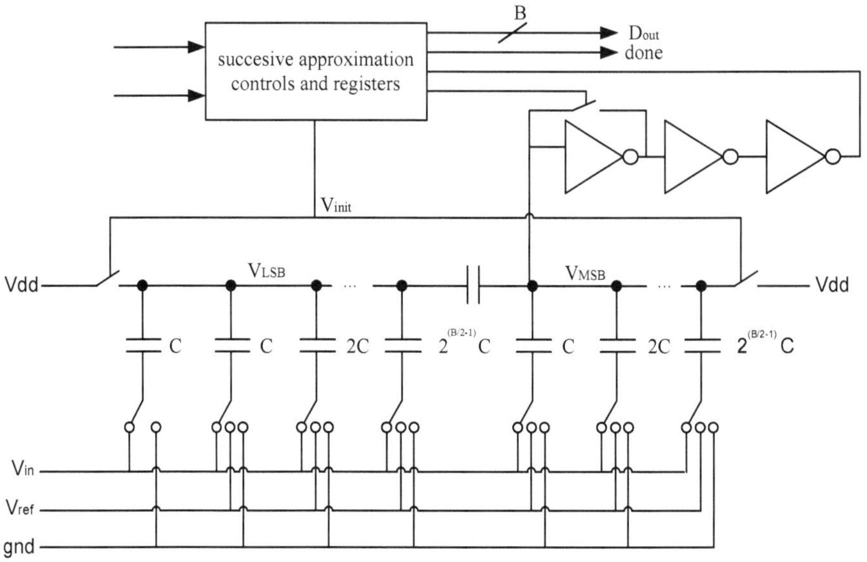

Fig. 16. SAR ADC circuit implementation for low power consumption

2.3 Demodulator and Clock Generator

Data transmission from the base station to the transponder is implemented using amplitude shift keying (ASK modulation). The shape of the antenna signal during communication is depicted in figure 17 ($V_{UHF,in}$). The amplitude of this signal varies with the distance between the transponder and the reader. A simple demodulator circuit (see figure 17) is used to extract the envelope and the average (or the delayed envelope) signal and to decide between the two logic states.

A small two stage multiplier that is loaded with a resistor extracts the envelope signal for the data frequency of 40 kHz. The circuit is similar to the main rectifier, but the diodes are very small so that the additional capacitive load that is presented to the antenna is not significantly raised. The envelope signal is fed into an additional low pass filter to extract the average value, which is different for each operating distance. The hysteretic comparator filters out noise and generates the logic signal for the digital part of the chip. The signals at the output of the envelope detector and the second low-pass filter are also shown in figure 17. Figure 18 shows the comparator circuit [7]. The unregulated supply voltage (VDC) is applied to the input stage to increase the common mode input range in close distance to the base station.

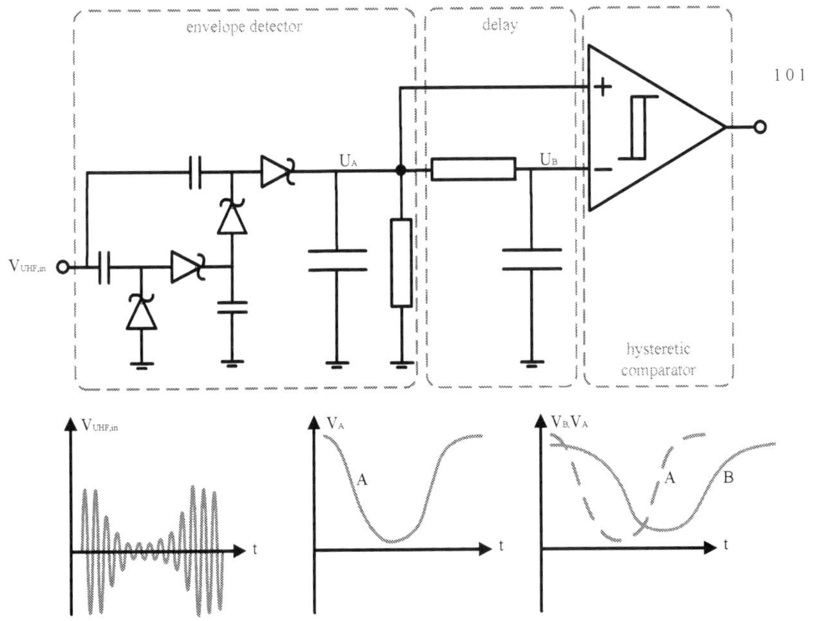

Fig. 17. Demodulator circuit

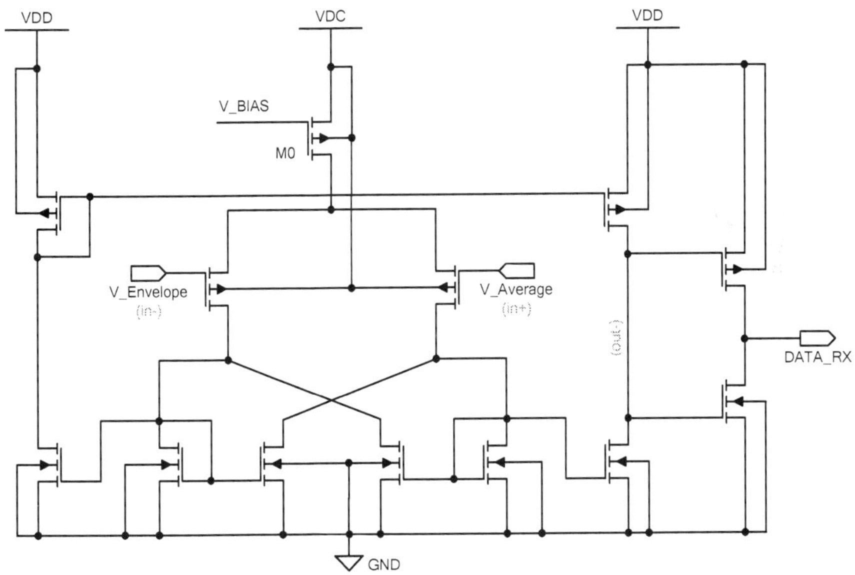

Fig. 18. hysteretic comparator circuit

In order to decode the data, the digital part also requires a clock signal that is several times faster than the data rate, but still at least two orders of magnitude smaller than the carrier frequency. This signal is generated by a local oscillator circuit. The relaxation type oscillator is shown in figure 19.

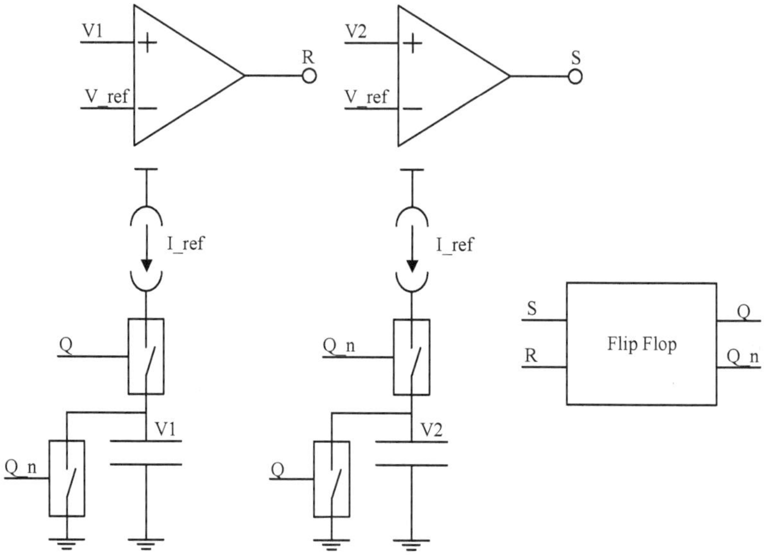

Fig. 19. low power relaxation type clock oscillator

Whenever the flip-flop is set, capacitor C2 is charged by a constant current while capacitor C1 is quickly discharged through transistor M1. The voltage at the input of comparator 2 rises linearly until it reaches V_ref. The flip-flop is reset, capacitor C2 is discharged through transistor M2 and capacitor C1 is charged. The advantage of this structure compared to a single capacitor design is that no hysteresis comparator is required in the oscillator and that the discharge time has no influence on the output frequency. The total current consumption of the oscillator is 400 nA at a frequency of 1 MHz. The frequency variation is mainly determined by the accuracy of the reference current, the capacitor accuracy, and the comparator delay. The comparator delay causes the capacitor voltage to peak above the reference voltage according to figure 20. This delay depends on process and temperature variation and is not controlled very well. When high frequencies are required the comparator current needs to be increased to reduce the delay in relation to the total oscillation period time. For high data rates above 100 kHz an oscillator frequency of 1.6 MHz is usually required. A ring oscillator circuit has the advantage of less current consumption at high frequencies. However, for the required data rate of 40 kHz, the relaxation design offers improved accuracy at lower oscillation frequency.

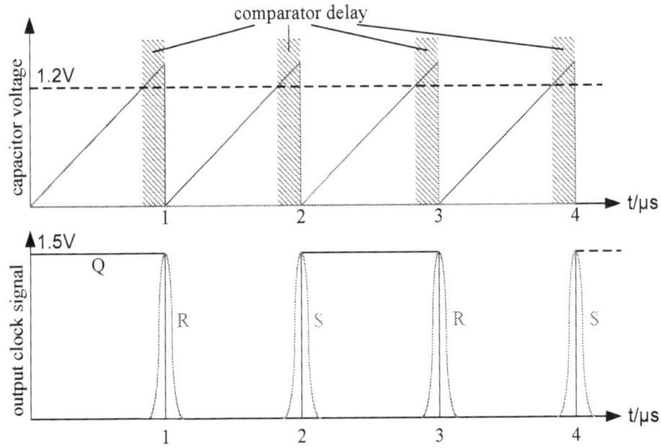

Fig. 20. internal oscillator waveforms

3. Conclusion

Wireless power transmission for sensor transponders in the electromagnetic far field is feasible. The architecture of the transponder circuits is more complex and requires more supply voltage and power than simple RFID transponders. The requirements for the air interface and the analog front-end are therefore more stringent. The voltage multiplier is the most critical circuit block concerning the power conversion efficiency and the maximum operating range.

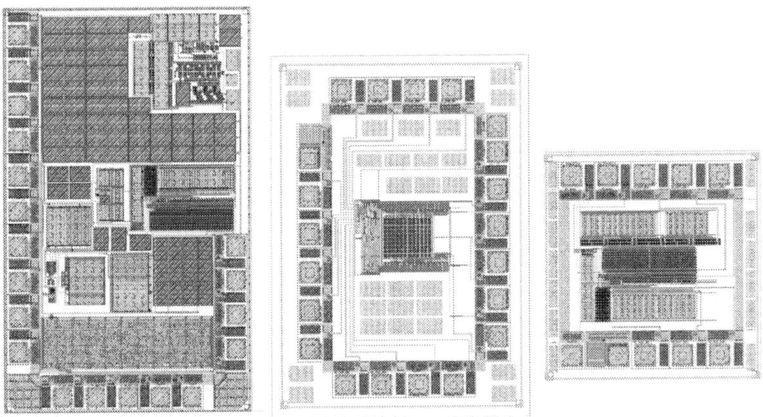

Fig. 21. layout of the analog front end, the ADC, and the sensor (left to right)

A rectifier has been presented that reduces the threshold voltage drop across rectifying devices. It uses a secondary unloaded voltage multiplier to generate a high DC voltage at low input amplitude. This voltage is used to generate the bias voltage for the rectifying transistors in the main power rectifier stack. The minimum required input voltage and the efficiency of the main rectifier is therefore reduced compared to the conventional Schottky diode rectifier. Three transponder test chips have been developed that contain all required analog circuits including the front-end, the sensor and a low power ADC.

4. References

Bechen, B. (2008). "Systematischer Entwurf analoger Low-Power Schaltugnen in CMOS anhand einer kapazitiven Sensorauslese" (German), Fraunhofer IRB Verlag, Germany, 2008

Curty, J.-P. ; Declercq, M. ; Dehollain, C. & Joehl, N. (2006). "Design and Optimization of Passive UHF RFID Systems", Springer 2006, Germany

Feldengut, T.; Wang, J. ; Kolnsberg, S. & Kokozinski, R. (2008). "An Analog Front End for a Passive UHF Transponder With Temperature Sensor" Proceedings of the Microwave Conference, EuMC, 38th European

Finkenzeller, K. (2003) RFID Handbook, Radio Frequency Identification Fundamentals and Applications. 2nd ed. New York: Wiley, 2003

Karthaus, U. & M. Fischer (2003). "Fully Integrated Passive UHF RFID Transponder With 16.7 µW Minimum RF Input Power" IEEE J. Solid State Circuits, vol. 38, no. 10, pp. 1602-1608, Oct 2003.

Lee, T. H. (1998) "The Design of CMOS Radion Frequency Integrated Circuits" , Cambridge University Press, Cambridge, UK

Nakamoto, H. ; Yamazaki, D. ; Yamamoto, T. ; Kurata, H. ; Yamada, S. ; Mukaida, K. ; Ninomiya, T. ; Ohkawa, T. ; Masui, S. & Gotoh, K. (2006). "A Passive UHF RFID Tag LSI with 36,6% Efficiency CMOS Only Rectifier and Current Mode Demodulator in 0,35 µm FeRAM Technology" Proceedings of the International Solid State Circuits Conference, ISSCC, Session 17, 2006

Razavi, B. (2001) "Design of Analog CMOS Integrated Circuits", McGraw-Hill, New York, NY

Umeda, T. ; Yoshida, H. ; Sekine, S. ; Fujita, Y. ; Suzuki, T. & Otaka, S. (2005). "A 950 MHz Rectifier Circuit for Sensor Networks with 10 m Distance Proceedings of the International Solid State Circuits Conference, ISSCC, Session 14, 2005.

21

Remote Characterization of Microwave Networks - Principles and Applications

Somnath Mukherjee
RB Technology
USA

1. Introduction

The present work deals with characterization of RF/Microwave one-port networks remotely, i.e. without any wired connections between the Device under Test (DUT) and the measuring apparatus. The objective is a frequency domain characterization of the complex reflection coefficient (or complex impedance) of the one-port so that its equivalent circuit model may be constructed. Finally, applications of this technique to ultra low-cost sensors and Radio Frequency Identification Devices (RFID) will be outlined

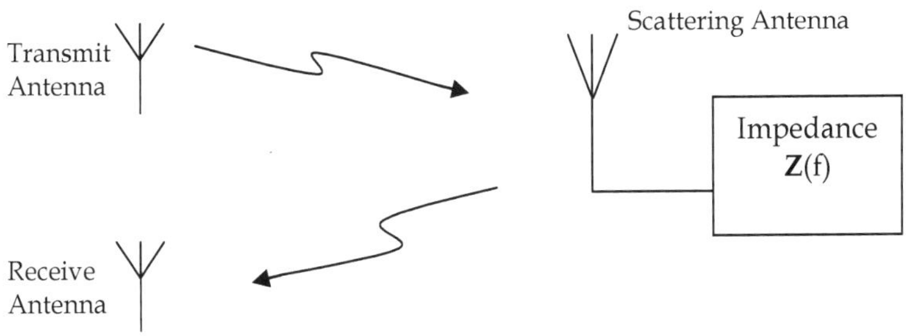

Fig. 1. Scattering Antenna in Bistatic Mode

Fig.1 diagrammatically depicts the problem, where the one-port under test, represented by the impedance **Z**(f) is used as a termination for the so-called Scattering Antenna. The Transmit Antenna is a part of the measurement system and is used to illuminate the Scattering Antenna with RF energy. The backscatter from the Scattering Antenna is received by the Receive Antenna - again part of the measurement system. The objective is to recover the frequency dependent complex impedance **Z**(f) by processing the backscatter. Fig. 1 illustrates the measurement in a bistatic arrangement, though monostatic implementations may be considered if necessary.
We would like to point out that the scattering antenna and the one-port may be devoid of any active electronics, including means to convert RF energy to DC.

If $\mathbf{Z}_A$ is the impedance of the scattering antenna at the point where the unknown one-port $\mathbf{Z}(f)$ is connected, then the reflection coefficient $\Gamma(f)$ at the scattering antenna terminals is

$$\Gamma(f) = \frac{Z(f) - Z_A}{Z(f) + Z_A} \tag{1}$$

For the special case of Z(f) being lossless (= X(f)) and $\mathbf{Z}_A$ being resistive ($=R_A$), we have

$$|\Gamma(f)| = 1 \tag{2a}$$

and,

$$\Gamma(f) = 1.\arg(\Gamma(f)) = -2\arctan(X(f)/R_A) \tag{2b}$$

Therefore, the task of characterizing **Z**(f) (i.e. Γ (f)) for a lossless one-port reduces to characterizing the phase-frequency profile of its reflection coefficient.

The fundamental advantage of frequency domain characterization is the ability to cover a large RF bandwidth while maintaining a narrow detection bandwidth, resulting in a favorable Carrier to Noise ratio (CNR). Furthermore, use of a large RF bandwidth enables spatial resolution of 'targets' - the theoretical limit to such resolution given by c/2B where c = speed of light, B = Bandwidth of the illuminating signal (Ulaby et al., 1982). The above principle behind spatial resolution is general and independent of the type of waveform. Ultra-wide band (UWB) radar is therefore attractive due to its ability to provide fine resolution, in addition to its robustness against multi-path and narrow-band interference (Yarovoy 2007).

Frequency domain measurements in bounded medium are routinely performed using a Vector Network Analyzer (VNA) and one is tempted to use the same for remote measurement as well. Use of VNA for polarimetric scattering measurements has been documented extensively such as (Ulaby et al., 1990). Such instruments have been extensively used to characterize ground, sea-ice and similar natural objects.

The present discussion is focused on the remote characterization of synthetic targets (i.e. a scattering antenna terminated by one-port) by coherent processing of backscatter - with applications to sensors and RFID. For such applications, the probing platform ('Reader') needs to be priced economically, as well be capable of functioning in presence of clutter and multipath. The transmit and the receive antennas are more often than not collocated, and therefore significant carrier leakage exist between the two. The above undesired phenomena termed impairments need to be addressed in a successful measurement scheme.

2. Scattering Antennas

The present section outlines the type of antennas suitable for remote measurement of one-port networks.

2.1 Salient Features of Scattering Antennas

When **Z**(f) in Fig.1 is lossless, intuitively speaking, all the power captured by the antenna should be re-radiated back as no energy is dissipated in the terminating reactance. Therefore, as described by eqn. 2(a) and (b), the scattering antenna should scatter back the same power irrespective of the type of lossless termination (including open and short), as they all produce a reflection coefficient of unity and therefore the radar cross-section (RCS) should be ideally constant. However, a counter-example in the form of 'minimum-scatter antennas' (Kahn 1965, Nikitin 2006) where no power is scattered back under open circuited condition demands further investigation. A typical example of 'minimum-scatter antennas' is the one-dimensional dipole without parasitic elements, where no current flows under open circuit condition when illuminated with RF energy, and RCS tends to disappear.

A Thevenin/Norton equivalent circuit is traditionally used to model antennas in transmit and receive modes (Balanis 1982). According to that model, under matched conditions, a receive antenna scatters back as much power as absorbed by the load. The limitations to that model have been discussed in several references such as Andersen 2003, Colin 2003, Pozar 2004, and in general, the model is not valid to interpret scattering related questions. The limitation is obvious for antennas containing parasitic elements, where the RCS might approach zero under matched conditions (Pozar 2004).

Let us now summarize some desirable characteristics of a scattering antenna.

1. Obviously, a 'minimum scatter antenna' is not a suitable candidate as RCS approaches zero under open circuit conditions.
2. It is important to realize the limitation of Thevenin/Norton model to study scattering by antennas that are not 'minimum scatter'.
3. The concept of antenna aperture may not be relevant as ideally no power is absorbed in a lossless termination. Complex RCS (equation (6)) appears to be a more suitable metric.
4. The physical mechanism through which an antenna absorbs power from an incident wave is destructive interference, i.e. the antenna scatters so as to cancel some of the incident fields (Kahn 1965). Ideally we would prefer no 'structural scattering' (Nikitin 2006, Hansen 1989) from a scattering antenna, i.e. no scattering when the said antenna is terminated by a matched load. This ensures that the energy incident on the antenna travels in its entirety to the terminal reactance and eventually gets reflected and re-radiated. Antennas with parasitic or passive elements have the potential to reach zero structural scattering (Pozar 2004) and can scatter even when the active element is open circuited (Andersen 2003), making them attractive as scattering antennas.

2.2 Two-port Representation of Scattering Antennas

Let us consider a Scattering Antenna operating in bistatic mode as in Fig.1. We propose that a non-minimum scattering antenna can be modeled as a simple two-port network (Fig.2) whose parameters can be extracted using traditional Vector Network measurement techniques. We assume far-field behavior throughout.

The incident wave from the transmit antenna **a** is modeled as coming from a generator of impedance Z0 where Z0= impedance of free space (120π). $\gamma l1$ represent the forward propagation coefficient from the Transmit Antenna to the Scattering Antenna, and $\gamma l2$ the return propagation coefficient from the Scattering Antenna to the Receive Antenna. The scattered wave **b** is generated due to reflection from the equivalent two-port network terminated by the unknown one-port network Γ (DUT). The objective is to de-embed the effect of the two-port network "error" network and determine Γ. In other words, this ensures characterizing the DUT alone, minus the effect of the antenna.

As discussed in section 2.1, a certain class of antennas terminated by a matched load results in low RCS. This phenomenon occurs only when the antenna is resonant, i.e. capable of radiating power in the transmit mode. Outside resonance, the antenna radiates negligible power in transmit mode, and scatters back finite energy even when terminated by the

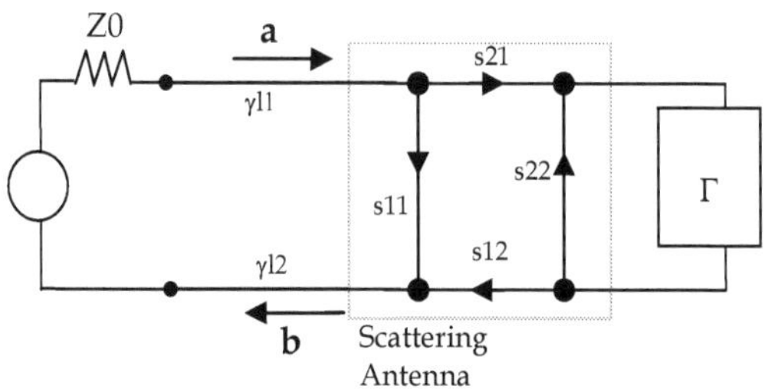

Fig. 2. Scattering Antenna modeled as a 2-port

characteristic impedance, most of the backscatter energy coming from "structural" scattering. As a result, the unknown one-port **Γ** will have negligible effect on the backscatter under these conditions. Therefore, under resonant condition, **s11** in Fig.2 is expected to become small, and outside resonance, |**s11**| may even approach unity. **s11** is therefore an important figure of merit to quantify structural scattering.

2.3 Example of Two-port Representation of a Scattering Antenna

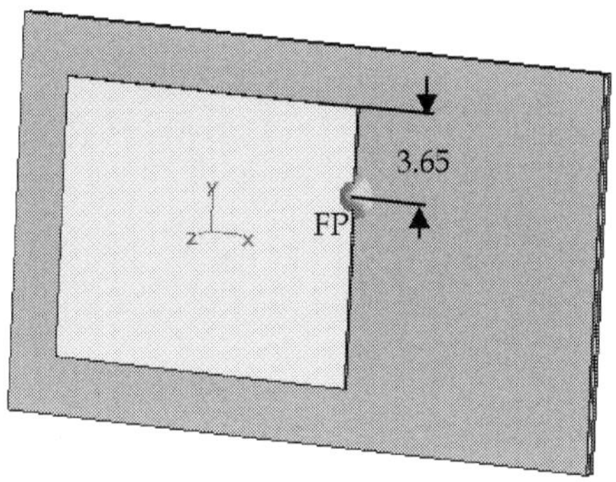

Fig. 3. Rectangular Microstrip Patch used as Scattering Antenna

A microstrip patch satisfies the requirements listed in Section 2.1, and was therefore considered a candidate for a scattering antenna. Simulation using commercially available Electromagnetic Simulator (CST Microwave Studio 2008 from Computer Simulation Technology) was tried on a rectangular patch on a finite ground as shown in Fig.3. The patch had dimensions of 11.5 X 15 mm and mounted on a ground plane 16.3 X 30 mm maintaining symmetry in the Y-direction. The ground plane was intentionally made larger in the X-direction than necessary to create 'structural scattering' or 'clutter'. Copper was selected as the conducting material for both ground plane and the patch, whereas the substrate was assumed lossless with a permittivity (ε_r) of 4.5. The thickness of the substrate was 0.8 mm. A feed point (FP), located 3.65 mm below the longer edge (Fig.3), was used to connect various one-port networks. The impedance at FP happened to be 50Ω, though feed points with other impedance levels could be used. Typical one-port networks used were open, short, termination (50Ω), lumped inductors and capacitors.

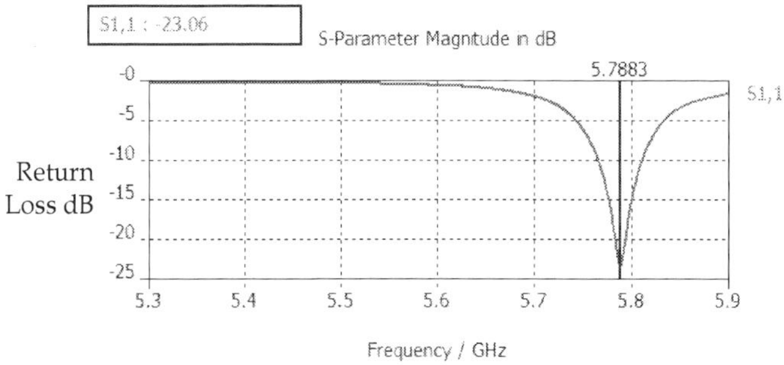

Fig. 4. Return Loss at FP in Transmit Mode

An approximate analysis (Balanis 1982) predicts the fundamental resonant frequencies of the patch to be 4.56 and 6.15 GHz. To investigate the scattering phenomenon, we decided to operate around the higher resonant frequency, i.e. when the longer sides are radiating and FP is located on a non-radiating edge. The non-radiating edge was selected as feed point so as to disturb the radiating field as little as possible. Simulation found the resonance to be close to 5.8 GHz in radiating mode (Fig.4), as depicted in the Return Loss at FP referenced to 50Ω. To investigate scattering, the patch was illuminated with linearly polarized wave parallel to Y-axis (Fig.3) traveling in negative Z-direction and the co-polarized backscatter was observed.

The scattered electric field vector **E**, under far-field conditions, at co-ordinates (r,θ,ϕ) can be expressed as:

$$\mathbf{E}(r,\theta,\phi) = \mathbf{e}(\theta,\phi).\ \exp(-j\beta r)/r \tag{3}$$

where β is the propagation constant.

The magnitude of RCS is defined as:

$$\sigma(\theta_i,\phi_i,\theta,\phi) = 4.\pi.[|\mathbf{e}(\theta,\phi)|^2/\ |\mathbf{E}_i(\theta_i,\phi_i)|^2] \tag{4}$$

where $\mathbf{E}_i(\theta_i, \phi_i)$ is the incident electric field vector in the θ_i,ϕ_i direction. Phase of the scattered field, using a convenient reference point (r = 1 meter was used in the simulation) can be expressed as

$$\Psi(\theta_i,\phi_i,\theta,\phi) = \arg[\mathbf{e}(\theta,\phi)/\ \mathbf{E}_i(\theta_i,\phi_i)] \tag{5}$$

Based on (4) and (5), we define a complex reflection coefficient as (Mukherjee 2008):

$$\Gamma(\theta_i,\phi_i,\theta,\phi) = \sqrt{\sigma(\theta_i,\phi_i,\theta,\phi)}.e^{i\Psi(\theta_i,\phi_i,\theta,\phi)} \tag{6}$$

The patch was successively terminated with an open (no termination), short and 50Ω termination at FP, and the magnitude and phase of the co-polarized Γ as defined by (6) were computed (Please note this Γ is different from one in Fig.2). Let Γo, Γs and Γl be the

complex reflection coefficient as defined in (6) due to open, short and load conditions for $\theta=\theta_t=0$ and $\phi=\phi_t=0$. The incident and backscatter waves are in negative and positive z-direction respectively, i.e. under bore-sight conditions. It was observed that $|\Gamma l|$ is significantly lower in the middle of the band compared to $|\Gamma s|$ and $|\Gamma o|$, indicating high amount of absorption by the termination resistor (Mukherjee 2008).
The following equations were used to calculate the parameters of the error network of Fig.1 from the calibration data, viz. Γo, Γs and Γl.

$$\gamma l1.\gamma l2.\mathbf{s11} = \Gamma l \tag{7}$$

$$\mathbf{s22} = (\Gamma o + \Gamma s - 2.\gamma l1.\gamma l2.\mathbf{s11})/(\Gamma o - \Gamma s) \tag{8}$$

$$\gamma l1.\gamma l2.\mathbf{s21}.\mathbf{s12} = (1\text{-}\mathbf{s22}).(\Gamma o - \gamma l1.\gamma l2.\mathbf{s11}) \tag{9}$$

We note that **s11** and **s21.s12** occur de-normalized in conjunction with the term γl1.γl2.

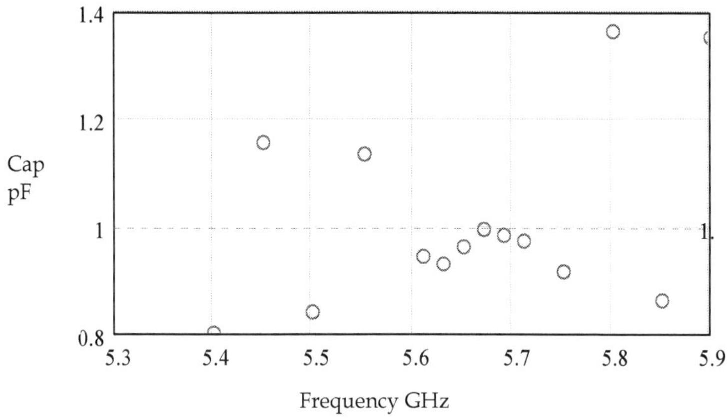

Fig. 5. De-embedded value of DUT (1 pF capacitor)

For an unknown one-port producing a backscattered reflection coefficient Γ, the de-embedded value of reflection coefficient of the DUT itself can be calculated according to

$$\Gamma corr = [\Gamma\text{-}\ \gamma l1.\gamma l2.\mathbf{s11}]/[\gamma l1.\gamma l2.\mathbf{s21}.\mathbf{s12} + \mathbf{s22}.(\Gamma - 1)] \tag{10}$$

Fig.5 shows the results of de-embedding the effect of scattering antenna (using eqn. (7), (8), (9),(10)) for a 1 pF lumped capacitor as DUT. In the vicinity of resonance, we find good agreement between the actual and the de-embedded values. However, as one moves away from resonance, the accuracy falls due to increase of structural scattering (increase in **s11** in the two-port representation of Fig.2). The simulation assumed negligible contribution from noise.
It is worth mentioning that the backscatter measured at angles away from bore-sight also allows determining the unknown device making the approach aspect independent within certain limits.

3. Measurement Technique

The practical applications of the present technique viz. in the RFID and sensors often require the probing platform to be low cost and portable; as a result, the transmit and receive antennas may need to be co-located. This brings forth the problem of detecting a backscatter signal in the same order of magnitude or lower than the leakage between the transmit and receive antennas. RFID devices containing chips solve this problem (Dobkin 2007) by generating a subcarrier at the tag level. Since chipless devices cannot afford that luxury, one has to resort to alternative probing methods. We present here a technique suggested by Mukherjee 2007.

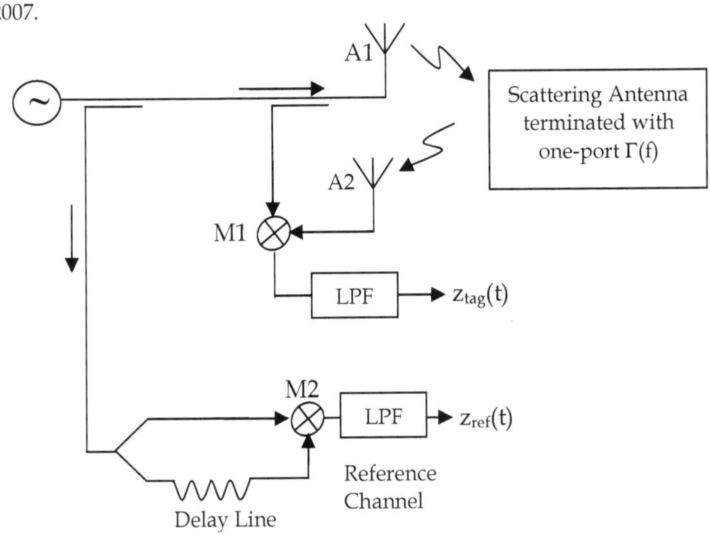

Fig. 6. Remote Measurement of Impedance

The technique bears a superficial similarity to the FMCW radar (Ulaby1982). A source generates a chirp signal that is radiated by the antenna A1 (Fig.6) to illuminate the 'target' (scattering antenna terminated by the unknown one-port $\Gamma(f)$), and another antenna A2 receives the backscattered signal. The transmitted and received signals are mixed by a mixer M1 and lowpass filtered (LPF) to generate the Intermediate Frequency (IF) signal $z_{tag}(t)$. The backscatter from the scattering antenna can be resolved spatially from the carrier leakage by filtering the IF signal. Henceforth, we consider the IF signal due to the scattering antenna only.

Simultaneously, internal to the Reader, a sample of the transmit signal is mixed with a delayed version of itself and lowpass filtered to generate a reference signal $z_{ref}(t)$.

The transmitted chirp signal can be expressed as (Brunfeldt 1991):

$$x(t) = a(t).\cos[2\pi f_0 t + K\int_0^t s(t)dt] \tag{11}$$

where a(t) is the incidental amplitude modulation of the source, and s(t) is the frequency modulating signal which is linear for a linear chirp. The instantaneous frequency f(t) is given by:

$$f(t) = f_0 + \frac{K}{2\pi}.s(t) \tag{12}$$

where f_0 is the start frequency.
The backscattered signal received at the Reader can be expressed as:

$$y(t) = L.\Gamma(f(t-\tau)).a(t-\tau).\cos\left[2\pi f_0(t-\tau) + K.\int_0^{t-\tau} s(t)dt + \psi(f(t-\tau))\right] \tag{13}$$

where τ is the round trip delay and L is the total loss (assumed constant over the frequency band), between antennas A1 and A2 through the scattering antenna.

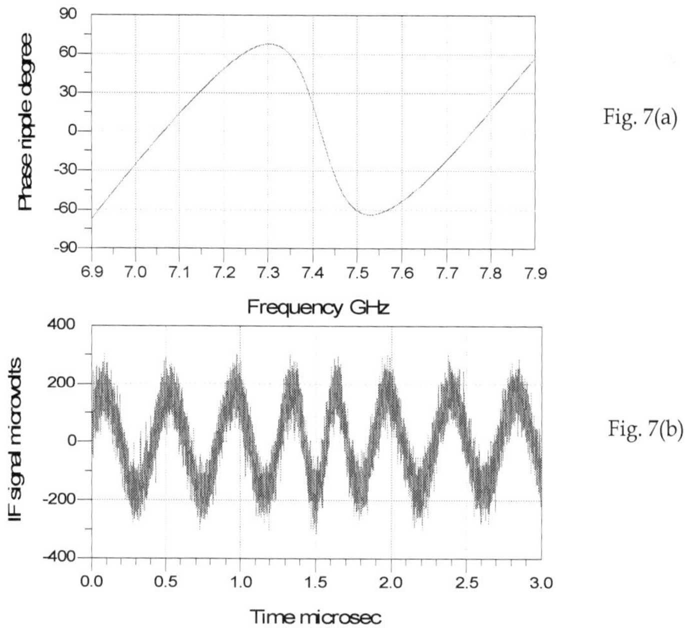

Fig. 7. (a) Phase Ripple of Γ(f)
Fig. 7. (b) Time Domain IF Waveform Depicting Phase Modulation

The output from the mixer M1, after filtering, can be approximated as:

$$z_{tag}(t) \approx \tfrac{1}{2}.L.a^2(t).\Gamma(f(t)).\cos[2\pi f_0\tau + K\tau.s(t) + \psi(f(t))] \tag{14}$$

provided the delay τ is small compared to the chirp duration.
For the special case of a linear chirp, (14) becomes:

$$z_{tag}(t) \cong \tfrac{1}{2}.L.a^2(t).\Gamma(f(t)).\cos[2\pi f_0\tau + K\tau.t + \psi(f(t))] \quad (15)$$

The IF signal in such a case is a nominal sine wave with frequency

$$K\tau/2\pi = (B/T_R).\tau \quad (16)$$

where B = chirp bandwidth and T_R = chirp duration
The IF is thus a modulated sine wave of "carrier frequency" $(B/T_R).\tau$ together with amplitude and phase modulation according to $\Gamma(f(t))$. In other words, a mapping occurs for the complex function $\Gamma(f)$ from the frequency to time domain. Therefore, demodulation of the IF signal in (15) should provide information on the reflection coefficient between frequencies f_0 and f_0 + B.
Fig.7(b) represents $z_{tag}(t)$ as generated by simulation with the following parameters: f_0 = 6.9 GHz, B=1 GHz, T_R= 3 μs with a round trip delay τ produced by propagation through 2 meters in air (i.e.1 meter in each direction). $\Gamma(f)$ was chosen to be a lossless one-port with a phase-frequency profile as in Fig.7(a).
It is interesting to observe the correlation in phase modulation in Fig.7(b) with the phase ripple of $\Gamma(f)$ in Fig.7(a).
The task of recovering the phase ripple of $\Gamma(f)$ (magnitude being unity) could be achieved through the use of the Reference Channel in Fig.6.
Following the same procedure as above, we have:

$$z_{ref}(t) \cong \tfrac{1}{2}.L_{ref}.a^2(t).\cos[2\pi f_0\tau_{ref} + K\tau_{ref}.t + \psi_{ref}] \quad (17)$$

where τ_{ref} is the delay in the reference channel.
We have made the reasonable assumption that the loss L_{ref} and phase shift ψ_{ref} in the reference channel are independent of frequency.
Each of the equations (15) and (17) can be converted from real time functions to complex time functions by application of Hilbert Transform. The complex functions can be expressed as:

$$\zeta tag(t) = \tfrac{1}{2}.L.a^2(t).\Gamma(f(t)).\exp\{j[2\pi f_0\tau + K\tau.t + \psi(f(t))]\} \quad (18a)$$

$$\zeta ref\ (t) = \tfrac{1}{2}.L_{ref}.a^2(t).\exp\{j[2\pi f_0\tau_{ref} + K\tau_{ref}.t + \psi_{ref}]\} \quad (18b)$$

Therefore,

$$\arg(\zeta tag(t)/\ \zeta ref(t)) = 2\pi f_0.(\tau-\tau_{ref}) + K.(\tau-\tau_{ref}).t + \psi(f(t)) + \psi_{ref} \quad (19)$$

We are usually interested in the phase ripple $\psi(f(t))$ and therefore the first and fourth terms in the right hand side of (19) can be scaled out. The second term indicates a linearly progressive phase shift with frequency that can be conveniently factored out by unwrapping the phase. Therefore, the desired phase ripple $\psi(f(t))$ can be determined.

There could be alternative ways to recover the phase ripple information and this method is not claimed to be the optimum one.
We observe that the detection bandwidth in the method as in Section 3 can be made sufficiently small as to reduce the thermal noise - while operating with large RF bandwidth at the same time.
The approach is essentially a broadband technique and therefore carries its usual advantages. Moreover, it need not operate over a continuous spectrum of frequencies - as it could maintain phase coherence in the reader while selectively shutting off parts of the chirp in time.

4. Applications

The present section will outline some applications of the remote measurement of impedance to RFID and sensors. The RFID would use the one-port as a vehicle for storing the coded information. And, a remotely monitored sensor could be constructed by utilizing a one-port that gets predictably affected by a physical parameter such as temperature, pressure, strain, magnetic field etc. In either case, they would operate without DC power (including one generated by a rectifying antenna) and be free of semiconductor components. This opens up the possibility of printed RF barcodes (with conducting ink on low-cost substrate such as paper or plastic) and disposable sensors. This approach has the potential to reduce the cost by orders of magnitude compared to existing technology including printed electronics.
Such devices should have a longer range compared to those utilizing rectifying antenna, where a significant fraction of the received RF power is converted to DC to operate the associated electronics. On the other hand, the chipless devices being designed lossless (ideally), could potentially re-radiate all the received power.

4.1 Application to RFID

Section 3 demonstrated that it is possible to recover remotely the phase-frequency profile of an unknown reactance (one-port) connected to an antenna port. And, we know that such phase-frequency profile is completely characterized by the poles and zeros of the one-port. With this background, we propose creation of multiple tag signatures by suitable placement of poles and zeros. As all information about the reactive network is embedded in the poles and zeros, identifying them identifies the network uniquely. The following discussion is an attempt to estimate a lower bound on the number of bits that can be encoded using this technique.
To start with, let us consider a pair consisting of a pole and zero as shown in Fig. 8 located inside a segment of bandwidth with start and stop frequencies f1 and f2. We are allowed to move the positions of poles and zeros within that segment following certain rules, and thereby calculate the number of distinct permissible states to ascertain the number of unique identification signatures.

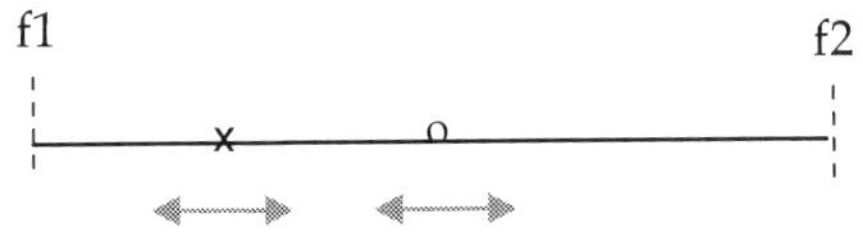

Fig. 8. Positioning of Poles and Zeros

The following rules are defined:

1. The minimum separation between a pole and its companion zero is θ. This separation is dictated by the quality factor (Q) of the resonators in the reactive network.
2. Poles or zeros can be moved in steps of $\geq \Delta p$ as dictated by measurement accuracy in presence of noise and various impairments.

To start with, let the pole be placed at f = f1 and the zero at f= f1+θ.
Next, the zero starts its journey from f1+θ and travels up to f2-θ in increments of Δp.
The i-th position of the zero is described as

$$f_i = f1 + \theta + i.\Delta p \tag{20a}$$

where i = 0.....N with

$$N = \frac{f2 - f1 - 2\theta}{\Delta p} \tag{20b}$$

Now, corresponding to the i-th position of the zero, the pole could start its journey at f1 in steps of Δp and commence at (f_i-θ).
Therefore, number of states the zero can assume for the i-th position of the pole is given by

$$\frac{f_i - \theta - f1}{\Delta p} \tag{21}$$

And, therefore corresponding to all the positions of the zero, the total number of pole-zero combinations is given by

$$\sum_{i=0}^{N} \frac{f_i - \theta - f1}{\Delta p} \tag{22}$$

Using (20a) and (20b), (22) is simplified to

$$\frac{(f2 - f1 - \theta).(f2 - f1 - \theta + \Delta p)}{2.\Delta p^2} \tag{23}$$

Now, the above exercise can be repeated for a scenario where the frequency of pole exceeds that of zero, and would generate an identical number of states. Therefore, the total number of states is given by

$$\Theta_{single} = \frac{(f2 - f1 - \theta).(f2 - f1 - \theta + \Delta p)}{\Delta p^2} \tag{24}$$

where the subscript 'single' implies number of states calculated over a single segment of bandwidth, viz. between f1 and f2 for a pole-zero pair.

Now, let us consider 'm' number of identical bandwidth segments following the same rules of positioning of poles and zeros. The total number of states can be given by

$$\Theta_m = \left(\frac{(f2 - f1 - \theta).(f2 - f1 - \theta + \Delta p)}{\Delta p^2} \right)^m \quad (25)$$

In deriving the above expression, we made some simplifying assumptions that did not account for some additional states as follows:

Each bandwidth segment always contained a pole zero pair. However, valid states are possible with just a single pole (zero) or none at all is present in a particular segment. The missing poles (zeros) could have migrated to other segments.

For a given total available bandwidth, i.e. m.(f2-f1), presence of m pole-zero pairs were assumed. However, additional states can be considered to be generated by single pair, double pair up to (m-1) pairs inhabiting the total available bandwidth. However, the number of states generated by pole zero pairs <m will be small compared to that generated from m pairs.

With the above premises, we can conclude that the number of encoded bits B is given by

$$B > \log_2(\Theta_m) \quad (26)$$

As an example, let us consider a total bandwidth of 3 GHz divided into 6 segments. Therefore, f2-f1= 500 MHz and m = 6. Let us further assume that θ =100 MHz and Δp = 25 MHz. This results in B > 48 bits, a number comparable with information content available from optical bar codes.

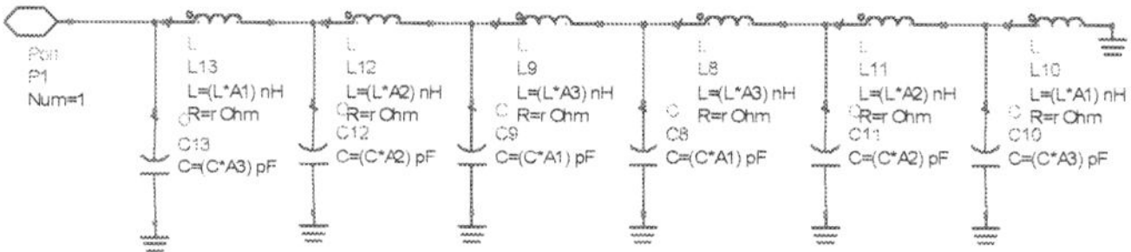

Fig. 9(a)

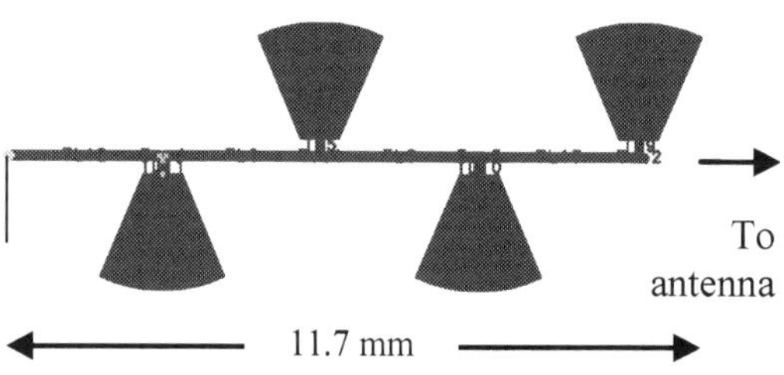

Fig. 9(b)

Fig. 9.(a) Example of Lumped Equivalent Circuit for X(f)
Fig. 9.(b) Microstrip Implementation of L-C Ladder

4.1.1 L-C Ladder as One-port

An example of the reactive one-port is a L-C ladder circuit as shown in Fig. 9(a), with its corresponding microstrip implementation - amenable to printing technique - in Fig. 9(b). The scattering antenna - not shown in Fig. 9(b) - need to possess properties outlined in Section 2.1. The narrow lines (Fig. 9(b)) represent the series inductors and the stubs work as shunt capacitors. By changing the values of these elements, the poles and zeros can be controlled as in Section 4.1 to generate RFID information bits.

4.1.2 Stacked Microstrip Patches as Scattering Structure

While the previous discussions premised on the separation of the scattering antenna and the one-port, we now present an example where the scattering structure does not require a distinguishable one-port.

Fig. 10. depicts a set of three (there could be more) stacked rectangular patches as a scattering structure where the upper patch resonates at a frequency higher than the middle patch. When the upper patch is resonant, the middle patch acts as a ground plane. Similarly, when the middle patch is resonant, the bottom patch acts as a ground plane (Bancroft 2004).

Fig. 10(a)

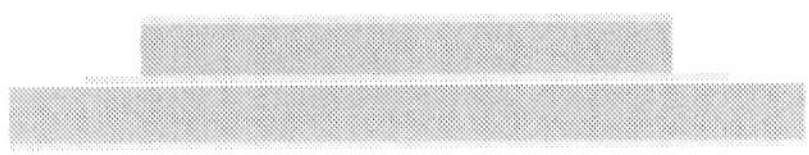

Fig. 10(b)

Fig. 10. (a) Stacked Rectangular Patches as Scattering Structure - Isometric
Fig. 10. (b) Stacked Rectangular Patches as Scattering Structure - Elevation

If the patches are perfectly conducting and the dielectric material is lossless, the magnitude of the RCS of the above structure could stay nominally fixed over a significant frequency range. As the frequency is swept between resonances, the structural scattering tends to maintain the RCS relatively constant over frequency - and therefore is not a reliable parameter for coding information. However, the phase (and therefore delay) undergoes significant changes at resonances. Fig. 11(a) and 11(b) illustrates this from simulation on the structure of Fig.10 (b). The simulation assumed patches to be of copper with conductivity

5.8. 10^7 S/m and the intervening medium had a dielectric constant =4.5 with loss tangent = 0.002. As a result of the losses, we see dips in amplitude at the resonance points.
Just like networks can be specified in terms of poles and zeros, it has been shown by numerous workers that the backscatter can be defined in terms of complex natural resonances (e.g. Chauveau 2007). These complex natural resonances (i.e. poles and zeros) will depend on parameters like patch dimension and dielectric constant. As a result, the principle of poles and zeros to encode information may be applied to this type of structure as well. However, being a multi-layer structure, the printing process may be more expensive than single layer (with ground plane) structures as in Fig. 9(b).

4.2 Application to Sensors

The principle of remote measurement of impedance could be used to convert a physical parameter (e.g. temperature, strain etc.) directly to quantifiable RF backscatter. As this method precludes the use of semi-conductor based electronics, it could be used in hazardous environments such as high temperature environment or for highly dense low cost sensors in Structural Health Monitoring (SHM) applications.

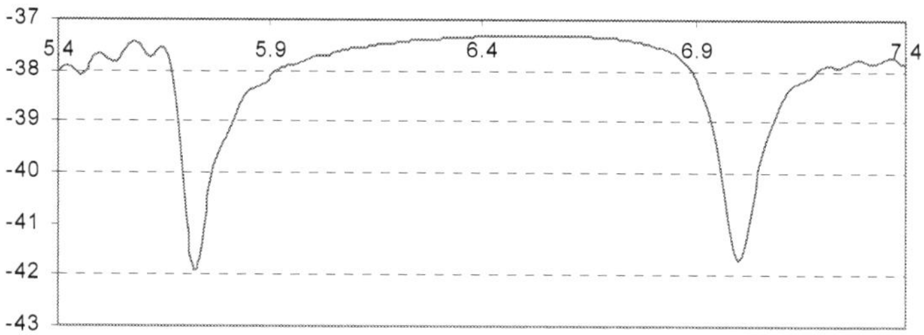

Fig. 11 (a)

Frequency GHz

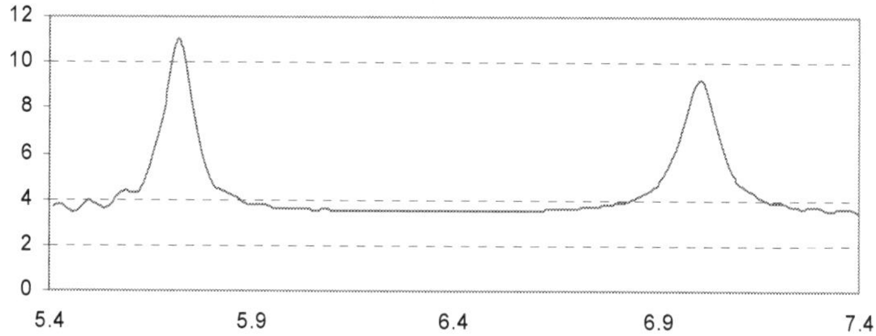

Fig. 11 (b)

Fig. 11. (a) Magnitude of Backscatter (dBV/m) from structure of Fig. 10 (a)
Fig. 11. (b) Group Delay (ns) of Backscatter from structure of Fig. 10 (a)

As an example, a temperature sensor using stacked microstrip patch has been proposed by

Mukherjee 2009. The space between a pair of patches could be constructed of temperature sensitive dielectric material whereas between the other pair could be of zero or opposite temperature coefficient. Fig.12 illustrates the movement of resonance peak in group delay for about 2.2% change in dielectric constant due to temperature.
Other types of sensors, such as strain gauge for SHM are under development.

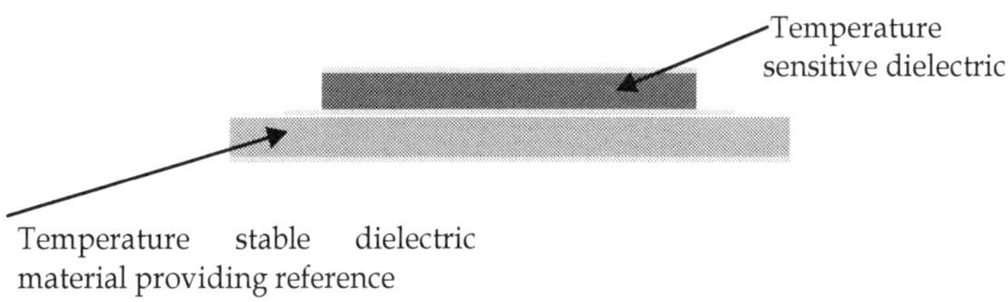

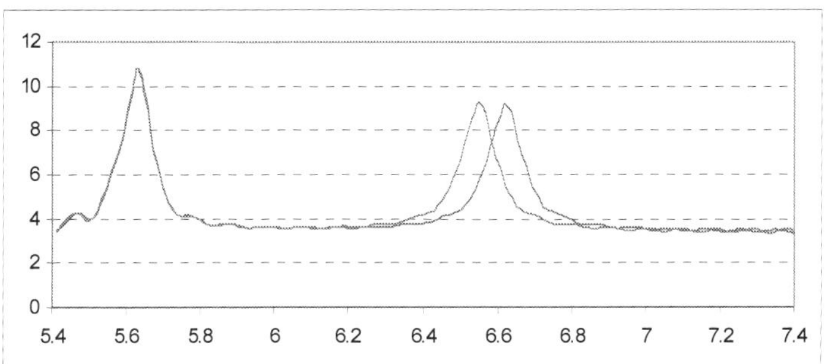

Fig. 12. Change in higher frequency resonance due to 2.2% change in ε_r

5. Impairment Mitigation

Cause of impairment is due to multipath and backscatter from extraneous objects - loosely termed clutter. The boundary between multipath and clutter is often vague, and so the term impairment seems to be appropriate. Mitigation of impairment is especially difficult in the present situation as there is no electronics in the scatterer to create useful differentiators like subcarrier, non-linearity etc. that separates the target from impairments. Impairment mitigation becomes of paramount importance when characterizing devices in a cluster of devices or in a shadowed region.
Fig. 13 illustrates with simulation data how impairments corrupt useful information. The example used the scatterer of Fig. 10 with associated clutter from a reflecting backplane, dielectric cylinder etc.
To mitigate the effect of impairments, we propose using a target scatterer with constant RCS but useful information in phase only (analogous to all-pass networks in circuits). In other words, the goal is to phase modulate the complex RCS in frequency domain while keeping

the amplitude constant. The 'modulating signal' is the information content for RFID or sensors - as the case may be. A lossless stacked microstrip patch has poles and zeros that are mirror images about the jω axis. When loss is added to the scatterer, the symmetry about jω axis is disturbed. Fig.14 illustrates the poles and zeros for the lossy scatterer described in Fig.10. The poles and zeros are not exactly mirror image about jω axis due to losses but close enough for identification purposes as long as certain minimum Q is maintained. We hypothesize that poles and zeros due to impairments will in general not follow this 'all-pass' property and therefore be distinguishable from target scatterers. Investigation using genetic algorithm is underway to substantiate this hypothesis. And, while the complex natural resonances from the impairments could be aspect dependent, the ones from the target will in general not be (Baev 2003).

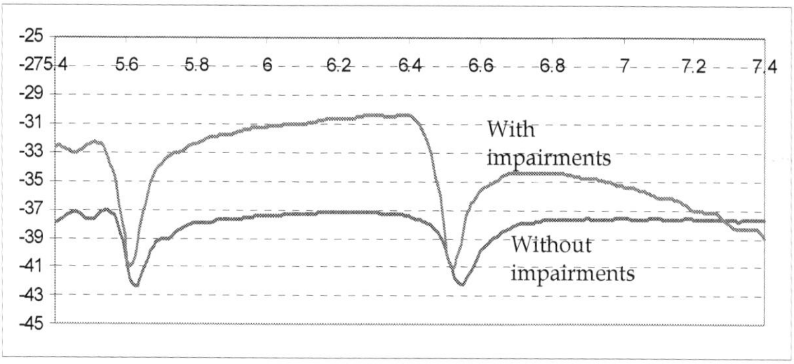

Fig. 13(a)

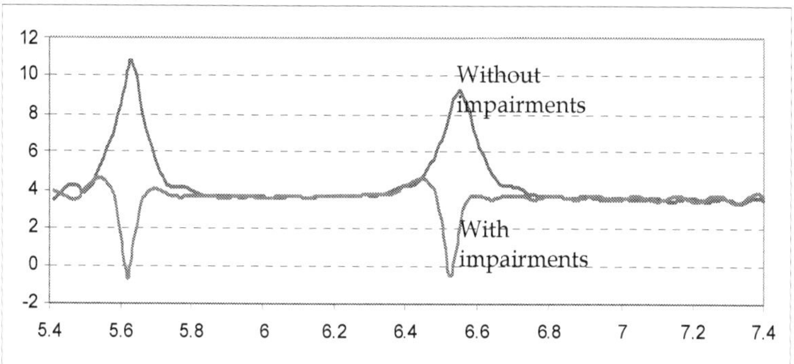

Fig. 13(b)

Fig. 13. (a) Magnitude of Backscatter (dBV/m) with and without impairments
Fig. 13. (b) Group Delay (ns) of Backscatter with and without impairments

6. Summary and Outlook

Several novel ideas have been introduced in this work - the foundation being remotely determining the complex impedance of a one-port. The above approach is next used for the development of chipless RFID and sensors. The approach has advantages like spatial resolution (due to large bandwidth), distance information, long range (lossless scatterer and low detection bandwidth), low cost (no semiconductor or printed electronics), ability to operate in non-continuous spectrum, potential to mitigate impairments (clutter, multipath) and interference and so on.

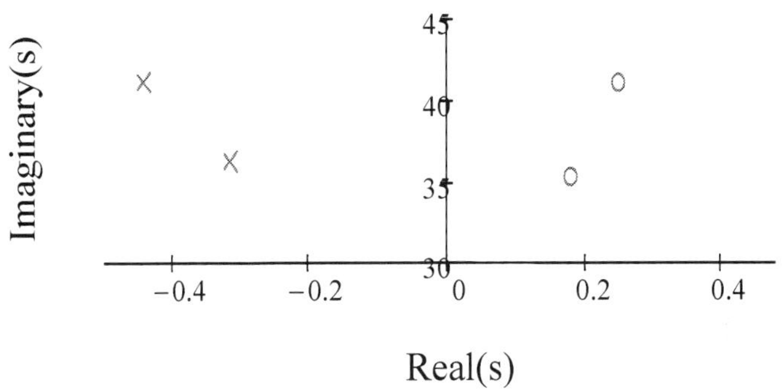

Fig. 14. Poles and Zeros of Stacked Microstrip Patches (Complex conjugate ones not shown)

The technique has the potential of providing sub-cent RF barcodes printable on low cost substrates like paper, plastic etc. It also has the potential to create sensors that directly convert a physical parameter to wireless signal without the use of associated electronics like Analog to Digital Converter, RF front-end etc.

To implement the approach, a category of antennas with certain specific properties has been identified. This type of antennas requires having low RCS with matched termination and constant RCS when terminated with a lossless reactance.

Next, a novel probing method to remotely measure impedance has been introduced. The method superficially resembles FMCW radar but processes signal differently.

Finally, a novel technique for the mitigation of impairments has been outlined. The mitigation technique is premised on the extraction of poles and zeros from frequency response data and separation of all-pass (target) from non all-pass (undesired) functions.

The work so far - based on mathematical analysis and computer simulation has produced encouraging results and therefore opens the path towards experimental verification.

There are certain areas that need further investigation e.g. development of various types of broadband 'all-pass' scattering structures with low structural scattering - or preferably, a general purpose synthesis tool to that effect. Another area is the development of broadband antennas that satisfy the scattering property mentioned earlier.

7. References

Andersen J.B. and Vaughan R.G. (2003) Transmitting, receiving and Scattering Properties of Antennas, *IEEE Antennas & Propagation Magazine*, Vol.45 No.4, August 2003.

Baev A., Kuznetsov Y. and Aleksandrov A. (2003) Ultra Wideband Radar Target Discrimination using the Signatures Algorithm, *Proceedings of the 33rd European Microwave Conference*, Munich 2003.

Balanis C.A. (1982) *Antenna Theory Analysis and Design*, Harper and Row

Bancroft R. (2004) *Microstrip and Printed Antenna Design*, Noble Publishing Corporation

Brunfeldt D.R. and Mukherjee S. (1991) A Novel Technique for Vector Measurement of Microwave Networks, *37th ARFTG Digest*, Boston, MA, June 1991.

Chauveau J., Beaucoudrey N.D. and Saillard J. (2007) Selection of Contributing Natural Poles for the Characterization of Perfectly Conducting Targets in Resonance Region, *IEEE Transactions on Antennas and Propagation, Vol. 55, No. 9, September 2007*

Collin R.E. (2003) Limitations of the Thevenin and Norton Equivalent Circuits for a Receiving Antenna, *IEEE Antennas and Propagation Magazine*, Vol.45, No.2, April 2003.

Dobkin D. (2007) *The RF in RFID Passive UHF RFID in Practice*, Elsevier

Hansen R.C. (1989) Relationship between Antennas as Scatterers and Radiators, *Proc. IEEE*, Vol.77, No.5, May 1989

Kahn W. and Kurss H. (1965) Minimum-scattering antennas, *IEEE Transactions on Antennas and Propagation*, vol. 13, No. 5, Sep. 1965

Mukherjee S. (2007) Chipless Radio Frequency Identification based on Remote Measurement of Complex Impedance, *Proc. 37th European Microwave Conference*, Munich, 2007

Mukherjee S. (2008) Antennas for Chipless Tags based on Remote Measurement of Complex Impedance, *Proc. 38th European Microwave Conference*, Amsterdam, 2008.

Mukherjee S., Das S.K and Das A.K. (2009) Remote Measurement of Temperature in Hostile Environment, *US Provisional Patent Application 2009.*

Nikitin P.V. and Rao K.V.S. (2006) Theory and Measurement of Backscatter from RFID Tags, *IEEE Antennas and Propagation Magazine*, vol. 48, no. 6, pp. 212-218, December 2006

Pozar D (2004) Scattered and Absorbed Powers in Receiving Antennas, *IEEE Antennas and Propagation Magazine*, Vol.46, No.1, February 2004.

Ulaby F.T., Moore R.K., and Fung A.K. (1982) *Microwave Remote Sensing, Active and Passive*, Vol. II, Addison-Wesley.

Ulaby F.T., Whitt M.W., and Sarabandi K. (1990) VNA Based Polarimetric Scatterometers, *IEEE Antennas and Propagation Magazine*, October 1990.

Yarovoy A. (2007) Ultra-Wideband Radars for High-Resolution Imaging and Target Classification, *Proceedings of the 4th European Radar Conference*, October 2007.

22

Solving Inverse Scattering Problems Using Truncated Cosine Fourier Series Expansion Method

Abbas Semnani & Manoochehr Kamyab
K. N. Toosi University of Technology
Iran

1. Introduction

The aim of inverse scattering problems is to extract the unknown parameters of a medium from measured back scattered fields of an incident wave illuminating the target. The unknowns to be extracted could be any parameter affecting the propagation of waves in the medium.

Inverse scattering has found vast applications in different branches of science such as medical tomography, non-destructive testing, object detection, geophysics, and optics (Semnani & Kamyab, 2008; Cakoni & Colton, 2004).

From a mathematical point of view, inverse problems are intrinsically ill-posed and nonlinear (Colton & Paivarinta, 1992; Isakov, 1993). Generally speaking, the ill-posedness is due to the limited amount of information that can be collected. In fact, the amount of independent data achievable from the measurements of the scattered fields in some observation points is essentially limited. Hence, only a finite number of parameters can be accurately retrieved. Other reasons such as noisy data, unreachable observation data, and inexact measurement methods increase the ill-posedness of such problems. To stabilize the inverse problems against ill-posedness, usually various kinds of regularizations are used which are based on a priori information about desired parameters. (Tikhonov & Arsenin, 1977; Caorsi, et al., 1995). On the other hand, due to the multiple scattering phenomena, the inverse-scattering problem is nonlinear in nature. Therefore, when multiple scattering effects are not negligible, the use of nonlinear methodologies is mandatory.

Recently, inverse scattering problems are usually considered in global optimization-based procedures (Semnani & Kamyab, 2009; Rekanos, 2008). The unknown parameters of each cell of the medium grid would be directly considered as the optimization parameters and several types of regularizations are used to overcome the ill-posedness. All of these regularization terms commonly use a priori information to confine the range of mathematically possible solutions to a physically acceptable one. We will refer to this strategy as the direct method in this chapter.

Unfortunately, the conventional optimization-based methods suffer from two main drawbacks. The first is the huge number of the unknowns especially in 2-D and 3-D cases which increases not only the amount of computations, but also the degree of ill-posedness. Another disadvantage is the determination of regularization factor which is not straightforward at all. Therefore, proposing an algorithm which reduces the amount of computations along with the sensitivity of the problems to the regularization term and initial guess of the optimization routine would be quite desirable.

2. Truncated cosine Fourier series expansion method

Instead of direct optimization of the unknowns, it is possible to expand them in terms of a complete set of orthogonal basis functions and optimize the coefficients of this expansion in a global optimization routine. In a general 3-D structure, for example the relative permittivity could be expressed as

$$\varepsilon_r(x,y,z) = \sum_{n=0}^{N-1} d_n f_n(x,y,z) \tag{1}$$

where f_n is the n_{th} term of the complete orthogonal basis functions.

It is clear that in order to expand any profile into this set, the basis functions must be complete. On the other hand, orthogonality is favourable because with this condition, a finite series will always represent the object with the best possible accuracy and coefficients will remain unchanged while increasing the number of expansion terms.

Because of the straightforward relation to the measured data and its simple boundary conditions, using harmonic functions over other orthogonal sets of basis functions is preferable. On the other hand, cosine basis functions have simpler mean value relation in comparison with sine basis functions which is an important condition in our algorithm.

We consider the permittivity and conductivity profiles reconstruction of lossy and inhomogeneous 1-D and 2-D media as shown in Fig. 1.

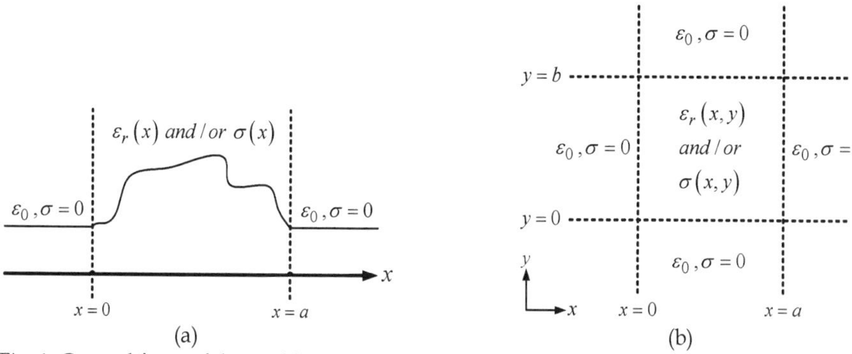

Fig. 1. General form of the problem, (a) 1-D case, (b) 2-D case

If cosine basis functions are used in one-dimensional cases, the truncated expansion of the permittivity profile along x which is homogeneous along the transverse plane could be expressed as

$$\varepsilon_r(x) = \sum_{n=0}^{N-1} d_n \cos\left(\frac{n\pi}{a}x\right) \tag{2}$$

where a is the dimension of the problem in the x direction and the coefficients, d_n, are to be optimized. In this case, the number of optimization parameters is N in comparison with conventional methods in which this number is equal to the number of discretized grid points. This results in a considerable reduction in the amount of computations. As another very important advantages of the expansion method, no additional regularization term is needed, because the smoothness of the cosine functions and the limited number of expansion terms are considered adequate to suppress the ill-posedness

In a similar manner for 2-D cases, the expansion of the relative permittivity profile in transverse x-y plane which is homogeneous along z can be written as

$$\varepsilon_r(x,y) = \sum_{n=0}^{N-1}\sum_{m=0}^{M-1} d_{nm} \cos\left(\frac{n\pi}{a}x\right)\cos\left(\frac{m\pi}{b}y\right) \tag{3}$$

where a and b are the dimensions of the problem in the x and y directions, respectively. Similar expansions could be considered for conductivity profiles in lossy cases.

The proposed expansion algorithm is shown in Fig. 2. According to this figure, based on an initial guess for a set of expansion coefficients, the permittivity and conductivity are calculated according to the expansion relations like (2) or (3). Then, an EM solver computes a trial electric and magnetic simulation fields. Afterwards, cost function which indicates the difference between the trial simulated and reference measured fields is calculated. In the next step, global optimizer is used to minimize this cost function by changing the permittivity and conductivity of each cell until the procedure leads to an acceptable predefined error.

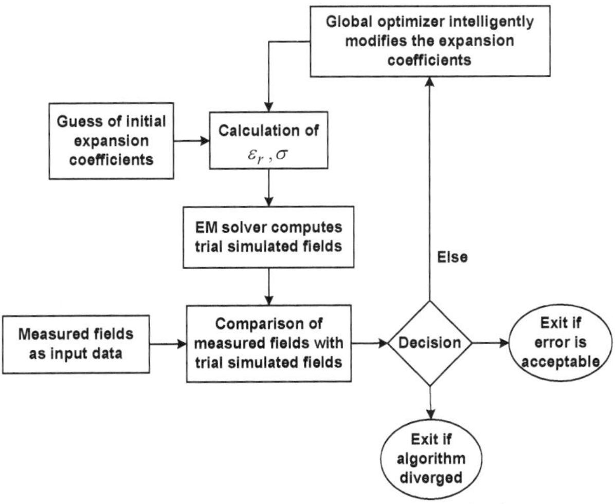

Fig. 2. Proposed algorithm for reconstruction by expansion method

3. Mathematical Considerations

As mentioned before, inverse problems are intrinsically ill-posed. Therefore, a priori information must be applied for stabilizing the algorithm as much as possible which is quite straightforward in direct optimization method. In this case, all the information can be applied directly to the medium parameters which are as the same as the optimization parameters. In the expansion algorithm, however, the optimization parameters are the Fourier series expansion coefficients and a priori information could not be considered directly. Hence, a useful indirect routine is vital to overcome this difficulty.
There are two main assumptions about the parameters of an unknown medium. For example, we may assume first that the relative permittivity and conductivity have limited ranges of variation, i.e.

$$1 \le \varepsilon_r \le \varepsilon_{r,\max} \tag{4}$$

and

$$0 \le \sigma \le \sigma_{max} \tag{5}$$

The second assumption is that the permittivity and conductivity profiles may not have severe fluctuations or oscillations. These two important conditions must be transformed in such a way to be applicable on the expansion coefficients in the initial guess and during the optimization process.
It is known that average of a function with known limited range is located within that limit, that is if

$$L_1 \le g(x) \le L_2, \quad a \le x \le b \tag{6}$$

Then

$$L_1 \le \frac{1}{b-a}\int_a^b g(x)dx \le L_2 \tag{7}$$

Thus, for 1-D permittivity profile expansion we have

$$1 \le d_0 \le \varepsilon_{r,\max} \tag{8}$$

For $x = 0$, (2) reduces to

$$\varepsilon_r(0) = \sum_{n=0}^{N-1} d_n \quad \Rightarrow \quad 1 \le \sum_{n=0}^{N-1} d_n \le \varepsilon_{r,\max} \tag{9}$$

and for $x = a$, we have

$$\varepsilon_r(a) = \sum_{n=0}^{N-1} (-1)^n d_n \;\Rightarrow\; 1 \le \sum_{n=0}^{N-1} (-1)^n d_n \le \varepsilon_{r,\max} \tag{10}$$

Using Parseval theorem, another relation between expansion coefficients and upper bound of permittivity may be written. For a periodic function $g(x)$ with period T, we have

$$\frac{1}{T}\int_T |g(x)|^2\,dx = \sum_{n=0}^{\infty} |d_n|^2 \tag{11}$$

Based on (2), (11) may be simplified to

$$1 \le \sum_{n=0}^{N-1} |d_n|^2 \le \varepsilon_{r,\max}^2 \tag{12}$$

It is possible to achieve the similar relations for 2-D cases.

$$1 \le d_{00} \le \varepsilon_{r,\max} \tag{13}$$

$$1 \le \sum_{n=0}^{N-1}\sum_{m=0}^{M-1} d_{nm} \le \varepsilon_{r,\max} \tag{14}$$

$$1 \le \sum_{n=0}^{N-1}\sum_{m=0}^{M-1} (-1)^{n+m} d_{nm} \le \varepsilon_{r,\max} \tag{15}$$

$$1 \le \sum_{n=0}^{N-1}\sum_{m=0}^{M-1} |d_{nm}|^2 \le \varepsilon_{r,\max}^2 \tag{16}$$

By using the above supplementary equations in the initial guess of the expansion coefficients and as a boundary condition (Robinson & Rahmat-Samii, 2004) during the optimization, the routine converges in a considerable faster rate. Similar conditions can be used for conductivity profiles in lossy cases.

4. Numerical Results

Proposed method stated above is utilized for reconstruction of some different 1-D and 2-D media. In each case, reconstruction by the proposed expansion method is compared with different number of expansion functions in terms of the amount of computations and reconstruction precision.

The objective of the proposed reconstruction procedure is the estimate of the unknowns by minimizing the cost function

$$C = \frac{\sum_{i=1}^{I}\sum_{j=1}^{J}\sum_{t=1}^{T}\left|E_{ij}^{meas}(t) - E_{ij}^{sim}(t)\right|^2}{\sum_{i=1}^{I}\sum_{j=1}^{J}\sum_{t=1}^{T}(E_{ij}^{meas}(t))^2} \tag{17}$$

where $\vec{E}^{sim}$ is the simulated field in each optimization iteration. $\vec{E}^{meas}$ is measured field, I and J are the number of transmitters and receivers, respectively and T is the total time of measurement.

To quantify the reconstruction accuracy, the reconstruction errors for example for relative permittivity in 1-D case is defined as

$$e(\varepsilon) = \sqrt{\frac{\sum_{i=1}^{M_x}\left|\varepsilon_{ri} - \varepsilon_{ri}^{o}\right|^2}{\sum_{i=1}^{M_x}(\varepsilon_{ri}^{o})^2}} \times 100 \tag{18}$$

where M_x is the number of subdivisions along x axis and " ^{o} " denotes the original scatterer properties.

In all reconstructions in this chapter, FDTD (Taflove & Hagness, 2005) and DE (Storn & Price, 1997) are used as forward EM solver and global optimizer, respectively.

4.1 One-dimensional case

Reconstruction of two 1-D cases is considered in this section. The first one is inhomogeneous and lossless and the second one is considered to be lossy. In the simulations of both cases, one transmitter and two receivers are used around the medium as shown in Fig. 3.

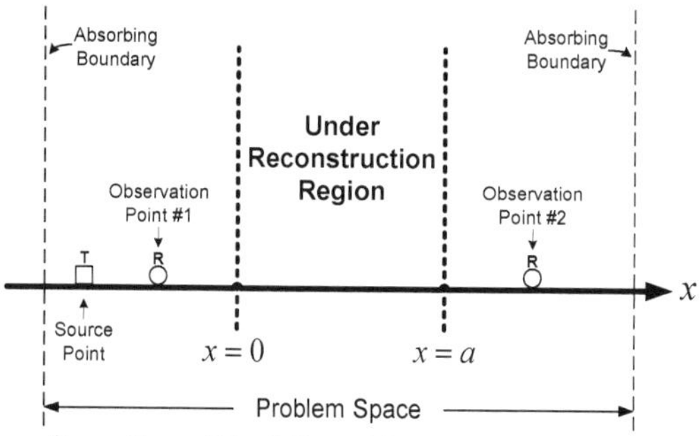

Fig. 3. Geometrical configuration of the 1-D problem

Test case #1: In the first sample case, we consider an inhomogeneous and lossless medium consisting 50 cells. Therefore, only the permittivity profile reconstruction is considered. In the expansion method, the number of expansion terms is set to 4, 5, 6 and 7 which results in a lot of reduction in the number of the unknowns in comparison with the direct method. The population in DE algorithm is chosen equal to 100 and the maximum iteration of optimization is considered to be 300. It must be noted that the initial populations in all reconstruction problems in this chapter are chosen completely random in the solution space. The exact profile and reconstructed ones by the expansion method with different number of expansion terms are shown in Fig. 4a. The variations of cost function (17) and reconstruction error (18) versus the iteration number are plotted in Figs. 4b and 4c, respectively.

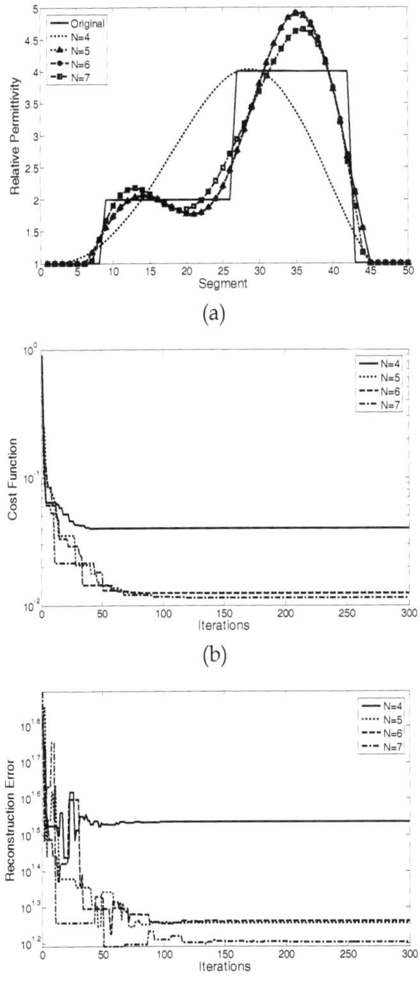

Fig. 4. Reconstruction of 1-D test case #1, (a) original and reconstructed profiles, (b) the cost function and (c) the reconstruction error

Test case #2: In this case, a lossy and inhomogeneous medium again with 50 cell length is considered. So, the number of unknowns in direct optimization method is equal to 100. In the expansion method for both permittivity and conductivity profiles expansion, N is chosen equal to 4, 5, 6 and 7. The optimization parameters are considered equal to the first sample case. The original and reconstructed profiles in addition of the variations of cost function and reconstruction error are presented in Fig. 5.

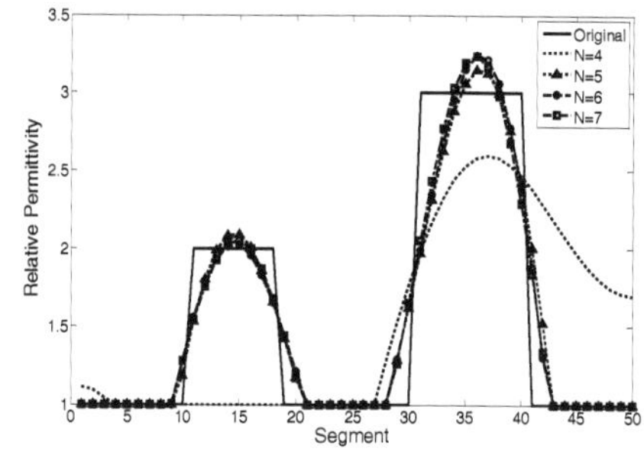

(a)

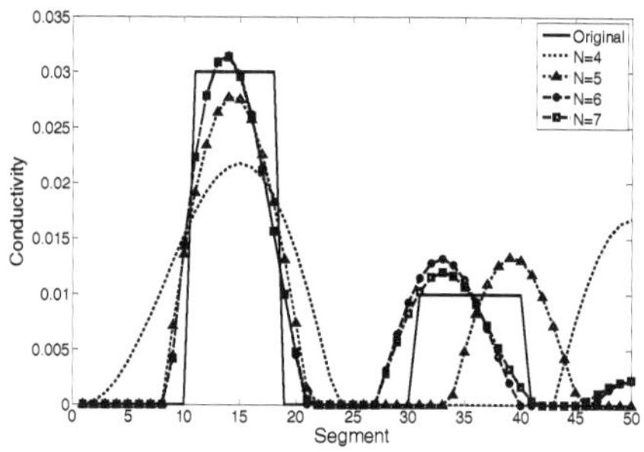

(b)

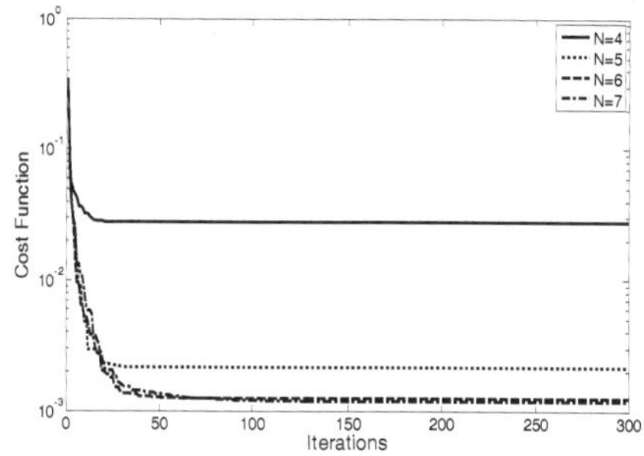

(c)

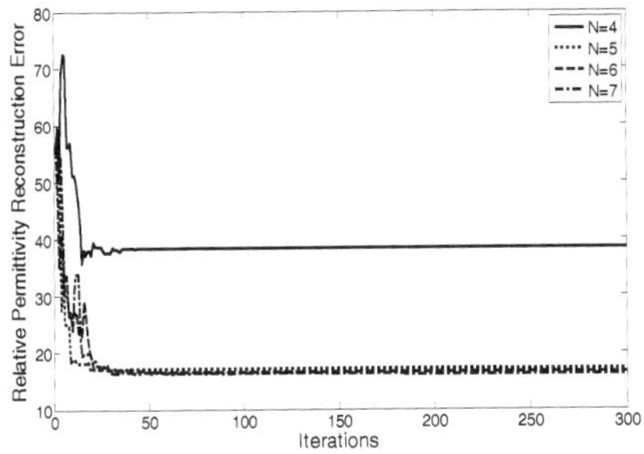

(d)

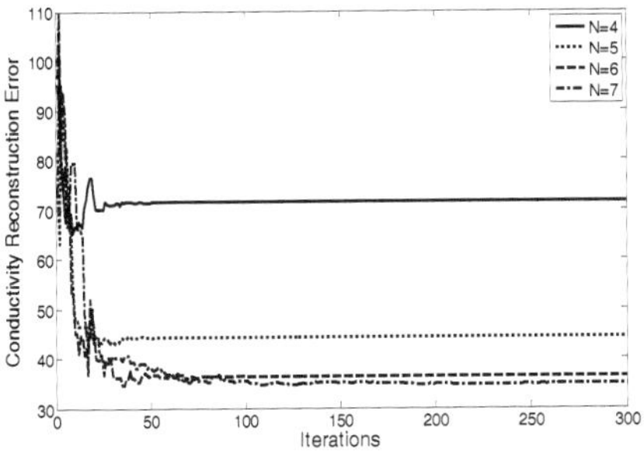

(e)

Fig. 5. Reconstruction of 1-D test case #2, (a) original and reconstructed permittivity profiles, (b) original and reconstructed conductivity profiles, (c) the cost function, (d) the permittivity reconstruction error and (e) the conductivity reconstruction error

4.2 Two-dimensional case

The proposed expansion method is also utilized for two 2-D cases. In the simulations of both cases, four transmitter and eight receivers are used as shown in Fig. 6. The population in DE algorithm is chosen equal to 100, the maximum iteration is considered to be 300.

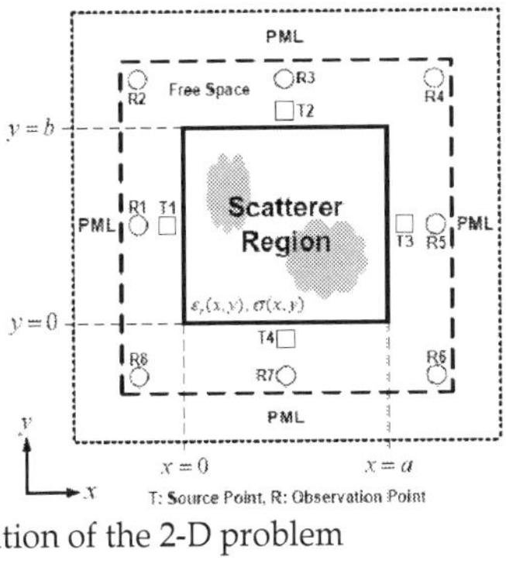

Fig. 6. Geometrical configuration of the 2-D problem

Case study #1: In the first sample case, we consider an inhomogeneous and lossless 2-D medium consisting 20*20 cells. Therefore, only the permittivity profile reconstruction is considered. In the expansion method, the number of expansion terms in both x and y directions are set to 4, 5, 6 and 7.

The original profile and reconstructed ones with the use of expansion method are shown in Fig. 7.

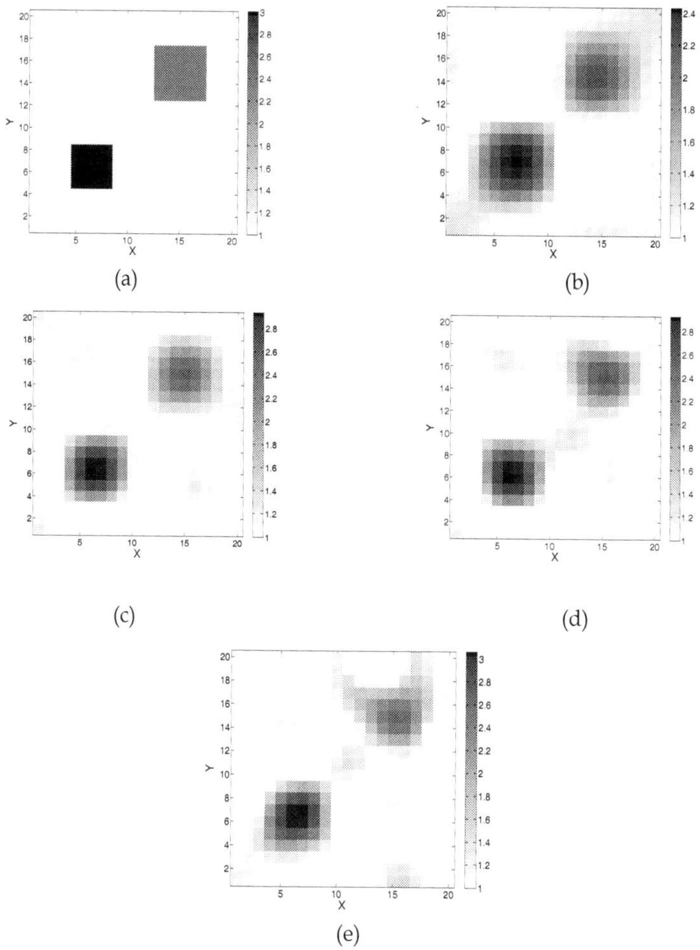

Fig. 7. Reconstruction of 2-D test case #1, (a) original profile, reconstructed profile with (b) N=M=4, (c) N=M=5, (d) N=M=6 and (e) N=M=7

The variations of cost function and reconstruction error versus the iteration number are graphed in Fig. 8.

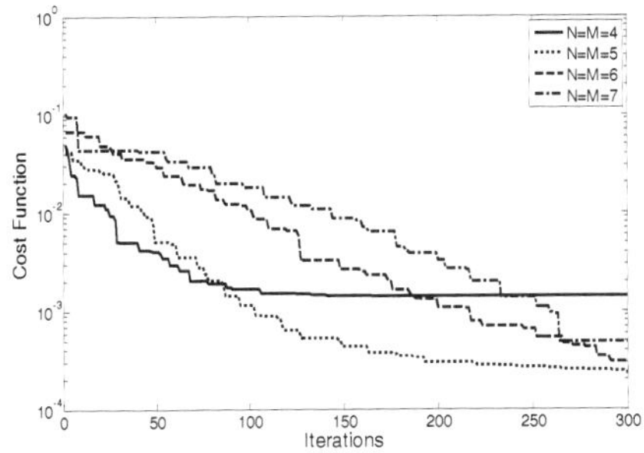

(a)

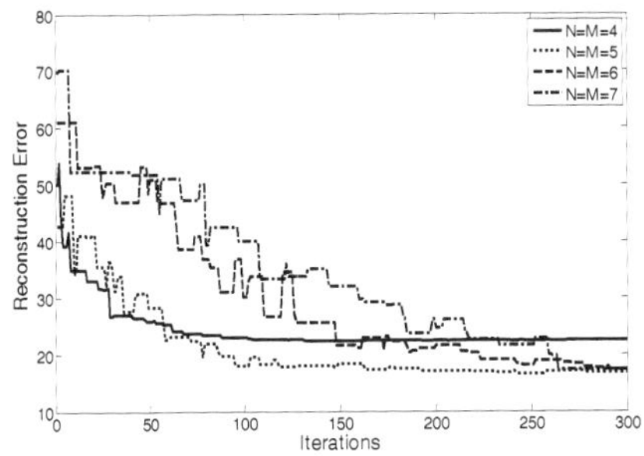

(b)

Fig. 8. Reconstruction of 2-D test case #1, (a) the cost function, (b) the reconstruction error

Case study #2: In this case, a lossy and inhomogeneous medium again with 20*20 cells is considered. Therefore, we have two expansions for relative permittivity and conductivity profiles and in both expansions, N and M are chosen equal to 4, 5, 6 and 7. It is interesting to note that the number of direct optimization unknowns in this case is equal to 800 which is really a large optimization problem. The reconstructed profiles of permittivity and conductivity are shown in Figs. 9 and 10, respectively.

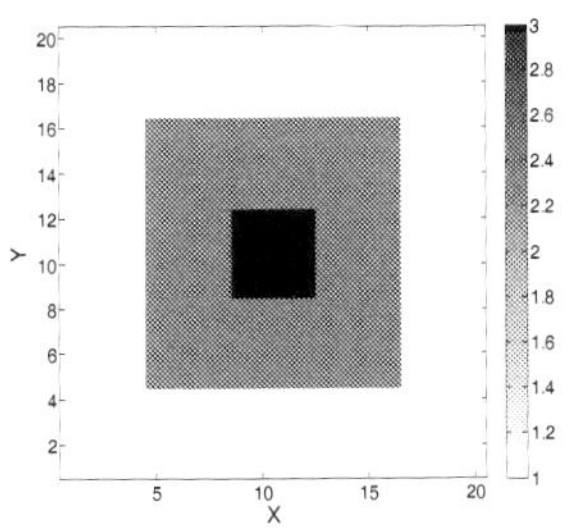

(a)

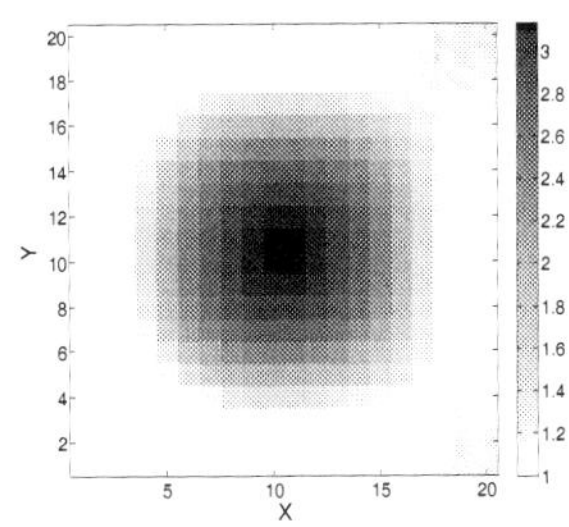

(b)

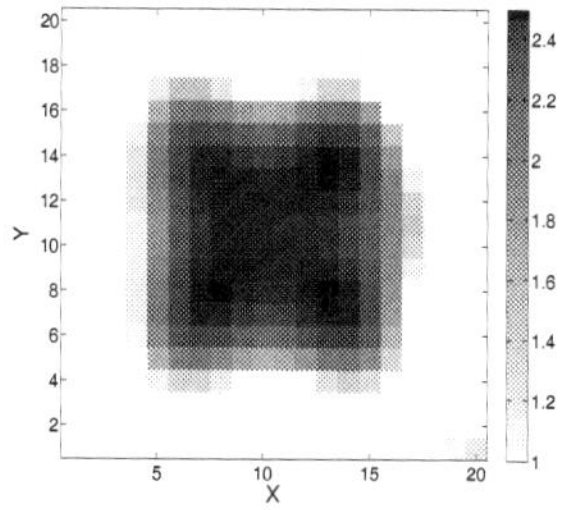

(c)

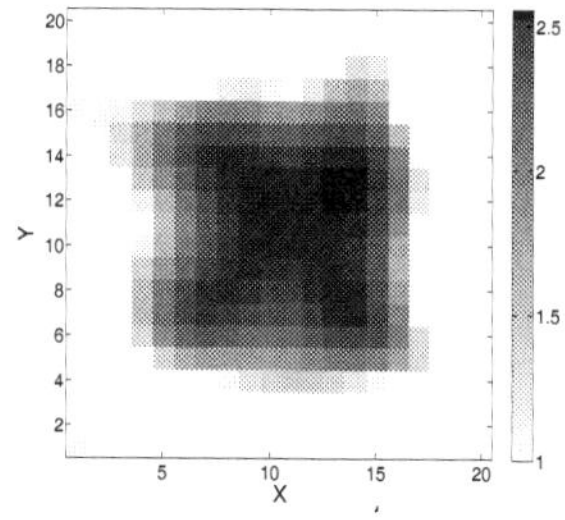

(d)

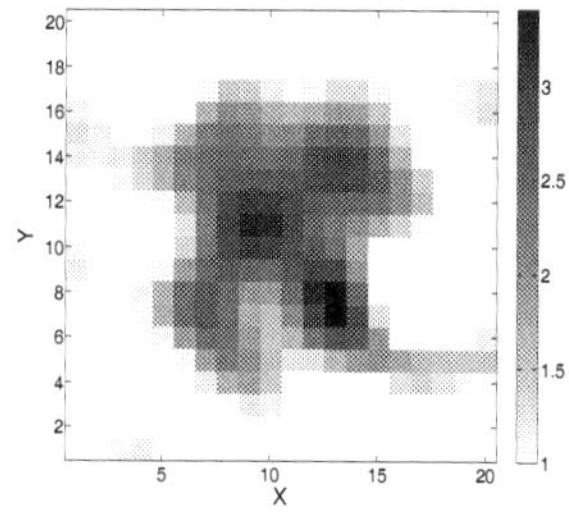

(e)

Fig. 9. Reconstruction of 2-D test case #2, (a) original permittivity profile, reconstructed permittivity profile with (b) N=M=4, (c) N=M=5, (d) N=M=6 and (e) N=M=7

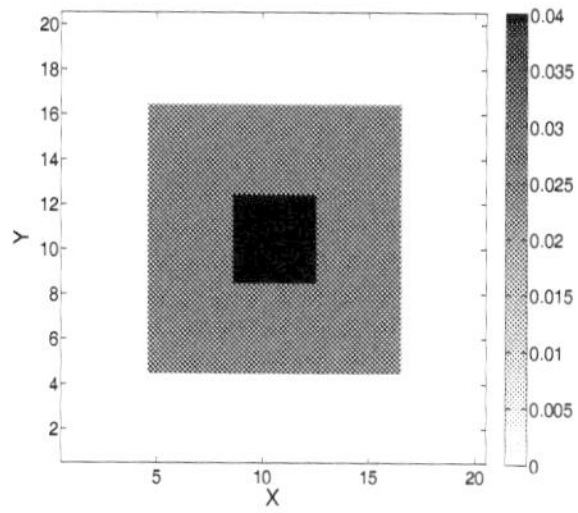

(a)

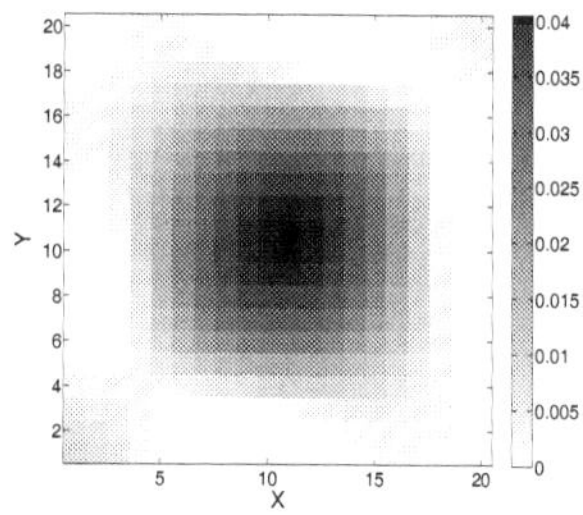

(b)

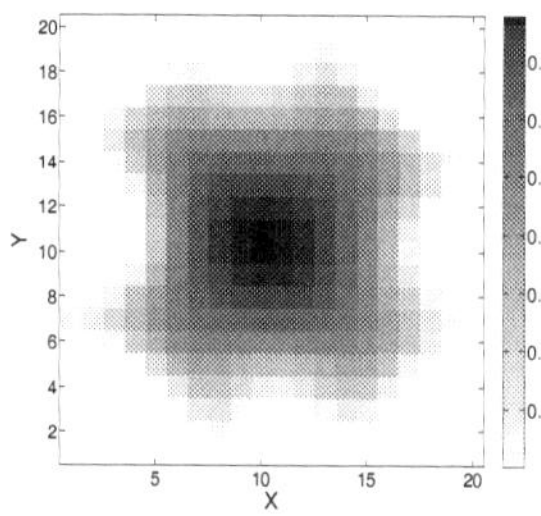

(c)

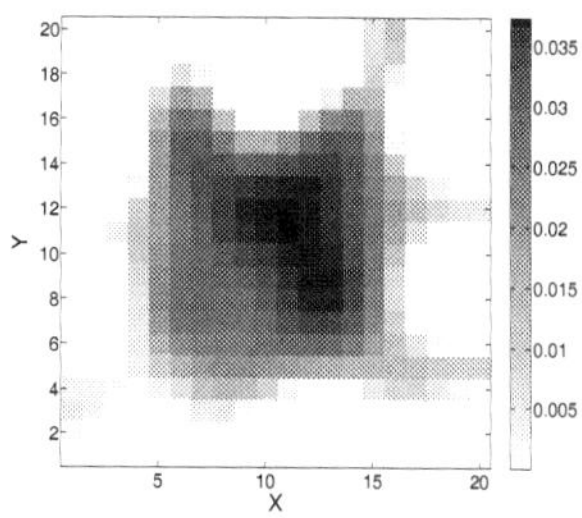

(d)

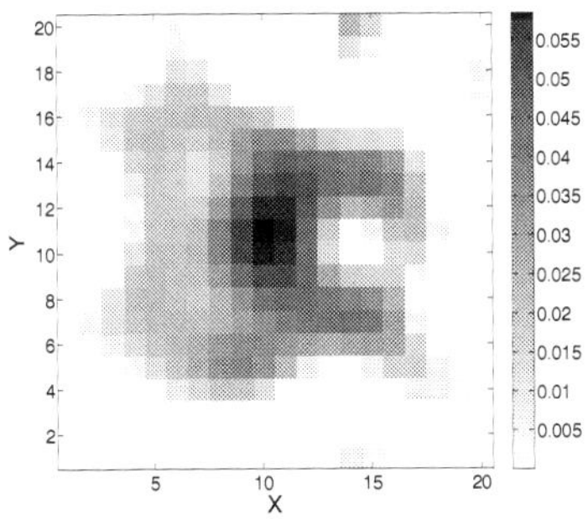

(e)

Fig. 10. Reconstruction of 2-D test case #2, (a) original conductivity profile, reconstructed conductivity profile with (b) N=M=4, (c) N=M=5, (d) N=M=6 and (e) N=M=7

The variations of cost function and reconstruction error are shown in Fig. 11.

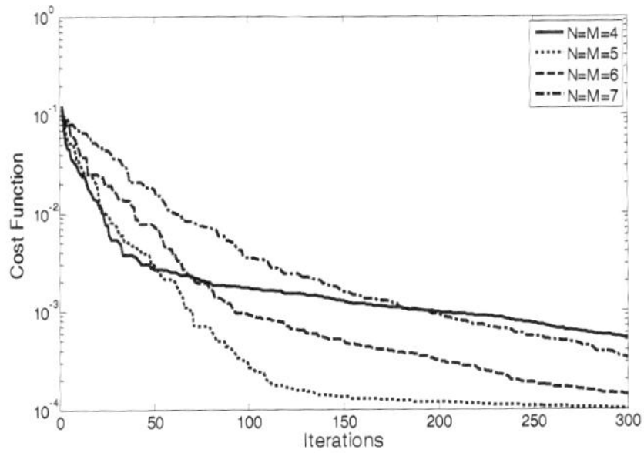

(a)

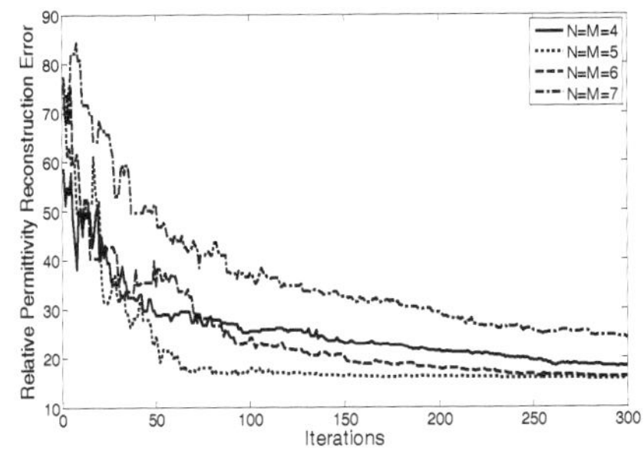

(b)

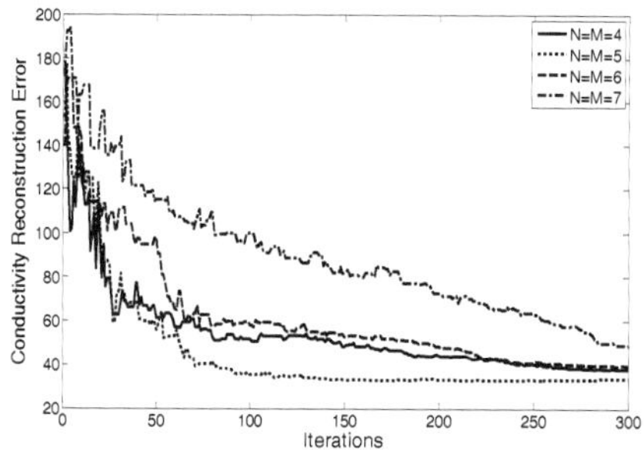

(c)

Fig. 11. Reconstruction of 2-D test case #2, (a) the cost function, (b) the permittivity reconstruction error and (c) the conductivity reconstruction error

The results of all 1-D and 2-D cases which are generally inhomogeneous and lossy or lossless media show that the proposed expansion method can tolerably reconstruct the unknown media with a considerable reduction in the amount of computations as compared to the conventional direct optimization of the unknowns.

5. Sensitivity Considerations

It is obvious that the performance of the expansion method directly depends on the number of expansion terms. Larger number of terms results in a more precise reconstruction at the expense of higher degree of ill-posedness. On the other hand, lower ones leads to a less accurate solution with higher probability of convergence of the inverse algorithm. Therefore, suitable selection of N has a notable impact on the convergence speed of the algorithm.
The reconstructed profiles of two 1-D cases with larger values of N are shown in Figs. 12 and 13 for test case #1 and #2, respectively.

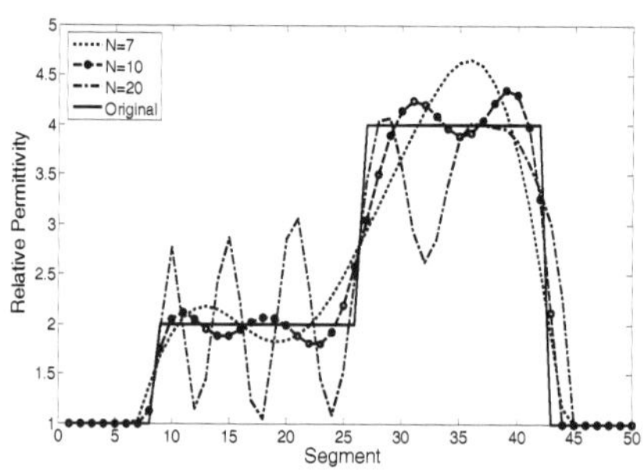

Fig. 12. Reconstruction of 1-D test case #1, the original profiles and reconstructed ones with N=7, 10 and 20

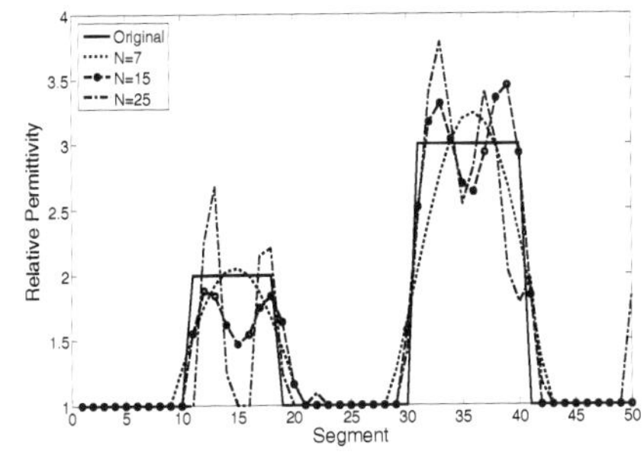

(a)

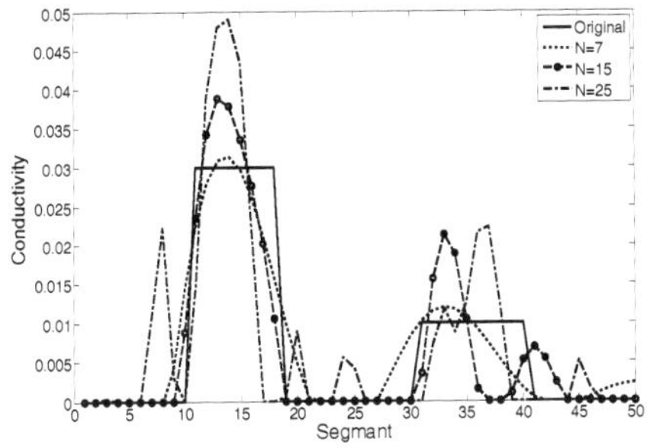

(b)

Fig. 13. Reconstruction of 1-D test case #2, the original profiles and reconstructed ones with N=7, 15 and 25, (a) permittivity profile and (b) conductivity profile

It is seen that increasing the number of expansion terms results oscillatory reconstruction because of the more ill-posedness of the problem.

We can come to similar conclusion for 2-D cases by comparing different parts of Figs. 7, 9 and 10.

Our experiences in studying various permittivity and conductivity profiles reconstruction show that choosing the number of expansion terms between 5 and 10 may be suitable for most of the reconstruction problems.

6. Conclusion

A computationally efficient method which is based on combination of the cosine Fourier series expansion, an EM solver and a global optimizer has been proposed for solving 1-D and 2-D inverse scattering problems. The mathematical formulations of the method have been derived completely and the algorithm has been examined for reconstruction of several inhomogeneous lossless and lossy cases. With a considerable reduction in the number of the unknowns and consequently the required number of populations and optimization iterations, along with no need to the regularization term, the relative permittivity and conductivity profiles have been reconstructed successfully. It has been shown by sensitivity analysis that for obtaining well-posedness as well as accurate reconstruction simultaneously, the number of expansion terms must be chosen intelligently.

7. References

Cakoni, F. & Colton, D. (2004). Open problems in the qualitative approach to inverse electromagnetic scattering theory. *Euro. Jnl. of Applied Mathematics*, Vol. 00, (1-15)

Caorsi, S.; Ciaramella, S.; Gragnani, G. L. & Pastorino, M. (1995). On the use of regularization techniques in numerical invere-scattering solutions for microwave imaging applications. *IEEE Trans. Microwave Theory Tech.*, Vol. 43, No. 3, (March). (632-640)

Colton, D. & Paivarinta, L. (1992). The uniqueness of a solution to an inverse scattering problem for electromagnetic waves. *Arc. Ration. Mech. Anal.*, Vol. 119, (59-70)

Isakov, V. (1993). Uniqueness and stability in multidimensional inverse problems," *Inverse Problems*, Vol. 9, (579-621)

Rekanos, I. T. (2008). Shape reconstruction of a perfectly conducting scatterer using differential evolution and particle swarm optimization. *IEEE Trans. Geosci. Remote Sens.*, Vol. 46, No. 7, (July). (1967-1974)

Robinson, J. & Rahmat-Samii, Y. (2004). Particle swarm optimization in electromagnetics. *IEEE Transactions on Antennas and Propagation*, Vol. 52, No. 2, (397-407)

Semnani, A. & Kamyab, M. (2008). Truncated cosine Fourier series expansion method for solving 2-D inverse scattering problems. *Progress In Electromagnetics Research*, Vol. 81, (73-97)

Semnani, A. & Kamyab, M. (2009). An enhanced hybrid method for solving inverse scattering problems. *IEEE Transaction on Magnetics*, Vol. 45, No. 3, (March). (1534-1537)

Storn, R. & Price, K. (1997). Differential evolution - A simple and efficient heuristic for global optimization over continuous space. *J. Global Optimization*, Vol. 11, No. 4, (Dec). (341-359)

Taflove, A. & Hagness, S. C. (2005). *Computational Electrodynamics: The finite-difference time-domain method*, Third Edition, Artech House

Tikhonov, A. N. & Arsenin, V. Y. (1977). *Solutions of Ill-Posed Problems*, Winston, Washington, DC

23

Electromagnetic Solutions for the Agricultural Problems

Hadi Aliakbarian[1], Amin Enayati[1], Maryam Ashayer Soltani[2], Hossein Ameri Mahabadi[3] and Mahmoud Moghavvemi[3]

[1] *Departement Elektrotechniek (ESAT), Katholieke Universiteit Leuven (KUL), Belgium*

[2] *Department of Bioprocess Engineering, Malaysia (UTM), Malaysia*

[3] *Department of Electrical Engineering, University of Malaya (UM), Malaysia*

1. The idea of electromagnetic waves in agricultural applications

1.1 Introduction

In the recent years, interactive relations between various branches of science and technology have improved interdisciplinary fields of science. In fact, most of the research activities take place somewhere among these branches. Therefore, a specialist from one branch usually can propose novel methods, whenever enters a new field, based on his previous knowledge.

Taking a look at the extensive problems in the field of agriculture, an expert in the field of Electromagnetic waves can easily suggest some innovative solutions to solve them. The major suffering problems with which a farmer faces are the damages caused by the harmful pests as well as the product freezing in unexpected cold weather. The promising available biological methods of treatment have decreased the need for new treatment methods effectively. However, some advantages of electromagnetic treatment is still without competitor. The environment-friendly methods we introduce in this chapter are to use electromagnetic waves to kill pest insects without killing the taste or texture of the food they infest.

1.2 Electromagnetic waves in agricultural applications

Electromagnetic waves as tools in the field of agriculture have been used in many applications such as remote sensing, imaging, quality sensing, and dielectric heating in a pre-harvest or post-harvest environment. However, the goal here is to discuss about applications which are directly related to the main electromagnetic wave effect which is warming. Among variable methods applicable in the agriculture section, Radio frequency (RF) power has been known as physical (non-chemical) thermal method. In this method, the general idea is the same as heating food products to kill bacteria. It can be used to disinfest various foods and non food materials including soil. On the other hand, there are applications of using radio frequency to measure soil parameters and soil salinity, as well.

1.3 Pest control and electromagnetic waves

Traditional agricultural producers usually use simple conventional chemical sprays to control pests. Despite the simplicity of use, these chemical fumigants such as Methyl Bromide have many disadvantages such as reducing the thickness of Ozone layer (Tang et al. 2003). Additionally, the probable international ban of methyl bromide for post-harvest treatments will increase the attention to other methods. Three other methods including ionizing radiation, cold treatments and conventional heating has been reviewed in (Wang & Tang, 2001). In ionizing radiation, the main problem is that it is not possible to shut of the radiation after ending the treatment. In addition, although there are still some road blocks to use irradiation effectively and also commercially. Cold treatments are not a complete method due to high price and relatively long required time. The drawback of the conventional heating methods originates from the fact that this kind of heating warms both pest and the agricultural product similarly which may destroy product's quality. To overcome these problems, some modern techniques such as genetic treatments, ultrasonic waves and electromagnetic treatments have been suggested in the literature.

The use of electromagnetic exposure, mainly electromagnetic heating has been started in 1952 by Frings (Frings, 1952) and then Thomas in 1952 (Thomas, 1952) and Nelson from 1966 (Nelson, 1966). But today there are vast applications for electromagentic waves are proposed at least to be an alternate treatment method. Formerly, the electromagnetic wave method was suggested as a post-harvest treatment, but recently, it has been suggested to be used as an in-the-field method for pest control or to prevent the agricultural product from getting freezed (Aliakbarian et al., 2007).

1.4 Challenging problems

Although the effectiveness of using radio waves to kill destructive insects in agricultural products has been known for 70 years, the technique has rarely been applied on a commercial scale because of the technical and market problems. There are at least six challenging problems against the vast implementation of electromagnetic waves use in agricultural applications: high electromagnetic power needed, probable human health effects, probable biological effects on the surrounding environment, finalized price, frequency allocation and system design complexity.

Power problem can be easily solved if the employed frequency is not more than the low-gigahertz range. Based on the fact that high power sources are now available in VHF and UHF frequencies the power problem can be solved.

The problem of price is also an economic topic that should be considered by investors. The enormous detriments of *pests* may motivate large companies in this investment. In addition, RF technology is already used commercially and has existed for about 40 years. Consequently, the machinery that delivers the RF blast will probably be affordable for the industry. However, it is still costly to pay $ 2,000/kW or higher in some bands. Although it is reported that the technology is already commercially applied to food products including biscuits and bakery products in 40 MHz (Clarck, 1997), researches of a team led by J. Tang since 2005 (Flores, 2003(2)) shows that we still need more researches to the economical industrial use of radio waves.

The problem of frequency allocation in some countries is crucial. However, shifting the frequency to the closest ISM bands can solve the frequency allocation problem. The Federal Communications Commission has allocated twelve industrial, scientific and medical, or

ISM, bands starting from 6.7 MHz to 245 GHz. For the outdoor environments, electromagnetic waves are needed for a few days in a year.

Another problem is to design such a proper controllable system to warm up pests uniformly. For example, in a complex environment, if a single power source is used, it will be difficult to cover the whole environment. Thus an array of sources should be designed. Moreover, the frequency of treatment must be selected in such a manner that the absorption of energy by pest be more than other materials available.

Today, electromagnetic wave is known as a potential hazard of health and biological effects such as cancer. It is tried to shield and protect the radiation space from the outside environment. On the other hand, in the outdoor problems, we reduce the hazard lowering the exposure time. Moreover, treatment environments are usually empty of human population. In spite of the health effect, biological effects of electromagnetic exposure should be evaluated to ensure that it does not have a harmful effect on the ecosystem.

2. Theory of electromagnetic selective warming

2.1 Introduction

There are various ideas about the mechanism of pest control using electromagnetic waves. Most of the researchers believe that the waves can only warm up the pests. This belief originates from the fact that these insects are mostly composed of water. Normally, the water percentage in their body is more than the other materials present in the surrounding environment. On the other hand, there are some claims expressing that not only do the electromagnetic waves heat the pest, but also they can interfere with their bodys' functionality with their none-thermal effects. (Shapovalenko et al., 2000). Fig.1 represents a practical tests of electromagnetic exposure which shows pests running away from the antenna. Their escape may be due to heating effect or due to some other colfict to their dody. Although attraction is also reported, a reapetable test has not been verified. However, none-thermal effects of electromangetic waves on living tissue has been confirmed (Geveke & Brunkhorst, 2006).

The imaginary part of the dielectric constant can be used to heat up a material remotely using radio waves. However the main goal is not just to heat a material (*i.e.* a flower) in the indoor or outdoor environment since it can be done using a heater or 2.4 GHz microwave source. The mission, here, is to warm a material while the surrounding materials are not affected. This can be done using the difference between the imaginary parts of the dielectric constants of two different material at a specified frequency. Taking into account that the dielectric constant of each material is frequency-dependent, there can be an appropriate frequency for which the electromagnetic energy is absorbed by the pest while the product or plant don't absorb the energy at this frequency. Concequently, this process will not affect the quality of the agricultural products, specially important for the products which are sensitive to the temperature increase.

Fig. 1. Repelling in response of *Sunne* pest to electromagnetic exposure

2.2 Electromagnetic heating

Dielectric materials, such as most plants, can store electric energy and convert electric energy into heat. Each material has a complex permittivity (ε) in general. According to measurements, usually this value is noticeably frequency dependent. The imaginary part (ε'') of this value is responsible for absorption of electromagnetic waves in each material. Eq.1 shows the general form of the first Maxwell's equations considering (ε'').

$$\nabla \times H = j\omega\varepsilon' E + J = j\omega\varepsilon' E + \sigma E \quad (1)$$
$$= j\omega(\varepsilon' - j\frac{\sigma}{\omega})E = j\omega(\varepsilon' - j\varepsilon'')E$$

As a consequence, total power absorption in a specific material is achieved if the second part of the equation is integrated over the material volume as follows.

$$P_{Loss} = \int_v E.JdV = \int_v \sigma |E|^2 dV = \omega \int_v \varepsilon'' |E|^2 dV \quad (2)$$

The basic idea is to use ε'' to warm up the selected materials which are located far from an electromagnetic source. On the other hand, the E-field distribution inside the absorbing material highly depends on the shape of absorber and surrounding scatterers, compared to wavelength and also the source of excitation. For instance, regarding the distribution of a cluster of walnuts inside an oven, the resulted electromagnetic wave inside one of them is a function of its shape, the shape and the position of other walnut as external scatterers and the oven and its exciting antenna as the source. Hence, not only the total absorbed energy resulted from (2) is important, but also the uniformity of the electromagnetic field distribution is crucial. If the dimensions of the exposed object are electrically small, we can assume the E-field distribution inside the object is uniform. For bigger objects, the penetration of the wave inside the object, i.e. skin depth, can be calculated using (3) depending on frequency used and the dielectric properties of the sample under test, especially conductivity.

$$\delta = \frac{1}{\sqrt{\pi\mu_o}}\sqrt{\frac{1}{\sigma\mu f}} \tag{3}$$

Wherein, δ is the skin depth in meter, f is frequency in Hz, μ is relative permeability, σ is conductivity ($S \cdot m^{-1}$). and μ_o is $4\pi \times 10^{-7}$. The increase in the temperature of a material by absorbin the electromagnetic energy can be expressed in Eq. (4) as stated in (Nelson, 1996)

$$\rho C \frac{\Delta T}{\Delta t} = 55.63 \times 10^{-12} f E^2 \varepsilon'' \tag{4}$$

where C is the specific heat capacity of the material ($J.kg^{-1}.°C^{-1}$), ρ is the density of the material (kg.m-3), E is the electric field intensity ($V.m^{-1}$), f is the frequency (Hz), ε'' is the dielectric loss factor (farad/m) of the material, Δt is the time duration (s) and ΔT is the temperature rise in the material (°C).
Thus, if the goal is to absorb as much as the energy to the victim, the optimized solution is to maximize P_{loss} or simply $\varepsilon''(f)$ in a predefined structure.
Thus, it is essential to measure the dielectric constant of the material. Fig.2 shows the measured dielectric constant and loss factor of tissue of fresh Navel Orange in terms of frequency measured at different temperatures (Nelson, 2004). The figure indicates that the dielectric constant relatively depends on the temperature of the material.

Fig. 2. Dielectric constant of fresh Navel Orange in different temperatures (Nelson, 2004)

2.2 Differential heating

The dielectric constant parameter for materials as a whole and for agricultural products specifically varies with frequency. For instance, ε" of water has a peak in 24 GHz frequency. The absorption frequency of water may help us in warming the water in the insects' bodies but probably all of the other water-composed materials in the nearby environment absorb the energy as well. Thus, to be more efficient and safe, the electromagnetic wave should have a frequency which maximizes the difference between temperature increment in pest on one side and the agricultural products on the other side. This goal can be reached by using the frequency dependent character of the dielectric constants of the two materials. Using (4), the function in (5) represents a goal function which should be maximized in the volume of an electrically small object.

$$\begin{aligned} Goal(f) &= \frac{(\Delta T_{pest}(f) - \Delta T_{Orchid}(f))}{\Delta t} \\ &= \frac{\alpha f E_{pest}^2(f)\varepsilon''_{pest}(f)}{\rho_{pest} C_{pest}} - \frac{\alpha f E_{Orchid}^2(f)\varepsilon''_{Orchid}(f)}{\rho_{Orchid} C_{Orchid}} \\ \alpha &= 55.63 \times 10^{-12} \end{aligned} \tag{5}$$

Using the assumption that specific heat capacities of the both materials are equal, goal function is reduced to (6).

$$Goal(f) = \frac{\alpha f}{\rho C}\left(E_{pest}^2(f)\varepsilon''_{pest}(f) - E_{Orchid}^2(f)\varepsilon''_{Orchid}(f)\right) \tag{6}$$

If we simply suppose that electric field is equal in pest and agricultural product regions, the goal function is reduced to (7)

$$Goal(f) = \frac{\alpha f E^2}{\rho C}(\varepsilon''_{pest}(f) - \varepsilon''_{Orchid}(f)) \tag{7}$$

Therefore, approximately, it can be stated that we are searching for a frequency at which the difference between ε″ (f) of the agricultural material and pest is the most possible value. In order to solve this problem, we are going to measure the effective permittivity of the agricultural products to find the optimum frequency in which the difference between ε″ of the pest and the agricultural product is the largest.

The above discussion assumes that the target object is a small one in terms of heat convection, while it is not the case almost all of the time. Therefore, to predict the temperature in a practical three dimensional domain, Maxwell equations and Navier-Stokes equations should be solved simultaneously in the presence of all products. Navier-Stokes equations are nonlinear partial differential equations describes the temperature and gas distribution in an environment. The combined equation can help us to predict the real situation inside a silo.

In conclusion, taking a look at equations (4) and (7), we can find out that there are two approaches based on maximum energy transfer and maximum differential heating which does not necessarily happen at the same frequency.

2.3 Measurement of the dielectric constant (Wang et al., 2003)

There are many methods for the measurement of the dielectric properties of materials. The best one for arbitrarily shaped materials is the open-ended coaxial probe which ended at the material under measurement with full contact. Using the method, we can measure the properties in a wide range of frequencies using reflection data. The more accurate one is the transmission line method, but it is necessary to fill a part of transmission line with the samples accurately. In order to measure very low loss materials, cavity method can be used. In this method, the sample is inserted in a cavity and the change in the reflection coefficient and the resonance frequency shift is measured. Using accurate perturbation formulas, the dielectric constant can be calculated in one fixed frequency. Many experimental data has been released for several foods and agricultural products (Wang et al., 2003) but yet few works has been done on pest's properties. The measurement results shows that the properties highly depends on frequency, temperature, moisture content and also state of the moisture, namely frozen, free or bound.

3. Proposed treatments

The proposed treatments can be categorized from different aspects. Here, we divide the applications in two categories of post-harvest and pre-harvest treatments. On the other hand, the vast range of frequencies from low RF to microwave and millimeter waves can be used which is mentioned.

The applications of electromagnetic waves in agriculture are not known without Tang and Wang's works. For many years, they have tried to replace fumigation with radio frequency treatment for export fruit quarantine applied on cherries and apples in Washington, citruses in Texas, and also walnuts and almonds in California (Flores et al., 2003(1)).

3.1. Post-harvest treatment

3.1.1. Walnuts treatment in ISM band

Keeping in mind that the two third of world nuts are supplied by US, the importance of quality improvement will be clear. The dielectric loss factors of nuts' pests are much higher than those of the nuts illustrated in Fig.3 (Wang & Tang, 2001). Within a 3 minutes of treatment, the Codling moth, which infests the walnuts, is killed due to the high absorption of energy compared to the walnut (Ikediala et al., 2000). On the other hand, the shell and the air inside it act as an insulator and protect the walnut from convectional heating, while the electromagnetic wave selects the victim inside the walnut to transfer the energy. The speed of temperature increase is approximately 10 times the hot air method.

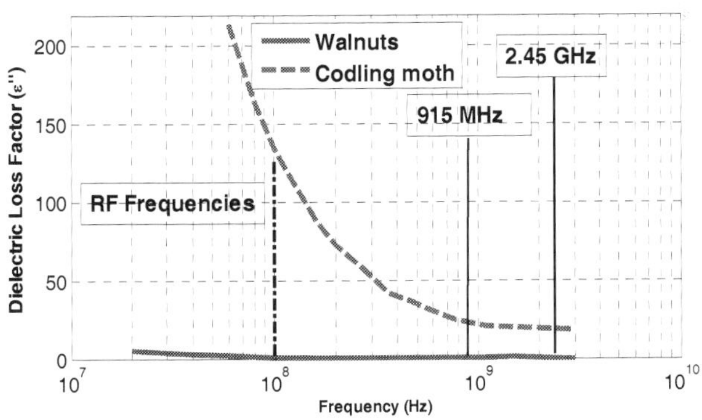

Fig. 3. Difference between the loss factor of codling moth and walnut (Wang & Tang, 2001)

Thus, the idea of combined methods is raised to remove some of the disadvantages of each one of them. If the RF heating is combined with hot air (Wang et al, 2002), the temperature drop during the holding period will be reduced and surface heating will be improved as well. The schematic of the system is described in Fig.4. 6 kW RF power in 27 MHz is supplied by an oscillator circuit, but the gap between electrodes is adjusted to expose 0.8 kW to the samples under treatment. From the other side, hot air is supplied using a tray drier.

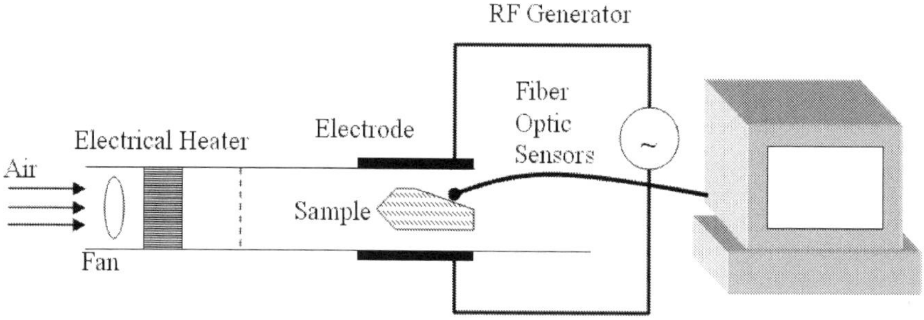

Fig. 4. Schematic of 27 MHz combined RF and hot air prototype

As another example of combined methods (Wang, 2001), RF heating can be combined with chemical fumigation. After fumigation, in-shell walnuts are washed and dried. During this process, walnuts are dropped into storage bins a number of times which may cause walnuts' shells to be cracked. With the use of RF waves to heat and dry the walnuts, we can effectively reduce damages, treatment time and required space. Yet, there are many practical problems such as the problem of different moisture content in walnuts. Moist material (basically water) has high dielectric constant. One of the reasons of the different moisture content is the different bleaching operations based on the customer (Wang et al, 2006). For example in US, 3% hypochlorite is used for Spain export while 6% hydrogen peroxide for Germany which are absorbed differently by the walnuts' tissues. Moreover, the absorption depends on the condition of the walnuts such as to be opened, closed, or cracked. A scaled pilot system is designed and implemented in 27 MHz (Wang, 2006). To overcome the problems of cost and quality, some solutions such as walnut orientation and intermittent mixing of the walnuts are suggested.

3.1.2 Thermal and none-thermal treatment of fruit Juice using low-frequency waves. (Geveke & Brunkhorst, 2006)

This work is to some extent similar to post-baking applications (Clarck, 1997) because both of them concerns food processing application rather than agriculture. On the other hand, this example is exceptional due to the use of none-thermal effects of electromagnetic waves. Use of radio waves to make safer fruit juices has been worked out by researchers for many years but has not been commercialized yet due to economical reasons. Conventional pasteurization is done using different heating techniques, but they can affect the nutrient composition and flavour of the fruit and vegetable juices. The new method is totally different. The radio frequency electric fields inactivate bacteria in apple juices without

heating them. According to Geveke and Brunkhors's work, the method has been used half century ago for pasteurization purposes but this is the first time that they could inactivate bacteria of fruit juice using this technique successfully.
However, using the combined method, namely the use of moderate heat in addition to the none-thermal method, has much greater effects than those of the either processes has alone. They have built a specially designed treatment to apply high-intensity radio frequency electric fields to apple juice. The schematic of the device shown in Fig.5 illustrates the juice flow passing the RF part in the center. It also highlights how it is tried to reduce required RF power to converge the juice flow into a narrow line. Current simulation is done using Quick field finite element software.

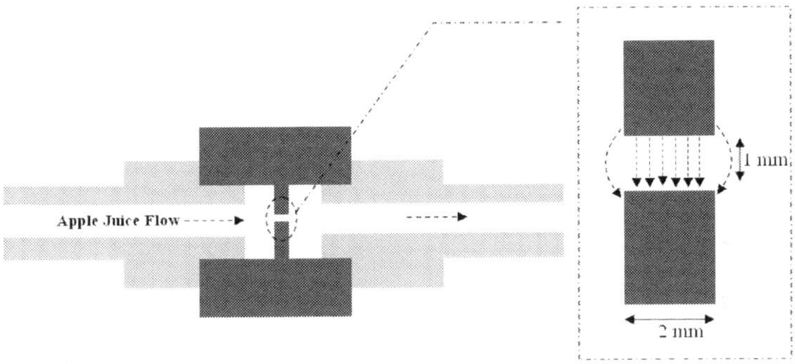

Fig. 5. Schematic of the treatment device for apple juice bacteria inactivation

The juice has been exposed to electrical field strengths of up to 20 kilovolts per centimeter and frequencies in the range of 15 to 70 kilohertz for 0.17 ms period, using a 4-kilowatt power supply. Based on the experiments, frequency increase as well as field strength and temperature increment enhances the inactivation. However, exposure above 16 Kilovolts intensity does not improve the inactivation performance and so do not the frequencies of more than 20 KHz. The experiment on different drinks in 18 MHz and intensity of 0.5 Kilovolts/cm does not show any none-thermal effect.

3.1.3 Indoor differential warming for wheat (Nelson & Tetson, 1974)

It is a key advantage if we can warm the infecting insects while not affecting the products. This decreases the undesirable effects of waves on the products especially when they can not tolerate temperature increment. As stated in the previous subchapters, this localized differential warming is based on the possible considerable difference between dielectric loss factors of the insects' body and products. Considering the fact that the insect objects are different in biological and physiological nature, they have different dielectric constants. The shapes and sizes are also different. Thus it is rather difficult to find a single optimized frequency for the differential warming.
Nelson (Nelson & Tetson, 1974) believes that treatment of the affected products with the lower frequency bands, namely 11-90 MHz, is much better and more efficient than those of the microwave bands such as 2450 MHz, meaning that pest can be controlled in lower

temperatures and using less power in the lower bands rather than microwave band. The complex dielectric constant of one kind of rice weevils and a kind of wheat in a wide range of frequencies are compared in Fig.6. It can be seen from the Fig.6 (a) that the band between 5 MHz and 100 MHz is the best option for differential heating.

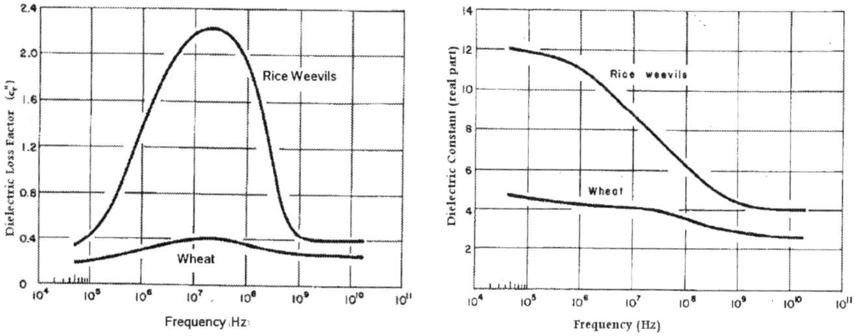

(a) (b)

Fig. 6. (a) Dielectric loss factor of rice weevil and wheat versus frequency (b) Dielectric constant of rice weevil and wheat versus frequency (Nelson & Tetson, 1974) © 1974 IEEE

The theory is also confirmed by measurements done in different frequencies shown in Fig.7. In this figure, insect mortality in terms of temperature is shown for two different bands of 39 MHz and 2450 MHz and different durations, 1 and 8 days. It is obvious that complete mortality for 39 MHz frequency is achieved with less temperature around 50° degrees compared to more than 80° degrees for 2450 MHz. Thus, it shows that the complete mortality is delayed to be achieved in higher frequencies. The point which is not mentioned is that how long does is take to increase the temperature to the required level. Moreover, for a fair claim of differential heating, the magnitude of RF power and the resulted temperature of exposed wheat should also be mentioned.

In some cases, during the treatment, while the temperature increases, the frequency of maximum absorption (relaxation frequency) shifts to higher frequencies as shown in Fig.8. This is due to a change in the biological tissue of the insects. In another word, the dielectric loss factor depends on the temperature. Consequently, it may be more efficient to change the frequency of exposure during the treatment. This can be done using a sweeper starting from the lower up to the upper frequency bound.

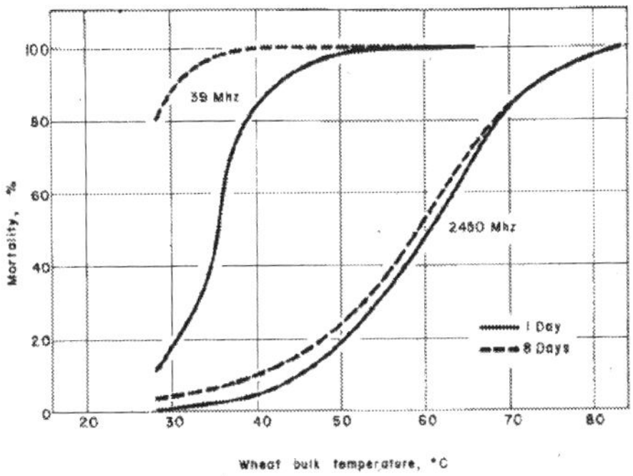

Fig. 7. Mortalities of adult rice weevils in different frequencies in terms of temperature (Mofidian et al., 2007) © 1974 IEEE

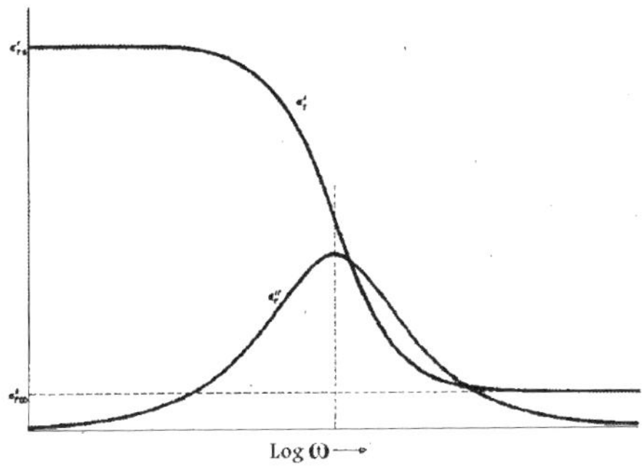

Fig. 8. Dispersion and absorption curves based on the Debye relaxation process for polar molecules (Mofidian et al., 2007) © 1974 IEEE

3.1.4 Millimeter wave pest killer (Halverson et al., 1998)

A practical device for stored-grains has been designed by Halverson presented in (Halverson et al., 1998). He has tried to assess the effectiveness and financial side of controlling stored-grain insects with microwave energy in millimeter wave and microwave band using the free-water relaxation frequency. It is worth pointing out that the crucial bottleneck of using these bands, which is the development of high-power microwave oscillators with tolerable price, has already been solved.

Another problem in using these bands is the poor penetration depth compared to low RF. The skin depth in a dense medium, mentioned in Equation (3), is inversely proportional to the frequency and the conductivity. Conductivity (σ) is also directly related to loss factor (ε'') according to Equation (1). Thus, a good compromise should be done between volume percentage of the gain in a mixture of air and grain when mass product rolls in. This calculation can help us to estimate the efficiency of maximum penetration of the energy into the flowing products. The 3 dB attenuation depth of energy (or similarly penetration depth) is then calculated using Equation (8) (Halverson & Bigelow, 2001).

$$\delta = 0.3466/(2\pi f\sqrt{\mu\varepsilon_o\varepsilon_r\cos(\tfrac{1}{2}\arctan(-\varepsilon_r''/\varepsilon_r' \pm 2\pi)+\pi/2)}) \tag{8}$$

And the ε_r of the mixture is calculated using Equation (9).

$$\varepsilon_r^{1/3} = v_2\varepsilon_{grain}^{1/3} + v_1\varepsilon_{air}^{1/3} \tag{9}$$

which υ_1 and υ_2 are the ratios of the volume of the air and infested product respectively. He has made several one-way path attenuation measurements on controlled air-grain mixtures of flowing soft white wheat, hard red wheat, and rice over a range of 18 to 50 GHz. Fig. 9. shows the semi-schematic for test fixture which performed attenuation tests. The grains are coming down from the hopper and the scalar network analyzer measures the insertion loss of receiver to transmitter link. The measurement results of maximum and minimum penetration depth for the three products, soft while wheat (SWW), hard red wheat (HRW) and rice, shown in Table 1, illustrate that the highest penetration depth occurs in the range of 18 to 26.5 GHz compared to that of the 26.5-40 GHz and 33-50 GHz frequency bands.

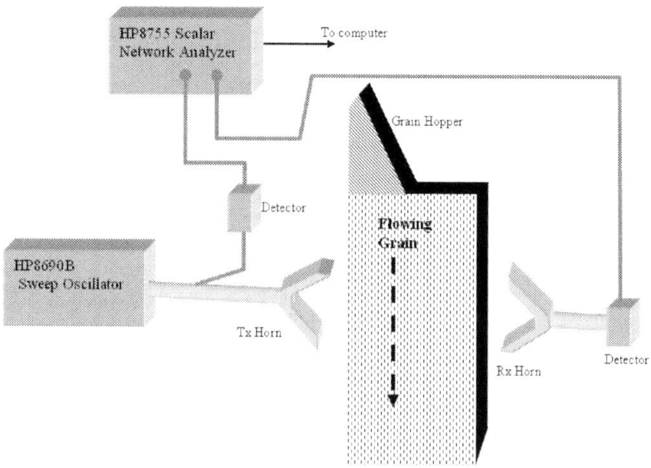

Fig. 9. Semi-schematic diagram of the one-way path attenuation measurement (Halverson et al., 1998).

Using measurement results, he designed the finalized version of his microwave/millimeter apparatus patented in 2001. The schematic of the device has been described in details in the patent (Halverson & Bigelow, 2001).

Grain	f (GHz)	L-3dB max (cm)	L-3dB min(cm)
HRW	18 to 26.5	501.39	75.70
	26.5 to 40	34.76	8.78
	33 to 50	101.95	13.68
SWW	18 to 26.5	139.66	35.48
	26.5 to 40	84.26	7.96
	33 to 50	100.94	12.36
Rice	18 to 26.5	180.97	46.34
	26.5 to 40	61.54	9.57
	33 to 50	131.26	10.22

Table 1. Maximum and minimum penetration depth corresponding to estimated attenuation (Halverson et al., 1998)

3.1.5 Microwave-protected silo (Mofidian et al., 2007)

The prototype system described here has used a bigger microwave oven to control insects of stored wheat. A 2.44 GHz magnetron source has been used to affect two kinds of existing harmful insects, *Sitophilus granarius* and *Tribolium*. This frequency band has been tried before (Andreuccetti, 1994) as the commercial high power low-price technology exists.

Most stored-product pests are killed within few minutes having temperature of 50° C or more shown in Table 2 (Mofidian et al., 2007). On the other hand, there are possible methods such as cutting down the insects' activities using a lower temperature increment which requires a lower power as well.

Mortality, as a general rule, depends on the duration that insects are exposed. However, during heat treatment, temperature can be different within structural profile of a storing facility. Hence, the essential time which insects are exposed to the lethal temperature can differ depending on their location within the facility. This is one of the main problems of the electromagnetic exposure systems.

Temprature(°C)	**Effect**	**Zone**
50-60	Death in minutes	Lethal
45	Death in hours	Lethal
35	Development stops	Suboptimum
33-35	Development slows	Suboptimum
25-33	Maximum rate of development	Optimum
13-25	Development slows	Suboptimum
3-13	Death in days (unacclimated) movement stops	Lethal
-5 to -10	Death in weeks to months if acclimated	Lethal
-25 to -15	Death in minutes, insects freeze	Lethal

Table 2. Response of stored product insects to various temperature zones

The practical scaled system was designed similar to a real wheat storing silo. The system has been modeled in CST Microwave Studio 5 shown in Fig.11 with more than 1 meter diameter and 70 centimeters height. The exposed wheat is located at the bottom of the silo and the insects are inserted in middle areas of wheat-filled section. As can be seen in the Fig.11, the exciting monopole antenna is positioned at the top of the silo's lid below the microwave source.

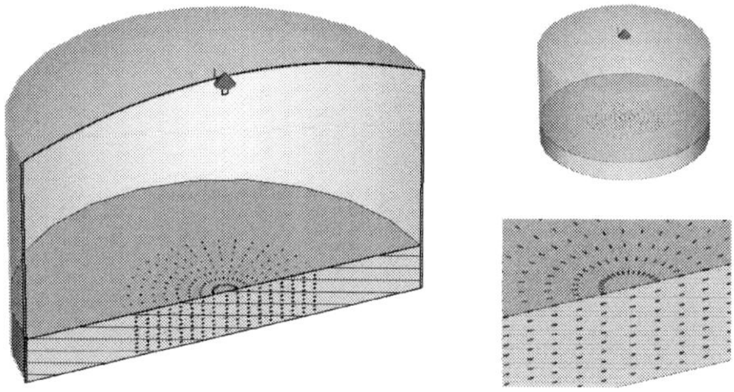

(a)

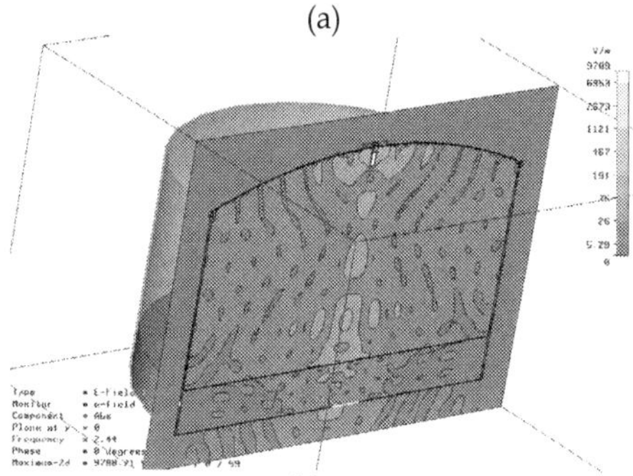

(b)

Fig. 11. Design and simulation of the silo structure using CST software (a) defined structure (b) simulated E-field (Mofidian et al., 2007) © 2007 IEEE

The simulation of the structure has been done using 2.44 GHz normalized microwave source. The resulted E-field in the simulated silo is illustrated in Fig.11(b). It is obvious that the power density is concentrated in the center part of the system, approximately ten times higher than near the wall. Therefore, we expect that the wheats located in the center of the bulk absorb much more power in comparison to the other areas. Fig. 12 presents the constructed system of the silo and the exciting circuit including 220/2000 volt transformer, voltage rectifier, antenna and the 600 watt magnetron with its feeding circuit.

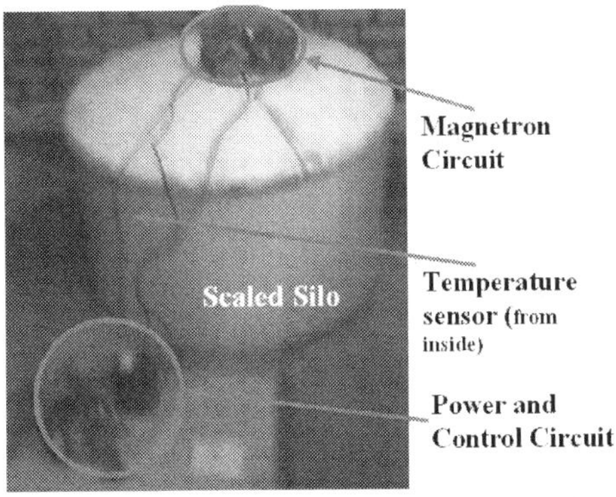

Fig. 12. The constructed scaled Silo using a 2.44 GHz source (Mofidian et al., 2007) © 2007 IEEE

Regarding the comparison of mortality rate between larvae and adults, it is evident that larvae are more susceptible than adults to high temperatures as larvae's mortality is more than adult's which can be seen in Fig.13 in terms of different exposed times. The susceptibilities of both insects are depicted in Fig.14. having similar behavior although *Sitophilus granarius* is quite more outstanding than *Tribolium*. The low mortality rate for 10-minute and 20-minute tests caused by the none-uniform distribution of the field. In the high intensity part, the pests are burnt while they are still alive in the low intensity parts. To improve the performance, the system antenna can be restructured using an array antenna. Obviously, we can reach a complete mortality of insects by increasing the exposure time or the power magnitude as well.

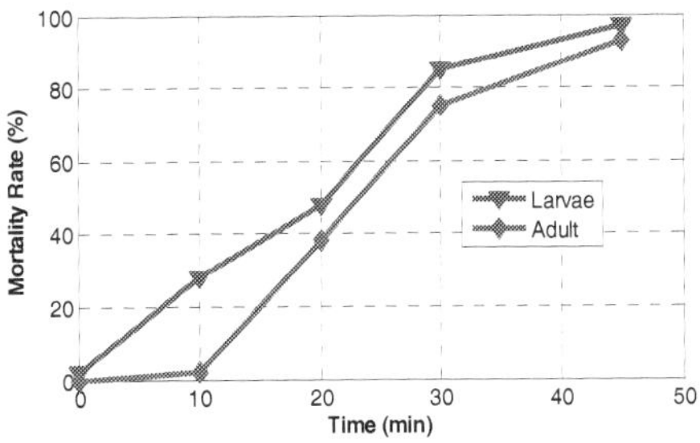

Fig. 13. Comparison between larvae and adults mortality rate of *Tribolium and Sitophilus Granarius* exposed to microwave radiation (2.44 GHz) in terms of exposed time

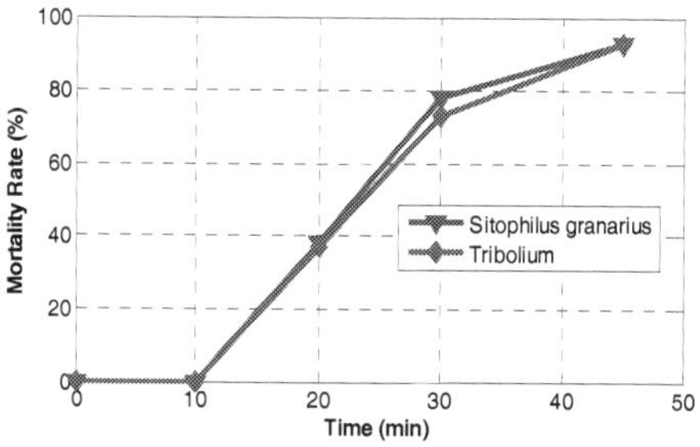

Eradicated pests

Fig. 14. Comparison between mortality rate of adult *Tribolium* and *Sitophilus Granarius* exposed to microwave radiation (2.44 GHz) in terms of exposed time

3.2. Pre-harvest treatment

3.2.1 Disinfection of soil (Lagunas-Solar, 2006)

Lagunas-Solar (Lagunas-Solar, 2006), at UC Davis, has used lower bands of the RF spectrum (few kHz to < 10 MHz) to disinfest the soil from pests. New RF systems has been designed and engineered based upon solid state electronics. A conceptual schematic of the system is illustrated in Fig.15. Its test is said to be relatively efficient for the control of fungi, nematodes and can compete with the other methods specially fumigation such as methyl bromide.

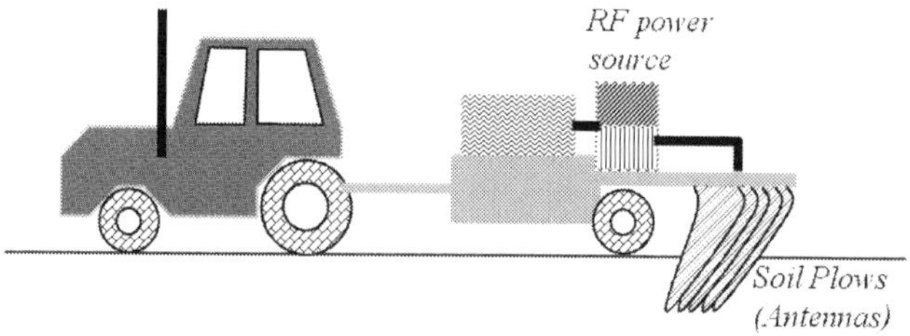

Fig. 15. Schematic for a Portable RF-Soil processing system.

The principle of the working is based on the relatively high electrical conductivity and heat capacity of the agricultural soils. Therefore, the RF oscillator of the system can transfer the energy to the soil, make it warm efficiently, and then it will retain the energy for a while. One of the biggest challenges for the efficient implementation of the system is the large volume (and mass) of soils and consequently the required energy. To save the energy inside the soil, some kind of covers on top can be used.

Using lower frequencies, the RF waves can easily penetrate the soil but do not affect the soil itself. Then, the absorbed energy in pests, mites, and mico-organisms annihilate them. Lagunas-Solar states that the using microwave frequencies may also cause permanent changes in the soil in contrary to the RF lower frequency waves' effects which are always reversible. Fig.16 shows the exposure time required for the soil to reach 60 degrees for different kind of soils. A comparison of soil disinfection using three methods is shown in Table.(2)

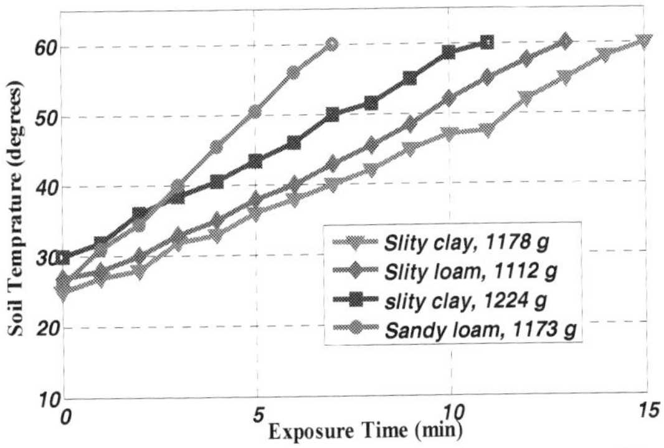

Fig. 16. Soil temperatures vs RF exposure time resulted from a 500 W test (Lagunas-Solar, 2006).

3.2.2 Pre-emptive attack (Aliakbarian et al., 2004),(Aliakbarian et al., 2007)

'Sunne' pest or *'Eurygaster integriceps'*, shown in Fig.1, is the most destroying sap-sucking bug of wheat in the middle east, central and east Europe and North Africa in about twenty countries. The most injures in wheat production in these regions is due to this pest. The proposed idea is to use electromagnetic exposure to control *Sunne* pest in winters before their migration and attack to wheat farms, not to kill them with heat. The method is based on interfering the biological organization of *Sunne* pests in their life period.

The lifetime of the *Sunne* pests is only one year. By the end of March, groups of *Sunne* pests start to migrate to wheat farms and hurt their widest harm to flourishing wheat in 15 to 30 days. Also they start reproducing and then they come back to mountains until the next year. Furthermore, they have a winter sleep when they are in mountains and use their stored energy until the next year. *Sunne* pest can fly about 30 kilometers to the farms and so all of their winter shelters are known and are smaller than wheat farms. Traditional method of spraying poison to kill them in the winter is unsuccessful because they take crucibles as their shelters during these days. *Sunne* pests, like many other insects, are too sensitive to temperature variations and hence they don't attack to warmer regions. In addition, *Sunne* pests are more sensitive to temperature variations in the period of sleep in winter and the variation of climate temperature in this period usually causes an immense damage on them. The reason is that they have *diapause* phenomenon in this time which does not allow them to reproduce and nourish, so it makes them resist against coldness in order to save their energy during winter. As a consequence, if we can heat them up to about 12 to 15 degrees they will wake up and their *diapause* will be broken. Consequently, they should fly, reproduce, and move but not eat because they don't have any food. These activities result in shedding their energy with impunity and probably they can't live until spring or if they can, they can't fly to wheat farms due to the lack of energy. If this heating up is exposed more, their lives will be threatened seriously. The work then is to find the proper frequency which now is more focused on RF ISM band.

3.2.3 Anti-freezing (Aliakbarian et al., 2007)

Sudden freeze of product in a cold day of spring is one of the most damaging agricultural events. In many desert areas, temperature reduction in a few days may cause huge economic injuries. These detriments will be more painful when occur for costly productions like pistachio. By the end of winter, at the beginning of spring, plants are about to flourish. Because of the fact that the weather is not stable, the temperature may fall all of a sudden. Therefore, the biological tissues of the budded pistachio or other products may be damaged. It has been found that if the temperature of the production is increased about two or three degrees, we can save them from being offended immensely.

The previous techniques of anti-freezing have been limited to physical, biophysical and genetic treatments. For instance, in some areas, farmers put a fan and a diesel heater under each tree. These methods are more expensive and hard to exploit than the solution which is suggested here. More over, they have some potential hazards for consumers. Additionally, they need much time than they can be exploited on demand when the weather gets colder. We must estimate the weather condition far before necessity while, with the use of electromagnetic waves, there is no need to an exact prediction of weather condition. Regarding these advantages, it seems that this method can find a suitable place among the other methods in anti-freezing application.

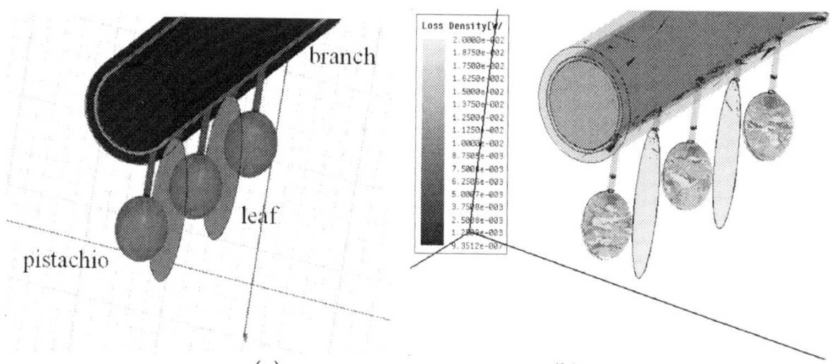

(a) (b)

Fig. 17. (a) Pistachio branch model (b) Volume loss density, the hatched lines show losses (Aliakbarian et al., 2007) © 2007 IEEE

The proposed idea is to warm up the pistachio remotely and selectively using electromagnetic exposure while the other materials of the environment are not warmed up. The most significant work is to find the optimum frequency in which the difference in the absorption rate of energy in pistachio and sensitive objects is the most. This frequency also depends on the electromagnetic characteristics of the objects and can be measured practically. We have done some primary simulations using approximate parameters. Fig.17 shows an HFSS model of a pistachio branch and the volume loss density caused by an incident electromagnetic wave respectively. The simulation in 2.4 GHz in Fig.17 shows that volume loss density in the pistachio is higher than leafs, branch lines, and stems due to difference in dielectric constant in the used frequency.

4. Conclusion

Electromagnetic waves have been suggested for use in a vast range of applications in agriculture and food processing society. Although there is sill a long way for them to be used commercially, the idea can help us to consider this method as an alternative solution for different problems. Heating effect of the waves, especially if used as differential heating, is an efficient way to keep the pests away from the valuable products. The treatment base on EM wave can be employed in indoor or outdoor environments. Nevertheless, it has already been used for indoor environments due to technical and environmental issues. Other topics such as anti-freezing and none-thermal treatment have also been discussed. In conclusion, the usage of the method in commercial scale is likely applicable in a near future .

5. References

Aliakbarian, H.; Enayati, A.; Ameri H.; Ashayer-Soltani, M. (2007). "Agricultural Applications for Electromagnetic Exposure", Proceeding of *Asia-Pacific Microwave Conf., 11-14 Dec. 2007, Bangkok.*

Aliakbarian, H.; Enayati, A.; Farsi, S.; Ajam, H.; Ameri, H.; Ashayer-Soltani, M.; (2004). "Pre-harvest Annihilation of *Sunne*-Pest in Winters Using Electromagnetic Exposure", *Proc. ICCEA, IEEE,* 2004, pp.446-448.

Andreuccetti, D.; Bini, M.; Ignesti, A.; Gambetta, A.; Olmi, R. (1994). "Microwave destruction of woodworm", *Journal of Microwave Power and Electromagnetic Energy,* vol.29, pp.153-160.

Clarck, D. (1997). "The current status of radio frequency post-baking drying technology", The 72nd Annual Technical Conference of the Biscuit and Cracker Manufacturers", Texas, 21 Oct. 1997.

Flores, A.; Suszkiw J., Wood, J.; (2003)1. "Radio Frequency Puts the Heat on Plant Pests", *Agricultural Research* magazine, Feb. 2003, Vol. 51, No. 2, pp.15-17.

Flores, A. (2003)2. "Radio Frequencies Used To Kill Agriculture Pests", Agricultural Research Service, USDA, Agricultural Research magazine.

Frings, H. (1952). "Factors determining the effects of radio-frequency electromagnetic fields and materials they infest", *Journal of economic entomology,* Vol. 45(3), pp. 396-408, 1952.

Geveke, D. J.; Brunkhorst, C.; (2006) "Inactivation of in Apple Juice by Radio Frequency Electric Fields", Journal of Food Science, Vol. 69, Issue 3, pp. FEP134 - FEP0138, May 2006.

Halverson, S. L.; Bigelow, T. S.; Lieber K.; (1998). "Penetration of Infested Stored-Products by EHF/SHF Microwave Energy"; *Annual Intern. Research Conf. on Methyl Bromide Alternatives and Emissions Reductions,* 1998.

Halverson, S. L.; Bigelow, T. S.; (2001). "Microwave and Millimeter method and apparatus for controlling insects in stored products", US Patent No.: 6,192,598 B1, 27 Feb. 2001.

Ikediala, J. N.; Tang, J.; Neven, L. G.; Drake, S. R.; (2000). "Dielectric Properties of apple culvitars and codling moth larve, Transaction of ASAE, vol. 43(5), pp.1175-1184, 2000.

Lagunas-Solar, M. C.; Zeng, N. X.; Essert, T. K.; Truong, T. D.; Piña, C. (2006). "Thermal Disinfection of Soils with Radiofrequency Power", *California Agriculture*, 60(4), pp.192-199, October-December 2006.

Mofidian, M. M., Aliakbarian, H.; Mofidian, M. A.; (2007). "Stored-Product Insects Protection Using Microwave Exposure", Applied Electromagnetics Conf., 2007, 19-20 Dec. 2007, pp.1 - 3.

Nelson, S. O. (1966). "Electromagnetic and sonic energy for insect control", *Transactions of the ASAE*, Vol. 9(3), pp. 398-404, 1966.

Nelson, S.; (2004). "Dielectric Spectroscopy Applications in Agriculture" , 3rd International Conference on Broadband Dielectric Spectroscopy and its Applications, 23-26 August 2004, Delft, Netherlands, pp.200.

Nelson, S.O.; Tetson, L. E.; (1974). "Possibilities for Controlling Insects with Microwaves and Lower Frequency RF Energy", *IEEE Trans. on MTT*, vol. 22, Dec 1974, pp. 1303 - 1305.

Nelson, S.O. (1996). "Review and assessment of radio-frequency and microwave energy for stored-grain insect control", *Trans. ASAE*, vol. 39, pp.1475–1484, 1996.

Shapovalenko, O.I.; Yanyuk, T.I.; Yanenko, A.F.; (2000). "The Influence of Microwave Radiation on the Quality of Wheat Germs", *Proc. Of 10th international Crimean Conf. 'The microwave and Telecomm. Technology*,2000.

Tang, J.; Wang, S. ; Hansen, J.; Johnson J.; Mitcham E.; Drake S.; Hallman G.; (2003). "Postharvest Control of Insect Pests in Nuts and Fruits Based on Radio Frequency Energy"; *ISHS Acta Horticultura*, 2003, ISSU 599, pp. 175-182.

Thomas, A.M. (1952). "Pest control by high-frequency electric fields critical resume", *Technical report W/T23. Leatherhead, Surrey, England: British electric and allied industries association*, 1952.

Wang, S.; Tang J.; (2001). "Radio Frequency and Microwave Alternative Treatments for Insect Control in Nuts: A Review"; *Agricultural Engineering Journal*, vol. 10 (3&4), pp.105-120, 2001.

Wang, S.; Tang, J.; Johnson, J. A.; Mitcham, E.; Hansen, J. D.; Hallman, G.; Drake, S. R.; Wang, Y.; (2003). "Dielectric Properties of Fruits and Insect Pests as related to Radio Frequency and Microwave Treatments"; *Bio-systems Engineering Journal*, pp.:201-212, April, 2003.

Wang, S.; Tang, J.; Johnson J.A.; Mitcham, E.; Hansen, J.D.; Cavalieri, R.P.; Bower, J.; Biasi, B. (2002). "Process protocols based on radio frequency energy to control field and storage pests in in-shell walnuts", *Post-harvest Biology And Technology*, May 20, 2002.

Wang, S.; Tang, J.;Sun, T.; Mitcham, E.; Koral, T.; Birla, S.L. (2006). "Considerations in design of commercial radio frequency treatments for postharvest pest control in in-shell walnuts", *Journal of Food Engineering*, August 10, 2006.